D1432002

Quantum Theory of Solids

SECOND REVISED PRINTING

NO LONGER
PROPERTY OF
OLIN LIBRARY
WASHINGTON UNIVERSITY

C. Kittel, *Department of Physics, University of California, Berkeley, California*

Solutions Appendix Prepared by C. Y. FONG,

Department of Physics, University of California, Davis, California

Quantum Theory of Solids

SECOND REVISED PRINTING

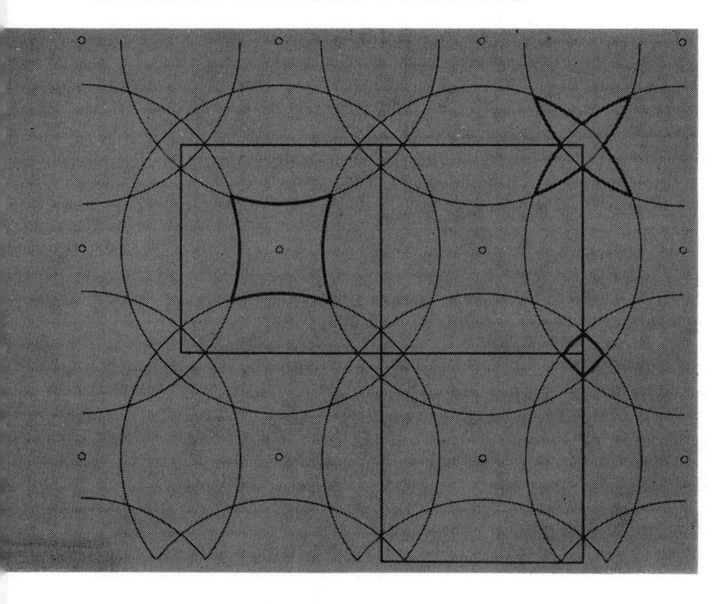

Phy
QC 176
K53
1987
c. 2

JOHN WILEY & SONS

New York • Chichester • Brisbane • Toronto • Singapore

WASHINGTON UNIVERSITY
PHYSICS LIBRARY
ST. LOUIS, MISSOURI 63130

Copyright © 1963, 1987 by John Wiley & Sons, Inc.

All rights reserved. Published simultaneously in Canada.

Reproduction, or translation of any part of
this work beyond that permitted by Sections
107 and 108 of the 1976 United States Copyright
Act without the permission of the copyright
owner is unlawful. Requests for permission
or further information should be addressed to
the Permissions Department, John Wiley & Sons.

Library of Congress Cataloging in Publication Data:

Kittel, Charles.
Quantum theory of solids.

Includes indexes.
1. Solid state physics. 2. Quantum theory.
I. Fong, C. Y. (Ching-yao) II. Title.
QC176.K53 1987 530.4'1 86-32478
ISBN 0-471-62412-8 (pbk.)

Printed in the United States of America

10 9 8 7 6 5 4

Preface

The object of this textbook is to present the central principles of the quantum theory of solids to theoretical physicists generally and to those experimental solid state physicists who have had a one year course in quantum mechanics. The book is intended for use in a one year graduate course, to follow a year of general education in the facts of solid state physics at the level at least of my earlier book *Introduction to Solid State Physics*, second edition. Much of the physical background to the theory is developed there, and it is hoped in the next edition to develop the physical interpretation up to the level needed here.

As far as possible I have tried to emphasize unifying principles. The first part of the book treats phonon, magnon, and electron fields and their interactions, culminating in the theory of superconductivity. The second part treats fermi surfaces and electron wavefunctions in metals, alloys, semiconductors, and insulators, with considerable attention to the theory of the important types of experiments which bear on our understanding. The third part deals with correlation functions and their application to time-dependent effects in solids, with a brief introduction to Green's functions. The order of the chapters, particularly within the second part, is not rigorously linear —I attempted to avoid an accumulation of material which did not challenge the imagination and participation of the reader. The first part of the book forms by itself a short course which has been observed to interest students of field theory and particle physics.

A number of important theoretical calculations in solid state physics are too long, intricate, fearsome, or tedious to present in full in a course; and a summary treatment often has no pedagogical value. To give several examples, the full theory of Bloch electrons in a magnetic field, by Blount, Roth, Wannier, Kohn, and others, has been omitted, although a less complete treatment is included. Equally, the work of Soviet physicists and Luttinger on the Landau theory of the fermi gas is clearly beyond the scope of the book, together with a number of other many-body problems. The subject of the properties of paramagnetic ions in various environments appears to me to be too specialized for a general text. While writing it became abundantly clear that there is no level at which a textbook like this could aim to be a complete treatment of all major aspects of the theory of solids; the field is simply too vast. It is natural not to report several detailed subjects which are available in existing texts—thus the material on standard transport theory in the books by A. H. Wilson and J. Ziman is not repeated here; the discussion in Peierls of phonon interactions and the book by Abragam on nuclear magnetism are treatments so complete that it would be ridiculous to paraphrase them. I renounced an attempt to do everything by the method of Green's functions, for then the contents would be almost entirely inaccessible to experimentalists at the present time. Given an opportunity, many students will choose to write a term paper on an application of Green's functions to a many-body problem. The quantum theory of transport processes is not treated here. Fortunately, for many subjects excellent monographs exist, particularly in the monumental series *Solid State Physics—Advances in Research and Applications*, edited by F. Seitz and D. Turnbull, and in the *Encyclopedia of Physics—Handbuch der Physik* series. Details far beyond the scope of this text can be found in these and other specialized works.

This book contains problems and is a textbook; it is not a history of the development of the subject. I have actively tried as a matter of policy to avoid proper names, assignment of priority, and allocation of credit. Detailed references and names are given only when it would be positively clumsy to omit them, or when the work is too recent to have been included in reviews. An adequate bibliography would be as long as the text itself. Many lists of references are readily available in the *Advances* and in the *Handbuch* series. It is increasingly clear that many active research workers cannot find time both to write a text and to give full historical credit to all their colleagues responsible for the development of a large subject.

A number of very important results are developed in the problems to be found at the end of most chapters. It is urgently recommended that the problems be read over in conjunction with the text, and it would be vastly better for the reader to solve the problems.

A note on notation: $[\,,\,]$ = commutator; $\{\,,\,\}$ = anticommutator; the symbols c, c^+ are usually reserved for fermion operators. The units

$\hbar = 1$ are employed throughout, but $\hbar$ is sometimes restored to the final result. A Bloch function of wavevector $\mathbf{k}$ is written as $|\mathbf{k}\rangle$. When convenient the volume Ω of the specimen is set equal to unity; N usually refers to the total number of particles and n to their concentration. The symbol Ψ usually denotes a field operator, and Φ usually denotes a state vector.

It is a pleasant duty to acknowledge here in part the wide assistance I have received. M. H. Cohen, W. A. Harrison, W. Kohn, H. Suhl, J. Friedel, A. Blandin, P. Argyres, B. Cooper, S. Silverstein, B. Dreyfus, J. W. Halley, G. Mahan, D. Mills, and F. Sheard have helpfully suggested improved treatments of various demonstrations. My colleague, J. Hopfield, alertly resolved innumerable apparent paradoxes which appeared during the writing. At Stanford, M. S. Sparks and his associates, including R. M. White, R. Adler, K. Nordtvedt, K. Motozuki, and I. Ortenberger, detected many lapses in the early versions of the manuscript. R. Peierls graciously agreed to share the title of his most useful book. The typing was done with perfection by Mrs. Eleanor Thornhill, without whose help few physics books would be written in Berkeley; Mrs. Sue Limoli kindly assisted with the proofs.

C. KITTEL

August 1963

Preface to the Second Revised Printing

The original preface suggested that "A number of very important results are developed in the problems... ." Experience has shown that the problem solutions can be really useful both for self-study and as a supplement to lectures. Indeed, many of the results are too important to be left to the chance that the reader will take time off to work them out. It is fortunate that Professor C. Y. Fong offered to prepare an appendix giving the solutions to selected problems. He also made a number of typographic and algebraic corrections in the text itself. Thus the revised edition is defined by the original text, the solutions, and the corrections.

C. KITTEL

November 1986

BOOKS BY CHARLES KITTEL

Introduction to Solid State Physics. Sixth Edition
Elementary Statistical Physics
Elementary Solid State Physics
Quantum Theory of Solids
Thermal Physics, Second Edition (Freeman)

Contents

1. *Mathematical Introduction, 1*

2. *Acoustic Phonons, 12*

3. *Plasmons, Optical Phonons, and Polarization Waves, 34*

4. *Magnons, 49*

5. *Fermion Fields and the Hartree-Fock Approximation, 75*

6. *Many-Body Techniques and the Electron Gas, 99*

7. *Polarons and the Electron-Phonon Interaction, 130*

8. *Superconductivity, 150*

9. *Bloch Functions—General Properties, 179*

10. *Brillouin Zones and Crystal Symmetry, 199*

11. *Dynamics of Electrons in a Magnetic Field: de Haas-van Alphen Effect and Cyclotron Resonance, 217*

12. *Magnetoresistance, 237*

13. *Calculation of Energy Bands and Fermi Surfaces, 249*

14. *Semiconductor Crystals: I. Energy Bands, Cyclotron Resonance, and Impurity States, 268*

15. *Semiconductor Crystals: II. Optical Absorption and Excitons, 294*

16. *Electrodynamics of Metals, 308*

17. *Acoustic Attenuation in Metals, 326*

18. *Theory of Alloys, 338*

19. *Correlation Functions and Neutron Diffraction by Crystals, 368*

20. *Recoilless Emission, 386*

21. *Green's Functions—Application to Solid State Physics, 397*

Appendix: Perturbation Theory and the Electron Gas, 417

Appendix: Solutions A-1–A-70

Index, I-1–I-9

References frequently cited

QUANTUM MECHANICS BACKGROUND

L. Schiff, *Quantum Mechanics*, McGraw-Hill, New York, second edition, 1955.

A. Messiah, *Quantum Mechanics*, North-Holland, Amsterdam, 1961–62, 2 vols.

L. Landau and E. Lifshitz, *Quantum Mechanics*, Addison-Wesley, 1958.

E. M. Henley and W. Thirring, *Elementary Quantum Field Theory*, McGraw-Hill, New York, 1962.

SOLID STATE BACKGROUND

Solid State Physics: *Advances in Research and Applications*, edited by F. Seitz and D. Turnbull, Academic Press, New York. This is cited as *Solid state physics*.

C. Kittel, *Introduction to Solid State Physics*, Wiley, New York, 1986, sixth edition. This is cited as *ISSP*.

Low Temperature Physics, edited by C. De Witt, B. Dreyfus, and P. G. de Gennes, Gordon and Breach, New York, 1962. This is cited as *LTP*.

MANY-BODY PROBLEMS

D. Pines, *Many-Body Problem*, Benjamin, New York, 1961. This is cited as Pines.

K. A. Brueckner, in *Many Body Problems*, edited by C. De Witt, Wiley, New York, 1959, pp. 47–242.

TABLES

H. B. Dwight, *Table of Integrals and Other Mathematical Data*, Macmillan, New York, 1957 3rd edition. Integrals in here are cited by their number.

Quantum Theory of Solids

SECOND REVISED PRINTING

1 Mathematical introduction

It is convenient to gather together here a number of definitions and results which are utilized throughout the text.

RECIPROCAL LATTICE

We recall several important properties of the reciprocal lattice. The basis vectors $\mathbf{a}^*$, $\mathbf{b}^*$, $\mathbf{c}^*$ of the reciprocal lattice are defined in terms of the primitive basis vectors, $\mathbf{a}$, $\mathbf{b}$, $\mathbf{c}$ of the direct lattice by the equations

$$(1) \quad \mathbf{a}^* = 2\pi \frac{\mathbf{b} \times \mathbf{c}}{\mathbf{a} \cdot \mathbf{b} \times \mathbf{c}}; \qquad \mathbf{b}^* = 2\pi \frac{\mathbf{c} \times \mathbf{a}}{\mathbf{a} \cdot \mathbf{b} \times \mathbf{c}}; \qquad \mathbf{c}^* = 2\pi \frac{\mathbf{a} \times \mathbf{b}}{\mathbf{a} \cdot \mathbf{b} \times \mathbf{c}};$$

this definition includes a factor 2π which is not usually present in the usual crystallographic definition followed by elementary texts. In treating the interaction of waves with periodic lattices we constantly encounter in the statements of wavevector conservation an additive term which is 2π times the crystallographic reciprocal lattice vector; thus we find it handy to include the 2π in the definition here. Otherwise our notation here is standard; the use of the asterisk superscript in no way implies "complex conjugate." All the basis vectors are real. We note that $\mathbf{a} \cdot \mathbf{a}^* = 2\pi$; $\mathbf{a} \cdot \mathbf{b}^* = 0$; etc.

By simple vector analysis it follows from (1) that

$$(2) \qquad V_c^* = \frac{(2\pi)^3}{V_c},$$

where V_c^* is the volume of the primitive cell in the reciprocal lattice and $V_c = \mathbf{a} \cdot \mathbf{b} \times \mathbf{c}$ is the volume of the primitive cell in the direct lattice. We note that the conversion of wavevector sums to integrals involves

$$(3) \qquad \boxed{\sum_{\mathbf{k}} \rightarrow \frac{\Omega}{(2\pi)^3} \int d^3k = (N/V_c^*) \int d^3k,}$$

where the direct volume Ω contains N primitive cells.

THEOREM. The vector $\mathbf{r}^*(hkl)$ to the point hkl of the reciprocal lattice is normal to the (hkl) plane of the direct lattice.

Proof: Note that

$$\frac{1}{h}\mathbf{a} - \frac{1}{k}\mathbf{b}$$

is a vector in the (hkl) plane of the direct lattice, by definition of the lattice indices. But

$$\mathbf{r}^* \cdot \left(\frac{1}{h}\mathbf{a} - \frac{1}{k}\mathbf{b}\right) = (k\mathbf{a}^* + k\mathbf{b}^* + l\mathbf{c}^*) \cdot \left(\frac{1}{h}\mathbf{a} - \frac{1}{k}\mathbf{b}\right) \tag{4}$$

$$= \mathbf{a}^* \cdot \mathbf{a} - \mathbf{b}^* \cdot \mathbf{b} = 0;$$

therefore $\mathbf{r}^*$ is normal to one vector in the plane. By the same argument $\mathbf{r}^*$ is normal to the second vector

$$\frac{1}{h}\mathbf{a} - \frac{1}{l}\mathbf{b}$$

in the plane, and thus $\mathbf{r}^*$ is normal to the plane.

THEOREM. The length of the vector $\mathbf{r}^*(hkl)$ is equal to 2π times the reciprocal of the spacing $d(hkl)$ of the planes (hkl) of the direct lattice.

Proof: If $\mathbf{n}$ is the unit normal to the plane, then $h^{-1}\mathbf{a} \cdot \mathbf{n}$ is the interplanar spacing. Now

$$\mathbf{n} = \mathbf{r}^*/|\mathbf{r}^*|, \tag{5}$$

so that the spacing $d(hkl)$ is

$$d(hkl) = \frac{1}{h}\mathbf{n} \cdot \mathbf{a} = \frac{\mathbf{r}^* \cdot \mathbf{a}}{h|\mathbf{r}^*|} = \frac{2\pi}{|\mathbf{r}^*|}. \tag{6}$$

We now go on to two important theorems about expansions of periodic functions.

THEOREM. A function $f(\mathbf{x})$ which is periodic with the period of the lattice may be expanded in a fourier series in the reciprocal lattice vectors $\mathbf{G}$.

Proof: Consider the series

$$f(\mathbf{x}) = \sum_{\mathbf{G}} a_{\mathbf{G}} e^{i\mathbf{G}\cdot\mathbf{x}}; \tag{7}$$

to show that this is periodic with the period of the lattice we increase $\mathbf{x}$ by a lattice vector:

$$\mathbf{x} \rightarrow \mathbf{x} + m\mathbf{a} + n\mathbf{b} + p\mathbf{c}, \tag{8}$$

where m, n, p are integers. Then

$$f(\mathbf{x} + m\mathbf{a} + n\mathbf{b} + p\mathbf{c}) = \Sigma\, a_{\mathbf{G}} e^{i\mathbf{G}\cdot\mathbf{x}} e^{i\mathbf{G}\cdot(m\mathbf{a}+n\mathbf{b}+p\mathbf{c})}; \tag{9}$$

but

$$\begin{aligned} \mathbf{G}\cdot(m\mathbf{a} + n\mathbf{b} + p\mathbf{c}) &= (h\mathbf{a}^* + k\mathbf{b}^* + l\mathbf{c}^*)\cdot(m\mathbf{a} + n\mathbf{b} + p\mathbf{c}) \\ &= 2\pi(hm + kn + lp), \end{aligned} \tag{10}$$

which is just an integer times 2π, so that

$$f(\mathbf{x} + m\mathbf{a} + n\mathbf{b} + p\mathbf{c}) = f(\mathbf{x}), \tag{11}$$

and the representation (7) has the required periodicity.

THEOREM. If $f(\mathbf{x})$ has the periodicity of the lattice,

$$\int d^3x\, f(\mathbf{x}) e^{i\mathbf{K}\cdot\mathbf{x}} = 0, \tag{12}$$

unless $\mathbf{K}$ is a vector in the reciprocal lattice.

Proof: This result is a direct consequence of the preceding theorem and is essentially a selection rule for interband ($\mathbf{G} \neq 0$) and intraband ($\mathbf{G} = 0$) transitions. By (7)

$$f(\mathbf{x}) = \sum_{\mathbf{G}} a_{\mathbf{G}} e^{i\mathbf{G}\cdot\mathbf{x}}, \tag{13}$$

and

$$\int d^3x\, f(\mathbf{x}) e^{i\mathbf{K}\cdot\mathbf{x}} = \sum_{\mathbf{G}} a_{\mathbf{G}} \int d^3x\, e^{i(\mathbf{K}+\mathbf{G})\cdot\mathbf{x}} = \Omega \sum_{\mathbf{G}} a_{\mathbf{G}} \Delta(\mathbf{K} + \mathbf{G}), \tag{14}$$

where Δ is the kronecker symbol; and Ω is the volume of the specimen; we also write $\Delta(\mathbf{K} + \mathbf{G})$ as $\delta_{\mathbf{K},-\mathbf{G}}$.

FOURIER LATTICE SERIES

Consider the series

$$q_r = N^{-\frac{1}{2}} \sum_k Q_k e^{ikr}. \tag{15}$$

We shall usually determine the allowed values of k by the periodic boundary condition $q_{r+N} = q_r$, whence $e^{ikN} = 1$; this condition is satisfied by $k = 2\pi n/N$, where n is any integer. Only N values of n give independent values of the N coordinates q_r. It is convenient to take N as even and to choose the values of n as $0, \pm 1, \pm 2, \cdots, \pm(\frac{1}{2}N - 1), \frac{1}{2}N$. We note that $\frac{1}{2}N$ and $-\frac{1}{2}N$ give identical values of e^{ikr} for all r, so that we need take only $\frac{1}{2}N$. The value $n = 0$ or

$k = 0$ is associated with what is called the *uniform mode* in which all q_r are equal, independent of r.

THEOREM. Given (15), then

$$Q_k = N^{-1/2} \sum_s q_s e^{-iks}. \tag{16}$$

Proof: Substitute (16) in (15):

$$q_r = N^{-1} \sum_{ks} q_s e^{ik(r-s)}. \tag{17}$$

If $s = r$, then the sum over k gives Nq_r, the desired result. If $s - r = \sigma$, some other integer,

$$\sum_n e^{ik\sigma} = \sum_n e^{i2\pi n\sigma/N} = \sum_{n=0}^{\frac{1}{2}N} e^{i2\pi n\sigma/N} + \sum_{n=1}^{\frac{1}{2}N-1} e^{-i2\pi n\sigma/N} \tag{18}$$

$$= \sum_{n=0}^{N-1} e^{i2\pi n\sigma/N} = \frac{1 - e^{i2\pi\sigma}}{1 - e^{i2\pi\sigma/N}} = 0$$

for $\sigma \neq 0$. Thus we have the orthogonality relation

$$\sum_k e^{ik(r-s)} = N\delta_{sr}. \tag{19}$$

This is the analog for discrete sums of the delta function representation

$$\int_{-\infty}^{\infty} e^{ikx}\, dk = 2\pi\delta(x). \tag{20}$$

Consider the series defined for $-\frac{1}{2}L < x < \frac{1}{2}L$:

$$q(x) = L^{-1/2} \Sigma\, Q_k e^{ikx}, \tag{21}$$

where k is any integer times $2\pi/L$.

THEOREM. Given (21), then

$$Q_k = L^{-1/2} \int_{-\frac{1}{2}L}^{\frac{1}{2}L} d\xi\, q(\xi) e^{-ik\xi}. \tag{22}$$

Proof: Substitute (21) in (22):

$$Q_k = L^{-1} \sum_{k'} Q_{k'} \int d\xi\, e^{i(k'-k)\xi} = \sum_{k'} Q_{k'}\delta_{kk'} = Q_k, \tag{23}$$

because

$$\int_{-\frac{1}{2}L}^{\frac{1}{2}L} d\xi\, e^{i(k'-k)\xi} = \frac{2 \sin \frac{1}{2}(k' - k)L}{i(k' - k)} = 0,$$

except for $k = k'$.

THEOREM. The potential $1/|\mathbf{x}|$ may be expanded in a fourier series as

$$\frac{1}{|\mathbf{x}|} = \frac{4\pi}{\Omega} \sum_{\mathbf{q}} \frac{1}{q^2} e^{i\mathbf{q}\cdot\mathbf{x}}, \tag{24}$$

where Ω is the volume of the crystal.

Proof: Following (22) consider, with $r = |\mathbf{x}|$,

$$\int d^3x \frac{e^{-\alpha r} e^{-i\mathbf{q}'\cdot\mathbf{x}}}{r} = 2\pi \int r\,dr \int_{-1}^{1} d\mu\, e^{-iq'r\mu} e^{-\alpha r} \tag{25}$$
$$\cong \frac{2\pi}{iq'} \int_0^\infty dr\, (e^{-i(q'-i\alpha)r} - e^{i(q'+i\alpha)r}) = \frac{4\pi}{q'^2 + \alpha^2}.$$

On taking the limit $\alpha \to +0$ we obtain (24).

SUMMARY OF QUANTUM EQUATIONS ($\hbar = 1$)

$$i\dot{\psi} = H\psi. \tag{26}$$

$$i\dot{F} = [F,H], \qquad \text{for an operator } F. \tag{27}$$

$$[f(\mathbf{x}),\mathbf{p}] = i\frac{\partial}{\partial \mathbf{x}} f(\mathbf{x}); \qquad \mathbf{p} = -i\,\mathrm{grad} - \frac{e}{c}\mathbf{A}. \tag{28}$$

$$[f(\mathbf{x}),p_x^2] = 2i\frac{\partial f}{\partial x} p_x + \frac{\partial^2 f}{\partial x^2}. \tag{29}$$

$$\sigma_x = \begin{pmatrix} 0 & 1 \\ 1 & 0 \end{pmatrix}; \quad \sigma_y = \begin{pmatrix} 0 & -i \\ i & 0 \end{pmatrix}; \quad \sigma_z = \begin{pmatrix} 1 & 0 \\ 0 & -1 \end{pmatrix}. \tag{30}$$

$$\sigma^+ = \begin{pmatrix} 0 & 2 \\ 0 & 0 \end{pmatrix}; \quad \sigma^- = \begin{pmatrix} 0 & 0 \\ 2 & 0 \end{pmatrix}. \tag{31}$$

For the harmonic oscillator,

$$\langle n|x|n+1\rangle = (2m\omega)^{-1/2}(n+1)^{1/2}; \qquad \langle n|p|n+1\rangle = -i(m\omega/2)^{1/2}(n+1)^{1/2}. \tag{32}$$

$$\mathrm{Tr}\{A[B,C]\} = \mathrm{Tr}\{[A,B]C\}; \qquad \mathrm{Tr}\{ABC\} = \mathrm{Tr}\{CAB\}. \tag{33}$$

$$\lim_{s\to+0} \frac{1}{x \pm is} = \mathcal{P}\frac{1}{x} \mp \pi i\delta(x); \qquad \mathcal{P} \equiv \text{principal value}. \tag{34}$$

Transition rate:

$$W(n \to m) = 2\pi|\langle m|H'|n\rangle|^2 \delta(\epsilon_m - \epsilon_n). \tag{35}$$

Density of states per unit energy range, free electrons:

$$\rho_E = \frac{\Omega}{2\pi^2}(2m)^{3/2}\varepsilon^{1/2}. \tag{36}$$

$$\int dx\, f(x)\delta(ax - y) = \frac{1}{|a|}f(y/a). \tag{37}$$

$$\delta(g(x)) = \sum_i \frac{1}{|g'(x_i)|}\delta(x - x_i), \tag{38}$$

where the x_i are the roots of $g(x) = 0$.

$$\int_{-\infty}^{\infty} dx\, e^{ixy} = 2\pi\delta(y). \tag{39}$$

For nondegenerate states,

$$|m\rangle^{(1)} = |m\rangle + \sum_k{}' \frac{|k\rangle\langle k|H'|m\rangle}{\epsilon_m - \epsilon_k}; \tag{40}$$

$$\varepsilon_m^{(2)} = \varepsilon_m^{(0)} + \langle m|H'|m\rangle + \sum_k{}' \frac{|\langle m|H'|k\rangle|^2}{\varepsilon_m - \varepsilon_k}.$$

$$[AB,C] = A[B,C] + [A,C]B. \tag{41}$$

GENERAL TIME-DEPENDENT PERTURBATION THEORY

We consider the hamiltonian

$$H = H_0 + V, \tag{42}$$

where V is called the perturbation. Even when H_0 and V are independent of time, important results of perturbation theory appear more naturally from time-dependent theory than from the usual time-independent perturbation theory. We assume that the lowest eigenstate Φ of H can be derived from the unperturbed lowest eigenstate Φ_0 of H_0 by adiabatically switching on the interaction V in the time interval $-\infty$ to 0. This assumption is not necessarily always true, and in particular it fails if the perturbation causes one or more bound states to appear below a continuum. The assumption is called the *adiabatic hypothesis.* We shall use (only in this development) the notation $|)$ to denote an eigenstate of H and $|\rangle$ to denote an eigenstate of H_0. The unperturbed ground state is $|0\rangle$ and the exact ground state is $|0)$. The same notation is used again at the end of Chapter 6 in the identical connection.

THEOREM. If E_0 is defined by

$$H_0|0\rangle = E_0|0\rangle, \tag{43}$$

and ΔE by

(44) $$(H_0 + V)|0) = (E_0 + \Delta E)|0),$$

then the exact shift in the ground-state energy caused by the perturbation is

(45) $$\boxed{\Delta E = \frac{(0|V|0)}{(0|0)}.}$$

Proof: The result (45) follows on subtracting

(46) $$(0|H_0|0) = E_0(0|0)$$

from

(47) $$(0|H_0 + V|0) = (E_0 + \Delta E)(0|0).$$

Then

(48) $$(0|V|0) = \Delta E(0|0). \qquad \text{Q.E.D.}$$

We now undertake to calculate $|0)$. We replace V by

(49) $$\lim_{s \to +0} e^{-s|t|}V, \qquad s > 0.$$

This defines the process of adiabatic switching, in which the interaction is switched on slowly between $t = -\infty$ and $t = 0$. Between $t = 0$ and $t = \infty$ the interaction is switched off slowly. We shall in the following always understand that the limit $s \to +0$ is to be carried out.

We work with the perturbation in the interaction representation:

(50) $$V(t) = e^{iH_0t}Ve^{-iH_0t}e^{-s|t|},$$

so that the time-dependent Schrödinger equation has the form

(51) $$i\frac{\partial \Phi}{\partial t} = V(t)\Phi,$$

with the boundary condition $\Phi(-\infty) = \Phi_0$. In the interaction representation

(52) $$\Phi(t) = e^{iH_0t}\Phi_s(t),$$

where Φ_s is in the Schrödinger representation. We confirm (51) by forming

(53) $$i\dot{\Phi} = -H_0e^{iH_0t}\Phi_s + ie^{iH_0t}\dot{\Phi}_s = Ve^{iH_0t}\Phi_s.$$

Then

$$i\dot{\Phi}_s = (H_0 + V)\Phi_s, \tag{54}$$

as required in the Schrödinger representation.

We now define the operator

$$U(t,t') \equiv \sum_{n=0}^{\infty} (-i)^n \int_{t'}^{t} \cdots \int_{t'}^{t_{n-2}} \int_{t'}^{t_{n-1}} dt_1 \cdots dt_n \, V(t_1)V(t_2) \cdots V(t_n), \tag{55}$$

which may be written as

$$U(t,t') \equiv \sum_n \frac{(-i)^n}{n!} \int_{t'}^{t} \cdots \int_{t'}^{t} dt_1 \cdots dt_n \, P\{V(t_1)V(t_2) \cdots V(t_n)\}, \tag{56}$$

where P is a time-ordering operator—the Dyson chronological operator —which orders all quantities on its right in order of decreasing time from left to right. Here $V(t)$ is in the interaction representation. We may also write

$$U(t,t') \equiv P\left\{\exp\left[-i\int_{t'}^{t} dt \, V(t)\right]\right\}. \tag{57}$$

THEOREM. With $U(t,t')$ as defined above, the exact ground state $|0)$ is given in terms of the unperturbed ground state $|0\rangle$ by

$$|0) = \frac{U(0,-\infty)|0\rangle}{\langle 0|U(0,-\infty)|0\rangle}, \tag{58}$$

with all states in the interaction representation.

Proof: If (58) is to be true,

$$\Phi_s(t) = e^{-iH_0t}U(t,-\infty)\Phi_0 \tag{59}$$

should satisfy the time-dependent Schrödinger equation (54). We form, from (59) and (57), and then (50),

$$\begin{aligned} i\frac{\partial}{\partial t}\Phi_s(t) &= H_0\Phi_s(t) + e^{-iH_0t}V(t)U\Phi_0 \\ &= (H_0 + Ve^{-s|t|})e^{-iH_0t}U(t,-\infty)\Phi_0 = H\Phi_s(t), \end{aligned} \tag{60}$$

as required. Thus, at $t = 0$, $\Phi_s = U(0, -\infty)\Phi_0$; the denominator in (58) follows if we normalize to make the mixed product

$$\langle 0|0\rangle = 1. \tag{61}$$

Equivalence to Time-Independent Perturbation Theory. From (50) and (56) we see that the lowest order term in the construction of a matrix element $\langle f|U(0, -\infty)|0\rangle$, using

$$U(0, -\infty) = \sum_n \frac{(-i)^n}{n!} \int_{-\infty}^{0} \cdots \int_{-\infty}^{0} dt_1 \cdots dt_n \, P\{V(t_1) \cdots V(t_n)\}, \tag{62}$$

is

$$\langle f|U_1(0, -\infty)|0\rangle = -i \int_{-\infty}^{0} dt_1 \, \langle f|V|0\rangle e^{i(E_f - E_0 - is)t_1} = -\frac{\langle f|V|0\rangle}{E_f - E_0 - is}, \tag{63}$$

exactly as in ordinary time-independent perturbation theory. Similarly, the second order term is

$$\begin{aligned} \langle f|U_2(0, -\infty)|0\rangle &= (-i)^2 \int_{-\infty}^{0} dt_1 \int_{-\infty}^{t_1} dt_2 \sum_p \langle f|V|p\rangle e^{i(E_f - E_p - is)t_1} \langle p|V|0\rangle e^{i(E_p - E_0 - is)t_2} \\ &= i \int_{-\infty}^{0} dt_1 \sum_p \langle f|V|p\rangle e^{i(E_f - E_p - is)t_1} \langle p|V|0\rangle \frac{e^{i(E_p - E_0 - is)t_1}}{E_p - E_0 - is} \\ &= \sum_p \frac{\langle f|V|p\rangle\langle p|V|0\rangle}{(E_f - E_0 - 2is)(E_p - E_0 - is)}. \end{aligned} \tag{64}$$

One advantage of the time-dependent formulation is that the prescription

$$\lim_{s \to +0} \frac{1}{x - is} = \mathcal{P}\frac{1}{x} + i\pi\delta(x) \tag{65}$$

tells us how to handle the poles which arise in ordinary perturbation theory; here $\mathcal{P}$ denotes principal part. Interesting consequences follow from a graphical analysis of the contributions to the U operator, as discussed in Chapter 6.

An important advantage of the time-dependent development makes it possible to separate immediately parts of the problem which refer to disconnected parts of the system. In ordinary perturbation theory the factorization is seen only at the end of the calculation. Suppose that a and b refer to different regions of space a and b with no physical connection. Let

$$H = H_a^0 + H_b^0 + V_a + V_b. \tag{66}$$

We form, noting that all commutators necessarily involve δ_{ab},

$$\begin{aligned} U(0,-\infty) &= P\left\{\exp\left[-i\int_{-\infty}^{0} dt\, V(t)\right]\right\} \\ &= P\left\{\exp\left[-i\right.\right. \\ &\qquad \left.\left.\int_{-\infty}^{0} dt\, e^{iH_a{}^0t}e^{iH_b{}^0t}(V_a + V_b)e^{-iH_a{}^0t}e^{-iH_b{}^0t}\right]\right\} \\ &= P\left\{\exp\left[-i\int_{-\infty}^{0} dt\, e^{iH_a{}^0t}V_ae^{-iH_a{}^0t}\right]\right. \\ &\qquad \left.\exp\left[-i\int_{-\infty}^{0} dt\, e^{iH_b{}^0t}V_be^{-iH_b{}^0t}\right]\right\} \\ &= U_a(0,-\infty)U_b(0,-\infty). \end{aligned} \tag{67}$$

This is a very useful result, one which is difficult to obtain in ordinary perturbation theory.

PROBLEMS

1. Show that

$$\int d^3k e^{i\mathbf{k}\cdot\mathbf{x}} = 4\pi\left(\frac{\sin k_Fr - k_Fr\cos k_Fr}{r^3}\right), \tag{68}$$

where the integral is taken over the sphere $0 < k < k_F$.

2. Show that

$$\int d^3x e^{i\mathbf{K}\cdot\mathbf{x}}\frac{x_{\mathbf{K}}}{r^3} = \frac{4\pi i}{|\mathbf{K}|}, \tag{69}$$

where $x_{\mathbf{K}}$ is the projection of $\mathbf{x}$ on $\mathbf{K}$.

3. Show that

$$\theta(t) = \lim_{s\to+0}\frac{i}{2\pi}\int_{-\infty}^{\infty} dx\,\frac{e^{-ixt}}{x+is}, \tag{70}$$

where

$$\theta(t) = \begin{matrix} 1 & t > 0 \\ 0 & t < 0. \end{matrix} \tag{71}$$

4. (*a*) Show that the commutator

$$[e^{-i\mathbf{k}\cdot\mathbf{x}},\mathbf{p}] = \mathbf{k}e^{-i\mathbf{k}\cdot\mathbf{x}}. \tag{72}$$

(b) Show that

$$[e^{-i\mathbf{k}\cdot\mathbf{x}},p^2] = e^{-i\mathbf{k}\cdot\mathbf{x}}(2\mathbf{k}\cdot\mathbf{p} - k^2). \tag{73}$$

2 Acoustic phonons

The duality of waves and particles is the dominant concept of modern physics. In crystals there are many fields which combine wave and particle aspects. Names have been given to the quantized unit of energy in these fields. Exactly as the word *photon* describes the particle aspect of the electromagnetic field in a vacuum, the words phonon, magnon, plasmon, polaron, and exciton describe several of the quantized fields in crystals. Phonons are associated with elastic excitations—acoustic phonons correspond to ordinary elastic waves. Magnons are the elementary excitations of the system of electron spins coupled together by exchange interactions. Plasmons are the collective coulomb excitations of the electron gas in metals. Excitons are neutral particles associated with the dielectric polarization field; and polarons are charged particles associated with the polarization field, usually in ionic crystals. Except for polarons, the particular particles given above act as bosons. The Cooper pairs of electrons in the Bardeen-Cooper-Schrieffer theory of superconductivity act to a certain extent as bosons. The quasiparticle composed of an electron and its interactions with the electron gas in a metal acts as a fermion.

Much of this book is concerned with these particles—their quantization, their spectra, and their interactions. The most convenient mathematical description of the particles uses the method of second quantization—the quantization of wave-particle fields. The method is quite easy to learn and to use: excellent introductions are given in standard textbooks on quantum mechanics, particularly those by Schiff (Chapter 13), Landau and Lifshitz, and Henley and Thirring. The nonrelativistic field problems we treat are good pedagogical illustrations of the more elementary content of quantum field theory.

The fields of interest to us can be developed for a lattice of discrete atoms or for a homogeneous continuum. Much of the time the particular model employed is a matter of indifference to us, but for the discrete lattice the dispersion relations and selection rules are more general than for the continuum. The lattice includes the continuum as a limiting case.

DISCRETE ELASTIC LINE

The transverse motion of an elastic line under tension is a simple example of a boson field. We exhibit the hamiltonian of a line composed of discrete points of unit mass at unit spacing under unit tension. The appropriate parameters will be restored later, after we have settled the mathematical preliminaries. We shall later impose periodic boundary conditions, with periodicity N. Let p_i denote the transverse momentum and q_i the transverse displacement of the mass at the point i; then for small displacements

$$H = \frac{1}{2}\sum_i [p_i^2 + (q_{i+1} - q_i)^2]. \tag{1}$$

This form is derived from the lagrangian

$$\begin{aligned} L &= \text{kinetic energy} - \text{potential energy} \\ &= \tfrac{1}{2}\Sigma\, \dot{q}_i^2 - \tfrac{1}{2}\Sigma\, (q_{i+1} - q_i)^2 \end{aligned} \tag{2}$$

by forming the canonical momenta

$$p_i = \partial L/\partial \dot{q}_i = \dot{q}_i, \tag{3}$$

and evaluating the hamiltonian

$$H = \Sigma\, p_i\dot{q}_i - L. \tag{4}$$

The theory is quantized by the usual condition

$$[q_r,p_s] = i\delta_{rs}, \tag{5}$$

in units with $\hbar = 1$.

We now find the eigenfrequencies and eigenvectors of (1). We transform to phonon or wave coordinates Q_k:

$$q_r = N^{-1/2}\sum_k Q_k e^{ikr}; \qquad Q_k = N^{-1/2}\sum_s q_s e^{-iks}. \tag{6}$$

These transformations are consistent: it was shown in Chapter 1 that $\sum_k e^{ik(s-r)} = N\delta_{sr}$. The coordinates q_r must be hermitian variables in quantum mechanics, so that q_r must be equal to its hermitian adjoint:

$$q_r = q_r^+ = N^{-1/2}\,\Sigma\, Q_k e^{ikr} = N^{-1/2}\,\Sigma\, Q_k^+ e^{-ikr}. \tag{7}$$

This relation is satisfied if

$$Q_k = Q^+_{-k}, \tag{8}$$

with Q^+ the hermitian adjoint operator to Q; we may write

$$q_r = \tfrac{1}{2}N^{-1/2} \sum_k (Q_k e^{ikr} + Q^+_k e^{-ikr}), \tag{9}$$

with the sum still taken over all allowed values of the wavevector k, positive and negative.

The allowed values of k are usually determined as in Chapter 1 by the periodic boundary condition

$$q_{r+N} = q_r, \tag{10}$$

whence

$$e^{ikN} = 1, \tag{11}$$

which is satisfied by

$$k = 2\pi n/N, \qquad n = 0, \pm 1, \pm 2, \cdots, \pm(N-1), N. \tag{12}$$

This is not the only possible choice for k, but it is the most useful choice.

It is a good practice to make the coordinate transformation (6) in the lagrangian, because we can then determine the momentum component P_k canonical to Q_k. We need the relations

$$\sum_r (\dot{q}_r)^2 = N^{-1} \sum_k \sum_{k'} \sum_r \dot{Q}_k \dot{Q}_{k'} e^{i(k+k')r} = \sum_k \dot{Q}_k \dot{Q}_{-k}. \tag{13}$$

Further,

$$\begin{aligned} \sum (q_{r+1} - q_r)^2 &= N^{-1} \sum_k \sum_{k'} \sum_r Q_k Q_{k'} e^{ikr}(e^{ik} - 1)e^{ik'r}(e^{ik'} - 1) \\ &= 2 \sum_k Q_k Q_{-k}(1 - \cos k). \end{aligned} \tag{14}$$

Thus

$$L = \tfrac{1}{2} \Sigma\, \dot{Q}_k \dot{Q}_{-k} - \Sigma\, (1 - \cos k) Q_k Q_{-k}; \tag{15}$$

$$P_k = \partial L/\partial \dot{Q}_k = \dot{Q}_{-k} = P^+_{-k}, \tag{16}$$

and

$$H = \tfrac{1}{2} \Sigma\, P_k P_{-k} + \Sigma\, (1 - \cos k) Q_k Q_{-k}. \tag{17}$$

In terms of the new coordinates,

$$P_k = N^{-1/2} \Sigma\, \dot{q}_s e^{iks} = N^{-1/2} \Sigma\, p_s e^{iks}; \tag{18}$$

$$p_r = N^{-1/2} \Sigma\, P_k e^{-ikr}. \tag{19}$$

The commutation relation becomes

(20) $$[Q_k,P_{k'}] = N^{-1}\left[\sum_r q_re^{-ikr}, \sum_s p_se^{ik's}\right]$$
$$= N^{-1}\sum_{rs}[q_r,p_s]e^{-i(kr-k's)} = i\delta_{kk'}.$$

Except for $k = 0$, the wavevector index k refers to internal coordinates and it has no connection with the total momentum of the system. The total momentum is $\Sigma\, p_r$, and by (18) we see that

(21) $$P_0 = N^{-\frac{1}{2}}\Sigma\, p_r.$$

The total momentum involves only the $k = 0$ mode, which is a uniform translation of the system. Many interaction processes in crystals proceed as if the total wavevector $\Sigma\,\mathbf{k}$ were conserved for the interacting particles, and for this reason $\mathbf{k}$ is often spoken of as the *crystal momentum* or as the quasimomentum. If a system is invariant under an infinitesimal translation, then the total wavevector is indeed rigorously conserved. A crystal lattice is not invariant under an infinitesimal translation, but under a translation by a multiple of the basis vectors **a**, **b**, **c** of the primitive cell. The conservation law becomes, as we shall see in later chapters,

(22) $$\Sigma\,\mathbf{k}_\nu = \mathbf{G},$$

where $\mathbf{G}$ is any reciprocal lattice vector, but this conservation law is distinct from the conservation of momentum of the center-of-mass.

The hamiltonian (17) is not quite in the form of a set of harmonic oscillators because of the mixture of terms in k and $-k$. Our object must be to get the hamiltonian into the harmonic oscillator form

(23) $$H = \Sigma\,\omega_k\hat{n}_k = \Sigma\,\omega_k a_k^+ a_k,$$

where $\hat{n}_k$ is the phonon population operator, and the a^+, a are boson creation and annihilation operators with the commutator

(24) $$[a_k,a_{k'}^+] = \delta_{kk'}.$$

Let us try the linear transformation

(25) $$\boxed{\begin{aligned} a_k^+ &= (2\omega_k)^{-\frac{1}{2}}(\omega_kQ_{-k} - iP_k); \\ a_k &= (2\omega_k)^{-\frac{1}{2}}(\omega_kQ_k + iP_{-k}), \end{aligned}}$$

using the conditions (8) and (16); here ω_k is defined as

(26) $$\omega_k = [2(1 - \cos k)]^{\frac{1}{2}}.$$

The ω_k are just the classical oscillation frequencies. For low k we have $\omega \propto k$, so that there is no dispersion. Dispersion enters when the wavelength $2\pi/k$ approaches the interatomic separation, so that the waves see the discrete nature of the lattice. Note that ω is a periodic function of k. Now

$$(27)\quad [a_k,a_{k'}^+] = (2\omega_k)^{-1}(-i\omega_k[Q_k,P_{k'}] + i\omega_k[P_{-k},Q_{-k'}]) = \delta_{kk'},$$

as required.

Let us form the sums over $\pm k$ in the hamiltonian (17):

$$(28)\quad \tfrac{1}{2}(P_kP_{-k} + P_{-k}P_k) + (1 - \cos k)(Q_kQ_{-k} + Q_{-k}Q_k) \\ = \tfrac{1}{2}\omega_k(a_k^+a_k + a_ka_k^+ + a_{-k}^+a_{-k} + a_{-k}a_{-k}^+).$$

Thus the hamiltonian may be written as

$$(29)\quad \boxed{H = \Sigma\, \omega_k(a_k^+a_k + \tfrac{1}{2}); \qquad \omega_k = [2(1 - \cos k)]^{1/2}.}$$

This contains the operator for the number of bosons in the state k:

$$(30)\quad \hat{n}_k = a_k^+a_k.$$

If a state Φ is specified by the occupancy eigenvalues n_k, then the Schrödinger equation is

$$(31)\quad H\Phi = E\Phi = (\Sigma\, n_k\omega_k)\Phi.$$

An increase of n_k by one is described as the excitation of one *phonon* of energy ω_k.

The inverse transformation to (25) is

$$(32)\quad P_k = i(\omega_k/2)^{1/2}(a_k^+ - a_{-k}); \qquad Q_k = (2\omega_k)^{-1/2}(a_k + a_{-k}^+).$$

We insert (32) for Q_k into (9) to obtain, with $\hbar$ and M restored,

$$(33)\quad \boxed{q_r = \Sigma\, (\hbar/2MN\omega_k)^{1/2}(a_ke^{ikr} + a_k^+e^{-ikr}).}$$

The modes of oscillation which we have quantized are identical with the classical modes of the system.

There are problems for which it is convenient to work only with real coordinates in the phonon expansion—the Q_k above are not real because $Q_k^+ \neq Q_k$. Real Q_k are particularly advantageous if we desire to exhibit the Schrödinger eigenfunctions $|Q_k\rangle$ of the normal modes. A simple set[1] of real coordinates is provided by standing waves:

$$(34)\quad q_r = (2/N)^{1/2} \sum_{k>0} \{Q_k^{(c)} \cos kr + Q_k^{(s)} \sin kr\},$$

[1] G. Leibfried, in *Handbuch der Physik*, VII/1, 104 (1955); see especially pp. 160, 165, 260; for a discussion of the phonon polarization vector, see p. 174.

where k assumes values $(2\pi/N)$ times a positive integer up to $N/2$. We note that the eigenvectors $(2/N)^{1/2}\cos kr$ and $(2/N)^{1/2}\sin kr$ are orthonormal. For example,

$$(35)\quad \sum_r (2/N)\cos kr\cos k'r = (1/N)\sum_r \{\cos (k+k')r + \cos (k-k')r\};$$

now

$$(36)\quad \sum_r \cos (k \pm k')r = \Re \sum_{r=0}^{N-1} e^{i2\pi nr/N} = \Re \frac{1-e^{i2\pi n}}{1-e^{i2\pi n/N}} = 0,$$

unless $n = 0$, in which case the sum is equal to N. Thus

$$(37)\quad \sum_r (2/N)\cos kr\cos k'r = \delta_{kk'};$$

and similarly

$$(38)\quad \Sigma\,(2/N)\sin kr\sin k'r = \delta_{kk'}.$$

The cross-product terms $\cos k'r\sin kr$ are always orthogonal.

The momentum is

$$(38a)\quad p_r = (2/N)^{1/2}\sum_{k>0}(P_k^{(c)}\cos kr + P_k^{(s)}\sin kr).$$

We wish to find the commutation relations of the P_k, Q_k:

$$(39)\quad Q_k^{(c)} = (2/N)^{1/2}\sum_r q_r\cos kr; \qquad Q_k^{(s)} = (2/N)^{1/2}\sum_r q_r\sin kr;$$

$$(40)\quad P_k^{(c)} = (2/N)^{1/2}\sum_r p_r\cos kr; \qquad P_k^{(s)} = (2/N)^{1/2}\sum_r p_r\sin kr.$$

Then

$$(41)\quad [Q_k^{(c)},P_{k'}^{(c)}] = (2/N)\sum_{rs}[q_r,p_s]\cos kr\cos k's = (2/N)\sum_r i\cos kr\cos k'r = i\delta_{kk'},$$

by (37). In general,

$$(42)\quad [Q_k^{(\alpha)},P_{k'}^{(\beta)}] = i\delta_{\alpha\beta}\delta_{kk'}.$$

The hamiltonian can be written as

$$(43)\quad H = \sum_{k\alpha}\left\{\frac{1}{2M}P_k^{(\alpha)^2} + \tfrac{1}{2}M\omega_k{}^2Q_k^{(\alpha)^2}\right\},$$

where α runs over c and s; we assume $\omega_k^{(c)} = \omega_k^{(s)}$. This is just the hamiltonian for a collection of harmonic oscillators of frequency ω_k.

Now Q_k is the amplitude of the oscillator k. The wavefunction $\varphi_{n_k}(Q_k) \equiv |Q_k n_k\rangle$ satisfies the Schrödinger equation

$$\left\{-\frac{1}{2M}\frac{\partial^2}{\partial Q_k{}^2}+\frac{1}{2}M\omega_k{}^2Q_k{}^2\right\}|Q_kn_k\rangle = (n_k+\tfrac{1}{2})\omega_k|Q_kn_k\rangle. \tag{44}$$

The wavefunction of the whole system may be written as the product $\Pi|Q_kn_k\rangle$.

In terms of creation and annihilation operators,

$$Q_k^{(\alpha)} = (2M\omega_k)^{-\frac{1}{2}}(a_{k\alpha} + a_{k\alpha}^+); \tag{45}$$

$$P_k^{(\alpha)} = i(M\omega_k/2)^{\frac{1}{2}}(a_{k\alpha} - a_{k\alpha}^+). \tag{46}$$

Then

$$H = \sum_{k\alpha} (\hat{n}_{k\alpha} + \tfrac{1}{2})\omega_k. \tag{47}$$

QUANTUM THEORY OF THE CONTINUOUS ELASTIC LINE

We consider an elastic line of linear density ρ under tension T. The classical lagrangian is $L = \int dx\, \mathfrak{L}$; from standard classical mechanics the lagrangian density

$$\mathfrak{L} = \tfrac{1}{2}\rho\dot{\psi}^2 - \tfrac{1}{2}T\left(\frac{\partial\psi}{\partial x}\right)^2, \tag{48}$$

where $\psi(x,t)$ is the displacement of the string from the equilibrium position. The result (48) follows directly from the lagrangian (2) for the discrete line, on taking appropriate limits.

The classical derivation of the lagrangian equations of motion for a field follow from the variational principle

$$\delta\int_{t_1}^{t_2} L\,dt = 0, \tag{49}$$

subject to $\delta\psi(x,t_1) = 0$; $\delta\psi(x,t_2) = 0$. Thus $\int\int \delta\mathfrak{L}\,dt\,dx = 0$, where

$$\delta\mathfrak{L} = \frac{\partial\mathfrak{L}}{\partial\psi}\delta\psi + \frac{\partial\mathfrak{L}}{\partial\dot{\psi}}\delta\dot{\psi} + \frac{\partial\mathfrak{L}}{\partial(\partial\psi/\partial x)}\delta\frac{\partial\psi}{\partial x} + \frac{\partial\mathfrak{L}}{\partial t}\delta t, \tag{50}$$

with $\delta\dot{\psi} = \dfrac{\partial}{\partial t}\delta\psi$; $\delta\dfrac{\partial\psi}{\partial x} = \dfrac{\partial}{\partial x}\delta\psi$. Then

$$\int_{t_1}^{t_2}\frac{\partial\mathfrak{L}}{\partial\dot{\psi}}\frac{\partial}{\partial t}\delta\psi\,dt = \left[\frac{\partial\mathfrak{L}}{\partial\dot{\psi}}\delta\psi\right]_{t_1}^{t_2} - \int_{t_1}^{t_2}\left(\frac{\partial}{\partial t}\frac{\partial\mathfrak{L}}{\partial\dot{\psi}}\right)\delta\psi\,dt, \tag{51}$$

where the first term on the right is zero because of the conditions on the variational principle. Further,

$$\int \frac{\partial \mathcal{L}}{\partial(\partial\psi/\partial x)} \frac{\partial}{\partial x} \delta\psi \, dx = - \int \left(\frac{\partial}{\partial x} \frac{\partial \mathcal{L}}{\partial(\partial\psi/\partial x)} \right) \delta\psi \, dx. \tag{52}$$

The extremum condition becomes

$$\frac{\partial \mathcal{L}}{\partial \psi} - \frac{\partial}{\partial t} \frac{\partial \mathcal{L}}{\partial \dot{\psi}} - \frac{\partial}{\partial x} \frac{\partial \mathcal{L}}{\partial(\partial\psi/\partial x)} = 0. \tag{53}$$

This is the equation of motion for the lagrangian density, and for the elastic line (48) leads to

$$\rho \frac{\partial^2 \psi}{\partial t^2} - T \frac{\partial^2 \psi}{\partial x^2} = 0, \tag{54}$$

a well-known wave equation.

The momentum density π is defined by

$$\pi \equiv \frac{\partial \mathcal{L}}{\partial \dot{\psi}}, \tag{55}$$

by analogy with the definition $p \equiv \partial L/\partial \dot{q}$ of the particle momentum. The hamiltonian density $\mathcal{H}$ is defined by

$$\mathcal{H} = \pi \dot{\psi} - \mathcal{L} = \frac{1}{2\rho} \pi^2 + \tfrac{1}{2} T \left(\frac{\partial \psi}{\partial x} \right)^2. \tag{56}$$

The definition (55) of momentum density is consistent with the limit of the discrete result $(\Delta\tau)\pi = p = \partial L/\partial \dot{q} = \partial(\Delta\tau)\mathcal{L}/\partial \dot{q}$ for a cell of volume $\Delta\tau$.

QUANTIZATION OF THE FIELD

The particle quantum condition is $[q_r, p_s] = i\delta_{rs}$. If the continuum is divided into cells of extension $\Delta\tau_s$, then the commutator becomes, with $\psi(r)$ written for q_r,

$$[\psi(r), \pi(s)\, \Delta\tau_s] = i\delta_{rs}, \tag{57}$$

or, in three dimensions,

$$[\psi(\mathbf{x}), \pi(\mathbf{x}')] = i\delta(\mathbf{x},\mathbf{x}'), \tag{58}$$

where $\delta(\mathbf{x},\mathbf{x}') = 1/\Delta\tau$ if $\mathbf{x}$, $\mathbf{x}'$ are in the same cell $\Delta\tau$ and zero otherwise. The function $\delta(\mathbf{x},\mathbf{x}')$ has the property that $\int f(\mathbf{x})\delta(\mathbf{x},\mathbf{x}')\, d\tau$ is equal to the average value of f for the cell in which $\mathbf{x}'$ is located; therefore in the limit as $\Delta\tau \to 0$,

$$\delta(\mathbf{x},\mathbf{x}') \to \delta(\mathbf{x} - \mathbf{x}'), \tag{59}$$

the delta function in three dimensions. In this limit the quantum condition becomes

$$[\psi(\mathbf{x},t),\pi(\mathbf{x}',t)] = i\delta(\mathbf{x} - \mathbf{x}'), \tag{60}$$

noting that our discussion has referred only to a common time t.

It is apparent that there are two ways of dealing further with the hamiltonian of the elastic line. One method is to solve the quantum equation of motion directly in the Heisenberg representation; in some problems one gains physical insight by working with the equations of motion. Often it is tedious to obtain these equations, and we may wish simply to transform the quantum operators in the hamiltonian to a form solvable on inspection.

In looking for suitable transformations of

$$\mathcal{H} = \frac{1}{2\rho}\pi^2 + \tfrac{1}{2}T\left(\frac{\partial\psi}{\partial x}\right)^2 \tag{61}$$

we may be guided by our experience with the discrete line. We apply periodic boundary conditions over a length L. We set

$$\psi(x) = L^{-\frac{1}{2}}\sum_k Q_k e^{ikx}, \qquad Q_k = Q_{-k}^+; \tag{62}$$

from Eq. (1.22),

$$Q_k = L^{-\frac{1}{2}}\int d\xi\, \psi(\xi)e^{-ik\xi}, \tag{63}$$

where the integral is taken between $-L/2$ and $L/2$. Then

$$\begin{aligned} \int (\partial\psi/\partial x)^2\, dx &= -(1/L)\sum_k\sum_{k'} kk'Q_kQ_{k'}\int e^{i(k+k')x}\, dx \\ &= \sum_k k^2 Q_kQ_{-k}. \end{aligned} \tag{64}$$

We define $P_k = P_{-k}^+$ such that

$$P_k = L^{-\frac{1}{2}}\int \pi(\xi)e^{ik\xi}\, d\xi; \qquad \pi(x) = L^{-\frac{1}{2}}\sum_k P_k e^{-ikx}. \tag{65}$$

Then

$$\int \pi^2\, dx = \sum_k P_kP_{-k}, \tag{66}$$

and the hamiltonian becomes

$$H = \int \mathcal{H}\, dx = \sum\left(\frac{1}{2\rho}P_kP_{-k} + \tfrac{1}{2}Tk^2Q_kQ_{-k}\right). \tag{67}$$

The quantum condition is readily found:

$$[Q_k,P_{k'}] = L^{-1} \iint [\psi(x),\pi(x')]e^{i(k'x'-kx)}\,dx\,dx' \tag{68}$$
$$= iL^{-1} \int e^{i(k'-k)x}\,dx = i\delta_{kk'}.$$

The form (68) is similar to (20), so that the subsequent transformation may be modeled after (25) and (32):

$$\begin{aligned} a_k^+ &= -i(2\rho\omega_k)^{-1/2}P_k + (T/2\omega_k)^{1/2}kQ_{-k}; \\ a_k &= i(2\rho\omega_k)^{-1/2}P_{-k} + (T/2\omega_k)^{1/2}kQ_k, \end{aligned} \tag{69}$$

where

$$\omega_k = (T/\rho)^{1/2}|k|. \tag{70}$$

LONG WAVELENGTH ACOUSTIC MODE PHONONS IN ISOTROPIC CRYSTALS

We let **R** denote the vector displacement operator in a solid,[2] with $\mathbf{R} = \mathbf{x}' - \mathbf{x}$, where $\mathbf{x}$ is the initial position of an atom or of a volume element of the solid, and $\mathbf{x}'$ is the position after deformation. We assume the crystal is isotropic elastically, that is, that the elastic energy associated with a given state of strain is independent of the orientation of the crystalline axes. It turns out that the eigensolutions may be classified rigorously as longitudinal or transverse when the elastic energy density is isotropic. Cubic crystals with large primitive cells, such as yttrium iron garnet, may tend to be isotropic elastically.

First we find expressions quadratic in the strain components $\partial R_\mu/\partial x_\nu$ and invariant under arbitrary rotation of axes. There are three such invariants: $(\text{div}\,\mathbf{R})^2$, $|\nabla\mathbf{R}|^2$, and $|\text{curl}\,\mathbf{R}|^2$. The deformation associated with curl **R** is a rotation, not a strain, so that this term will not appear in the elastic energy. The strain energy density U in the quadratic approximation may then be written

$$U = \tfrac{1}{2}\alpha(\text{div}\,\mathbf{R})^2 + \tfrac{1}{2}\beta|\nabla\mathbf{R}|^2 = \tfrac{1}{2}\alpha\frac{\partial R_\mu}{\partial x_\mu}\frac{\partial R_\nu}{\partial x_\nu} + \tfrac{1}{2}\beta\frac{\partial R_\mu}{\partial x_\nu}\frac{\partial R_\mu}{\partial x_\nu}, \tag{71}$$

where α, β are constants related to elastic moduli. We are to sum over repeated indices. The coordinate axes x_μ are assumed to be orthogonal. The term in α is the square of the trace of the strain tensor; the term in β is the sum of the squares of the tensor components.

[2] *ISSP*, Chapter 4. The vector **R** now plays the role of the scalar ψ in the preceding section on the elastic line. In *ISSP* the symbol $\boldsymbol{\rho}$ was used.

The hamiltonian density of the isotropic elastic continuum is

$$\mathcal{H} = \frac{1}{2\rho}\Pi_\mu\Pi_\mu + \tfrac{1}{2}\alpha\frac{\partial R_\mu}{\partial x_\mu}\frac{\partial R_\nu}{\partial x_\nu} + \tfrac{1}{2}\beta\frac{\partial R_\mu}{\partial x_\nu}\frac{\partial R_\mu}{\partial x_\nu}, \tag{72}$$

where ρ is the density and Π_μ are the components of the momentum density. We assume cyclic boundary conditions over a unit cube and define the transformation to phonon variables $Q_{\mathbf{k}}^\mu$, $\mu = x, y, z$:

$$Q_{\mathbf{k}}^\mu = \int d^3x\, R_\mu(\mathbf{x})e^{-i\mathbf{k}\cdot\mathbf{x}}; \qquad R_\mu(\mathbf{x}) = \sum_{\mathbf{k}} Q_{\mathbf{k}}^\mu e^{i\mathbf{k}\cdot\mathbf{x}}, \tag{73}$$

with the condition $Q_{\mathbf{k}}^\mu = (Q_{-\mathbf{k}}^\mu)^+$. We have

$$\int d^3x\, \frac{\partial R_\mu}{\partial x_\mu}\frac{\partial R_\nu}{\partial x_\nu} = \sum_{\mathbf{k}} k_\mu k_\nu Q_{\mathbf{k}}^\mu Q_{-\mathbf{k}}^\nu. \tag{74}$$

We define momentum components $P_{\mathbf{k}}^\mu$ such that

$$\Pi_\mu = \sum_{\mathbf{k}} P_{\mathbf{k}}^\mu e^{-i\mathbf{k}\cdot\mathbf{x}}; \qquad P_{\mathbf{k}}^\mu = (P_{-\mathbf{k}}^\mu)^+; \tag{75}$$

$$P_{\mathbf{k}}^\mu = \int d^3x\, \Pi_\mu(\mathbf{x})e^{i\mathbf{k}\cdot\mathbf{x}} \tag{76}$$

then

$$\int d^3x\, \Pi_\mu\Pi_\mu = \sum_{\mathbf{k}} P_{\mathbf{k}}^\mu P_{-\mathbf{k}}^\mu, \tag{77}$$

and the hamiltonian becomes

$$H = \frac{1}{2\rho}\sum_{\mathbf{k}} P_{\mathbf{k}}^\mu P_{-\mathbf{k}}^\mu + \tfrac{1}{2}\alpha\sum_{\mathbf{k}} k_\mu k_\nu Q_{\mathbf{k}}^\mu Q_{-\mathbf{k}}^\nu + \tfrac{1}{2}\beta\sum_{\mathbf{k}} k_\nu^2 Q_{\mathbf{k}}^\mu Q_{-\mathbf{k}}^\mu. \tag{78}$$

The hamiltonian was constructed to be invariant under rotation of crystal axes, and we may choose the directions of the coordinate axes as we wish. Because different **k**'s, outside the pair $\pm\mathbf{k}$, are not mixed in (78), we may choose different axes for each **k**. It is convenient to choose one of the axes, x_l, parallel to **k**. Then

$$H = \frac{1}{2\rho}\sum_{\mathbf{k}\mu} P_{\mathbf{k}}^\mu P_{-\mathbf{k}}^\mu + \sum_{\mathbf{k}} k^2\left(\tfrac{1}{2}\alpha Q_{\mathbf{k}}^l Q_{-\mathbf{k}}^l + \tfrac{1}{2}\beta\sum_\mu Q_{\mathbf{k}}^\mu Q_{-\mathbf{k}}^\mu\right). \tag{79}$$

Now introduce the boson operators

$$\begin{aligned} a_{\mathbf{k}\mu}^+ &= -i(2\rho\omega_{\mathbf{k}\mu})^{-1/2}P_{\mathbf{k}}^\mu + \omega_{\mathbf{k}\mu}^{-1/2}(\alpha\delta_{\mu l} + \beta)^{1/2}kQ_{-\mathbf{k}}^\mu; \\ a_{\mathbf{k}\mu} &= i(2\rho\omega_{\mathbf{k}\mu})^{-1/2}P_{-\mathbf{k}}^\mu + \omega_{\mathbf{k}\mu}^{-1/2}(\alpha\delta_{\mu l} + \beta)^{1/2}kQ_{\mathbf{k}}^\mu. \end{aligned} \tag{80}$$

In (80) repeated indices are not summed; the index μ has now the significance of a polarization index, relating the particle displacement to the wavevector. We verify readily that the energy is

$$E = \sum_{\mathbf{k}\mu} \omega_{\mathbf{k}\mu}(n_{\mathbf{k}\mu} + \tfrac{1}{2}); \qquad \omega_{\mathbf{k}\mu} = [(\alpha\delta_{\mu l} + \beta)/\rho]^{1/2}k, \tag{81}$$

where l denotes the longitudinal phonon ($Q_{\mathbf{k}} \parallel \mathbf{k}$) and the other two choices of μ denote transverse phonons. The two transverse phonons for a given $\mathbf{k}$ are degenerate in an isotropic solid, but for a general direction in a cubic crystal there are three modes for each $\mathbf{k}$, all nondegenerate, and none exactly longitudinal or transverse. The velocities of sound given by (81) are

$$v_\mu = \partial\omega_\mu/\partial k = [(\alpha\delta_{\mu l} + \beta)/\rho]^{1/2}. \tag{82}$$

The lattice displacement operator $\mathbf{R}$ is found from (73) and (80):

$$\boxed{\mathbf{R}(\mathbf{x}) = \sum_{\mathbf{k}} \mathbf{e}_{\mathbf{k}\mu}(2\rho\omega_{\mathbf{k}\mu})^{-1/2}(a_{\mathbf{k}\mu}e^{i\mathbf{k}\cdot\mathbf{x}} + a_{\mathbf{k}\mu}^{+}e^{-i\mathbf{k}\cdot\mathbf{x}}),} \tag{83}$$

where $\mathbf{e}_{\mathbf{k}\mu}$ is a unit vector in the direction of the polarization of the phonon. The dilation operator is

$$\Delta(\mathbf{x}) = \partial R_\mu/\partial x_\mu = i \sum_{\mathbf{k}} (2\rho\omega_{\mathbf{k}l})^{-1/2}k(a_{\mathbf{k}l}e^{i\mathbf{k}\cdot\mathbf{x}} - a_{\mathbf{k}l}^{+}e^{-i\mathbf{k}\cdot\mathbf{x}}); \tag{84}$$

classically, Δ is the fractional volume change $\delta V/V$.

PHONONS IN A CONDENSED BOSON GAS

We want to show by a method of Bogoliubov how phonons may arise in a system of weakly interacting particles. We consider a system of a large number of weakly interacting bosons described by the hamiltonian

$$H = \sum_{\mathbf{k}} \varepsilon_{\mathbf{k}}a_{\mathbf{k}}^{+}a_{\mathbf{k}} + \tfrac{1}{2}\sum V(\mathbf{k}_1 - \mathbf{k}_1{}')a_{\mathbf{k}_1}^{+}a_{\mathbf{k}_2}^{+}\,a_{\mathbf{k}_2{}'}a_{\mathbf{k}_1{}'}\Delta(\mathbf{k}_1 + \mathbf{k}_2 - \mathbf{k}_1{}' - \mathbf{k}_2{}'), \tag{85}$$

where the Kronecker delta symbol Δ assures conservation of wavevector. If $V = 0$, the ground state has all particles condensed because of the bose statistics in the lowest one-particle state, normally $\mathbf{k} = 0$. With a weak and short-range interaction most of the particles will be in the ground state, and a few ($n \ll N$) will be in excited states $k > 0$. It will make little difference in the calculation of macroscopic quantities whether we use states with N or $N + n$ particles, but the expectation value of the total number of particles will be well defined. In liquid He^4 the potential is not so very weak, and it is believed from the interpretation of neutron diffraction studies that in the ground state less than 0.1 of the atoms are in the zero momentum state [see A. Miller, D. Pines, and P. Nozières, *Phys. Rev.* **127,** 1452 (1962)]. The model we treat probably bears only a qualitative relation to real He^4.

Then (85) may be written, assuming $V_{\mathbf{k}} = V_{-\mathbf{k}}$,

$$\text{(86)} \quad H = \sum_{\mathbf{k}} \varepsilon_{\mathbf{k}} a_{\mathbf{k}}^{+} a_{\mathbf{k}} + \tfrac{1}{2} N_0{}^2 V_0 + N_0 V_0 \sum_{\mathbf{k}}' a_{\mathbf{k}}^{+} a_{\mathbf{k}} + N_0 \sum_{\mathbf{k}}' V_{\mathbf{k}} a_{\mathbf{k}}^{+} a_{\mathbf{k}} + \tfrac{1}{2} N_0 \sum_{\mathbf{k}}' V_{\mathbf{k}} (a_{\mathbf{k}} a_{-\mathbf{k}} + a_{\mathbf{k}}^{+} a_{-\mathbf{k}}^{+}) + \text{terms of higher order.}$$

Here $N_0 = a_0^{+} a_0$ and the summations do not include $\mathbf{k} = 0$. Reading from left to right, the terms retained in the hamiltonian (86) are:

(*a*) Kinetic energy: $\varepsilon_{\mathbf{k}} a_{\mathbf{k}}^{+} a_{\mathbf{k}}$.

(*b*) Interactions in the ground state: $a_0^{+} a_0^{+} a_0 a_0$.

(*c*) One particle not excited in the ground state: $a_0^{+} a_{\mathbf{k}}^{+} a_{\mathbf{k}} a_0$ and $a_{\mathbf{k}}^{+} a_0^{+} a_0 a_{\mathbf{k}}$.

(*d*) Exchange of one particle in the ground state: $a_{\mathbf{k}}^{+} a_0^{+} a_{\mathbf{k}} a_0$ and $a_0^{+} a_{\mathbf{k}}^{+} a_0 a_{\mathbf{k}}$.

(*e*) Both initial or both final particles in the ground state: $a_0^{+} a_0^{+} a_{\mathbf{k}} a_{-\mathbf{k}}$ and $a_{\mathbf{k}}^{+} a_{-\mathbf{k}}^{+} a_0 a_0$.

Terms with three ground-state operators are excluded by momentum conservation.

We now take the expectation value of $N_0 + \sum_{\mathbf{k}}' a_{\mathbf{k}}^{+} a_{\mathbf{k}}$ to be the number of particles N in the system. We collect terms and rewrite (86) as the reduced hamiltonian

$$\text{(87)} \quad H_{\text{red}} = \tfrac{1}{2} N^2 V_0 + \sum_{\mathbf{k}}' (\varepsilon_{\mathbf{k}} + N V_{\mathbf{k}}) a_{\mathbf{k}}^{+} a_{\mathbf{k}} + \tfrac{1}{2} N \sum_{\mathbf{k}}' V_{\mathbf{k}} (a_{\mathbf{k}} a_{-\mathbf{k}} + a_{-\mathbf{k}}^{+} a_{\mathbf{k}}^{+}) + \cdots.$$

This is a bilinear form in the boson operators and may be diagonalized exactly. The reduced hamiltonian does not commute with the operator for the number of particles, but for a large system this introduces no perceptible error (see Pines, p. 335). The Green's function method discussed in Chapter 21 allows us to introduce in a natural way matrix elements connecting states with different numbers of particles.

But first let us consider a perturbation-theory approach, as discussed in detail in Brueckner, pp. 205–241. It is natural to treat $\Sigma\, \varepsilon_{\mathbf{k}} a_{\mathbf{k}}^{+} a_{\mathbf{k}}$ as the unperturbed energy and $\tfrac{1}{2} \Sigma\, V_{\mathbf{k}} (2 a_{\mathbf{k}}^{+} a_{\mathbf{k}} + a_{\mathbf{k}} a_{-\mathbf{k}} + a_{-\mathbf{k}}^{+} a_{\mathbf{k}}^{+})$ as the perturbation. The perturbation may easily be shown to lead to divergent terms in higher orders of the perturbation calculation.

Bogoliubov calls the divergent terms "dangerous diagrams." It is actually possible to sum the divergent terms to obtain a convergent result, but it is best to avoid perturbation theory by diagonalizing exactly the reduced hamiltonian (87), as we do now.

We carry out the diagonalization by a method using the equations of motion. The method is a systematic way of finding transformations such as (25), which was produced full-blown with no derivation. We look for new boson operators $\alpha_{\mathbf{k}}^{+}$, $\alpha_{\mathbf{k}}$ such that

$$(88)\qquad [\alpha_{\mathbf{k}}^{+},H] = -\lambda\alpha_{\mathbf{k}}^{+}; \qquad [\alpha_{\mathbf{k}},H] = \lambda\alpha_{\mathbf{k}}; \qquad [\alpha_{\mathbf{k}},\alpha_{\mathbf{k}'}^{+}] = \delta_{\mathbf{k}\mathbf{k}'}.$$

The first two relations are satisfied if the hamiltonian can be written in the diagonal form

$$(89)\qquad H = \sum_{\mathbf{k}} \lambda_{\mathbf{k}}\alpha_{\mathbf{k}}^{+}\alpha_{\mathbf{k}} + \text{constant}.$$

Let us write (87) as

$$(90)\qquad H_{\text{red}} - \tfrac{1}{2}N^2V_0 = \Sigma\, H_{\mathbf{k}};$$

$$(91)\qquad H_{\mathbf{k}} = \omega_0(a_{\mathbf{k}}^{+}a_{\mathbf{k}} + a_{-\mathbf{k}}^{+}a_{-\mathbf{k}}) + \omega_1(a_{\mathbf{k}}a_{-\mathbf{k}} + a_{-\mathbf{k}}^{+}a_{\mathbf{k}}^{+}),$$

with

$$(92)\qquad \omega_0 = \varepsilon_{\mathbf{k}} + NV_{\mathbf{k}}; \qquad \omega_1 = NV_{\mathbf{k}}.$$

Note that $i\dot{a}_{\mathbf{k}} = \omega_0 a_{\mathbf{k}} + \omega_1 a_{-\mathbf{k}}^{+}$ couples $a_{\mathbf{k}}$ and $a_{-\mathbf{k}}^{+}$. We make the transformation

$$(93)\qquad \alpha_{\mathbf{k}} = u_{\mathbf{k}}a_{\mathbf{k}} - v_{\mathbf{k}}a_{-\mathbf{k}}^{+}; \qquad \alpha_{\mathbf{k}}^{+} = u_{\mathbf{k}}a_{\mathbf{k}}^{+} - v_{\mathbf{k}}a_{-\mathbf{k}},$$

where $u_{\mathbf{k}}$, $v_{\mathbf{k}}$ are real functions of $\mathbf{k}$. The commutator

$$(94)\qquad [\alpha_{\mathbf{k}},\alpha_{\mathbf{k}}^{+}] = u_{\mathbf{k}}^2 - v_{\mathbf{k}}^2,$$

and we can choose $u_{\mathbf{k}}$, $v_{\mathbf{k}}$ to make this equal to one. With this choice $a_{\mathbf{k}} = u_{\mathbf{k}}\alpha_{\mathbf{k}} + v_{\mathbf{k}}\alpha_{-\mathbf{k}}^{+}$.

Using (91) and (93),

$$(95)\qquad [\alpha_{\mathbf{k}}^{+},H_{\mathbf{k}}] = u_{\mathbf{k}}(-\omega_0 a_{\mathbf{k}}^{+} - \omega_1 a_{-\mathbf{k}}) - v_{\mathbf{k}}(\omega_0 a_{-\mathbf{k}} + \omega_1 a_{\mathbf{k}}^{+}),$$

which, according to (88), we want to equal

$$-\lambda(u_{\mathbf{k}}a_{\mathbf{k}}^{+} - v_{\mathbf{k}}a_{-\mathbf{k}}).$$

Thus

$$(96)\qquad \begin{aligned} \omega_0 u_{\mathbf{k}} + \omega_1 v_{\mathbf{k}} &= \lambda u_{\mathbf{k}}; \\ \omega_1 u_{\mathbf{k}} + \omega_0 v_{\mathbf{k}} &= -\lambda v_{\mathbf{k}}. \end{aligned}$$

These equations have a solution if

$$\begin{vmatrix} \omega_0 - \lambda & \omega_1 \\ \omega_1 & \omega_0 + \lambda \end{vmatrix} = 0, \tag{97}$$

or

$$\lambda^2 = \omega_0{}^2 - \omega_1{}^2 = (\varepsilon_{\mathbf{k}} + NV_{\mathbf{k}})^2 - (NV_{\mathbf{k}})^2. \tag{98}$$

Now $\varepsilon_{\mathbf{k}} = k^2/2M$, so that

$$\lambda(k \to 0) = (NV_0/M)^{1/2}k; \tag{99}$$

this limit is the dispersion relation for a phonon with velocity

$$v_s = (NV_0/M)^{1/2}. \tag{100}$$

We have assumed $V_{\mathbf{k}} \cong V_0$ for small k. Obviously, the result (99) is applicable only for positive V, which corresponds to a repulsive potential. For high k,

$$\lambda(k \to \infty) \cong \varepsilon_{\mathbf{k}} + NV_{\mathbf{k}}, \tag{101}$$

which is a particle-like dispersion relation. The frequency in (98) is a monotonically increasing function of the wavevector, as we see immediately below. The quasiparticle excitation spectrum of liquid helium has been determined directly by studies of inelastic neutron scattering, with results shown in Fig. 1. The dip in the dispersion relation near $k = 1.9\ \text{A}^{-1}$ agrees quite well with the calculations of Feynman and Cohen;[3] see also Problem 6 in this chapter. On this simplified model the dip can result from a k-dependence of $V_{\mathbf{k}}$. The portion of the excitation spectrum near the dip is known as the *roton* spectrum.

Superfluidity.[3,4] The elementary excitation described by the operator $\alpha_{\mathbf{k}}^{+}$ is a collective property of the system, particularly for low $\mathbf{k}$; this is shown most directly by an extension of the argument of Problem 5 to a state with one elementary excitation. But there is an important difference between the present phonons and those carried by a crystal lattice. In a lattice, the phonons are referred to a relative coordinate system and do not carry momentum, as we saw in the discussion related to (21). In a gas, the phonon excitations do carry momentum, and from this we can construct a criterion for superfluidity.

Suppose a body of mass M is moving with velocity $\mathbf{v}$ through a bath of liquid helium at rest at 0°K. The body will be slowed down if it can

[3] See R. P. Feynman in *Progress in Low Temperature Physics,* **1,** 17 (1955); R. P. Feynman and M. Cohen, *Phys. Rev.* **102,** 1189 (1956).

[4] N. Bogoliubov, *J. Phys. USSR* **11,** 23 (1947); also reprinted in Pines, p. 292.

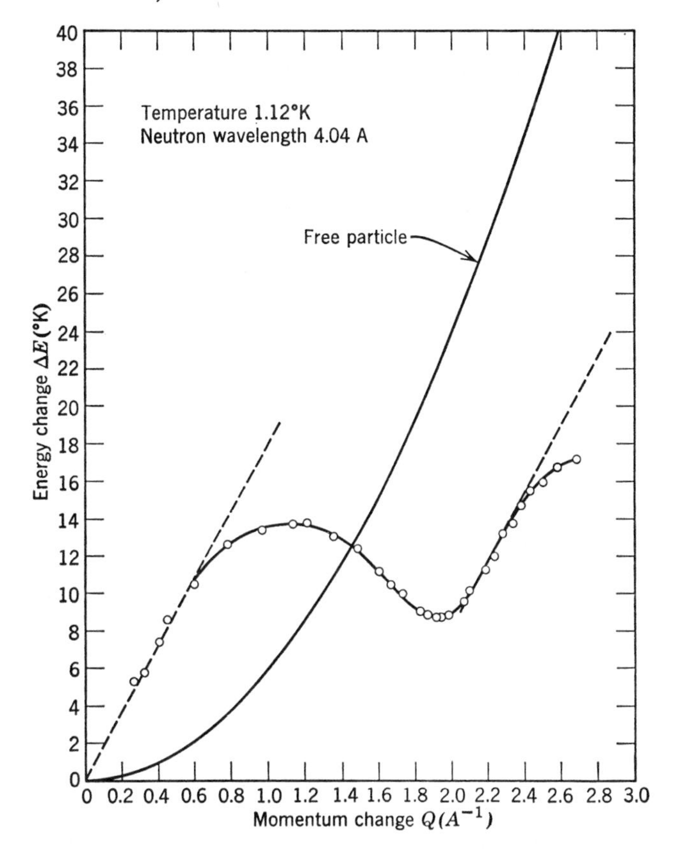

FIG. 1. Spectrum of elementary excitations in liquid helium at 1.12°K at normal vapor pressure, as determined by Henshaw and Woods by inelastic neutron scattering. The broken curve through the origin corresponds to a velocity of sound of 2.37×10^4 cm/sec; the parabolic curve through the origin is calculated for *free* helium atoms.

generate elementary excitations. In an event in which an elementary excitation $\omega_{\mathbf{k}}$ is created, we must have

$$\frac{1}{2}Mv^2 = \frac{1}{2}Mv'^2 + \omega_{\mathbf{k}}; \tag{102}$$

$$M\mathbf{v} = M\mathbf{v}' + \mathbf{k}, \tag{103}$$

from energy and momentum conservation. Combining these two equations,

$$0 = -\mathbf{v} \cdot \mathbf{k} + (2M)^{-1}k^2 + \omega_{\mathbf{k}}. \tag{104}$$

The lowest value of $\mathbf{v}$ for which this equation can be satisfied will occur for $\mathbf{v} \parallel \mathbf{k}$; this critical velocity is then

$$v_c = \min\left(\frac{k}{2M} + \frac{\omega_{\mathbf{k}}}{k}\right). \tag{105}$$

When $M \to \infty$, the critical velocity is determined by the minimum of $\omega_{\mathbf{k}}/k$. If there is an energy gap, that is, all $\omega_{\mathbf{k}} > 0$, we will have $v_c > 0$. The situation portrayed in Fig. 1 also leads to a positive minimum of $\omega_{\mathbf{k}}/k$. If the elementary excitations had a free particle dispersion relation—$\omega \propto k^2$—then the critical velocity would be zero. What has been shown here is that for $v < v_c$ the body will move at 0°K without loss of energy to the fluid; we say that the fluid acts as a superfluid. In actual liquid helium the critical velocities observed are often two or three orders of magnitude lower than the calculated values, probably because of the possibility of creating other low-lying excitations of the form of vortex lines. At finite temperatures the moving body does encounter resistance; one source of resistance comes from Raman processes in which a thermal phonon is scattered inelastically by the moving body.

SECOND SOUND IN CRYSTALS

We have treated acoustic phonons in crystals and in liquid helium. The acoustic phonons couple at a crystal/gas or liquid/gas interface with ordinary sound waves in the gas. We denote propagation by acoustic phonons as ordinary or *first sound.*

The acoustic phonons themselves possess many of the attributes of particles. These particles interact weakly with each other by virtue of anharmonic terms in the lattice potential. At low temperatures, the collision events among phonons conserve both energy and wavevector, but not necessarily the number of phonons. At higher temperatures, the total wavevector of the colliding particles may change by a reciprocal lattice vector; our discussion does not apply if such umklapp processes are frequent. The dispersion relation of the phonons at sufficiently long wavelengths is $\omega = c_1 k$, where c_1 is the velocity of first sound.

We are interested in the possibility that collective waves may propagate in the phonon gas—modes in which the local excitation density or phonon concentration is modulated in a wavelike way. Such waves are known as *second sound.* They surely exist in liquid helium II, but have not yet been reported in crystals. The problem of second sound in liquid helium differs in important respects from the problem of second sound in crystals, and we discuss only the latter.

We assume that the phonon mean free path is much shorter than the wavelength of second sound. In the case of ordinary conduction of heat in a crystal there also exists a gradient in the phonon density, yet no wavelike solution ordinarily results because wavevector conservation is destroyed by umklapp processes.

We now derive an expression for the velocity of second sound in a phonon gas in a crystal. For simplicity we consider only longitudinal phonons. Let

$$f(\mathbf{k},\mathbf{x})\, d^3k\, d^3x$$

be the number of phonons of wavevector in d^3k at $\mathbf{k}$ and of position in d^3x at $\mathbf{x}$. According to the boltzmann transport equation[5]

$$\frac{\partial f}{\partial t} + \mathbf{v} \cdot \frac{\partial f}{\partial \mathbf{x}} = \Delta_c f, \tag{106}$$

where $\Delta_c f$ denotes the rate of change of f from collisions. We have omitted the acceleration term $\boldsymbol{\alpha} \cdot (\partial f/\partial \mathbf{v})$ from the equation because the acceleration $\boldsymbol{\alpha} \equiv 0$ in the absence of external forces. For long wavelength phonons the ν component of velocity is $v_\nu = c_1 k_\nu / k$, so that (106) becomes

$$\frac{\partial f}{\partial t} + c_1 \frac{k_\nu}{k} \frac{\partial f}{\partial x_\nu} = \Delta_c f. \tag{107}$$

Now form the crystal momentum density $\mathbf{P}$ and energy density U:

$$\mathbf{P} = \int d^3k\, \mathbf{k} f; \qquad U = \int d^3k\, c_1 k f. \tag{108}$$

Because we have assumed wavevector and energy are conserved in all collisions,

$$\int d^3k\, \mathbf{k}\, \Delta_c f = 0; \qquad \int d^3k\, c_1 k\, \Delta_c f = 0. \tag{109}$$

Thus on multiplying (107) by k_μ and integrating:

$$\frac{\partial P_\mu}{\partial t} + \frac{\partial}{\partial x_\nu} \int d^3k\, \frac{c_1 k_\nu k_\mu}{k} f = 0; \tag{110}$$

and on multiplying by $c_1 k$ and integrating:

$$\frac{\partial U}{\partial t} + \frac{\partial}{\partial x_\nu} \int d^3k\, c_1{}^2 k_\nu f = \frac{\partial U}{\partial t} + c_1{}^2 \frac{\partial P_\nu}{\partial x_\nu} = 0. \tag{111}$$

[5] For a discussion of the boltzmann equation see, for example, C. Kittel, *Elementary statistical physics*, Wiley, New York, 1958, pp. 192–196.

If f is only slightly perturbed from an isotropic distribution,

$$\int d^3k\, \frac{c_1 k_\nu k_\mu}{k} f = \delta_{\mu\nu} \tfrac{1}{3} U, \tag{112}$$

and (110) may be written

$$\frac{\partial P_\mu}{\partial t} + \frac{1}{3}\frac{\partial U}{\partial x_\mu} = 0. \tag{113}$$

On combining (111) with (113),

$$\frac{\partial^2 U}{\partial t^2} - \frac{c_1{}^2}{3}\frac{\partial^2 U}{\partial x_\nu\, \partial x_\nu} = 0, \tag{114}$$

which is the equation of a wave having velocity $c_1/3^{1/2}$. The ratio $c_1/c_2 = 3^{1/2}$ of the velocities of first and second sound is approximately satisfied in liquid helium as $T \rightarrow 0$, but the attainable experimental accuracy is poor because second sound becomes highly damped at very low temperatures as the phonon mean free path increases relative to the wavelength of the second sound.

The waves of second sound are periodic variations of excitation energy associated with periodic variations of phonon concentration. The above calculation is due to J. C. Ward and J. Wilks, *Phil. Mag.* **43,** 48 (1952). Our result for $c_2{}^2$ does not include their factor $[1 - (\rho_n/\rho)]$ in the ratio of the effective mass density ρ_n of the excitations to the actual density ρ of the medium. This factor is present in liquid helium, but it is not evident that it should be present in crystals. In any event $\rho_n/\rho \rightarrow 0$ as $T \rightarrow 0$ for phonons in liquid helium and in crystals. A discussion of second sound in liquid helium is given by R. B. Dingle, *Advances in Physics* **1,** 112–168 (1952).

FREQUENCY DISTRIBUTION FOR PHONONS

The phonon frequency distribution function $g(\omega)$ of a crystal is defined as the number of phonon frequencies per unit frequency range, divided by the total number of frequencies. The frequency distribution determines an important part of the thermodynamic properties of the crystal. The distribution may be calculated numerically from the dispersion relation,[6] usually with considerable labor. An exact solution for the two-dimensional square lattice has been obtained by Montroll,[7] who found that the distribution function has a logarithmically infinite peak.

[6] For a review of this area, see J. de Launay, in *Solid state physics* **2,** 268 (1956).
[7] E. Montroll, *J. Chem. Phys.* **15, 575 (1947).**

Such singularities are of great importance for the thermodynamic properties. A general investigation of the singularities in the distribution function has been given by Van Hove,[8] who utilizes a topological theorem of M. Morse. The theorem states that any function of more than one independent variable which, as $\omega(\mathbf{k})$, is periodic in all its variables has at least a certain number of saddle points. This number is determined by topological considerations and depends only on the number of independent variables. A saddle point usually leads to a singularity in the distribution function. In two dimensions the singularity is logarithmic; in three dimensions $g(\omega)$ is continuous, but $dg/d\omega$ has infinite discontinuities.

The fact that $\omega(\mathbf{k})$ is periodic is shown in the chapter on Bloch functions; we have seen an example of the periodicity in (26) above. The density of states $g(\omega)\,d\omega$ is directly proportional to $\int d\mathbf{k}$, where the integral is over the volume in $\mathbf{k}$ space bounded by the constant-energy surfaces at ω and $\omega + d\omega$. The thickness of the shell in the direction normal to a bounding surface is given by

$$|\nabla_{\mathbf{k}}\omega|\,dk_n = d\omega, \tag{115}$$

so that

$$g(\omega) = \frac{V_0}{Zl}\int_{S(\omega)} \frac{dS}{|\nabla_{\mathbf{k}}\omega|}, \tag{116}$$

where V_0 is the volume of the primitive cell; Z denotes the number of atoms per cell; and l is the dimensionality of the space. The integration is over the surface S of constant ω. We expect a singularity when

$$|\nabla_{\mathbf{k}}\omega| = \left[\left(\frac{\partial\omega}{\partial k_x}\right)^2 + \left(\frac{\partial\omega}{\partial k_y}\right)^2\right]^{1/2} = 0, \tag{117}$$

written for two dimensions. It is apparent that the singularities of $g(\omega)$ originate from the critical points of $\omega(\mathbf{k})$, where all the derivatives in (117) vanish.

PROBLEMS

1. From the equation of motion for the field operator ψ in the Heisenberg representation, $i\dot{\psi} = [\psi,H]$, show for the elastic line (56) that $\dot{\psi} = \pi/\rho$, in agreement with the classical equation. We note that ψ and any function of ψ,

[8] L. Van Hove, *Phys. Rev.* **89,** 1189 (1953); see also H. P. Rosenstock, *Phys. Rev.* **97,** 290 (1955).

such as $\partial\psi/\partial x$, commute, so the term in T does not contribute to $\dot{\psi}$. The term

$$(118)\quad [\psi(\mathbf{r}), \int \pi(\mathbf{r}')^2\, d\tau'] = \int [\psi(\mathbf{r}),\pi(\mathbf{r}')]\pi(\mathbf{r}')\, d\tau' + \int \pi(\mathbf{r}')[\psi(\mathbf{r}),\pi(\mathbf{r}')]\, d\tau',$$

which may be reduced, using (60).

2. Show that for the elastic line the quantum equation of motion gives

$$(119)\quad \dot{\pi} = T\frac{\partial^2\psi}{\partial x^2},$$

which may be combined with $\ddot{\psi} = \dot{\pi}/\rho$ to give the wave equation in the field operator ψ:

$$(120)\quad \rho\ddot{\psi} = T\frac{\partial^2\psi}{\partial x^2}.$$

Note that

$$(121)\quad \dot{\pi} = -\tfrac{1}{2}iT\int dx'\left\{\left[\pi(x), \frac{\partial\psi(x')}{\partial x'}\right]\frac{\partial\psi(x')}{\partial x'} + \frac{\partial\psi(x')}{\partial x'}\left[\pi(x), \frac{\partial\psi(x')}{\partial x'}\right]\right\},$$

where

$$(122)\quad \left[\pi(x), \frac{\partial\psi(x')}{\partial x'}\right] = -i\frac{\partial}{\partial x'}\,\delta(x - x') = i\frac{\partial}{\partial x}\,\delta(x - x').$$

For theorems on derivatives of delta functions, see Messiah, pp. 469–471.

3. Consider a continuous elastic line of length L with fixed ends, so that $\psi(0) = \psi(L) = 0$. Develop $\psi(x,t) = \sum_k Q_k(t)u_k(x)$, where

$$(123)\quad u_k = (2/L)^{1/2}\sin kx,$$

with $k = n\pi/L$, $n = 1, 2, 3, \cdots$. (*a*) Diagonalize the hamiltonian for this problem and indicate the form of the ground state Schrödinger wavefunction in the Q representation. (*b*) Calculate the mean square fluctuation of ψ in the ground state of the line, averaged over the length of the line.

4. Show that on the Debye model the low temperature ($T \ll \Theta$) heat capacity per unit volume of an isotropic monatomic solid containing n atoms per unit volume is

$$(124)\quad C = (12\pi^4 nk_B/5)(T/\Theta)^3;$$

here k_B is the boltzmann constant and

$$(125)\quad \frac{3}{\Theta^3} = \frac{1}{\Theta_l^3} + \frac{2}{\Theta_t^3},$$

where $k_B\Theta_{l,t} = \hbar v_{l,t}(6\pi^2 n)^{1/3}$.

5. The normalization of the commutator (94) is assured if we write

$$(126)\quad u_{\mathbf{k}} = \cosh\chi_{\mathbf{k}};\qquad v_{\mathbf{k}} = \sinh\chi_{\mathbf{k}}.$$

(*a*) Show that $H_{\mathbf{k}}$ in (91) is diagonal if

$$(127)\quad \tanh 2\chi_{\mathbf{k}} = -\frac{NV_{\mathbf{k}}}{\varepsilon_{\mathbf{k}} + NV_{\mathbf{k}}}.$$

(*b*) Show that

$$(128)\quad \overset{+}{a}_{\mathbf{k}} a_{\mathbf{k}} = u_{\mathbf{k}}^2 \overset{+}{\alpha}_{\mathbf{k}} \alpha_{\mathbf{k}} + v_{\mathbf{k}}^2 + v_{\mathbf{k}}^2 \alpha_{-\mathbf{k}}^+ \alpha_{-\mathbf{k}} + u_{\mathbf{k}} v_{\mathbf{k}} (\overset{+}{\alpha}_{\mathbf{k}} \alpha_{-\mathbf{k}}^+ + \alpha_{-\mathbf{k}} \alpha_{\mathbf{k}}).$$

(*c*) The ground state Φ_0 has the property

$$(129)\qquad \alpha_{\mathbf{k}} \Phi_0 = 0$$

for all $\alpha_{\mathbf{k}}$, whence show that the mixture of excitations **k** in the ground state is given by

$$(130)\qquad \langle \overset{+}{a}_{\mathbf{k}} a_{\mathbf{k}} \rangle_0 = \langle \Phi_0 | \overset{+}{a}_{\mathbf{k}} a_{\mathbf{k}} | \Phi_0 \rangle = v_{\mathbf{k}}^2 = \tfrac{1}{2}(\cosh 2\chi_{\mathbf{k}} - 1).$$

Make a rough plot of $\langle \overset{+}{a}_{\mathbf{k}} a_{\mathbf{k}} \rangle_0$ versus $|\mathbf{k}|$, assuming $V_{\mathbf{k}}$ = constant. Note that

$$(131)\qquad \cosh 2\chi_{\mathbf{k}} = \frac{\varepsilon_{\mathbf{k}} + NV_{\mathbf{k}}}{\{(\varepsilon_{\mathbf{k}} + NV_{\mathbf{k}})^2 - N^2 V_{\mathbf{k}}^2\}^{1/2}}.$$

6. Return to this problem after studying Chapter 6 and particularly Problems 6.9 and 6.10. Suppose that the interactions in our condensed boson gas are such that at low **k** only quasiparticle excitations exist and have the dispersion relation $\omega(\mathbf{k})$. Then at low **k** we approximate the dynamic structure factor $\mathcal{S}(\omega\mathbf{k})$ by

$$(132)\qquad \mathcal{S}(\omega\mathbf{k}) \cong N\mathcal{S}(\mathbf{k})\delta(\omega - \omega(\mathbf{k})),$$

where $\mathcal{S}(\mathbf{k})$ is the liquid structure factor used in Chapter 6. This equation satisfies the sum rule of Problem 6.9. Show, using the sum rule of Problem 6.10, that the dispersion relation is related to the liquid structure factor by the Feynman relation

$$(133)\qquad \omega(\mathbf{k}) = \frac{k^2}{2M\mathcal{S}(\mathbf{k})}.$$

The liquid structure factor measured by neutron diffraction [D. G. Henshaw, *Phys. Rev.* **119,** 9 (1960), Fig. 2] has a strong peak near $k = 2.0$ A^{-1}, in good agreement with the observed position of the minimum in the elementary excitation dispersion relation.

7. Diagonalize $H = \omega a^+ a + \varepsilon(ab^+ + ba^+)$, where the a, b are boson operators.

3 Plasmons, optical phonons, and polarization waves

In this chapter we study simple examples of several important effects:

(*a*) The phonon excitation spectrum will contain $3s$ branches if there are s atoms or ions in a primitive cell of a crystal. The three branches whose frequency approaches zero as the wavevector approaches zero are called the acoustical phonon branches. The remaining $3s - 3$ branches are called optical phonon branches, and these have finite eigenfrequencies as $\mathbf{k} \rightarrow 0$.

(*b*) Consider a crystal having two ions per primitive cell, with equal charges of opposite sign. The long range of the coulomb interaction will increase considerably the frequency of the longitudinal optical branch with respect to the two transverse optical branches. The plasma frequency in an electron gas with a positive background charge is a limiting case of this effect.

(*c*) In certain circumstances the long wavelength transverse optical modes will couple strongly and be mixed with the electromagnetic radiation field. There are then strong effects on the dispersion relations.

Thus we are concerned in the following with fields for which electric charges of one sign are displaced relative to charges of the opposite sign. Our first example is the electron gas in a metal; this problem will be developed in much greater detail in Chapters 5 and 6. We also discuss optical phonons in ionic crystals, both without and with coupling to the electromagnetic field.

PLASMONS

We consider a continuum model of an electron gas with a rigid fixed background of positive charge. The continuum model is only an approximation, but it allows us to see certain central features of the eigenfrequency problem in the presence of the long-range coulomb interaction. The electron gas is contained in unit volume, with n electrons. The volume is also filled uniformly with a rigid background of positive charge of density $\rho_0 = n|e|$, equal and opposite to the average charge density of the electrons. In our approximation there is no restoring force on shear waves because such waves do not change the local charge neutrality of the system. The transverse mode eigenfrequencies are therefore zero. Longitudinal waves in the plasma cause dilations and contractions in the electron gas, thereby destroying neutrality and bringing into play the powerful coulomb restoring forces. The longitudinal eigenfrequencies, called plasma frequencies, are relatively high frequencies.

The hamiltonian density is given in analogy with (2.72), with electrostatic terms included and shear terms omitted:

$$\mathcal{H} = \frac{1}{2nm}\Pi_\mu\Pi_\mu + \tfrac{1}{2}\alpha\frac{\partial R_\mu}{\partial x_\mu}\frac{\partial R_\nu}{\partial x_\nu} + \tfrac{1}{2}(\rho - \rho_0)\varphi(\mathbf{x}), \tag{1}$$

where nm is the mass density of the gas; the dilation Δ is $\partial R_\mu/\partial x_\mu$; α is the appropriate bulk modulus, as if the gas were uncharged; φ is the electrostatic potential, derived from the poisson equation

$$\nabla^2\varphi = -4\pi(\rho - \rho_0). \tag{2}$$

The factor $\frac{1}{2}$ in the last term on the right of (1) enters because the electrostatic term is just the self-energy of the electron gas. The positive background is exactly canceled by the uniform ($\mathbf{k} = 0$) component of the negative charge density, as is shown in detail in Chapter 5.

For small local dilations of the electron gas, the charge density fluctuation $\delta\rho = \rho - \rho_0$ is given by

$$\frac{\delta\rho}{\rho} = -\Delta(\mathbf{x}), \tag{3}$$

where Δ is the dilation. Thus

$$\delta\rho = -\rho\Delta = -ne\frac{\partial R_\mu}{\partial x_\mu}. \tag{4}$$

Using (2.73) for longitudinal waves ($Q_\mathbf{k} \parallel \mathbf{k}$),

$$R(\mathbf{x}) = \sum_\mathbf{k} Q_\mathbf{k} e^{i\mathbf{k}\cdot\mathbf{x}}, \tag{5}$$

and

$$\delta\rho = -ine \sum_{\mathbf{k}} kQ_{\mathbf{k}} e^{i\mathbf{k}\cdot\mathbf{x}}. \tag{6}$$

Let us write

$$\varphi = \sum_{\mathbf{k}} \varphi_{\mathbf{k}} e^{i\mathbf{k}\cdot\mathbf{x}}, \tag{7}$$

so that

$$\nabla^2\varphi = -\Sigma\, k^2 \varphi_{\mathbf{k}} e^{i\mathbf{k}\cdot\mathbf{x}}, \tag{8}$$

and the poisson equation becomes

$$\varphi_{\mathbf{k}} = 4\pi ineQ_{\mathbf{k}}/k. \tag{9}$$

The electrostatic term in (1) is, after integration over the volume,

$$\begin{aligned} \int d^3x\, \tfrac{1}{2}(\rho - \rho_0)\varphi &= -\sum_{\mathbf{k}\mathbf{k}'} \int d^3x\, 2\pi n_0{}^2 e^2 Q_{\mathbf{k}} Q_{\mathbf{k}'} e^{i(\mathbf{k}+\mathbf{k}')\cdot\mathbf{x}} (k/k') \\ &= 2\pi n^2 e^2 \sum_{\mathbf{k}} Q_{\mathbf{k}} Q_{-\mathbf{k}}. \end{aligned} \tag{10}$$

Then, with the momentum density components given by (2.75),

$$H = \tfrac{1}{2} \sum_{\mathbf{k}} \left(\frac{1}{nm} P_{\mathbf{k}} P_{-\mathbf{k}} + \alpha k^2 Q_{\mathbf{k}} Q_{-\mathbf{k}} + 4\pi n^2 e^2 Q_{\mathbf{k}} Q_{-\mathbf{k}} \right). \tag{11}$$

In direct analogy to the solution (2.81) we have

$$\boxed{\omega_{\mathbf{k}}{}^2 = \omega_p{}^2 + (\alpha/nm)k^2; \qquad \omega_p{}^2 = 4\pi ne^2/m.} \tag{12}$$

This is the dispersion relation for plasmons; ω_p is usually referred to as the plasma frequency. The excitation of one mode with energy $\omega_{\mathbf{k}}$ is described as the excitation of a plasmon. In the limit $e^2 \to 0$ the electrostatic effects disappear and we have left the usual dispersion relation $\omega = (\alpha/nm)^{1/2}k$ for phonons in a gas. For electrons in an alkali metal $n \sim 10^{23}$ cm^{-3}; $m \sim 10^{-27}g$, whence $\omega_p \sim 10^{16}$ sec^{-1}. If we substitute for $(\alpha/nm)^{1/2}$ the velocity of sound in a solid and take k near its probable maximum of 10^8 cm^{-1}, the term $(\alpha/nm)k^2$ in (12) is still negligible in comparison with the term $\omega_p{}^2$.

The frequency ω_p of the uniform ($\mathbf{k} = 0$) mode is derived easily by a direct argument. Let the electron gas be displaced by x; the dielectric polarization is $P = nex$. If the plasma is contained in a flat slab with the displacement normal to the slab, the polarization P will give rise to a depolarization field $E_d = -4\pi P = -4\pi nex$. The equation of motion of an electron in the plasma is

$$m\ddot{x} = eE_d = -4\pi ne^2x, \tag{13}$$

so that the resonance frequency is

$$\omega^2 = \frac{4\pi ne^2}{m} \equiv \omega_p^2. \tag{14}$$

The resonance frequency of the uniform mode depends on the geometrical form of the container. Once the wavelength of the plasmon becomes small in comparison with the dimensions of the container the shape effects disappear.

We now consider briefly the dielectric constant associated with the uniform plasma mode. If the electric field E of frequency ω is applied parallel to the surface of the slab within which the plasma is contained, there is no depolarization field. Then

$$m\ddot{x} = eE; \qquad x = -\frac{eE}{m\omega^2}; \qquad P = -\frac{ne^2E}{m\omega^2}; \tag{15}$$

and the dielectric constant is

$$\epsilon = 1 + 4\pi\frac{P}{E} = 1 - \frac{4\pi ne^2}{m\omega^2} = 1 - \frac{\omega_p^2}{\omega^2}. \tag{16}$$

When $\omega \geqq \omega_p$, we see that ϵ is positive and the refractive index $n = \epsilon^{1/2}$ is real; thus the threshold for the transparency of metals in the ultraviolet is given by $\omega = \omega_p$. This neglects ion core polarization, which is not really a plasma effect. If the electric field E is applied normal to the slab, with E_{int} to denote the internal field, $E_{\text{int}} = E - 4\pi P = E - 4\pi\chi E_{\text{int}}$, and

$$m\ddot{x} = eE_{\text{int}}; \qquad E_{\text{int}} = E/(1 + 4\pi\chi) = E/\epsilon \tag{17}$$

from (13); then

$$x = -\frac{eE}{m(\omega^2 - \omega_p^2)}. \tag{18}$$

The free oscillations of the system parallel (transverse) to the slab are given by the poles of ϵ; the free oscillations normal (longitudinal) to the slab are given by the zeros of ϵ.

LONG-WAVELENGTH OPTICAL PHONONS IN ISOTROPIC CRYSTALS[1]

There are three interesting points to be made in a discussion of optical phonons.

[1] General reference: M. Born and K. Huang, *Dynamical theory of crystal lattices*, Clarendon Press, Oxford, 1954.

(*a*) A crystal with s nonequivalent ions per primitive cell will have three branches (one mainly longitudinal, two mainly transverse) of its vibrational spectrum whose frequencies approach zero as $\mathbf{k} \rightarrow 0$. These branches are called *acoustical modes*. There will be in addition $3(s - 1)$ branches each having a finite limiting frequency as $\mathbf{k} \rightarrow 0$; such branches are called *optical modes*. The optical modes are suppressed in the usual macroscopic analysis, such as we carried out in Chapter 2 following (2.71). In an ionic crystal having two ions per primitive cell, such as NaCl, the three optical branches at long wavelengths may be classified approximately into one longitudinal and two transverse branches.

(*b*) The limiting frequency ω_l of the longitudinal branch as $\mathbf{k} \rightarrow 0$ is appreciably higher because of electrostatic effects than the limiting frequency ω_t of the transverse branches: the approximate theoretical connection is

$$\omega_l^2 = (\epsilon_0/\epsilon_\infty)\omega_t^2, \tag{19}$$

where ϵ_0 is the static dielectric constant and ϵ_∞ is the square of the optical refractive index.

(*c*) The electromagnetic coupling between photons and phonons is particularly marked for long-wavelength transverse optical phonons, and there results a forbidden frequency gap between ω_t and ω_l in which a thick crystal does not transmit energy. There is a strong optical reflection band in the region of the frequency gap.

We now consider these points, but we do not enter into the details to be found in the book by Born and Huang and in the paper by Lyddane and Herzfeld.[2] The normal modes at $\mathbf{k} = 0$ are simple in form: the corresponding ions in every cell, by definition of $\mathbf{k} = 0$, move with identical amplitudes and phases. There are $3s$ equations of motion of the s ions in a primitive cell. Three of the modes correspond to the uniform undistorted translation of the cell as a whole, and thus have zero frequency. This is the limiting frequency of the acoustic modes. The other $3(s - 1)$ modes are the $\mathbf{k} = 0$ optical modes; they represent motions of the ions relative to one another in the same cell or rotations of the group within a cell. None of these $3(s - 1)$ frequencies at $\mathbf{k} = 0$ will in general be zero.

The frequency difference of the long-wavelength transverse and longitudinal optical modes in a cubic ionic crystal can be appreciated by an elementary argument. A transverse optical phonon in a crystal

[2] R. H. Lyddane and K. F. Herzfeld, *Phys. Rev.* **54**, 846 (1938).

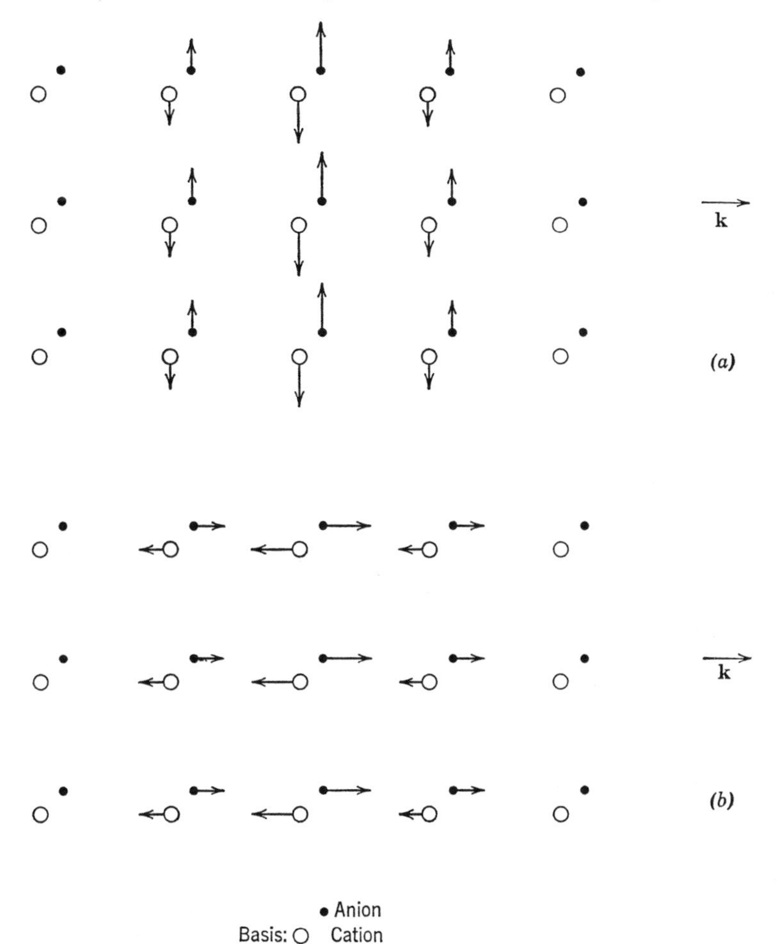

FIG. 1. Optical modes of oscillation: (a) transverse and (b) longitudinal. The directions of the displacements of individual ions are shown.

with two ions per cell, such as NaCl or CsCl, is pictured in Fig. 1*a*. If

$$u = u_+ - u_- \tag{20}$$

is the relative displacement of the positive and negative ion lattices, the equation of motion for the transverse oscillation is

$$M\ddot{u} + M\omega_t^2 u = 0, \tag{21}$$

where M is the reduced mass and ω_t^2 is the transverse resonance frequency. We assume that the wavelength is much smaller than the dimensions of the specimen, although long in comparison with atomic dimensions.

A longitudinal optical mode is shown in Fig. 1*b*. The restoring force on an ion is not $-\omega_t^2 Mu$, but is

$$-\omega_l^2 Mu = -\omega_t^2 Mu + E_i e, \tag{22}$$

where E_i is the internal electric field developed in the deformation. The field satisfies $D = 0 = E_i + 4\pi P$, so that $E_i = -4\pi P$. If we neglect the induced polarization on the ions, treating them internally as rigid, then $P = neu$, where n is the number of cells per unit volume. Thus the restoring force will be $-\omega_t^2 Mu - 4\pi n^2 eu$, so that

$$\omega_l^2 = \omega_t^2 + 4\pi ne^2/M. \tag{23}$$

Taking $n = 10^{22}$ cm^{-3}, $M = 10^{-22} g$, $e^2 \cong 25 \times 10^{-20}(\text{esu})^2$, we have $(4\pi ne^2/M)^{1/2} \approx 10^{13}$ sec^{-1}, of the same order of magnitude as ω_t. The effect on nonrigid ions is sometimes represented schematically by writing an effective charge e^* for e in (23).

The macroscopic theory of the optical modes of diatomic crystals which we now present was developed by Huang. For the optical mode at $\mathbf{k} = 0$ we introduce the coordinate

$$\mathbf{w} = (\mathbf{u}^+ - \mathbf{u}^-)/(Mn)^{1/2}, \tag{24}$$

where $\mathbf{u}^+ - \mathbf{u}^-$ is the relative displacement of the positive and negative ionic lattices; $M = M_+ M_-/(M_+ + M_-)$ is the reduced mass; and n is the number of cells per unit volume. Note that Mn is effectively the reduced mass density. The kinetic energy density associated with motion of one lattice against the other is $\frac{1}{2}\dot{\mathbf{w}}^2$. The potential energy density may contain terms in $\mathbf{w}^2$, $\mathbf{E}^2$, and $\mathbf{w} \cdot \mathbf{E}$, where $\mathbf{E}$ is the macroscopic internal electric field. We consider the general case where the ions themselves are polarizable (nonrigid). The lagrangian density is

$$\mathcal{L} = \tfrac{1}{2}\dot{\mathbf{w}}^2 - (\tfrac{1}{2}\gamma_{11}\mathbf{w}^2 - \gamma_{12}\mathbf{w} \cdot \mathbf{E} - \tfrac{1}{2}\gamma_{22}\mathbf{E}^2), \tag{25}$$

where γ_{11}, γ_{12}, γ_{22} are constants to be determined below. From (25) and the lagrangian equations we have the equation of motion

$$\ddot{\mathbf{w}} + \gamma_{11}\mathbf{w} - \gamma_{12}\mathbf{E} = 0. \tag{26}$$

The momentum density conjugate to $\mathbf{w}$ is $\mathbf{\Pi} = \partial\mathcal{L}/\partial\dot{\mathbf{w}} = \dot{\mathbf{w}}$, and the hamiltonian density is

$$\mathcal{H} = \tfrac{1}{2}\Pi^2 + \tfrac{1}{2}\gamma_{11}\mathbf{w}^2 - \gamma_{12}\mathbf{w} \cdot \mathbf{E} - \tfrac{1}{2}\gamma_{22}\mathbf{E}^2. \tag{27}$$

The polarization $\mathbf{P}$ is given by

$$\mathbf{P} = -\partial\mathcal{H}/\partial\mathbf{E} = \gamma_{12}\mathbf{w} + \gamma_{22}\mathbf{E}. \tag{28}$$

We now solve for γ_{11}, γ_{12}, γ_{22} in terms of the accessible constants ϵ_0, ϵ_∞, and ω_t. The transverse optical mode does not generate a depolarization field $\mathbf{E}$, so that (26) becomes $\ddot{\mathbf{w}}_t + \gamma_{11}\mathbf{w}_t = 0$. This must be identical with $\ddot{\mathbf{w}} + \omega_t^2\mathbf{w} = 0$, whence

$$\gamma_{11} = \omega_t^2. \tag{29}$$

Under static conditions $\ddot{\mathbf{w}} = 0$; an applied static electric field creates a value of $\mathbf{w}$ given by $\gamma_{11}\mathbf{w} - \gamma_{12}\mathbf{E} = 0$, according to (26). Substituting this value $\mathbf{w} = (\gamma_{12}/\gamma_{11})\mathbf{E}$ in (28), we have

$$\mathbf{P} = [(\gamma_{12}^2/\gamma_{11}) + \gamma_{22}]\mathbf{E} = \frac{\epsilon_0 - 1}{4\pi}\mathbf{E}, \tag{30}$$

for static conditions, where ϵ_0 is the static dielectric constant. At very high frequencies $\mathbf{w}$ approaches zero and

$$\mathbf{P} = \gamma_{22}\mathbf{E} = \frac{\epsilon_\infty - 1}{4\pi}\mathbf{E}, \tag{31}$$

where we understand ϵ_∞ to include the electronic polarizability. Thus

$$\gamma_{22} = \frac{\epsilon_\infty - 1}{4\pi}, \tag{32}$$

and

$$\gamma_{12} = \left(\frac{\epsilon_0 - \epsilon_\infty}{4\pi}\right)^{1/2}\omega_t. \tag{33}$$

The long-wavelength longitudinal optical modes are characterized by $\mathbf{D} = \mathbf{E} + 4\pi\mathbf{P} = 0$, in the absence of external fields. Thus (28) gives

$$E + 4\pi P = (1 + 4\pi\gamma_{22})E + 4\pi\gamma_{12}w_l = 0, \tag{34}$$

and the equation of motion (26) becomes

$$\ddot{w}_l + \left(\gamma_{11} + \frac{4\pi\gamma_{12}^2}{1 + 4\pi\gamma_{22}}\right)w_l = 0, \tag{35}$$

or

$$\omega_l^2 = \gamma_{11} + \frac{4\pi\gamma_{12}^2}{1 + 4\pi\gamma_{22}} = \frac{\epsilon_0}{\epsilon_\infty}\omega_t^2. \tag{36}$$

This result was first derived by Lyddane, Sachs, and Teller; it is consistent with (23) above as derived for $\epsilon_\infty = 1$. For NaCl, using the observed $\omega_t = 3.09 \times 10^{13}\ \text{sec}^{-1}$, one calculates $\omega_l = 4.87 \times 10^{13}\ \text{sec}^{-1}$, with $\epsilon_0 = 5.02$ and $\epsilon_\infty = 2.25$.

INTERACTION OF OPTICAL PHONONS WITH PHOTONS

The preceding treatment of the optical lattice mode problem at $\mathbf{k} = 0$ has neglected the interaction of the optical phonons with the photons of the electromagnetic field. This interaction is particularly important when the frequencies and wavevectors of the phonon and photon fields coincide—near the crossover of the dispersion relations even weak coupling of two fields can have drastic effects. We do not discuss here the **k**-dependence of the uncoupled optical phonons, but it is not usually very strong. The uncoupled solutions for a one-dimensional problem are indicated in *ISSP*, Fig. 5.3, p. 111. The dispersion relation for photons is $\omega = ck$, where c is the velocity of light; it is evident that this relation will cross at some point the dispersion relation for each optical phonon branch.

Our problem is to solve the maxwell equations simultaneously with the lattice equations:

$$\text{curl}\,\mathbf{H} = \frac{1}{c}(\dot{\mathbf{E}} + 4\pi\dot{\mathbf{P}});\qquad \text{curl}\,\mathbf{E} = -\frac{1}{c}\dot{\mathbf{H}};$$

$$\text{div}\,\mathbf{H} = 0;\qquad \text{div}\,(\mathbf{E} + 4\pi\mathbf{P}) = 0;$$

$$\ddot{\mathbf{w}} + \gamma_{11}\mathbf{w} - \gamma_{12}\mathbf{E} = 0;\qquad \mathbf{P} = \gamma_{12}\mathbf{w} + \gamma_{22}\mathbf{E}.$$

First we look for transverse solutions $\mathbf{E} \perp \mathbf{k}$ as a photon field. We try

$$E_x = E_x(0)e^{i(\omega t - kz)};\qquad P_x = P_x(0)e^{i(\omega t - kz)};$$

$$w_x = w_x(0)e^{i(\omega t - kz)};\qquad H_y = H_y(0)e^{i(\omega t - kz)}.$$

Then the differential equations become

$$ikH_y = (i\omega/c)(E_x + 4\pi P_x);\qquad -ikE_x = -(i\omega/c)H_y;$$

$$(-\omega^2 + \gamma_{12})w_x = \gamma_{12}E_x;\qquad P_x = \gamma_{12}w_x + \gamma_{22}E_x.$$

These equations have a nontrivial solution only if the determinant of the coefficients of E_x, H_y, P_x, w_x vanishes:

$$\begin{vmatrix} \omega/c & 4\pi\omega/c & -k & 0 \\ k & 0 & -\omega/c & 0 \\ \gamma_{12} & 0 & 0 & \omega^2 - \gamma_{11} \\ \gamma_{22} & -1 & 0 & \gamma_{12} \end{vmatrix} = 0, \tag{37}$$

which may be written as

$$\omega^4\epsilon_\infty - \omega^2(\omega_t^2\epsilon_0 + c^2k^2) + \omega_t^2c^2k^2 = 0, \tag{38}$$

where ω_t^2 now is to be viewed as a symbol defined by (29); ϵ_∞ and ϵ_0 are given by (32) and (33). The solutions are

$$\omega^2 = \frac{1}{2\epsilon_\infty}(\omega_t^2\epsilon_0 + c^2k^2) \pm \left[\frac{1}{4\epsilon_\infty^2}(\omega_t^2\epsilon_0 + c^2k^2)^2 - \omega_t^2k^2\left(\frac{c^2}{\epsilon_\infty}\right)\right]^{1/2}. \tag{39}$$

For $k \to 0$ the transverse solutions are

$$\boxed{\omega^2 = \omega_t^2(\epsilon_0/\epsilon_\infty) = \omega_l^2,} \tag{40}$$

and

$$\boxed{\omega^2 = (c^2/\epsilon_0)k^2.} \tag{41}$$

The twofold degeneracy in each of these solutions reflects the two independent orientations of **E** in the plane normal to **k**. For high **k** the solutions are

$$\omega^2 = c^2k^2/\epsilon_\infty; \qquad \omega^2 = \omega_t^2. \tag{42}$$

We see in Fig. 2 that the lower branch is photon-like at low **k** and phonon-like at high **k** with frequency ω_t. The upper branch is phonon-like at low **k** with frequency ω_l, even though the phonon is transverse, and becomes photon-like at high **k**.

There are no transverse solutions for frequencies between ω_t and

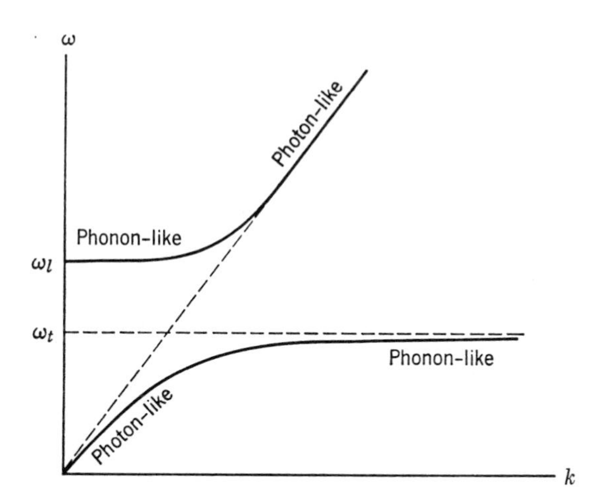

FIG. 2. Coupled modes of photons and transverse optical phonons in an ionic crystal. The broken lines show the spectra without interaction.

ω_l; further, there is no longitudinal mode solution in this region because the photon field is purely transverse in isotropic media. There is thus a forbidden band of frequencies between ω_t and ω_l in which it is impossible to transmit energy through the crystal. This is found experimentally, with appropriate correction for damping. The forbidden optical band appears as a frequency region of high reflectivity.

The longitudinal solutions are found by trying E_z, P_z, $w_z \sim e^{i(\omega t - kz)}$. Then $E_z + 4\pi P_z = 0$, and $\mathbf{H} = 0$. We have just (35) and (36), exactly as in the absence of retardation. Thus there is no effective coupling between the photon and the longitudinal phonon fields.

QUANTUM THEORY OF A CLASSICAL DIELECTRIC[3]

We treat the same problem by the methods of quantum field theory as an exercise in handling the quantization of coupled fields. No specifically quantum effects are discussed in detail. Near the crossover of the dispersion relations (ω versus $\mathbf{k}$) of the uncoupled photon and optical phonon fields we have seen that a weak coupling has drastic effects in mixing together the mechanical and electromagnetic fields. Such effects in isotropic or cubic crystals occur only for transverse optical phonons, as only these couple with an electromagnetic field, which is always transverse in isotropic media.

The lagrangian density for an electromagnetic field in an infinite classical dielectric of rigid ions may be written

$$(43) \quad \mathcal{L} = \frac{1}{8\pi}\left(\frac{1}{c}\dot{\mathbf{A}} + \text{grad}\,\varphi\right)^2 - \frac{1}{8\pi}(\text{curl}\,\mathbf{A})^2 + \frac{1}{2\chi}(\dot{\mathbf{P}}^2 - \omega_0^2\mathbf{P}^2) - \mathbf{P}\cdot\left(\frac{1}{c}\dot{\mathbf{A}} + \text{grad}\,\varphi\right),$$

where χ is a constant which in gaussian units has the dimensions of (frequency)^2. The lagrangian equations of motion for the field variables $\mathbf{A}$, φ, and $\mathbf{P}$ are equivalent to the maxwell equations plus the constitutive equation

$$(44) \quad -\frac{\partial\mathcal{L}}{\partial P_\alpha} + \frac{\partial}{\partial t}\frac{\partial\mathcal{L}}{\partial\dot{P}_\alpha} + \frac{\partial}{\partial X_\mu}\frac{\partial\mathcal{L}}{\partial(\partial P_\alpha/\partial X_\mu)} = \frac{1}{\chi}(\ddot{P}_\alpha + \omega_0^2 P_\alpha) + \left(\frac{1}{c}\dot{\mathbf{A}} + \text{grad}\,\varphi\right)_\alpha = 0,$$

or, with $\mathbf{E} \equiv -\frac{1}{c}\dot{\mathbf{A}} - \text{grad}\,\varphi$,

$$(45) \quad \ddot{P}_\alpha + \omega_0^2 P_\alpha = \chi E_\alpha.$$

[3] J. J. Hopfield, *Phys. Rev.* **112,** 1555 (1958); U. Fano, *Phys. Rev.* **103,** 1202 (1956).

The equation of motion for φ gives, in the coulomb gauge with div $\mathbf{A} = 0$,

$$(46) \qquad -\operatorname{div}(\operatorname{grad}\varphi - 4\pi\mathbf{P}) = \operatorname{div}(\mathbf{E} + 4\pi\mathbf{P}) = 0.$$

The equation of motion for A_α is, in the coulomb gauge,

$$(47) \qquad \nabla^2 A_\alpha - \frac{1}{c^2}\ddot{A}_\alpha = \frac{1}{c}[(\nabla\dot{\varphi})_\alpha - 4\pi\dot{P}_\alpha].$$

With the definition $\mathbf{H} = \operatorname{curl}\mathbf{A}$, this completes the derivation from the lagrangian of the maxwell equation and the constitutive equation. The latter is equivalent to a dielectric dispersion law appropriate to a rigid-ion model (that is, no electronic polarizability):

$$(48) \qquad \epsilon = 1 + \frac{4\pi\chi}{\omega_0^2 - \omega^2}.$$

We have seen in the preceding section that the longitudinal modes of the dielectric polarization field do not couple with photons and therefore do not concern us here. We may drop φ; the lagrangian density now reduces to

$$(49) \quad \mathfrak{L} = \frac{1}{8\pi c^2}(\dot{\mathbf{A}})^2 - \frac{1}{8\pi}(\operatorname{curl}\mathbf{A})^2 + \frac{1}{2\chi}(\dot{\mathbf{P}}^2 - \omega_0^2\mathbf{P}^2) - \mathbf{P}\cdot\frac{1}{c}\dot{\mathbf{A}}.$$

The momentum $\mathbf{M}$ conjugate to $\mathbf{A}$ is defined by

$$(50) \qquad M_\alpha \equiv \frac{\partial\mathfrak{L}}{\partial\dot{A}_\alpha} = \frac{1}{4\pi c^2}\dot{A}_\alpha - \frac{1}{c}P_\alpha;$$

and similarly for $\mathbf{\Pi}$:

$$(51) \qquad \Pi_\alpha = \frac{\partial\mathfrak{L}}{\partial\dot{P}_\alpha} = \frac{1}{\chi}\dot{P}_\alpha.$$

The hamiltonian density is given by

$$(52) \quad \mathcal{H} = M_\alpha\dot{A}_\alpha + \Pi_\alpha\dot{P}_\alpha - \mathfrak{L} = 2\pi c^2\mathbf{M}^2 + \frac{1}{8\pi}(\operatorname{curl}\mathbf{A})^2 + \frac{\chi}{2}\mathbf{\Pi}^2 + \left(2\pi + \frac{\omega_0^2}{2\chi}\right)\mathbf{P}^2 + 4\pi c\mathbf{M}\cdot\mathbf{P}.$$

We now expand, with periodic boundary conditions over unit volume,

$$(53) \qquad \mathbf{A} = \sum_{\mathbf{k}\lambda}\left(\frac{2\pi c}{k}\right)^{1/2}\mathbf{e}_{\mathbf{k}\lambda}(a_{\mathbf{k}\lambda}e^{i\mathbf{k}\cdot\mathbf{x}} + a_{\mathbf{k}\lambda}^+e^{-i\mathbf{k}\cdot\mathbf{x}});$$

$$(54) \qquad \mathbf{P} = \sum_{\mathbf{k}\lambda}\left[\frac{\chi}{2(4\pi\chi + \omega_0^2)^{1/2}}\right]^{1/2}\mathbf{e}_{\mathbf{k}\lambda}(b_{\mathbf{k}\lambda}e^{i\mathbf{k}\cdot\mathbf{x}} + b_{\mathbf{k}\lambda}^+e^{-i\mathbf{k}\cdot\mathbf{x}});$$

here $\boldsymbol{\varepsilon}_{\mathbf{k}\lambda}$ is a unit vector in the direction of the polarization of the wave; λ is the polarization index. The a, a^+ and b, b^+ have the properties of boson operators. The details of the calculation are given in reference 3. The hamiltonian becomes, with $\beta = \chi/\omega_0{}^2$,

$$(55)\quad H = \sum_{\mathbf{k}\lambda} \{ck(a^+_{\mathbf{k}\lambda}a_{\mathbf{k}\lambda} + \tfrac{1}{2}) + \omega_0(1 + 4\pi\beta)^{1/2}(b^+_{\mathbf{k}\lambda}b_{\mathbf{k}\lambda} + \tfrac{1}{2}) + i[\pi ck\beta\omega_0/(1 + 4\pi\beta)^{1/2}](a^+_{\mathbf{k}\lambda}b_{\mathbf{k}\lambda} - a_{\mathbf{k}\lambda}b^+_{\mathbf{k}\lambda} - a_{-\mathbf{k}\lambda}b_{\mathbf{k}\lambda} + a^+_{-\mathbf{k}\lambda}b^+_{\mathbf{k}\lambda})\}.$$

We have now set up the problem, but shall only sketch the solution. The hamiltonian is diagonalized by the introduction of annihilation operator

$$(56)\qquad \alpha_{\mathbf{k}} = wa_{\mathbf{k}} + xb_{\mathbf{k}} + ya^+_{-\mathbf{k}} + zb^+_{-\mathbf{k}},$$

where w, x, y, z are chosen to satisfy the relation

$$(57)\qquad [\alpha_{\mathbf{k}},H] = \omega_{\mathbf{k}}\alpha_{\mathbf{k}}.$$

The solution of the eigenvalue problem (57) is the dispersion relation

$$(58)\qquad \omega_k{}^4 - \omega_k{}^2[(1 + 4\pi\beta)\omega_0{}^2 + c^2k^2] + \omega_0{}^2c^2k^2 = 0,$$

identical with (38).

INTERACTION OF MAGNETIZATION AND THE ELECTROMAGNETIC FIELD

We solve the maxwell equations and the spin resonance equations simultaneously in a infinite medium. This is an interesting and striking example of coupling with the electromagnetic field. For $\epsilon = 1$,

$$(59)\qquad \text{curl }\mathbf{H} = \frac{1}{c}\dot{\mathbf{E}};\qquad \text{curl }\mathbf{E} = -\frac{1}{c}(\dot{\mathbf{H}} + 4\pi\dot{\mathbf{M}});$$

$$(60)\qquad \dot{\mathbf{M}} = \gamma\mathbf{M} \times \mathbf{H},$$

where γ is the magnetomechanical ratio $ge/2mc$, and $\mathbf{M}$ is the magnetization. The first two combine to give

$$(61)\quad \nabla^2\mathbf{H} = c^{-2}(\ddot{\mathbf{H}} + 4\pi\ddot{\mathbf{M}});\qquad -k^2H^+ = -c^{-2}\omega^2(H^+ + 4\pi M^+),$$

with $H^+ = H_x + iH_y$; $M^+ = M_x + iM_y$. The linearized resonance equation may be written, with $\omega_0 = \gamma H_0$ and $\omega_s = \gamma M_z$ for a static field H_0 in the z direction:

$$(62)\qquad i\omega M^+ = -i(\omega_0M^+ - \omega_sH^+),$$

so that

$$(63)\qquad M^+ = \frac{\omega_sH^+}{\omega + \omega_0}.$$

Using (63) in (61),

$$c^2k^2 = \omega^2\left(1 + \frac{4\pi\omega_s}{\omega + \omega_0}\right); \tag{64}$$

this has a branch at $\mathbf{k} = 0$ when

$$-\omega = \omega_0 + 4\pi\omega_s = \gamma(H_0 + 4\pi M_s) = \gamma B. \tag{65}$$

Thus the magnon branch (Fig. 3) comes in to the origin $\mathbf{k} = 0$ at $\omega = \gamma B$, rather than at $\omega = \gamma H$. This result has nothing to do with demagnetizing fields, but arises solely from the effect of displacement currents. The result is the analog of the Lyddane-Sachs-Teller relation (36) in ionic crystals.

However, magnetic resonance is observed in flat plates (with $H \perp$ surface) at $\omega = \gamma H_i$, where H_i is the internal field corrected for surface demagnetization, and not at $\omega = \gamma B_i$. The point is that in *thin* plates the displacement current term is not able to shift the solution from H_i to B_i. This effect is well known in ionic crystals and has been demonstrated by P. Pincus [*J. Appl. Phys.* **33,** 553 (1962)] for the magnetic resonance problem by solving exactly for the surface impedance of a plate as a function of the thickness. Roughly speaking, the displacement current shift in the resonance frequency develops when the plate contains a wavelength or more of the radiation.

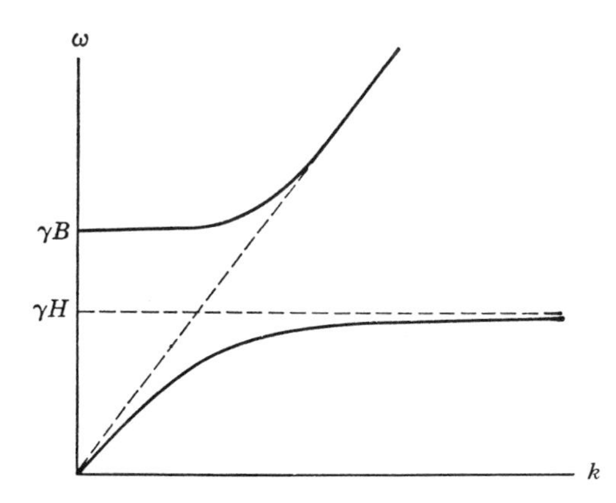

FIG. 3. Dispersion relation for coupled photon and magnetization fields, without exchange.

PROBLEMS

1. Suppose that an electron gas has imbedded in it n_a atoms per unit volume, each of atomic polarizability χ_a. Show that in (15)

$$P_{\|} = \left(-\frac{ne^2}{m\omega^2} + \chi_a\right) E. \tag{66}$$

Calculate for metallic silver the effect of χ_a on the value of the frequency for which $\epsilon_{\|} = 0$, using as a rough estimate for χ_a the polarizability of the Ag^+ ion in silver halides, as found in the literature.

2. Show that the eigenfrequencies of a plasma enclosed in a sphere are given by

$$\omega_L^2 = \omega_p^2 \frac{L}{2L + 1}, \tag{67}$$

where $\omega_p^2 = 4\pi ne^2/m$ and L is the order of the legendre polynomial in the exterior and interior potentials

$$\varphi_e \propto r^{-(L+1)} P_L^m(\cos\theta) e^{im\varphi}; \qquad \varphi_i \propto r^L P_L^m(\cos\theta) e^{im\varphi}. \tag{68}$$

We assume displacement current effects may be neglected.

Hint: the intrinsic dielectric constant of a plasma is

$$\epsilon = 1 - \frac{\omega_p^2}{\omega^2}. \tag{69}$$

This problem is the analog of the magnetostatic mode problem treated by L. R. Walker, *Phys. Rev.* **105**, 390 (1957).

4 Magnons

The low-lying energy states of spin systems coupled by exchange interactions are wavelike, as shown originally by Bloch for ferromagnets. The waves are called spin waves; the energy of a spin wave is quantized, and the unit of energy of a spin wave is called a *magnon*. Spin waves have been studied for all types of ordered spin arrays, including ferromagnetic, ferrimagnetic, antiferromagnetic, canted, and spiral arrays. We shall study the ferromagnetic and antiferromagnetic spin waves, first with an atomic hamiltonian and later with a macroscopic hamiltonian.

FERROMAGNETIC MAGNONS

The simplest hamiltonian of interest is the sum of nearest-neighbor exchange and zeeman contributions:

$$H = -J \sum_{j\boldsymbol{\delta}} \mathbf{S}_j \cdot \mathbf{S}_{j+\boldsymbol{\delta}} - 2\mu_0 H_0 \sum_j S_{jz}, \tag{1}$$

where the vectors $\boldsymbol{\delta}$ connect atom j with its nearest neighbors on a bravais lattice; J is the exchange integral and is assumed to be positive; $\mu_0 = (g/2)\mu_B$ is the magnetic moment; $\mathbf{S}_j$ is the spin angular momentum operator of the atom at j; H_0 is the intensity of a static magnetic field directed along the z axis. We take $H_0 > 0$ to make the magnetic moments line up along the positive z axis when the system is in the ground state.

The constants of the motion of the hamiltonian (1) include the total spin $\mathcal{S}^2 = \left(\sum_j \mathbf{S}_j\right)^2$ and the z component $\mathcal{S}_z = \sum_j S_{jz}$ of the total spin. The ground state $|0\rangle$ of a system of N identical atoms of spin S has

$$\mathcal{S}^2|0\rangle = NS(NS+1)|0\rangle; \qquad \mathcal{S}_z|0\rangle = NS|0\rangle. \tag{2}$$

HOLSTEIN-PRIMAKOFF TRANSFORMATION

The hamiltonian involves the three components S_{jx}, S_{jy}, S_{jz} of each spin $\mathbf{S}_j$. The components are not independent, but are connected by the identity $\mathbf{S}_j \cdot \mathbf{S}_j = S(S+1)$. It is more convenient to work with two operators which are independent. The Holstein-Primakoff transformation[1] to boson creation and annihilation operators a_j^+, a_j is defined by

$$S_j^+ = S_{jx} + iS_{jy} = (2S)^{1/2}(1 - a_j^+a_j/2S)^{1/2}a_j; \tag{3}$$

$$S_j^- = S_{jx} - iS_{jy} = (2S)^{1/2}a_j^+(1 - a_j^+a_j/2S)^{1/2}; \tag{4}$$

here we require

$$[a_j, a_l^+] = \delta_{jl}, \tag{5}$$

in order that the S^+, S^- satisfy the correct commutation relations.

From (3), (4), and (5) we may calculate the transformation for S_z. Now, dropping the subscript j,

$$\begin{aligned} S_z^2 &= S(S+1) - S_x^2 - S_y^2 = S(S+1) - \tfrac{1}{2}(S^+S^- + S^-S^+) \\ &= S(S+1) - S[(1 - a^+a/2S)^{1/2}aa^+(1 - a^+a/2S)^{1/2} \\ &\qquad + a^+(1 - a^+a/2S)a]. \end{aligned} \tag{6}$$

Using $[a^+a, a^+a] = 0$ and $[a^+a, a] = -a$, we develop (6) to find

$$\begin{aligned} S_z^2 &= S(S+1) - S[2a^+a(1 - a^+a/2S) \\ &\qquad + (1 - a^+a/2S) + a^+a/2S] \\ &= (S - a^+a)^2, \end{aligned} \tag{7}$$

whence

$$S_{jz} = S - a_j^+a_j. \tag{8}$$

It is convenient to make a transformation from the atomic a_j^+, a_j to the magnon variables $b_\mathbf{k}^+, b_\mathbf{k}$ defined by

$$b_\mathbf{k} = N^{-1/2} \sum_j e^{i\mathbf{k}\cdot\mathbf{x}_j}a_j; \qquad b_\mathbf{k}^+ = N^{-1/2} \sum_j e^{-i\mathbf{k}\cdot\mathbf{x}_j}a_j^+; \tag{9}$$

here $\mathbf{x}_j$ is the position vector of the atom j. The inverse transformation is then given by

$$a_j = N^{-1/2} \sum_\mathbf{k} e^{-i\mathbf{k}\cdot\mathbf{x}_j}b_\mathbf{k}; \qquad a_j^+ = N^{-1/2} \sum_\mathbf{k} e^{i\mathbf{k}\cdot\mathbf{x}_j}b_\mathbf{k}^+. \tag{10}$$

The signs of the exponents $\pm i\mathbf{k} \cdot \mathbf{x}_j$ have been chosen to agree with

[1] T. Holstein and H. Primakoff, *Phys. Rev.* **58,** 1098 (1940).

those adopted by Holstein and Primakoff. The commutator satisfies the boson commutation relation:

$$(11)\qquad [b_{\mathbf{k}},b_{\mathbf{k}'}^{+}] = N^{-1}\sum_{jl} e^{i\mathbf{k}\cdot\mathbf{x}_j}e^{-i\mathbf{k}'\cdot\mathbf{x}_l}[a_j,a_l^{+}] = N^{-1}\sum_{j} e^{i(\mathbf{k}-\mathbf{k}')\cdot\mathbf{x}_j} = \delta_{\mathbf{k}\mathbf{k}'},$$

and

$$(12)\qquad [b_{\mathbf{k}},b_{\mathbf{k}'}] = [b_{\mathbf{k}}^{+},b_{\mathbf{k}'}^{+}] = 0.$$

The operator $b_{\mathbf{k}}^{+}$ creates a magnon of wavevector **k**, and the operator $b_{\mathbf{k}}$ destroys a magnon of wavevector **k**. The discrete values of **k** summed over are those obtained from periodic boundary conditions.

We now wish to express S_j^{+}, S_j^{-}, and S_{zj} in terms of the spin-wave variables. We will be concerned chiefly with low-lying states of the system such that the fractional spin reversal is small:

$$(13)\qquad \langle a_j^{+}a_j\rangle/S = \langle n_j\rangle/S \ll 1,$$

so that it is pertinent to expand the square roots in (3) and (4). Then

$$(14)\qquad S_j^{+} = (2S)^{1/2}[a_j - (a_j^{+}a_ja_j/4S) + \cdots] = (2S/N)^{1/2}\Big[\sum_{\mathbf{k}} e^{-i\mathbf{k}\cdot\mathbf{x}_j}b_{\mathbf{k}} - (4SN)^{-1}\sum_{\mathbf{k},\mathbf{k}',\mathbf{k}''} e^{i(\mathbf{k}-\mathbf{k}'-\mathbf{k}'')\cdot\mathbf{x}_j}b_{\mathbf{k}}^{+}b_{\mathbf{k}'}b_{\mathbf{k}''} + \cdots\Big];$$

$$(15)\qquad S_j^{-} = (2S/N)^{1/2}\Big[\sum_{\mathbf{k}} e^{i\mathbf{k}\cdot\mathbf{x}_j}b_{\mathbf{k}}^{+} - (4SN)^{-1}\sum_{\mathbf{k},\mathbf{k}',\mathbf{k}''} e^{i(\mathbf{k}+\mathbf{k}'-\mathbf{k}'')\cdot\mathbf{x}_j}b_{\mathbf{k}}^{+}b_{\mathbf{k}'}^{+}b_{\mathbf{k}''} + \cdots\Big];$$

$$(16)\qquad S_{jz} = S - a_j^{+}a_j = S - N^{-1}\sum_{\mathbf{k}\mathbf{k}'} e^{i(\mathbf{k}-\mathbf{k}')\cdot\mathbf{x}_j}b_{\mathbf{k}}^{+}b_{\mathbf{k}'}.$$

We note that the total spin operator for the whole system is

$$NS - \mathcal{S}_z = NS - \sum_j S_{jz} = \sum_j a_j^{+}a_j;$$

$$(17)\qquad \boxed{\mathcal{S}_z = NS - N^{-1}\sum_{j\mathbf{k}\mathbf{k}'} e^{i(\mathbf{k}-\mathbf{k}')\cdot\mathbf{x}_j}b_{\mathbf{k}}^{+}b_{\mathbf{k}'} = NS - \sum_{\mathbf{k}} b_{\mathbf{k}}^{+}b_{\mathbf{k}}.}$$

This is exact.

Thus $b_{\mathbf{k}}^{+}b_{\mathbf{k}}$ may be viewed as the occupation number operator for the magnon state **k**; the eigenvalues of $b_{\mathbf{k}}^{+}b_{\mathbf{k}}$ are the positive integers $n_{\mathbf{k}}$.

We emphasize that the a's and b's act like boson amplitudes, despite the fact that electrons are fermions. This is no more surprising than having phonons act like bosons, even though every fundamental particle in the system (electrons, protons, neutrons) is a fermion. All field amplitudes which are macroscopically observable are boson fields: the field amplitude of a fermion state is restricted severely by the occupation rule 0 or 1 and so cannot be measured accurately.

HAMILTONIAN IN SPIN-WAVE VARIABLES

Using the transformation of S^+, S^-, S_z to spin-wave variables, the hamiltonian

$$H = -J \sum_{j\delta} \mathbf{S}_j \cdot \mathbf{S}_{j+\delta} - 2\mu_0 H_0 \sum_j S_{jz}$$

becomes, if there are z nearest neighbors,

$$H = -JNzS^2 - 2\mu_0 H_0 NS + \mathcal{H}_0 + \mathcal{H}_1, \tag{18}$$

where the term bilinear in spin-wave variables is

$$\mathcal{H}_0 = -(JS/N) \sum_{j\delta\mathbf{k}\mathbf{k}'} \{e^{-i(\mathbf{k}-\mathbf{k}')\cdot\mathbf{x}_j} e^{i\mathbf{k}\cdot\delta} b_\mathbf{k} b_{\mathbf{k}'}^+ + e^{i(\mathbf{k}-\mathbf{k}')\cdot\mathbf{x}_j} \times e^{-i\mathbf{k}'\cdot\delta} b_\mathbf{k}^+ b_{\mathbf{k}'} - e^{i(\mathbf{k}-\mathbf{k}')\cdot\mathbf{x}_j} b_\mathbf{k}^+ b_{\mathbf{k}'} - e^{-i(\mathbf{k}-\mathbf{k}')\cdot(\mathbf{x}_j+\delta)} b_\mathbf{k}^+ b_{\mathbf{k}'}\} + (2\mu_0 H_0/N) \sum_{j\mathbf{k}\mathbf{k}'} e^{i(\mathbf{k}-\mathbf{k}')\cdot\mathbf{x}_j} b_\mathbf{k}^+ b_{\mathbf{k}'}, \tag{19}$$

which becomes, on summing over j,

$$\mathcal{H}_0 = -JzS \sum_\mathbf{k} \{\gamma_\mathbf{k} b_\mathbf{k} b_\mathbf{k}^+ + \gamma_{-\mathbf{k}} b_\mathbf{k}^+ b_\mathbf{k} - 2 b_\mathbf{k}^+ b_\mathbf{k}\} + 2\mu_0 H_0 \sum_\mathbf{k} b_\mathbf{k}^+ b_\mathbf{k}, \tag{20}$$

where

$$\gamma_\mathbf{k} = z^{-1} \sum_\delta e^{i\mathbf{k}\cdot\delta} \tag{21}$$

over the z nearest neighbors. We note that $\sum_\mathbf{k} \gamma_\mathbf{k} = 0$. If there is a center of symmetry $\gamma_\mathbf{k} = \gamma_{-\mathbf{k}}$; then

$$\boxed{\mathcal{H}_0 = \sum_\mathbf{k} \{2JzS(1 - \gamma_\mathbf{k}) + 2\mu_0 H_0\} b_\mathbf{k}^+ b_\mathbf{k}.} \tag{22}$$

The term $\mathcal{H}_1$ in (18) contains fourth and higher order terms in magnon operators, and it may be neglected when the excitation is low.

We can write (22) as

$$\mathcal{H}_0 = \sum_\mathbf{k} \hat{n}_\mathbf{k} \omega_\mathbf{k}; \qquad \omega_\mathbf{k} = 2JSz(1 - \gamma_\mathbf{k}) + 2\mu_0 H_0. \tag{23}$$

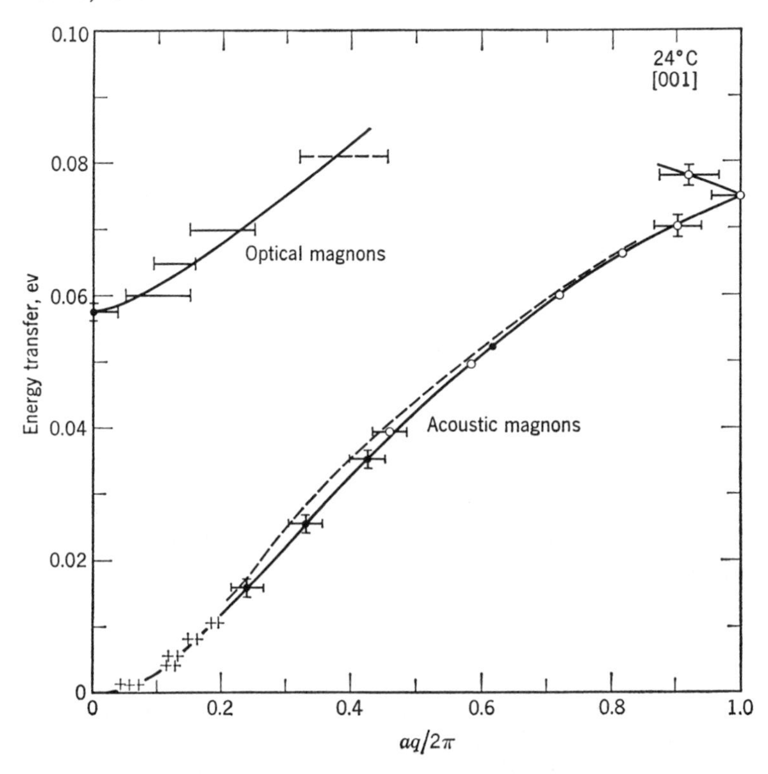

FIG. 1. Acoustical and optical magnon dispersion relations in magnetite, as determined from inelastic neutron scattering by Brockhouse and Watanabe (IAEA Symposium, Chalk River, Ontario, 1962).

This is the dispersion relation for magnons in a spin system forming a bravais lattice. Experimental curves for magnetite are given in Fig. 1; the primitive spin cell in Fe_3O_4 contains several ions, so that there is one acoustic magnon branch and several optical magnon branches. The terminology is borrowed from the corresponding phonon problem.

For $|\mathbf{k} \cdot \boldsymbol{\delta}| \ll 1$,

$$z(1 - \gamma_{\mathbf{k}}) \cong \tfrac{1}{2} \sum_{\boldsymbol{\delta}} (\mathbf{k} \cdot \boldsymbol{\delta})^2, \tag{24}$$

and

$$\omega_{\mathbf{k}} \cong 2\mu_0 H_0 + JS \sum_{\boldsymbol{\delta}} (\mathbf{k} \cdot \boldsymbol{\delta})^2, \tag{25}$$

which reduces for sc, bcc, and fcc lattices of lattice constant a to

$$\boxed{\omega_{\mathbf{k}} = 2\mu_0 H_0 + 2JS(ka)^2.} \tag{26}$$

We note that the exchange contribution to the magnon frequency is of the form of the de Broglie dispersion relation for a free particle of mass m^*, namely

$$\omega_{\mathbf{k}} = \frac{1}{2m^*} k^2, \tag{27}$$

if we set $2JSa^2 = 1/2m^*$, or

$$m^* = 1/4JSa^2. \tag{28}$$

For conventional ferromagnets with curie points at or above room temperature the observed dispersion relations lead to m^* of the order of ten times the electronic mass.

MAGNON INTERACTIONS

The exchange hamiltonian is only diagonal in the spin-wave variables if we neglect the interaction $\mathcal{H}_1$ in (18). The leading term in $\mathcal{H}_1$ is biquadratic and leads to coupling between spin waves. The collision cross section for two spin waves $\mathbf{k}_1$, $\mathbf{k}_2$ has been calculated by F. J. Dyson [*Phys. Rev.* **102,** 1217 (1956)] and is of the order of $(k_1a)^2(k_2a)^2a^2$, where a is the lattice constant. For magnons having microwave frequencies, $ka \sim 10^{-2}$ to 10^{-3}, so that the magnon-magnon exchange scattering cross section is of the order of 10^{-25} cm^2, which is very small for an atomic process. A physical interpretation of the Dyson result has been given by F. Keffer and R. Loudon, *J. Appl. Phys.* (*Supplement*) **32,** 2 (1961).

Using (14), (15), and (16), we find, after a simple but tedious enumeration and rearrangement of terms up to fourth order,

$$\mathcal{H}_1 = (zJ/4N) \sum_{1234} b_1^+ b_2^+ b_3 b_4 \Delta(\mathbf{k}_1 + \mathbf{k}_2 - \mathbf{k}_3 - \mathbf{k}_4)\{2\gamma_1 + 2\gamma_3 - 4\gamma_{1-3}\}, \tag{29}$$

where $\Delta(x) = 1$ for $x = 0$ and $\Delta(x) = 0$ otherwise. For $|\mathbf{k} \cdot \boldsymbol{\delta}| \ll 1$,

$$2\gamma_1 + 2\gamma_3 - 4\gamma_{1-3} \cong \sum_{\boldsymbol{\delta}} \frac{1}{2z} \{2(\mathbf{k}_1 \cdot \boldsymbol{\delta})^2 + 2(\mathbf{k}_3 \cdot \boldsymbol{\delta})^2 - 8(\mathbf{k}_1 \cdot \boldsymbol{\delta})^2(\mathbf{k}_3 \cdot \boldsymbol{\delta})^2\}. \tag{30}$$

We see that a transition probability involving $|\langle|\mathcal{H}|\rangle|^2$ will be proportional to $(ka)^4$, where a is the lattice constant. Thus the degree of diagonalization of the exchange interaction by the spin-wave transformation is extremely good for long-wavelength spin waves ($ka \ll 1$), which are the dominant excitations at low temperatures.

A discussion of the effect of magnon-magnon exchange interactions on the renormalization of magnon energies is given from Eq. (131) on.

MAGNON HEAT CAPACITY

We set $H = 0$, neglect magnon-magnon interactions, and assume $ka \ll 1$; then (26) gives

(31) $$\omega_{\mathbf{k}} = Dk^2; \qquad D \equiv 2SJa^2.$$

The internal energy of unit volume of the magnon gas in thermal equilibrium at temperature T is given by, with $\tau \equiv k_BT$,

(32) $$U = \sum_{\mathbf{k}} \omega_{\mathbf{k}} \langle n_{\mathbf{k}} \rangle_T = \sum_{\mathbf{k}} \omega_{\mathbf{k}} \cdot \frac{1}{e^{\omega_{\mathbf{k}}/\tau} - 1}$$
$$= \frac{1}{(2\pi)^3} \int d^3k \; Dk^2 \frac{1}{e^{Dk^2/\tau} - 1},$$

or, with $x = Dk^2/\tau$,

(33) $$U = \frac{\tau^{5/2}}{4\pi^2 D^{3/2}} \int_0^{x_m} dx \; x^{3/2} \frac{1}{e^x - 1},$$

where the upper limit may be taken as ∞ if we are interested in the region $\tau \ll \omega_{\max}$. Then the integral has the value $\Gamma(\frac{5}{2})\zeta(\frac{5}{2};1)$, where $\Gamma(x)$ is the gamma function, and $\zeta(S,a)$ is the Riemann zeta function (see Whittaker and Watson, *Modern analysis*, Chapter 13). Now $\Gamma(\frac{5}{2}) = 3\pi^{1/2}/4$ and $\zeta(\frac{5}{2};1) = 1.341$, according to the Jahnke-Ende tables. Thus

(34) $$U = \frac{3\tau^{5/2}\zeta(\frac{5}{2})}{2(4\pi D)^{3/2}} \cong \frac{0.45\tau^{5/2}}{\pi^2 D^{3/2}};$$

for the heat capacity of unit volume,

(35) $$\boxed{C = dU/dT = 0.113 k_B (k_BT/D)^{3/2}.}$$

If the heat capacity is composed solely of a magnon part $\propto T^{3/2}$ and a phonon part $\propto T^3$, a plot of $CT^{3/2}$ versus $T^{3/2}$ will be a straight line,

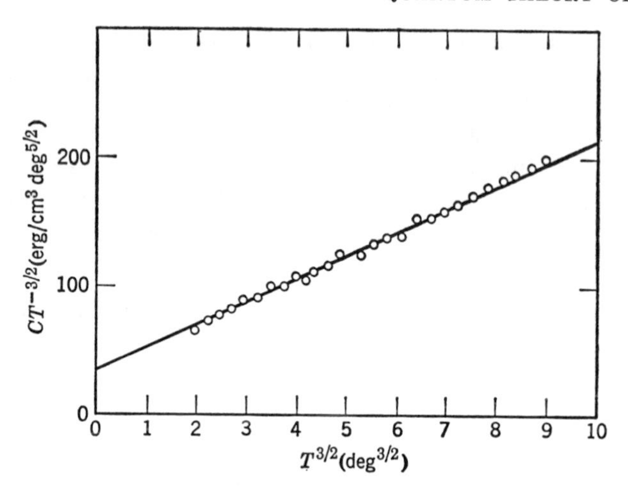

FIG. 2. Heat capacity of yttrium iron garnet, showing magnon and phonon contributions. (After Shinozaki.) The intercept at $T = 0$ measures the magnon contribution.

as in Fig. 2, with the intercept at $T = 0$ giving the magnon contribution and the slope giving the phonon contributions. Two specimens of yttrium iron garnet measured by Shinozaki gave $D = 0.81 \times 10^{-28}$ erg-cm^2 and 0.85×10^{-28} erg-cm^2, leading to $m^*/m \sim 6$.

MAGNETIZATION REVERSAL

The number of reversed spins is given by the ensemble average of the spin wave occupancy numbers. Thus for the saturation magnetization, with the same assumptions as for the heat capacity, and taking unit volume,

$$M_s = 2\mu_0 S_z = 2\mu_0(NS - \Sigma\, b_{\mathbf{k}}^+ b_{\mathbf{k}}), \tag{36}$$

and

$$M_s(0) - M_s(T) \equiv \Delta M = 2\mu_0 \sum_{\mathbf{k}} \langle n_{\mathbf{k}} \rangle = \frac{2\mu_0}{(2\pi)^3} \int d^3k \, \frac{1}{e^{Dk^2/\tau} - 1}. \tag{37}$$

At low temperatures such that $Dk_{\max}{}^2 \gg \tau$,

$$\Delta M = \frac{\mu_0}{2\pi^2} \left(\frac{\tau}{D}\right)^{3/2} \int_0^\infty dx \, x^{1/2} \frac{1}{e^x - 1}, \tag{38}$$

where the integral is equal to $\Gamma(\frac{3}{2})\zeta(\frac{3}{2};1)$. Therefore

$$\boxed{\Delta M = 0.117\mu_0(k_B T/D)^{3/2} = 0.117(\mu_0/a^3)(k_B T/2SJ)^{3/2},} \tag{39}$$

recalling that $\mu_0 = (g/2)\mu_B$, where μ_B is the bohr magneton. We note that $2S\mu_0/a^3 = (1,\frac{1}{2},\frac{1}{4})M_s(0)$ for sc, bcc, and fcc lattices, respectively.

Dyson [*Phys. Rev.* **102,** 1230 (1956)] has considered the terms in ΔM in higher orders in k_BT/J. Terms with exponents $\frac{5}{2}$ and $\frac{7}{2}$ arise from the use of full $\gamma_{\mathbf{k}}$ rather than the leading k^2 term in the expansion of $1 - \gamma_{\mathbf{k}}$. However, if one includes the $T^{5/2}$, $T^{7/2}$ terms, then for numerical significance the integral over d^3k should be taken only over the actual range of **k** space and not to ∞. The first term in ΔM arising from the nonideal aspect of the magnon gas, that is, from magnon-magnon exchange interactions, is of order $(k_BT/J)^4$.

Keffer and Loudon have given a simple picture of the origin of the T^4 term; they give a dynamical argument that in (39) one should use $\bar{S}$ for S, where $\bar{S}$ is the ensemble average projection of a spin on its nearest neighbor: $\bar{S} = \langle \mathbf{S}_j \cdot \mathbf{S}_{j+\delta} \rangle / S$. Thus $S - \bar{S} \propto T^{5/2}$ from (34); on expanding the leading term in the magnetization reversal $(k_BT/2\bar{S}J)^{3/2}$, we get a further term in $(k_BT/2SJ)^{3/2}(k_BT/2SJ)^{5/2} \propto T^4$. We see that magnon-magnon interactions have little effect on the temperature-dependence of the saturation magnetization, except near the curie temperature; for the high temperature region see M. Bloch, *Phys. Rev. Letters* **9,** 286 (1962).

We do not consider the effects which may arise in ferromagnetic metals because of the thermal redistribution of electron states within the bands. Herring and Kittel [*Phys. Rev.* **81,** 869 (1951)] have shown that there is no inconsistency for a metal to have, as observed, a low-temperature heat capacity $\propto T$ dominated by excitation of one-electron states and a magnetization reversal $\propto T^{3/2}$ dominated by magnon excitation.

Reviews of some aspects of the literature on ferromagnetic magnons are:

F. Keffer, in *Encyclopedia of Physics* (Springer) (in press).

A. I. Akhiezer, V. G. Bar'yakhtar, and M. I. Kaganov, *Soviet Physics-Uspekhi* **3,** 567 (1961); original in Russian, *Usp. Fiz. Nauk* **71,** 533 (1960).

J. Van Kranendonk and J. H. Van Vleck, *Revs. Mod. Phys.* **30,** 1 (1958).

C. Kittel, in *Low Temperature Physics,* Gordon and Breach, New York, 1962.

There are a considerable number of experiments in which selected magnons are excited and detected, including spin-wave resonances in thin films; parallel pumping; excitation by magnon-phonon coupling; inelastic neutron scattering; and magnon pulse propagation in discs.

ANTIFERROMAGNETIC MAGNONS[2]

We consider the hamiltonian

$$(40)\qquad H = J\sum_{j\delta} \mathbf{S}_j \cdot \mathbf{S}_{j+\delta} - 2\mu_0 H_A \sum_j S^a_{jz} + 2\mu_0 H_A \sum_j S^b_{jz};$$

here J is the nearest-neighbor exchange integral and with the new choice of sign is positive for an antiferromagnet. We omit everywhere in this section the effect of next-nearest-neighbor interactions, although these may be important in real antiferromagnets. We assume the spin structure of the crystal may be divided into two interpenetrating sublattices a and b with the property that all nearest neighbors of an atom on a lie on b, and *vice versa*. This simple subdivision is not possible in all structures. The quantity H_A is positive and is a fictitious magnetic field, which approximates the effect of the crystal anisotropy energy, with the property of tending for positive μ_0 to align the spins on a in the $+z$ direction and the spins on b in the $-z$ direction. We introduce H_A chiefly to stabilize the spin arrays along a preferred axis, the z axis. We shall see that the configuration with each sublattice saturated is not the true ground state.

We make the Holstein-Primakoff transformation:

$$(41)\qquad S^+_{aj} = (2S)^{1/2}(1 - a^+_j a_j/2S)^{1/2} a_j; \qquad S^-_{aj} = (2S)^{1/2} a^+_j (1 - a^+_j a_j/2S)^{1/2};$$

$$(42)\qquad S^+_{bl} = (2S)^{1/2} b^+_l (1 - b^+_l b_l/2S)^{1/2}; \qquad S^-_{bl} = (2S)^{1/2}(1 - b^+_l b_l/2S)^{1/2} b_l.$$

Here b^+_l, b_l are creation and annihilation operators which refer to the lth atom on sublattice b; they are not magnon variables. We have further

$$(43)\qquad S^a_{jz} = S - a^+_j a_j; \qquad -S^b_{lz} = S - b^+_l b_l,$$

using the other choice of sign permitted by (7). The motivation is obvious for the association of b^+ with S^+_b and a with S^+_a.

We introduce the spin-wave variables

$$(44)\qquad c_{\mathbf{k}} = N^{-1/2}\sum_j e^{i\mathbf{k}\cdot\mathbf{x}_j} a_j; \qquad c^+_{\mathbf{k}} = N^{-1/2}\sum_j e^{-i\mathbf{k}\cdot\mathbf{x}_j} a^+_j;$$

$$(45)\qquad d_{\mathbf{k}} = N^{-1/2}\sum_l e^{-i\mathbf{k}\cdot\mathbf{x}_l} b_l; \qquad d^+_{\mathbf{k}} = N^{-1/2}\sum_l e^{i\mathbf{k}\cdot\mathbf{x}_l} b^+_l.$$

[2] Nagamiya, Yosida, and Kubo, *Advances in Physics* **4,** 97 (1955); J. Ziman, *Proc. Phys. Soc. (London)* **65,** 540, 548 (1952); P. W. Anderson, *Phys. Rev.* **86,** 694 (1952); R. Kubo, *Phys. Rev.* **87,** 568 (1952); T. Nakamura, *Prog. Theo. Phys.* **7,** 539 (1952).

The sum for c is over the N atoms j on sublattice a, and for d over atoms l on b. The leading terms in the expansion of (41) and (42) are

$$(46)\qquad \begin{aligned} S_{aj}^{+} &= (2S/N)^{1/2}\left(\sum_{\mathbf{k}} e^{-i\mathbf{k}\cdot\mathbf{x}_j}c_{\mathbf{k}} + \cdots\right); \\ S_{aj}^{-} &= (2S/N)^{1/2}\left(\sum_{\mathbf{k}} e^{i\mathbf{k}\cdot\mathbf{x}_j}c_{\mathbf{k}}^{+} + \cdots\right); \end{aligned}$$

$$(47)\qquad \begin{aligned} S_{bl}^{+} &= (2S/N)^{1/2}\left\{\sum_{\mathbf{k}} e^{-i\mathbf{k}\cdot\mathbf{x}_l}d_{\mathbf{k}}^{+} + \cdots\right\}; \\ S_{bl}^{-} &= (2S/N)^{1/2}\left\{\sum_{\mathbf{k}} e^{i\mathbf{k}\cdot\mathbf{x}_l}d_{\mathbf{k}} + \cdots\right\}; \end{aligned}$$

$$(48)\qquad \begin{aligned} S_{jz}^{a} &= S - N^{-1}\sum_{\mathbf{k}\mathbf{k}'} e^{i(\mathbf{k}-\mathbf{k}')\cdot\mathbf{x}_j}c_{\mathbf{k}}^{+}c_{\mathbf{k}'}; \\ S_{lz}^{b} &= -S + N^{-1}\sum_{\mathbf{k}\mathbf{k}'} e^{-i(\mathbf{k}-\mathbf{k}')\cdot\mathbf{x}_l}d_{\mathbf{k}}^{+}d_{\mathbf{k}'}. \end{aligned}$$

The hamiltonian transformed to magnon variables is, if there are z nearest neighbors,

$$(49)\qquad H = -2NzJS^2 - 4N\mu_0H_AS + \mathcal{H}_0 + \mathcal{H}_1,$$

where the term bilinear in magnon variables is

$$(50)\qquad \mathcal{H}_0 = 2JzS\sum_{\mathbf{k}}\left[\gamma_{\mathbf{k}}(c_{\mathbf{k}}^{+}d_{\mathbf{k}}^{+} + c_{\mathbf{k}}d_{\mathbf{k}}) + (c_{\mathbf{k}}^{+}c_{\mathbf{k}} + d_{\mathbf{k}}^{+}d_{\mathbf{k}})\right] + 2\mu_0H_A\sum_{\mathbf{k}}(c_{\mathbf{k}}^{+}c_{\mathbf{k}} + d_{\mathbf{k}}^{+}d_{\mathbf{k}}),$$

with

$$(51)\qquad \gamma_{\mathbf{k}} = z^{-1}\sum_{\boldsymbol{\delta}} e^{i\mathbf{k}\cdot\boldsymbol{\delta}} = \gamma_{-\mathbf{k}},$$

assuming a center of symmetry. We neglect $\mathcal{H}_1$, which contains higher order terms.

We now look for a transformation to diagonalize $\mathcal{H}_0$. We transform to new creation and annihilation operators α^{+}, α; β^{+}, β with $[\alpha_{\mathbf{k}},\alpha_{\mathbf{k}}^{+}] = 1$; $[\beta_{\mathbf{k}},\beta_{\mathbf{k}}^{+}] = 1$; $[\alpha_{\mathbf{k}},\beta_{\mathbf{k}}] = 0$; etc. The problem becomes identical with the Bogoliubov problem discussed in Chapter 2 if in (91) there we make the identification

$$(52)\qquad a_{\mathbf{k}} \to c_{\mathbf{k}}; \qquad a_{\mathbf{k}}^{+} \to c_{\mathbf{k}}^{+}; \qquad a_{-\mathbf{k}} \to d_{\mathbf{k}}; \qquad a_{-\mathbf{k}}^{+} \to d_{\mathbf{k}}^{+};$$

$$(53)\qquad \omega_0 \to 2JzS + 2\mu_0H_A; \qquad \omega_1 \to 2JzS\gamma_{\mathbf{k}}.$$

Then the transformation is defined by

$$\begin{aligned} \alpha_{\mathbf{k}} &= u_{\mathbf{k}} c_{\mathbf{k}} - v_{\mathbf{k}} d_{\mathbf{k}}^{+}; \qquad & \alpha_{\mathbf{k}}^{+} &= u_{\mathbf{k}} c_{\mathbf{k}}^{+} - v_{\mathbf{k}} d_{\mathbf{k}}; \\ \beta_{\mathbf{k}} &= u_{\mathbf{k}} d_{\mathbf{k}} - v_{\mathbf{k}} c_{\mathbf{k}}^{+}; & \beta_{\mathbf{k}}^{+} &= u_{\mathbf{k}} d_{\mathbf{k}}^{+} - v_{\mathbf{k}} c_{\mathbf{k}}; \end{aligned} \tag{54}$$

here $u_{\mathbf{k}}$, $v_{\mathbf{k}}$ are real and satisfy $u_{\mathbf{k}}^2 - v_{\mathbf{k}}^2 = 1$.

By analogy with (2.98) the magnon eigenfrequencies $\omega_{\mathbf{k}}$ are given by

$$\boxed{\omega_{\mathbf{k}}^2 = (\omega_e + \omega_A)^2 - \omega_e^2 \gamma_{\mathbf{k}}^2,} \tag{55}$$

with

$$\omega_e \equiv 2JzS; \qquad \omega_A \equiv 2\mu_0 H_A. \tag{56}$$

The result is the dispersion relation for antiferromagnetic magnons. The bilinear hamiltonian becomes, with $\omega_{\mathbf{k}}$ taken to be positive,

$$\mathcal{H}_0 = -N(\omega_e + \omega_A) + \sum_{\mathbf{k}} \omega_{\mathbf{k}} (\alpha_{\mathbf{k}}^{+} \alpha_{\mathbf{k}} + \beta_{\mathbf{k}}^{+} \beta_{\mathbf{k}} + 1), \tag{57}$$

on actually using the inverse of (54) to rewrite (50). There are two degenerate modes for each **k**, one associated with the α operators and one with the β operators. The total hamiltonian (49) is

$$H = -2NzJS(S+1) - 4\pi\mu_0 H_A (S + \tfrac{1}{2}) + \sum_{\mathbf{k}} \omega_{\mathbf{k}} (n_{\mathbf{k}} + \tfrac{1}{2}) + \mathcal{H}_1, \tag{58}$$

where each value of **k** is to be counted twice, because of the double degeneracy; $n_{\mathbf{k}}$ is a positive integer.

If we neglect ω_A and take $ka \ll 1$, then $\{1 - \gamma_{\mathbf{k}}^2\}^{1/2} \approx 3^{-1/2} ka$ for a simple cubic lattice, and we have

$$\boxed{\omega_{\mathbf{k}} \cong 4(3)^{1/2} JSka.} \tag{59}$$

This is the dispersion law for antiferromagnets in the long wavelength limit, provided $\omega_{\mathbf{k}}/\omega_0 \gg 1$. For the uniform mode of antiferromagnetic resonance $\gamma_0 = 1$ and

$$\omega_0 = [(2\omega_e + \omega_A)\omega_A]^{1/2}, \tag{60}$$

which is the standard result.

If a uniform external magnetic field H is applied parallel to the axis of magnetization, the resonance frequencies are easily shown to become

$$\omega_{\mathbf{k}'} = \omega_{\mathbf{k}} \pm \omega_H, \qquad \omega_H = (ge/2mc)H; \tag{61}$$

here ω_H is the larmor frequency corresponding to H.

ZERO-POINT ENERGY

In the spin-wave approximation, neglecting H_A and $\mathcal{H}_1$, the exchange energy of the antiferromagnetic system at absolute zero is

$$E_0 = -2NzJS(S+1) + \sum_{\mathbf{k}} \omega_{\mathbf{k}}, \tag{62}$$

now counting each value of $\mathbf{k}$ once. We recall that the total number of spins in the system is $2N$, counting both sublattices. It is usual to express E_0 in terms of a constant β defined by

$$E_0 = -2NzJS(S + \beta z^{-1}). \tag{63}$$

Using (55),

$$\beta z^{-1} = N^{-1} \sum_{\mathbf{k}} [1 - (1 - \gamma_{\mathbf{k}}^2)^{1/2}]. \tag{64}$$

For a simple cubic lattice the value of β is 0.58.

ZERO-POINT SUBLATTICE MAGNETIZATION

A striking feature of the theory of antiferromagnetics is the departure of the sublattice magnetic moment at absolute zero from the value $2NS\mu_0$ corresponding to the elementary picture of a saturated sublattice. There is some experimental evidence from studies of nuclear magnetic resonance that the actual reduction in sublattice magnetic moment is rather less than is expected from the following calculation.

From (48) and (54),

$$\mathcal{S}_z^a = \sum_j S_{jz}^a = NS - \sum_{\mathbf{k}} c_{\mathbf{k}}^+ c_{\mathbf{k}} = NS - \sum_{\mathbf{k}} (u_{\mathbf{k}}^2 \alpha_{\mathbf{k}}^+ \alpha_{\mathbf{k}} + v_{\mathbf{k}}^2 \beta_{\mathbf{k}} \beta_{\mathbf{k}}^+ + \text{off-diagonal terms}). \tag{65}$$

Here we have used the inverse of (54):

$$c_{\mathbf{k}} = u_{\mathbf{k}} \alpha_{\mathbf{k}} + v_{\mathbf{k}} \beta_{\mathbf{k}}^+; \qquad d_{\mathbf{k}} = u_{\mathbf{k}} \beta_{\mathbf{k}} + v_{\mathbf{k}} \alpha_{\mathbf{k}}^+. \tag{66}$$

At absolute zero all $n_{\mathbf{k}} = 0$, and, counting each $\mathbf{k}$ once,

$$\Delta\mathcal{S}_z = NS - \langle \mathcal{S}_z \rangle = \sum_{\mathbf{k}} v_{\mathbf{k}}^2 = \tfrac{1}{2} \sum_{\mathbf{k}} (\cosh 2\chi_{\mathbf{k}} - 1) = \tfrac{1}{2} \sum_{\mathbf{k}} [(1 - \gamma_{\mathbf{k}}^2)^{-1/2} - 1], \tag{67}$$

for $\omega_A = 0$ and using the transcription of the result of Problem 5, Chapter 2. Therefore, for unit volume,

$$\Delta\mathcal{S}_z = -\tfrac{1}{2}N + \frac{1}{2(2\pi)^3} \int d^3k \, (1 - \gamma_{\mathbf{k}}^2)^{-1/2}, \tag{68}$$

where the integral is taken over the allowed values of $\mathbf{k}$. For a simple cubic lattice the integration gives

$$\Delta \mathcal{S}_z = 0.078N, \tag{69}$$

after Anderson.

Temperature Dependence of Sublattice Magnetization. From (65), with $\omega_A = 0$, and counting each $\mathbf{k}$ once,

$$\langle \mathcal{S}_z(0) \rangle - \langle \mathcal{S}_z(T) \rangle = \sum_{\mathbf{k}} \langle n_{\mathbf{k}} \rangle \cosh 2\chi_{\mathbf{k}} = \sum_{\mathbf{k}} \langle n_{\mathbf{k}} \rangle (1 - \gamma_{\mathbf{k}}{}^2)^{-\frac{1}{2}}, \tag{70}$$

where

$$\langle n_{\mathbf{k}} \rangle = \frac{1}{e^{\omega_{\mathbf{k}}/k_BT} - 1}. \tag{71}$$

In an appropriate temperature range $\omega_{\mathbf{k}}$ is given by (59), which may be written as

$$\omega_{\mathbf{k}} \cong k_B \Theta_N k / k_{\max} \tag{72}$$

for the appropriate range of $\mathbf{k}$; here Θ_N is a temperature of the order of the Néel temperature. For a simple cubic lattice $k_{\max} = \pm \pi/a$ along any cube edge. It is convenient to redefine $k_{\max}$ as for a Debye phonon spectrum:

$$n = \frac{1}{(2\pi)^3} \frac{4\pi}{3} k_{\max}^3, \tag{73}$$

where n is the number of atoms on one sublattice per unit volume. The temperature-dependent part (70) of the sublattice spin density is

$$\frac{3^{\frac{1}{2}}}{\pi^2 a} \int_0^{k_m} dk \cdot k \cdot \{e^{(k/k_m)(\Theta_N/T)} - 1\}^{-1} \cong \frac{3^{\frac{1}{2}}\Omega}{\pi^2 a k_m{}^2} \left(\frac{T}{\Theta_N}\right)^2 \int_0^\infty \frac{x\,dx}{e^x - 1}, \tag{74}$$

for $T \ll \Theta_N$. The sublattice magnetization decreases as $(T/\Theta_N)^2$. Kubo, *Phys. Rev.* **87,** 568 (1952), gives the numerical constants for NaCl and CsCl type lattices.

Heat Capacity. The heat capacity of an antiferromagnet having the dispersion relation

$$\omega_{\mathbf{k}} = [(\omega_e + \omega_A)^2 - \omega_e{}^2 \gamma_{\mathbf{k}}{}^2]^{\frac{1}{2}}$$

will be at low temperatures essentially exponential in $-1/T$ until $k_BT > \omega_0$, where $\omega_0 = [(2\omega_e + \omega_A)\omega_A]^{\frac{1}{2}}$. At higher temperatures (but not too high) the dispersion relation will be of the form (72). There are two antiferromagnetic magnons for every value of $\mathbf{k}$ instead of three phonons. Reducing the standard Debye result for the phonon heat capacity by a factor $\frac{2}{3}$, we have for the magnon contribution to the heat capacity per unit volume

$$C_{\text{mag}} = (4\pi^4/S)(2nk_B)^3(T/\Theta_N)^3, \tag{75}$$

at $T \ll \Theta_N$, but $T \gg \omega_0/k_B$. Bear in mind that Θ_N is defined by (72) and is not identical with the Néel temperature.

FURTHER TOPICS—FERROMAGNETIC MAGNONS

MACROSCOPIC MAGNON THEORY

For many purposes in ferromagnetic resonance and relaxation studies it is more convenient or more physical to work directly with the magnetization as a field $\mathbf{M}(\mathbf{x})$, rather than deal with individual spins $\mathbf{S}_j$. The range of usefulness of the macroscopic field theory is limited to regions of $\mathbf{k}$ space well away from the boundaries of the brillouin zone. We can only use the macroscopic theory where $ka \ll 1$, where a is the lattice parameter. The advantages of the macroscopic theory are that it is not based explicitly on a model in which each electron is attached to a particular atom, and we can easily introduce phenomenological constants relating to anisotropy, magnetoelastic, and magnetostatic energy.

We consider the vector spin density operator $\mathbf{s}(\mathbf{x})$ defined by

$$\mathbf{s}(\mathbf{x}) = \tfrac{1}{2}\sum_j \boldsymbol{\sigma}_j\delta(\mathbf{x} - \mathbf{x}_j), \tag{76}$$

where $\boldsymbol{\sigma}_j$ is the pauli matrix for the jth electron at position $\mathbf{x}_j$. The operator $\mathbf{s}(\mathbf{x})$ represents the spin moment density at $\mathbf{x}$, because

$$\int_\Omega d^3x\, \mathbf{s}(\mathbf{x})$$

is the total spin in the volume Ω. We examine the commutator, using the relation $[\sigma_{jx},\sigma_{ly}] = 2i\sigma_{jz}\delta_{jl}$:

$$\begin{aligned}[s_x(\mathbf{x}),s_y(\mathbf{x}')] &= \frac{i}{2}\sum_{jl}\sigma_{jz}\delta(\mathbf{x} - \mathbf{x}_j)\delta(\mathbf{x}' - \mathbf{x}_l) \\ &= \frac{i}{2}\sum_j \sigma_{jz}\delta(\mathbf{x} - \mathbf{x}_j)\delta(\mathbf{x}' - \mathbf{x}_j) \\ &= is_z(\mathbf{x})\delta(\mathbf{x} - \mathbf{x}'),\end{aligned} \tag{77}$$

on using the identity

$$\delta(\mathbf{x} - \mathbf{x}_j)\delta(\mathbf{x}' - \mathbf{x}_j) = \delta(\mathbf{x} - \mathbf{x}_j)\delta(\mathbf{x} - \mathbf{x}').$$

We introduce the magnetization by the relation

$$\mathbf{M}(\mathbf{x}) = g\mu_B\mathbf{s}(\mathbf{x}) = 2\mu_0\mathbf{s}(\mathbf{x}); \qquad \mu_0 = g\mu_B/2. \tag{78}$$

Then

$$[M_x(\mathbf{x}),M_y(\mathbf{x}')] = i2\mu_0 M_z(\mathbf{x})\delta(\mathbf{x} - \mathbf{x}'), \tag{79}$$

whence, with $M^{\pm}(\mathbf{x}) = M_x(\mathbf{x}) \pm iM_y(\mathbf{x})$,

$$[M^+(\mathbf{x}),M^-(\mathbf{x}')] = 4\mu_0 M_z(\mathbf{x})\delta(\mathbf{x} - \mathbf{x}'). \tag{80}$$

Now transform $\mathbf{M}(\mathbf{x})$ to Holstein-Primakoff field variables $a(\mathbf{x})$, $a^+(\mathbf{x})$, assumed to satisfy the commutation relation

$$[a(\mathbf{x}),a^+(\mathbf{x}')] = \delta(\mathbf{x} - \mathbf{x}'). \tag{81}$$

If we write, with $M = |\mathbf{M}(\mathbf{x})|$,

$$M^+(\mathbf{x}) = (4\mu_0 M)^{1/2}\{1 - (\mu_0/M)a^+(\mathbf{x})a(\mathbf{x})\}^{1/2}a(\mathbf{x}); \tag{82}$$

$$M^-(\mathbf{x}) = (4\mu_0 M)^{1/2}a^+(\mathbf{x})\{1 - (\mu_0/M)a^+(\mathbf{x})a(\mathbf{x})\}^{1/2}; \tag{83}$$

$$M_z(\mathbf{x}) = M - 2\mu_0 a^+(\mathbf{x})a(\mathbf{x}), \tag{84}$$

it is easy for the reader to verify that the commutation relations (79) or (80) on the components of $\mathbf{M}(\mathbf{x})$ are satisfied. For macroscopic purposes we may set $M = M_s$, a constant.

The transformation from $a^+(\mathbf{x})$, $a(\mathbf{x})$ to magnon field variables $b_{\mathbf{k}}^+$, $b_{\mathbf{k}}$ is defined by, for unit volume,

$$a(\mathbf{x}) = \sum_{\mathbf{k}} e^{-i\mathbf{k}\cdot\mathbf{x}}b_{\mathbf{k}}; \qquad a^+(\mathbf{x}) = \sum_{\mathbf{k}} e^{i\mathbf{k}\cdot\mathbf{x}}b_{\mathbf{k}}^+; \tag{85}$$

or

$$b_{\mathbf{k}} = \int d^3x\, a(\mathbf{x})e^{i\mathbf{k}\cdot\mathbf{x}}; \qquad b_{\mathbf{k}}^+ = \int d^3x\, a^+(\mathbf{x})e^{-i\mathbf{k}\cdot\mathbf{x}}. \tag{86}$$

The commutator is

$$\begin{aligned} [b_{\mathbf{k}},b_{\mathbf{k}'}^+] &= \int d^3x\, d^3x'\, [a(\mathbf{x}),a^+(\mathbf{x}')]e^{i(\mathbf{k}\cdot\mathbf{x}-\mathbf{k}'\cdot\mathbf{x}')} \\ &= \int d^3x\, e^{i(\mathbf{k}-\mathbf{k}')\cdot\mathbf{x}} = \delta_{\mathbf{k}\mathbf{k}'}. \end{aligned} \tag{87}$$

The magnetization components (82) to (84) in terms of the $b_{\mathbf{k}}^+$, $b_{\mathbf{k}}$ are

$$\begin{aligned} M^+(\mathbf{x}) = (4\mu_0 M)^{1/2} \Big[&\sum_{\mathbf{k}} e^{-i\mathbf{k}\cdot\mathbf{x}}\, b_{\mathbf{k}} \\ &- (\mu_0/2M) \sum_{\mathbf{k}\mathbf{k}'\mathbf{k}''} e^{i(\mathbf{k}-\mathbf{k}'-\mathbf{k}'')\cdot\mathbf{x}} b_{\mathbf{k}}^+ b_{\mathbf{k}'} b_{\mathbf{k}''} + \cdots \Big]; \end{aligned} \tag{88}$$

$$(89)\quad M^-(\mathbf{x}) = (4\mu_0 M)^{1/2}\Big[\sum_{\mathbf{k}} e^{i\mathbf{k}\cdot\mathbf{x}} b_{\mathbf{k}}^+ - (\mu_0/2M)\sum_{\mathbf{k}\mathbf{k}'\mathbf{k}''} e^{i(-\mathbf{k}+\mathbf{k}'+\mathbf{k}'')\cdot\mathbf{x}} b_{\mathbf{k}} b_{\mathbf{k}'}^+ b_{\mathbf{k}''}^+ + \cdots\Big];$$

$$(90)\quad M_z(\mathbf{x}) = M - 2\mu_0 \sum_{\mathbf{k}\mathbf{k}'} e^{i(\mathbf{k}-\mathbf{k}')\cdot\mathbf{x}} b_{\mathbf{k}}^+ b_{\mathbf{k}'}.$$

It is shown below that in a cubic crystal the macroscopic form of the exchange energy density must contain as the leading term

$$(91)\quad \boxed{\mathcal{H}_{\text{ex}} = C\,\frac{\partial M_\nu}{\partial x_\mu}\frac{\partial M_\nu}{\partial x_\mu},}$$

where C is a constant; repeated Greek subscripts are to be summed over x, y, z. To bilinear terms

$$(92)\quad \frac{\partial M^+}{\partial x_\mu}\frac{\partial M^-}{\partial x_\mu} = 4\mu_0 M \sum_{\mathbf{k}\mathbf{k}'} (\mathbf{k}\cdot\mathbf{k}') e^{i(\mathbf{k}'-\mathbf{k})\cdot\mathbf{x}} b_{\mathbf{k}} b_{\mathbf{k}'}^+; \qquad \frac{\partial M_z}{\partial x_\mu}\frac{\partial M_z}{\partial x_\mu} = 0.$$

Therefore the exchange energy density (91) is

$$(93)\quad \mathcal{H}_{\text{ex}}^{(1)} = 2C\mu_0 M \sum_{\mathbf{k}\mathbf{k}'} e^{i(\mathbf{k}'-\mathbf{k})\cdot\mathbf{x}} (b_{\mathbf{k}} b_{\mathbf{k}'}^+ + b_{\mathbf{k}'}^+ b_{\mathbf{k}})(\mathbf{k}\cdot\mathbf{k}')$$

to first order. The next highest order is

$$(94)\quad \mathcal{H}_{\text{ex}}^{(2)} = 2C\mu^2 \Sigma\, (k^2 + k'^2 - 4\mathbf{k}\cdot\mathbf{k}')\Delta(-\mathbf{k}+\mathbf{k}'-\mathbf{k}''+\mathbf{k}''') \cdot b_{\mathbf{k}}^+ b_{\mathbf{k}'} b_{\mathbf{k}''}^+ b_{\mathbf{k}'''},$$

of the same form as (29) derived on the localized spin or Heisenberg model.

The zeeman energy density is, for $\mathbf{H}$ along the z axis,

$$(95)\quad \mathcal{H}_Z = -H_0 M_z = 2\mu_0 H_0 \sum_{\mathbf{k}\mathbf{k}'} e^{i(\mathbf{k}'-\mathbf{k})\cdot\mathbf{x}} b_{\mathbf{k}'}^+ b_{\mathbf{k}},$$

dropping the constant term $-H_0 M$.

For directions near an easy axis of magnetization taken as the z axis the anisotropy energy density may be written

$$(96)\quad \mathcal{H}_K = (K/M_s^2)(M_x^2 + M_y^2) = \tfrac{1}{2}(K/M_s^2)(M^+M^- + M^-M^+) = 2\mu_0(K/M_s)\sum_{\mathbf{k}\mathbf{k}'} e^{i(\mathbf{k}'-\mathbf{k})\cdot\mathbf{x}} (b_{\mathbf{k}'}^+ b_{\mathbf{k}} + b_{\mathbf{k}} b_{\mathbf{k}'}^+).$$

Important effects arise from the demagnetizing field of the spin waves. A spin wave with $\mathbf{k} \parallel z$ gives no demagnetizing field to first order in the magnon amplitude, but other directions of $\mathbf{k}$ give a field

$\mathbf{H}_d$. We look for a solution of

$$\text{(97)} \qquad \operatorname{div} \mathbf{H} = -4\pi \operatorname{div} \mathbf{M}; \qquad \operatorname{curl} \mathbf{H} = 0.$$

If

$$\text{(98)} \qquad \mathbf{M} = \mathbf{M}_s + \Delta\mathbf{M}_0 e^{-i(\omega_{\mathbf{k}}t - \mathbf{k}\cdot\mathbf{x})}; \qquad \mathbf{H}_d = \mathbf{H}_d{}^0 e^{-i(\omega_{\mathbf{k}}t - \mathbf{k}\cdot\mathbf{x})},$$

then (97) is satisfied if

$$\text{(99)} \qquad \mathbf{k}\cdot\mathbf{H}_d{}^0 = -4\pi\mathbf{k}\cdot\Delta\mathbf{M}_0,$$

which is equivalent to

$$\text{(100)} \qquad \mathbf{H}_d = -\frac{4\pi(\mathbf{k}\cdot\Delta\mathbf{M})}{k^2}\,\mathbf{k}.$$

For the magnon system

$$\text{(101)} \qquad \mathbf{H}_d = -2\pi(4\mu_0 M)^{1/2} \sum_{\mathbf{k}} (k^- e^{-i\mathbf{k}\cdot\mathbf{x}} b_{\mathbf{k}} + k^+ e^{i\mathbf{k}\cdot\mathbf{x}} b_{\mathbf{k}}^+) k^{-2}\mathbf{k},$$

where $k^\pm = k_x \pm ik_y$. The demagnetizing energy density is, with a factor $\frac{1}{2}$ appropriate to a self-energy,

$$\text{(102)} \qquad \mathcal{H}_d = -\tfrac{1}{2}\mathbf{H}_d\cdot\mathbf{M};$$

it is convenient to specialize this to $\mathbf{k} \parallel \hat{\mathbf{x}}$, for which

$$\text{(103)} \qquad \mathcal{H}_{dx} = 2\pi\mu_0 M \sum_{\mathbf{k}\mathbf{k}'} (e^{-i\mathbf{k}\cdot\mathbf{x}} b_{\mathbf{k}} + e^{i\mathbf{k}\cdot\mathbf{x}} b_{\mathbf{k}}^+)(e^{-i\mathbf{k}'\cdot\mathbf{x}} b_{\mathbf{k}'} + e^{i\mathbf{k}'\cdot\mathbf{x}} b_{\mathbf{k}'}^+).$$

We integrate the several energy densities over the unit volume and drop the zero-point contributions. The bilinear terms are

$$\text{(104)} \qquad \mathcal{H}_0 = \int d^3x\,(\mathcal{H}_{\text{ex}}^{(1)} + \mathcal{H}_Z + \mathcal{H}_K + \mathcal{H}_{dx}) = \sum_{\mathbf{k}} \{A_{\mathbf{k}} b_{\mathbf{k}}^+ b_{\mathbf{k}} + B_{\mathbf{k}}^+ b_{\mathbf{k}}^+ b_{-\mathbf{k}}^+ + B_{\mathbf{k}} b_{\mathbf{k}} b_{-\mathbf{k}}\},$$

where for $\mathbf{k} \parallel \hat{\mathbf{x}}$,

$$\text{(105)} \qquad A_{\mathbf{k}} = A_{-\mathbf{k}} = 2\mu_0 M_s[2Ck^2 + (H/M_s) + (2K/M_s{}^2) + 2\pi]; \qquad B_{\mathbf{k}} = B_{-\mathbf{k}} = 2\pi\mu_0 M_s.$$

The diagonalization problem of (104) is again just the Bogoliubov problem (91) of Chapter 2. To vary the procedure slightly, form the equations of motion

$$\text{(106)} \qquad i\dot{b}_{\mathbf{k}} = [b_{\mathbf{k}},\mathcal{H}_0] = A_{\mathbf{k}} b_{\mathbf{k}} + 2B_{\mathbf{k}} b_{-\mathbf{k}}^+;$$

$$i\dot{b}_{-\mathbf{k}}^+ = [b_{-\mathbf{k}}^+,\mathcal{H}_0] = -A_{\mathbf{k}} b_{\mathbf{k}}^+ - 2B_{\mathbf{k}} b_{\mathbf{k}}.$$

We look for solutions with time dependence $e^{-i\omega_{\mathbf{k}}t}$. The eigenfrequencies are the roots of

$$\begin{vmatrix} \omega_{\mathbf{k}} - A_{\mathbf{k}} & -2B_{\mathbf{k}} \\ 2B_{\mathbf{k}} & \omega_{\mathbf{k}} + A_{\mathbf{k}} \end{vmatrix} = 0,$$

or

$$\omega_{\mathbf{k}} = (A_{\mathbf{k}}^2 - 4B_{\mathbf{k}}^2)^{1/2}. \tag{107}$$

Now consider several special cases. If **k** is small and the anisotropy $K = 0$, we have

$$\omega_0 = g\mu_B[H_0(H_0 + 4\pi M)]^{1/2}, \tag{108}$$

in agreement with the classical result. If the terms in k^2 and H_0 are dominant,

$$\omega_{\mathbf{k}} \cong g\mu_B(H_0 + 2CM_sk^2), \tag{109}$$

of the form of the atomic result (26). We note that $2JSa^2 = 2g\mu_BM_sC = 4\mu_0M_sC = D$, from (26) and (31). For general angles $\theta_{\mathbf{k}}$ between **k** and the z axis, the secular equation is given by (107) with

$$A_{\mathbf{k}} = 2\mu_0M_s\{2Ck^2 + (H_0/M_s) + (2K/M_s^2) + 2\pi \sin^2 \theta_{\mathbf{k}}\}; \qquad B_{\mathbf{k}} = 2\pi\mu_0M_s \sin^2 \theta_{\mathbf{k}}. \tag{110}$$

Here, as above, H_0 is the applied external static field H_a corrected for the static demagnetizing field of the specimen. Thus in a sphere we are to use $H_0 = H_a - (4\pi/3)M_s$. For $\theta_{\mathbf{k}} = 0$,

$$\omega_{\mathbf{k}} = g\mu_B[2CM_sk^2 + H_0 + (2K/M_s)]. \tag{111}$$

Representation of the Exchange Energy—Eq. (91). We need to know the form of the exchange energy associated with nonuniform macroscopic distributions of the direction of local magnetization. In general to know this requires a detailed quantitative theory of the exchange interaction in the solid, exactly as a general theory of the elastic deformation of a solid requires a detailed solution of the cohesive energy problem. We know, however, that we may treat many elastic problems in terms of macroscopic elastic constants if the characteristic wavelength of the deformation is long in comparison with the atomic spacing and if the relative deformation or strain is small. Similarly, we may treat magnetic deformations in terms of macroscopic exchange constants, subject to the same restrictions as the elastic constants. The fact that in some ferromagnetic metals the magnetic carriers may be mobile is no more of a restriction on the validity of the macroscopic approach than is the high mobility of the conduction electrons in the alkali or noble metals a restriction on the applicability of elastic constants.

The expression for the macroscopic isotropic exchange energy density must be invariant with respect to spin rotations; noninvariant contributions to the energy are included in the magnetocrystalline anisotropy energy. The desired expression must be invariant under change of sign of the magnetization deformation components; otherwise the state of uniform magnetization could not be the ground state of the system. We look for an expression of the lowest order in the derivatives of **M** compatible with the symmetry of the crystal. In an isotropic medium there are three quantities quadratic in the derivatives of **M** and invariant under rotation of the coordinate system: $(\mathrm{div}\,\mathbf{M})^2$, $(\mathrm{curl}\,\mathbf{M})^2$, and $|\mathrm{grad}\,\mathbf{M}|^2$. For magnetization directed in concentric circles, $\mathrm{div}\,\mathbf{M} = 0$, and thus the $(\mathrm{div}\,\mathbf{M})^2$ form may be ruled out. If **M** is radial, curl **M** is zero, and thus $(\mathrm{curl}\,\mathbf{M})^2$ may be ruled out. We have left

$$|\mathrm{grad}\,\mathbf{M}|^2 = (\nabla M_x)^2 + (\nabla M_y)^2 + (\nabla M_z)^2, \tag{112}$$

which is a satisfactory choice. Thus in an isotropic medium

$$\mathcal{H}_{\mathrm{ex}} = C \frac{\partial M_\alpha}{\partial x_\mu} \frac{\partial M_\alpha}{\partial x_\mu}, \tag{113}$$

where summation over repeated indices is implied. The form (113) is invariant under the operations of the cubic point group. For a general crystal symmetry

$$\mathcal{H}_{\mathrm{ex}} = C_{\mu\nu} \frac{\partial M_\alpha}{\partial x_\mu} \frac{\partial M_\alpha}{\partial x_\nu}, \tag{114}$$

where $C_{\mu\nu}$ is a tensor with the symmetry of the crystal. Just as $M_\alpha M_\alpha$ is invariant, so this form is manifestly invariant with respect to rotations of the entire spin system.

In antiferromagnets we must take into account the existence of separate sublattices, with exchange interactions within and between them. Kaganov and Tsukernik[3] have given the generalization of (114) to antiferromagnets

$$\mathcal{H}_{\mathrm{ex}} = C_{iklm}^{ss'} \frac{\partial M_{si}}{\partial x_k} \frac{\partial M_{s'l}}{\partial x_m}, \tag{115}$$

where s, s' are sublattice indices. We may compare (114) with the

[3] M. I. Kaganov and V. M. Tsukernik, *Soviet Physics—JETP* **34,** 73 (1958).

elastic energy density

$$(116) \qquad \mathcal{H}_{\mathrm{el}} = c_{\alpha\mu\beta\nu} \frac{\partial u_\alpha}{\partial x_\mu} \frac{\partial u_\beta}{\partial x_\nu};$$

here **u** is the particle displacement and $c_{\alpha\mu\beta\nu}$ is a component of the elastic stiffness tensor.

EXCITATION OF FERROMAGNETIC MAGNONS BY PARALLEL PUMPING

A small sphere of a ferromagnetic dielectric is placed in a magnetic field $H = H_0 + h \sin 2\omega t$, both fields parallel to the z axis. Here H_0 is a static field. We calculate the energy absorbed from the r-f field $h \sin 2\omega t$ by a particular magnon mode and find that the net power absorption in a certain approximation increases without limit when h exceeds a threshold value h_c. There is no resonant connection between ω or 2ω and the field H_0.

Consider a particular standing wave mode with wavevector **k** along the x axis:

$$(117) \qquad M_x = m_1 \sin kx \sin \omega t; \qquad M_y = m_2 \sin kx \cos \omega t.$$

The process consists of the absorption of one photon of frequency 2ω and the emission of two magnons **k** and $-$**k** each of frequency ω. Two magnons of equal but opposite wavevector give a standing wave. Now

$$(118) \qquad \operatorname{div} \mathbf{H} = \frac{\partial H_x}{\partial x} = -4\pi \operatorname{div} M = -4\pi m_1 k \cos kx \sin \omega t;$$

$$(119) \qquad H_x = -4\pi m_1 \sin kx \sin \omega t; \qquad H_y = 0.$$

The z component of the torque equation $\dot{\mathbf{M}} = \gamma \mathbf{M} \times \mathbf{H}$ is

$$(120) \qquad \dot{M}_z = \gamma(M_x H_y - M_y H_x) = 4\pi\gamma m_1 m_2 \sin^2 kx \sin \omega t \cos \omega t$$
$$= 2\pi\gamma m_1 m_2 \sin^2 kx \sin 2\omega t.$$

The mean rate of power absorption by a specimen of volume Ω is

$$(121) \qquad \mathcal{P} = \mathbf{H} \cdot \dot{\mathbf{M}}\Omega = 2\pi\gamma h m_1 m_2 \Omega \sin^2 kx \sin^2 2\omega t.$$

The time average of $\sin^2 2\omega t$ is $\frac{1}{2}$ and the volume average of $\sin^2 kx$ is $\frac{1}{2}$; thus

$$(122) \qquad \mathcal{P} = \frac{\pi}{2} \gamma h m_1 m_2 \Omega.$$

It is convenient to express $m_1 m_2$ in terms of the excitation quantum number $n_{\mathbf{k}}$ of the mode considered. We have, neglecting zero-point motion,

$$n_{\mathbf{k}}g\mu_B = \langle (M_s - M_z)\Omega\rangle_{\mathbf{k}}, \tag{123}$$

because in the approximation $m_1 \cong m_2$ each magnon excited reduces the magnetic moment by $g\mu_B$. The angular brackets indicate the space and time average with only the mode **k** excited. Now, writing $m_1 \cong m_2 \cong m$,

$$\begin{aligned} \langle M_z\Omega\rangle &= \Omega\langle [M_s^2 - M_x^2 - M_y^2]^{1/2}\rangle \\ &\cong M_s\Omega\left(1 - \frac{\langle M_x^2 + M_y^2\rangle}{2M_s^2}\right) = M_s\Omega\left(1 - \frac{m^2}{4M_s^2}\right), \end{aligned} \tag{124}$$

so that

$$n_k \cong m^2\Omega/4gM_s\mu_B, \tag{125}$$

and (122) becomes

$$\mathcal{P} = 2\pi\gamma M_s h g\mu_B n_{\mathbf{k}}. \tag{126}$$

The energy balance in the mode **k** is expressed by

$$\frac{dE_{\mathbf{k}}}{dt} = -\frac{1}{T_{\mathbf{k}}}(E_{\mathbf{k}} - \bar{E}_{\mathbf{k}}) + \mathcal{P}, \tag{127}$$

where $\bar{E}_{\mathbf{k}}$ is the thermal average of the energy $E_{\mathbf{k}}$ of the mode **k** and $T_{\mathbf{k}}$ is the relaxation time of the mode. Writing $E_{\mathbf{k}} = n_{\mathbf{k}}\omega_{\mathbf{k}}$ and using (126), we have

$$\omega_{\mathbf{k}}\frac{dn_{\mathbf{k}}}{dt} = -\frac{\omega_{\mathbf{k}}}{T_{\mathbf{k}}}(n_{\mathbf{k}} - \bar{n}_{\mathbf{k}}) + 2\pi\gamma M_s h g\mu_B n_{\mathbf{k}}. \tag{128}$$

In the steady state, $dn_{\mathbf{k}}/dt = 0$, so that

$$n_{\mathbf{k}} = \frac{\bar{n}_{\mathbf{k}}}{1 - 2\pi\gamma M_s h g\mu_B T_{\mathbf{k}}/\omega_{\mathbf{k}}}; \tag{129}$$

this expression has a singularity at

$$h_c = \frac{\omega_{\mathbf{k}}}{2\pi\gamma M_s g\mu_B T_{\mathbf{k}}}. \tag{130}$$

Thus a determination of h_c is equivalent to a measurement of $T_{\mathbf{k}}$. We have assumed that the lowest threshold h_c occurs for spin waves making an angle $\theta_{\mathbf{k}} = \pi/2$ with the z axis.

TEMPERATURE DEPENDENCE OF THE EFFECTIVE EXCHANGE

The problem is treated by evaluating the diagonal four-operator terms in the exchange hamiltonian. The problem is quite similar to that treated in Chapter 2 for liquid helium, except there the unperturbed system was in the ground state. We have from (29),

$$(131)\quad \mathcal{H}_1 = (Jz/4N) \sum_{1234} b_1^+ b_2^+ b_3 b_4 \Delta(\mathbf{k}_1 + \mathbf{k}_2 - \mathbf{k}_3 - \mathbf{k}_4)[2\gamma_1 + 2\gamma_3 - 4\gamma_{1-3}].$$

The off-diagonal terms give rise to magnon-magnon scattering; the diagonal terms renormalize the energy. The diagonal terms involve only two $\mathbf{k}$'s, which we denote as $\mathbf{k}_a$, $\mathbf{k}_b$.

There are two types of diagonal terms in (131):

$$\mathbf{k}_1 = \mathbf{k}_3 = \mathbf{k}_a; \qquad \mathbf{k}_2 = \mathbf{k}_4 = \mathbf{k}_b: \qquad \Sigma\,(4\gamma_{\mathbf{a}} - 4\gamma_0)n_a n_b,$$

$$\mathbf{k}_1 = \mathbf{k}_4 = \mathbf{k}_a; \qquad \mathbf{k}_2 = \mathbf{k}_3 = \mathbf{k}_b: \qquad \Sigma\,(2\gamma_{\mathbf{a}} + 2\gamma_{\mathbf{b}} - 4\gamma_{\mathbf{a-b}})n_a n_b.$$

Thus the diagonal part of (131) is

$$(132)\quad E_1 = (Jz/N) \sum_{\mathbf{ab}} (\gamma_{\mathbf{a}} + \gamma_{\mathbf{b}} - \gamma_0 - \gamma_{\mathbf{a-b}})n_a n_b = \tfrac{1}{2} \sum_{\mathbf{k}} \varepsilon_{1\mathbf{k}},$$

where

$$(133)\quad \varepsilon_{1\mathbf{k}} = n_{\mathbf{k}}(2Jz/N) \sum_{\mathbf{b}} (\gamma_{\mathbf{k}} + \gamma_{\mathbf{b}} - \gamma_0 - \gamma_{\mathbf{k-b}})n_b;$$

here we have collected all terms in which $n_{\mathbf{k}}$ occurs. The energy of magnon mode $\mathbf{k}$ is therefore

$$(134)\quad \varepsilon_{\mathbf{k}} = n_{\mathbf{k}}[\omega_{\mathbf{k}} + (2Jz/N) \sum_{\mathbf{b}} (\gamma_{\mathbf{k}} + \gamma_{\mathbf{b}} - \gamma_0 - \gamma_{\mathbf{k-b}})n_{\mathbf{b}}];$$

here $\omega_{\mathbf{k}}$ is the energy of the mode when all other modes are in their ground state.

Now to $O(k^4)$ for a lattice with a center of symmetry at each spin, using the definition of the γ's,

$$(135)\quad \gamma_a + \gamma_b - \gamma_0 - \gamma_{a-b} \cong -(\tfrac{1}{36})k_a{}^2 k_b{}^2 \delta^4,$$

so that

$$(136)\quad \varepsilon_{\mathbf{k}} = n_{\mathbf{k}}\Big[\omega_{\mathbf{k}} - (Jz/18N)k^2\delta^4 \sum_b k_b{}^2 n_b\Big].$$

We see that the energy is lowered by an amount proportional to k^2 and to $\Sigma\, k_b{}^2 n_b$, which is of the form of the total spin wave energy, to lowest order. In fact,

$$(137)\quad \sum_b k_b{}^2 \langle n_b \rangle \cong U_T/D,$$

where U_T is the thermal magnon energy (33) and D is the constant in the relation $\omega_{\mathbf{k}} = DK^2 = \frac{1}{3}SJz\delta^2k^2 = 2SJak^2$. Thus, if we write $\epsilon_{\mathbf{k}} = n_{\mathbf{k}}\omega_{\mathbf{k}}(\text{eff})$, the renormalized energy is given by

$$\omega_{\mathbf{k}}(\text{eff}) \cong [2SJa^2 - \delta^2 U_T/6N)]k^2. \tag{138}$$

For the special case of a simple cubic lattice one finds

$$\omega_{\mathbf{k}}(\text{eff}) = \omega_{\mathbf{k}}[1 - (12JNS^2)^{-1} \sum_{\mathbf{k}'} n_{\mathbf{k}'}\omega_{\mathbf{k}'}], \tag{139}$$

for all $\mathbf{k}$. Using (138),

$$D(T) \cong D_0\left(1 - \frac{U_T}{6ND_0/\delta^2}\right)k^2 = D_0\left(1 - \frac{U_T}{2U_0}\right), \tag{140}$$

where $U_0 = JNzS^2$.

We have set $z\delta^2 = 6a^2$, as for sc, bcc, and fcc lattices. The result (140) demonstrates that $D(T)$ scales as the magnon energy U_T, not as the saturation moment. The result as obtained applies to nearest-neighbor interactions within one lattice.

Note that we have not considered in the partial diagonalization of (131) terms of the form $a_0^+a_0^+a_{\mathbf{k}}a_{-\mathbf{k}}$ and $a_{\mathbf{k}}^+a_{\mathbf{k}}^+a_0a_0$ which were considered in the analogous problem for liquid helium. In the helium problem we can treat $N_0 + 2$ as nearly equal to N_0, where N_0 is of the order of the total number of particles in the system. In our present problem the number of magnons in the uniform mode is not a very large number: $N_0 \sim k_BT/\omega_{\mathbf{k}}$, and a change of 2 in this value is not obviously negligible. Further, N_0 is exceedingly small in comparison with the total number of spins in the system, and the effect on the dispersion relation of N_0 alone will be negligible. But actually other terms, such as $a_{\mathbf{k}'}^+a_{\mathbf{k}'}^+a_{\mathbf{k}}a_{\mathbf{k}}$ must be treated on an equal footing, and we see that we should begin to worry about the result (134) if the number of magnons is not negligible in comparison with the total number of particles.

MAGNETOSTATIC MODES

It follows directly from the torque equation $\dot{\mathbf{M}} = \gamma\mathbf{M} \times \mathbf{H}$ that for a static magnetic field in the z direction the r-f permeability of a ferromagnet is described in the absence of exchange by the equations

$$B_x = \mu H_x + \xi H_y; \qquad B_y = -\xi H_x + \mu H_y, \tag{141}$$

where μ, ξ are determined by H_0, ω, and M_s. With $\mathbf{H} = \nabla\varphi$, the equation div $\mathbf{B} = 0$ becomes

$$\mu\left(\frac{\partial^2\varphi}{\partial x^2} + \frac{\partial^2\varphi}{\partial y^2}\right) + \frac{\partial^2\varphi}{\partial z^2} = 0, \tag{142}$$

inside the specimen, whereas $\nabla^2\varphi = 0$ outside. The boundary conditions are that the tangential component of **H** and the normal component of **B** should be continuous across the boundary of the specimen.

Solutions of (142) are considered for several geometries in the following papers:

L. R. Walker, *Phys. Rev.* **105,** 390 (1957).
P. Fletcher and C. Kittel, *Phys. Rev.* **120,** 2004 (1960).
R. Damon and J. Eshbach, *Phys. Chem. Solids* **19,** 308 (1961).

PROBLEMS

1. Prove that $[\mathcal{S}^2,H] = 0$ and that $[\mathcal{S}_z,H] = 0$, where H is given by (1).
2. From (3), (4), and (5), show that

$$[S_x,S_y] = iS_z. \tag{143}$$

3. Show that, for the total spin $\mathcal{S}$,

$$\mathcal{S}^2 \cong (NS)^2 + NS - 2NS \sum_{\mathbf{k}\neq 0} b_{\mathbf{k}}^+ b_{\mathbf{k}}, \tag{144}$$

observing that $\sum_j S_j^- \cong (2SN)^{1/2} b_0^+$. Discuss the fact that excitation of $\mathbf{k} = 0$ magnons does not change $\mathcal{S}^2$.
4. Derive an expression for the velocity of second sound in a magnon gas, assuming $\omega = Dk^2$, where D is a constant.
5. Construct a spin function to represent a ferromagnetic spin system with one magnon excited.
6. Using the hamiltonian

$$H = -J \sum_{j\delta} \mathbf{S}_j \cdot \mathbf{S}_{j+\delta} - 2\mu_0 H_0 \sum_j S_{jz}, \tag{145}$$

find the quantum equation of motion for $\mathbf{S}_j$, using $i\dot{\mathbf{S}}_j = [\mathbf{S}_j,H]$ and recalling that $\mathbf{S}_j \times \mathbf{S}_j = i\mathbf{S}_j$. Form the difference equations for S_j^+ and S_j^-. Interpreting the spin operators as classical vectors, solve for the eigenfrequency of spin waves in the limit of small amplitude $(S^+/S \ll 1)$. Show that for long wavelengths $(ka \ll 1)$ the classical difference equation reduces to a partial differential equation, which for a simple cubic lattice is

$$\dot{\mathbf{S}} = 2Ja^2\mathbf{S} \times \nabla^2\mathbf{S} + 2\mu_0\mathbf{S} \times \mathbf{H}, \tag{146}$$

as in *ISSP*, Appendix *O*.
7. Show that for antiferromagnetic magnons

$$\mathcal{S}^2 = NS \left\{\frac{H_A}{H_A + 2H_E}\right\} (n_0^\alpha + n_0^\beta + 1) + \text{terms in } (n_k)^2. \tag{147}$$

8. Show that for a linear lattice $\beta = 0.726$, using (64). Note that $\gamma_k = \frac{1}{2}(e^{ika} + e^{-ika}) = \cos ka$ and $(1 - \gamma_k^2) = \sin^2 ka$; thus

$$\beta = (2/N) \sum_k (1 - |\sin ka|) = (4a/\pi) \int_0^{\pi/2a} dk(1 - \sin ka). \tag{148}$$

9. Show from $\sum_j S_{jz}^a + \sum_l S_{lz}^b$ that the excitation of an antiferromagnetic magnon is accompanied by a change of ± 1 in the z component of the total spin.

10. Consider the magnon-phonon hamiltonian

$$H = \sum_k \{\omega_k^m a_k^+ a_k + \omega_k^p b_k^+ b_k + c_k(a_k b_k^+ + a_k^+ b_k)\}, \tag{149}$$

where c_k is the coupling coefficient and a^+, a; b^+, b are magnon and phonon creation and annihilation operators. Show that the transformations

$$a_k = A_k \cos \theta_k + B_k \sin \theta_k; \qquad b_k = B_k \cos \theta_k - A_k \sin \theta_k, \tag{150}$$

with θ real diagonalize the hamiltonian if

$$\tan 2\theta_k = \frac{2c_k}{\omega_k^p - \omega_k^m}. \tag{151}$$

Show that the nominal crossover of the dispersion relations has

$$\omega_A = \omega_k - c_k; \qquad \omega_B = \omega_k + c_k; \qquad a^+ = \frac{A^+ + B^+}{\sqrt{2}}; \qquad b^+ = \frac{B^+ - A^+}{\sqrt{2}}.$$

5 Fermion fields and the Hartree-Fock approximation

The essential distinction between an assembly of fermi particles and an assembly of bose particles is the requirement of the pauli principle that the eigenfunctions describing the fermions must be antisymmetric under interchange of any two particles. The antisymmetrized eigenfunctions of a system of independent fermions may be written as Slater determinants of the one-electron wavefunctions. For some purposes the Slater determinant description is convenient: it is direct and explicit. However, it is tedious to indicate the subscripts distinguishing the individual indistinguishable electrons and to indicate the permutation operators. There exists a more elegant, flexible, and concise representation in terms of the second quantization theory of fermion fields. The theory is closely analogous to that for boson fields. The theory, just as the determinantal description, usually contemplates an assembly of more-or-less independent particles, coupled by weak interactions.

Suppose we have a set of orthonormal solutions of some one-particle wave equation; the equation may be typically the Hartree or Hartree-Fock equation with particle interactions taken into account in some average sense, or the equation may be for free particles. We write a solution of a one-particle wave equation as $\varphi_j(\mathbf{x})$, where

$$H\varphi_j(\mathbf{x}) = \varepsilon_j\varphi_j(\mathbf{x}). \tag{1}$$

Note that we do not label the electron with a number index, such as ν in $\mathbf{x}_\nu$. The eigenfunction label j will include the specification of the spin state, for example, $\uparrow$ or $\downarrow$, or α or β.

Next form the *field operator*

$$\Psi(\mathbf{x}) = \sum_j c_j\varphi_j(\mathbf{x}); \qquad \Psi^+(\mathbf{x}) = \sum_j c_j^+\varphi_j^*(\mathbf{x}), \tag{2}$$

where c_j is an operator with properties to be specified below; $\varphi_j(\mathbf{x})$ remains an eigenfunction and not an operator, so that it is essentially a c-number function of the coordinate $\mathbf{x}$. The field operator $\Psi(\mathbf{x})$ and the fermion operators c_j operate on a *state vector* which we write as Φ. This state vector is in the space of the occupation numbers of the one-electron states. Thus

$$\Phi_{\text{vac}} = |000 \cdots 0 \cdots\rangle = |\text{vac}\rangle \tag{3}$$

is the vacuum state in which all the occupation numbers n_j of the one-electron states are zero—no particles are present in the system. The unperturbed ground state of a system of N fermions in the independent-particle approximation will be written as Φ_0, where

$$\Phi_0 = |1_1 1_2 1_3 \cdots 1_N 0_{N+1} 0_{N+2} \cdots 0 \cdots\rangle, \tag{4}$$

where the states are numbered in order of increasing energy. We note (Problem 4) that $\Psi^+(\mathbf{x})$ is an operator which adds a particle to the system at $\mathbf{x}$.

The requirements of the pauli principle are satisfied if the fermion operators c, c^+ satisfy the anticommutation relations

$$c_l c_m^+ + c_m^+ c_l = \delta_{lm}; \qquad c_l c_m + c_m c_l = 0; \qquad c_l^+ c_m^+ + c_m^+ c_l^+ = 0. \tag{5}$$

We write these as

$$\{c_l,c_m^+\} = \delta_{lm}; \qquad \{c_l,c_m\} = 0; \qquad \{c_l^+,c_m^+\} = 0. \tag{6}$$

The $\{,\}$ will denote anticommutator; $[,]$ will continue to denote commutator. Another common notation for anticommutator is $[,]_+$.

The anticommutation relations may be satisfied uniquely by a representation in terms of 2×2 matrices, the Jordan-Wigner matrices. For a system with only a single state we represent c^+ and c in the following way:

$$c^+ = \begin{pmatrix} 0 & 1 \\ 0 & 0 \end{pmatrix} = \tfrac{1}{2}(\sigma_x + i\sigma_y); \qquad c = \begin{pmatrix} 0 & 0 \\ 1 & 0 \end{pmatrix} = \tfrac{1}{2}(\sigma_x - i\sigma_y), \tag{7}$$

where the σ's are the pauli matrices. Thus

$$c^+c + cc^+ = \begin{pmatrix} 0 & 1 \\ 0 & 0 \end{pmatrix}\begin{pmatrix} 0 & 0 \\ 1 & 0 \end{pmatrix} + \begin{pmatrix} 0 & 0 \\ 1 & 0 \end{pmatrix}\begin{pmatrix} 0 & 1 \\ 0 & 0 \end{pmatrix} = \begin{pmatrix} 1 & 0 \\ 0 & 1 \end{pmatrix}. \tag{8}$$

Further,

$$
(9)\qquad
\begin{aligned}
cc + cc &= 2\begin{pmatrix}0 & 0\\ 1 & 0\end{pmatrix}\begin{pmatrix}0 & 0\\ 1 & 0\end{pmatrix} = \begin{pmatrix}0 & 0\\ 0 & 0\end{pmatrix};\\
c^+c^+ + c^+c^+ &= 2\begin{pmatrix}0 & 1\\ 0 & 0\end{pmatrix}\begin{pmatrix}0 & 1\\ 0 & 0\end{pmatrix} = \begin{pmatrix}0 & 0\\ 0 & 0\end{pmatrix}.
\end{aligned}
$$

The 2×2 matrices are to be understood as operating on a two-component state vector in the occupation number of the particle:

$$
(10)\qquad |1\rangle = \begin{pmatrix}1\\ 0\end{pmatrix}; \qquad |0\rangle = \begin{pmatrix}0\\ 1\end{pmatrix},
$$

corresponding to the possible fermion state occupancy numbers, 1 and 0.

We have the property

$$
(11)\qquad c^+c|1\rangle = \begin{pmatrix}1 & 0\\ 0 & 0\end{pmatrix}\begin{pmatrix}1\\ 0\end{pmatrix} = 1\begin{pmatrix}1\\ 0\end{pmatrix} = 1|1\rangle;
$$

and

$$
(12)\qquad c^+c|0\rangle = \begin{pmatrix}1 & 0\\ 0 & 0\end{pmatrix}\begin{pmatrix}0\\ 1\end{pmatrix} = 0\begin{pmatrix}0\\ 1\end{pmatrix} = 0|0\rangle.
$$

Thus

$$
(13)\qquad \hat{n} = c^+c
$$

is the number operator and has the eigenvalues 1 and 0 for the eigenstates $|1\rangle$ and $|0\rangle$, respectively.

We see that c^+ is a particle creation operator:

$$
(14)\qquad c^+|0\rangle = \begin{pmatrix}0 & 1\\ 0 & 0\end{pmatrix}\begin{pmatrix}0\\ 1\end{pmatrix} = \begin{pmatrix}1\\ 0\end{pmatrix} = |1\rangle;
$$

and c is a particle annihilation operator:

$$
(15)\qquad c|1\rangle = \begin{pmatrix}0 & 0\\ 1 & 0\end{pmatrix}\begin{pmatrix}1\\ 0\end{pmatrix} = \begin{pmatrix}0\\ 1\end{pmatrix} = |0\rangle; \qquad c|0\rangle = 0.
$$

A state cannot be occupied by more than one fermion:

$$
(16)\qquad c^+|1\rangle = \begin{pmatrix}0 & 1\\ 0 & 0\end{pmatrix}\begin{pmatrix}1\\ 0\end{pmatrix} = 0.
$$

We construct a state Φ in which a one-particle state $\mathbf{k}$ is occupied by one particle by

$$
(17)\qquad \Phi = c_{\mathbf{k}}^+\Phi_{\text{vac}} = c_{\mathbf{k}}^+|00\cdots 0_{\mathbf{k}}\cdots\rangle = |00\cdots 1_{\mathbf{k}}\cdots\rangle.
$$

Similarly, the ground state of an unperturbed fermi sea is

$$\Phi_0 = \Big(\prod_{|\mathbf{k}|<k_F} c_{\mathbf{k}}^+ \Big) \Phi_{\text{vac}} = c_1^+ c_2^+ \cdot \cdot \cdot c_{\mathbf{k}}^+ \cdot \cdot \cdot c_{\mathbf{k}_F}^+ \Phi_{\text{vac}}. \tag{18}$$

If there is more than one particle present, the results of the operation $c_j^+ c_j$ are unchanged, but the results of the operations c_j^+ or c_j may alternate in sign according to the number and ordering in Φ of the other occupied states. We can see this best from a simple example. Consider the state

$$\Phi = c_1^+ c_2^+ \Phi_{\text{vac}} \tag{19}$$

under the operation c_2:

$$\begin{aligned} c_2 \Phi &= c_2 c_1^+ c_2^+ \Phi_{\text{vac}} = -c_1^+ c_2 c_2^+ \Phi_{\text{vac}} \\ &= -c_1^+ (1 - c_2^+ c_2) \Phi_{\text{vac}} = -c_1^+ \Phi_{\text{vac}}, \end{aligned} \tag{20}$$

whereas under the operation c_1 we have

$$c_1 \Phi = c_1 c_1^+ c_2^+ \Phi_{\text{vac}} = (1 - c_1^+ c_1) c_2^+ \Phi_{\text{vac}} = c_2^+ \Phi_{\text{vac}}. \tag{21}$$

The difference in sign between (20) and (21) is the consequence of the anticommutation of c_1^+ and c_2. *We must in fact put in a minus sign for every occupied state i which occurs to the left of the state j on which we operate with c_j^+ or c_j.* The statement "to the left" assumes a definite order has been adopted for the order of the one-particle states in the state vector Φ:

$$\Phi = c_1^+ c_2^+ \cdot \cdot \cdot c_j^+ \cdot \cdot \cdot c_N^+ \Phi_{\text{vac}}. \tag{22}$$

It is convenient in dealing with products of creation and annihilation operators to rearrange them into *normal product* form in which all creation operators stand to the left of any annihilation operators which may be present. If in a normal product there do occur one or more annihilation operators, we know at once that the result of operating on the vacuum state with the normal product is identically zero.

The general result is

$$\boxed{\begin{aligned} c_j | \cdot \cdot \cdot n_j \cdot \cdot \cdot \rangle &= n_j \theta^j | \cdot \cdot \cdot 0_j \cdot \cdot \cdot \rangle; \qquad (23) \\ c_j^+ | \cdot \cdot \cdot n_j \cdot \cdot \cdot \rangle &= (1 - n_j) \theta^j | \cdot \cdot \cdot 1_j \cdot \cdot \cdot \rangle, \qquad (24) \end{aligned}}$$

where

$$\theta^j = (-1)^{p_j}; \tag{25}$$

here p_j is the number of occupied states to the left of j in the state vector Φ. We may omit θ^j if we redefine the matrices representing

c_j^+, c_j as

$$c_j^+ = T_1 \cdots T_{j-1} \begin{pmatrix} 0 & 1 \\ 0 & 0 \end{pmatrix}; \tag{26}$$

$$c_j = T_1 \cdots T_{j-1} \begin{pmatrix} 0 & 0 \\ 1 & 0 \end{pmatrix}, \tag{27}$$

where

$$T = \begin{pmatrix} -1 & 0 \\ 0 & 1 \end{pmatrix} = -\sigma_z. \tag{28}$$

We include a factor T in (26), (27) for every state to the left of j, occupied or not.

We observe that

$$\{\Psi(\mathbf{x}),\Psi^+(\mathbf{x}')\} = \delta(\mathbf{x} - \mathbf{x}'), \tag{29}$$

because

$$\{\Psi(\mathbf{x}),\Psi^+(\mathbf{x}')\} = \sum_{jl} \{c_j^+,c_l\}\varphi_j(\mathbf{x})\varphi_l^*(\mathbf{x}') = \sum_j \varphi_j(\mathbf{x})\varphi_j^*(\mathbf{x}'), \tag{30}$$

and by closure

$$\sum_j \varphi_j(\mathbf{x})\varphi_j^*(\mathbf{x}') = \delta(\mathbf{x} - \mathbf{x}'). \tag{31}$$

The particle density operator is

$$\rho(\mathbf{x}) = \int d^3x'\, \Psi^+(\mathbf{x}')\delta(\mathbf{x} - \mathbf{x}')\Psi(\mathbf{x}') = \Psi^+(\mathbf{x})\Psi(\mathbf{x}) = \sum_{ij} c_i^+ c_j \varphi_i^*(\mathbf{x})\varphi_j(\mathbf{x}); \tag{32}$$

this is also known as the one-particle density matrix operator. Notice that if $|\rangle$ is an eigenstate of the occupancy operator $\hat{n}_i$, then

$$\langle|\rho(\mathbf{x})|\rangle = \Sigma\, n_i\varphi_i^*(\mathbf{x})\varphi_i(\mathbf{x}) = \Sigma\, n_i\rho_i(\mathbf{x}), \tag{33}$$

where $\rho_i(\mathbf{x}) \equiv \varphi_i^*(\mathbf{x})\varphi_i(\mathbf{x})$.

The hamiltonian is obtained in the second quantization representation by the general theorem that quantum operators are obtained directly from their classical analogs. For example, the kinetic energy is

$$H = \int d^3x\, \Psi^+(\mathbf{x}) \frac{p^2}{2m} \Psi(\mathbf{x}) = \int d^3x \sum_{jl} c_j^+ c_l \varphi_j^*(\mathbf{x}) \frac{p^2}{2m} \varphi_l(\mathbf{x}), \tag{34}$$

where $\mathbf{p}$ is the momentum operator. For free particles, with $\mathbf{p} = -i\,\mathrm{grad}$,

$$H = \sum_j \left(\frac{k_j^2}{2m}\right) c_j^+ c_j, \tag{35}$$

where the factor $c_j^+c_j$ automatically arranges that we count the energy of occupied states only in a representation in which the $c_j^+c_j$ are diagonal.

PARTICLE FIELD EQUATION OF MOTION METHOD FOR THE HARTREE-FOCK EQUATION

We consider a system of electrons described by the field operator

$$\Psi(\mathbf{x}) = \sum_j c_j \varphi_j(\mathbf{x}), \tag{36}$$

where c_j is a fermion operator and $\varphi_j(\mathbf{x})$ is a one-particle eigenfunction. Our object is to find approximate solutions of the equation of motion $i\dot{\Psi} = -[H,\Psi]$. The prescription for the hamiltonian in this representation is to write the mean energy in terms of individual particle wavefunctions and then replace the wavefunctions by the field operator $\Psi(\mathbf{x}')$. Thus

$$H = \int d^3x'\ \Psi^+(\mathbf{x}') \left[\frac{1}{2m} p^2 + v(\mathbf{x}')\right] \Psi(\mathbf{x}') + \tfrac{1}{2} \int d^3x'\ d^3y\ \Psi^+(\mathbf{x}')\Psi^+(\mathbf{y}) V(\mathbf{x}' - \mathbf{y})\Psi(\mathbf{y})\Psi(\mathbf{x}'), \tag{37}$$

where $V(\mathbf{x}' - \mathbf{y})$ is the interaction energy of two particles at $\mathbf{x}'$ and $\mathbf{y}$. The factor $\frac{1}{2}$ arises because of the self-energy. The order of terms is significant because $\Psi(\mathbf{x}')\Psi(\mathbf{y}) = -\Psi(\mathbf{y})\Psi(\mathbf{x}')$.

For convenience we let $v(\mathbf{x}') = 0$ and write

$$H = \int d^3x'\ \Psi^+(\mathbf{x}') \frac{1}{2m} p^2\Psi(\mathbf{x}') + \tfrac{1}{2} \int d^3x'\ d^3y\ \Psi^+(\mathbf{x}')\Psi^+(\mathbf{y}) V(\mathbf{x}' - \mathbf{y})\Psi(\mathbf{y})\Psi(\mathbf{x}'), \tag{38}$$

where

$$\int d^3x'\ \Psi^+(\mathbf{x}')\Psi(\mathbf{x}') = \int d^3x' \sum_{j,l} c_j^+ c_l \varphi_j^*(\mathbf{x}')\varphi_l(\mathbf{x}') = \sum_j c_j^+ c_j = \hat{N} \tag{39}$$

is the operator for the total number of particles. The first term in the commutator $[H,\Psi(\mathbf{x})]$ is, with $\mathbf{p}$ operating on $\Psi(\mathbf{x}')$,

$$\frac{1}{2m} \int d^3x'\ [\Psi^+(\mathbf{x}')p^2\Psi(\mathbf{x}'),\Psi(\mathbf{x})] = -\frac{1}{2m} \int d^3x'\ \{\Psi^+(\mathbf{x}'),\Psi(\mathbf{x})\} p^2\Psi(\mathbf{x}'), \tag{40}$$

because the anticommutator

$$\{\Psi(\mathbf{x}'),\Psi(\mathbf{x})\} = 0. \tag{41}$$

Note the mixture of commutators and anticommutators in (40). By (29),

$$\begin{aligned}(42)\quad -\frac{1}{2m}\int d^3x'\ \{\Psi^+(\mathbf{x}'),\Psi(\mathbf{x})\}p^2\Psi(\mathbf{x}') \\ = -\frac{1}{2m}\int d^3\mathbf{x}'\ \delta(\mathbf{x}'-\mathbf{x})p^2\Psi(\mathbf{x}') \\ = -\frac{1}{2m}p^2\Psi(\mathbf{x}).\end{aligned}$$

The second term in the commutator $[H,\Psi(\mathbf{x})]$ is

$$\begin{aligned}(43)\quad &\tfrac{1}{2}\int d^3x'\ d^3y\ [\Psi^+(\mathbf{x}')\Psi^+(\mathbf{y})V(\mathbf{x}'-\mathbf{y})\Psi(\mathbf{y})\Psi(\mathbf{x}'),\Psi(\mathbf{x})] \\ &= \tfrac{1}{2}\int d^3x'\ d^3y\ V(\mathbf{x}'-\mathbf{y})(\Psi^+(\mathbf{x}')\{\Psi(\mathbf{x}),\Psi^+(\mathbf{y})\}\Psi(\mathbf{y})\Psi(\mathbf{x}') \\ &\qquad - \delta(\mathbf{x}-\mathbf{x}')\Psi^+(\mathbf{y})\Psi(\mathbf{y})\Psi(\mathbf{x}')) \\ &= \tfrac{1}{2}\int d^3x'\ V(\mathbf{x}'-\mathbf{x})\Psi^+(\mathbf{x}')\Psi(\mathbf{x})\Psi(\mathbf{x}') \\ &\qquad - \tfrac{1}{2}\int d^3y\ V(\mathbf{x}-\mathbf{y})\Psi^+(\mathbf{y})\Psi(\mathbf{y})\Psi(\mathbf{x}) \\ &= -\int d^3y\ V(\mathbf{y}-\mathbf{x})\Psi^+(\mathbf{y})\Psi(\mathbf{y})\Psi(\mathbf{x}).\end{aligned}$$

Here

$$\begin{aligned}(44)\quad -\int d^3y\ V(\mathbf{y}-\mathbf{x})\Psi^+(\mathbf{y})\Psi(\mathbf{y})\Psi(\mathbf{x}) \\ = -\sum_{klm} c_k^+c_lc_m\int d^3y\ V(\mathbf{y}-\mathbf{x})\varphi_k^*(\mathbf{y})\varphi_l(\mathbf{y})\varphi_m(\mathbf{x}).\end{aligned}$$

This expression involves products of three operators.

In the lowest or Hartree-Fock approximation we consider only terms in a single operator times the number operator $c_k^+c_k$. We thus keep the terms $c_k^+c_kc_m$ and $c_k^+c_lc_k = -c_lc_k^+c_k$, whence

$$\begin{aligned}(45)\quad -\int d^3y\ V(\mathbf{y}-\mathbf{x})\Psi^+(\mathbf{y})\Psi(\mathbf{y})\Psi(\mathbf{x}) \\ \cong -\int d^3y\ V(\mathbf{y}-\mathbf{x})\langle\Psi^+(\mathbf{y})\Psi(\mathbf{y})\rangle\Psi(\mathbf{x}) \\ + \int d^3y\ \Psi(\mathbf{y})V(\mathbf{y}-\mathbf{x})\langle\Psi^+(\mathbf{y})\Psi(\mathbf{x})\rangle,\end{aligned}$$

where we have resumed the series for $\Psi(\mathbf{x})$ and $\Psi(\mathbf{y})$; the angular brackets indicate the expectation value in the ground state, that is, only terms of the form $c_k^+c_k$ are retained within the angular brackets, where $c_k^+c_k$ is evaluated for the ground state. The first term on the right-hand side of (45) is the direct coulomb term, and the second term is the exchange term.

Collecting (42) and (45),

$$(46)\quad [H,\Psi(\mathbf{x})] \cong \left[-\frac{1}{2m}p^2 - \int d^3y\, V(\mathbf{y}-\mathbf{x})\langle\Psi^+(\mathbf{y})\Psi(\mathbf{y})\rangle\right]\Psi(\mathbf{x}) + \int d^3y\, \Psi(\mathbf{y})V(\mathbf{y}-\mathbf{x})\langle\Psi^+(\mathbf{y})\Psi(\mathbf{x})\rangle,$$

or

$$(47)\quad [H,\Psi(\mathbf{x})] \cong -\sum_j c_j\left[\left(\frac{1}{2m}p^2 + \int d^3y\, V(\mathbf{y}-\mathbf{x})\langle\Psi^+(\mathbf{y})\Psi(\mathbf{y})\rangle\right)\varphi_j(\mathbf{x}) - \int d^3y\, \varphi_j(\mathbf{y})V(\mathbf{y}-\mathbf{x})\langle\Psi^+(\mathbf{y})\Psi(\mathbf{x})\rangle\right].$$

Suppose now that the $\varphi_j(\mathbf{x})$ are eigenfunctions of the operator in the square brackets on the right-hand side of (47), with the eigenvalues ε_j; then the equation of motion is

$$(48)\qquad [H,\Psi(\mathbf{x})] = -i\sum_j \dot{c}_j\varphi_j(\mathbf{x}) = -\sum_j \varepsilon_j c_j\varphi_j(\mathbf{x}),$$

where

$$(49)\quad \varepsilon_j\varphi_j(\mathbf{x}) = \left(\frac{p^2}{2m} + \int d^3y\, V(\mathbf{y}-\mathbf{x})\langle\Psi^+(\mathbf{y})\Psi(\mathbf{y})\rangle\right)\varphi_j(\mathbf{x}) - \int d^3y\, \varphi_j(\mathbf{y})V(\mathbf{y}-\mathbf{x})\langle\Psi^+(\mathbf{y})\Psi(\mathbf{x})\rangle.$$

This is the Hartree-Fock equation. In the usual form we see that $\varphi_j(\mathbf{x})$ is determined by a certain *averaged* potential:

$$(50)\quad \boxed{\varepsilon_j\varphi_j(\mathbf{x}) = \left(\frac{p^2}{2m} + \int d^3y\, V(\mathbf{y}-\mathbf{x})\sum_i n_i\varphi_i^*(\mathbf{y})\varphi_i(\mathbf{y})\right)\varphi_j(\mathbf{x}) - \int d^3y\, \varphi_j(\mathbf{y})V(\mathbf{y}-\mathbf{x})\sum_i n_i\varphi_i^*(\mathbf{y})\varphi_i(\mathbf{x}),}$$

where n_i is the occupancy 0 or 1 of the state i. The integration is understood to include taking the spin inner product. The term $i = j$ may be included in the two sums in (50) because for it the direct and exchange contributions cancel. In the second term on the right-hand side, the sum is over all states and all spin orientations; in the third term (the exchange term) on the right-hand side only states i with spin parallel to j will remain because we take the spin inner product in the integration over d^3y.

KOOPMANS'S THEOREM

This important theorem states that the energy parameter ϵ_l in the Hartree-Fock equation for $\varphi_l(\mathbf{x})$ is just the negative of the energy required to remove the electron in the state l from the solid, *provided* that the φ's are extended functions of the Bloch type and that the electronic system is very large.

Because the electron charge is spread throughout the entire crystal, the φ's will be essentially identical for the problem with or without an electron in the state l. This is our central assumption. The work done in removing the electron from this state is the difference

$$\langle\Phi_l|H|\Phi_l\rangle - \langle\Phi|H|\Phi\rangle,$$

where Φ_l does not have an electron in the state l; in other respects Φ_l is identical with Φ.

If the $\varphi_m(\mathbf{x})$ in $\Psi(\mathbf{x}) = \Sigma\, c_m\varphi_m(\mathbf{x})$ are solutions of the Hartree-Fock equation, then on taking the inner product of (50) with φ_j^+ we have, with the energy referred to the ground state, and writing l for j,

$$\varepsilon_l = \left\langle l\left|\frac{1}{2m}p^2\right|l\right\rangle - \sum_m n_m(\langle lm|V|lm\rangle - \langle lm|V|ml\rangle). \tag{51}$$

Here $\langle lm|V|lm\rangle \equiv \int d^3x\, d^3y\, \varphi_l^*(\mathbf{x})\varphi_m^*(\mathbf{y})V\varphi_l(\mathbf{x})\varphi_m(\mathbf{y})$ and $\langle lm|V|ml\rangle \equiv \int d^3x\, d^3y\, \varphi_l^*(\mathbf{x})\varphi_m^*(\mathbf{y})V\varphi_m(\mathbf{x})\varphi_l(\mathbf{y})$.

Now from (37)

$$\begin{aligned}\langle\Phi|H|\Phi\rangle &= \left\langle\int d^3x\, \Psi^+(\mathbf{x})\,\frac{1}{2m}\,p^2\Psi(\mathbf{x})\right\rangle \\ &\quad + \left\langle\tfrac{1}{2}\int d^3x\, d^3y\, \Psi^+(\mathbf{x})\Psi^+(\mathbf{y})V(\mathbf{y}-\mathbf{x})\Psi(\mathbf{y})\Psi(\mathbf{x})\right\rangle,\end{aligned} \tag{52}$$

where the carets on the right-hand side indicate the diagonal matrix element in the state Φ in the HF representation. Then

$$\begin{aligned}\langle\Phi|H|\Phi\rangle &= \sum_m n_m\left\langle m\left|\frac{1}{2m}p^2\right|m\right\rangle \\ &\quad + \tfrac{1}{2}\sum_{mp} n_mn_p(\langle mp|V|mp\rangle - \langle mp|V|pm\rangle),\end{aligned} \tag{53}$$

and the change in the energy on removing the lth particle is

$$\left\langle l\left|\frac{1}{2m}p^2\right|l\right\rangle - \sum_m n_m(\langle lm|V|lm\rangle - \langle lm|V|ml\rangle),$$

which is just the value of ε_l given by (51). We note that the derivation assumes invariance of the $\varphi_m(\mathbf{x})$ under the removal of the particle from the lth state; thus the theorem cannot apply to small systems.

FERMION QUASIPARTICLES

The low energy excitations of a quantum-mechanical system having a large number of degrees of freedom may be often approximately described in terms of a number of elementary excitations or quasiparticles. In some situations the description of the system in terms of a sum over quasiparticles is exact; in other situations the quasiparticle is a wavepacket of exact eigenstates. The width in energy of the eigenstates comprising the packet determines the lifetime of the packet and thus the range of validity of the quasiparticle concept. The quasiparticles of an ionic crystal lattice are phonons; of a spin lattice, magnons; of a free-electron gas, excitations which resemble one-electron excitations.

In treating the electron system it is particularly convenient to redefine the vacuum state as the filled fermi sea, rather than the state with no particles present. With the filled fermi sea as the vacuum we must provide separate fermion operators for processes which occur above or below the fermi level. The removal of an electron below the fermi level is described in the new scheme as the creation of a hole. We consider first a system of N free noninteracting fermions having the hamiltonian

$$H_0 = \sum_{\mathbf{k}} \varepsilon_{\mathbf{k}} c_{\mathbf{k}}^+ c_{\mathbf{k}}, \tag{54}$$

where $\varepsilon_{\mathbf{k}}$ is the energy of a single particle having $\varepsilon_{\mathbf{k}} = \varepsilon_{-\mathbf{k}}$. We agree to measure $\varepsilon_{\mathbf{k}}$ from the fermi level ε_F.

In the ground state of the system Φ_0 defined by (4) all one-particle states are filled to the energy ε_F and above ε_F all states are empty. We regard the state Φ_0 as the vacuum of the problem: it is then convenient to represent the annihilation of an electron in the fermi sea as the creation of a hole. Thus we deal only with electrons (for states $k > k_F$) and holes (for states $k < k_F$). The act of taking an electron from $\mathbf{k}'$ within the sea to $\mathbf{k}''$ outside the sea involves the creation of an electron-hole pair. The language of the theory has a similarity to positron theory, and there is a complete formal similarity between particles and holes.

We introduce the *electron operators* α^+, α by

$$\boxed{\alpha_{\mathbf{k}}^+ = c_{\mathbf{k}}^+; \qquad \alpha_{\mathbf{k}} = c_{\mathbf{k}}, \qquad \text{for } \varepsilon_{\mathbf{k}} > \varepsilon_F,} \tag{55}$$

and the hole operators β^+, β by

$$\boxed{\beta_{\mathbf{k}}^+ = c_{-\mathbf{k}}; \qquad \beta_{\mathbf{k}} = c_{-\mathbf{k}}^+, \qquad \text{for } \varepsilon_{\mathbf{k}} < \varepsilon_F.} \tag{56}$$

The $-\mathbf{k}$ introduced for the holes is a convention which gives correctly the net change of wavevector or momentum: the annihilation $c_{-\mathbf{k}}$ of an electron at $-\mathbf{k}$ leaves the fermi sea with a momentum $\mathbf{k}$. Thus $\beta_{\mathbf{k}}^{+} \equiv c_{-\mathbf{k}}$ creates a hole of momentum $\mathbf{k}$. The total momentum, referred to Φ_0 as a state of zero momentum, is

$$\mathbf{P} = \sum_{\mathbf{k}} \mathbf{k}(\alpha_{\mathbf{k}}^{+}\alpha_{\mathbf{k}} - \beta_{\mathbf{k}}^{+}\beta_{\mathbf{k}}). \tag{57}$$

The number operator for excited electrons is

$$\hat{N}_e = \sum_{\mathbf{k}} \alpha_{\mathbf{k}}^{+}\alpha_{\mathbf{k}}, \qquad (k > k_F) \tag{58}$$

and for holes

$$\hat{N}_h = \sum_{\mathbf{k}} \beta_{\mathbf{k}}^{+}\beta_{\mathbf{k}}, \qquad (k < k_F). \tag{59}$$

The hamiltonian for the noninteracting fermi gas is

$$H_0 = \sum_{k>k_F} \varepsilon_{\mathbf{k}}\alpha_{\mathbf{k}}^{+}\alpha_{\mathbf{k}} + \sum_{k<k_F} \varepsilon_{\mathbf{k}}\beta_{\mathbf{k}}^{+}\beta_{\mathbf{k}}, \tag{60}$$

with $\varepsilon_{\mathbf{k}}$ referred to ε_F as zero; thus $\varepsilon_{\mathbf{k}} < 0$ if $k < k_F$.

The ground state Φ_0 of the fermi sea has the property

$$\alpha_{\mathbf{k}}\Phi_0 = 0; \qquad \beta_{\mathbf{k}}\Phi_0 = 0, \tag{61}$$

each within the appropriate range of $\mathbf{k}$. The true vacuum state Φ_{vac} satisfies $c_{\mathbf{k}}\Phi_{\text{vac}} = 0$ for all $\mathbf{k}$. The state $\alpha_{\mathbf{k}''}^{+}\beta_{\mathbf{k}'}^{+}\Phi_0$ contains an electron-hole pair, as shown in Fig. 1.

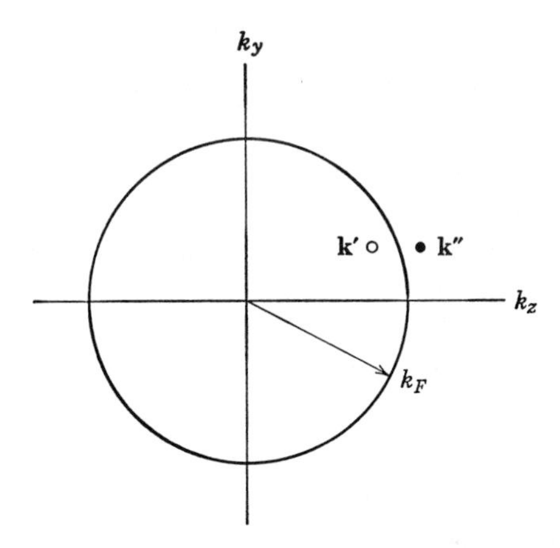

FIG. 1. Excitation of an electron-hole pair: the state is $\alpha_{\mathbf{k}''}^{+}\beta_{\mathbf{k}'}^{+}\Phi_0$.

ELECTRON GAS IN THE HARTREE AND HARTREE-FOCK APPROXIMATIONS

We consider the physical properties of a free fermion gas of N electrons of charge e in a volume Ω. To insure the neutrality of the system we add to the electrons a uniform background of positive charge having a charge density equal to the average charge density of the electrons.

In the Hartree independent-particle approximation we look for the product-type wavefunction of the form

$$\varphi(\mathbf{x}_1, \cdots, \mathbf{x}_N) = \prod_{j=1}^{N} \varphi_j(\mathbf{x}_j), \tag{62}$$

which minimizes the energy. The Hartree solutions satisfy the same equations as the Hartree-Fock solutions, but without the exchange term [the last term on the right-hand side of (50)].

$$\left[\frac{1}{2m} p^2 + \int d^3y \, \frac{e^2}{|\mathbf{x}-\mathbf{y}|} \sum_m{}' \varphi_m^*(\mathbf{y})\varphi_m(\mathbf{y}) - \int d^3y \, \rho_0^{(+)}|e| \frac{1}{|\mathbf{x}-\mathbf{y}|}\right] \varphi_j(\mathbf{x}) = \varepsilon_j \varphi_j(\mathbf{x}), \tag{63}$$

where $\rho_0^{(+)} = N|e|/\Omega$ arises from the positive background. This describes an electron moving in the average potential of all other particles. The method is said to be self-consistent if all φ's are eigenfunctions of this equation. The sum in (63) is over all occupied states except j.

We show now that the set of plane waves

$$\varphi_{\mathbf{k}}(\mathbf{x}) = \Omega^{-1/2} e^{i\mathbf{k}\cdot\mathbf{x}} \tag{64}$$

are self-consistent solutions of this equation. The electron charge density in (63) is constant for a product (62) of plane waves:

$$e \sum_m{}' \varphi_m^* \varphi_m = e\rho_0^{(-)} = \sum_m{}' \Omega^{-1} \cdot e \cdot 1_m = (N-1)e/\Omega, \tag{65}$$

which cancels $e\rho_0^{+}$ except for a trivial term arising from the difference between N and $N-1$: the associated energy is of the order of $e^2/\Omega^{1/3} \sim 10^{-19}$ ergs for $\Omega \sim 1 \text{ cm}^3$. To this order the coulomb interaction term in (63) drops out and the Hartree problem is just the free-electron problem:

$$\frac{1}{2m} p^2 \varphi_j = \varepsilon_j \varphi_j. \tag{66}$$

The energy of the electron gas in the Hartree approximation is purely kinetic and hence exactly the same as for free particles. The energy has the value at absolute zero

$$\langle \varepsilon_F \rangle = \tfrac{3}{5} \cdot \frac{1}{2m} k_F^2, \tag{67}$$

per particle, where the factor $\frac{3}{5}$ is from the average of k^2 over the volume of a sphere. The fermi momentum k_F is determined by

$$\frac{2\Omega}{(2\pi)^3} \cdot \frac{4\pi}{3} k_F^{\,2} = N, \tag{68}$$

with the factor of 2 coming from the spin. If we define a mean radius per particle by

$$\Omega = N \frac{4\pi}{3} r_0^{\,3}, \tag{69}$$

then (68) becomes

$$\frac{2}{(2\pi)^3} \left(\frac{4\pi}{3}\right)^2 (k_F r_0)^3 = 1, \tag{70}$$

or

$$\boxed{k_F = \frac{1}{\alpha r_0}; \qquad \alpha = (4/9\pi)^{1/3} = 0.52,} \tag{71}$$

and

$$\langle \varepsilon_F \rangle = \frac{3}{10\alpha^2 m r_0^{\,2}}. \tag{72}$$

It is often useful to express r_0 in terms of the bohr radius $a_H = 0.529$ A. We introduce the dimensionless parameter r_s as

$$\boxed{r_s = r_0/a_H = (me^2/\hbar^2) r_0.} \tag{73}$$

The region of actual metallic densities is $2 < r_s < 5$. With (73),

$$\langle \varepsilon_F \rangle = \frac{3}{10} \frac{me^4}{\hbar^2} \frac{1}{\alpha^2 r_s^{\,2}}, \tag{74}$$

or, in rydbergs $\frac{1}{2}(me^4/\hbar^2) = 13.60$ ev,

$$\langle \varepsilon_F \rangle = \frac{3}{5\alpha^2 r_s^{\,2}} \text{ ry} \cong \frac{2.21}{r_s^{\,2}} \text{ ry}. \tag{75}$$

This is the total energy per electron in the Hartree approximation. It turns out that the Hartree approximation does not lead to metallic cohesion—the electrons spend too much time in regions of repulsive potential energy. We note (on reflection) that in the Hartree approximation the coulomb self-energy of the positive background plus that of the electron gas just cancels the interaction energy of the electrons with the positive background.

Modified Hartree Model. We have seen that the coulomb energy vanishes for the Hartree solutions in a uniform background of positive charge. We now modify the model: keeping the previous uniform electron distribution, we collect the positive charge background into point charges $|e|$. There is one point per atomic volume Ω/N. To a close approximation the coulomb energy of this model without exchange is obtained by calculating the energy of a point charge $|e|$ interacting electrostatically with a uniform negative charge distribution throughout the volume of the sphere of radius r_0; we may consider also the electrostatic interaction of the electron distribution with itself, although this term is not present if only one electron is present in the sphere, rather than the N^{-1} part of each of N electrons. That is, there is no self-energy contribution from an electron interacting with itself, so we have to decide whether or not one individual electron is localized within the cell. We make the calculation as if each electron is distributed throughout the specimen.

The contribution of the point charge interacting with the negative charge distribution is

$$\mathcal{E}_1 = -e^2 \frac{3}{4\pi r_0{}^3} \int_0^{r_0} 4\pi r\, dr = -\frac{3}{2}\frac{e^2}{r_0}, \tag{76}$$

and the self-energy contribution of the electron distribution with itself is

$$\mathcal{E}_2 = e^2 \left(\frac{3}{4\pi r_0{}^3}\right)^2 \int_0^{r_0} d\xi\, \tfrac{1}{3}(4\pi)^2 \xi^4 = \frac{3}{5}\frac{e^2}{r_0}, \tag{77}$$

so that the total energy is, on this model,

$$\mathcal{E} = \mathcal{E}_1 + \mathcal{E}_2 + \langle \mathcal{E}_F \rangle = -\frac{9e^2}{10r_0} + \frac{3}{10\alpha^2 m r_0{}^2} = -\frac{1.80}{r_s} + \frac{2.21}{r_s{}^2}\ \text{ry}, \tag{78}$$

using (72) and (75). The equilibrium value of r_s is 2.45 bohr units or 1.30 A; the equilibrium value is obtained by setting $d\mathcal{E}/dr_s = 0$.

Hartree-Fock Approximation. We have derived the Hartree-Fock equation (50); on writing the spin variables s, s' explicitly,

$$\begin{aligned} \mathcal{E}_j \varphi_{js}(\mathbf{x}) = &\left(\frac{1}{2m} p^2 + v(\mathbf{x}) \right. \\ &\left. + \sum_{ls'} n_{ls'} \int d^3y\, \varphi_{ls'}^*(\mathbf{y}) \varphi_{ls'}(\mathbf{y}) V(\mathbf{x}-\mathbf{y}) \right) \times \varphi_{js}(\mathbf{x}) \\ &- \sum_l n_{ls} \left(\int d^3y\, \varphi_{ls}^*(\mathbf{y}) \varphi_{js}(\mathbf{y}) V(\mathbf{x}-\mathbf{y}) \right) \varphi_{ls}(\mathbf{x}). \end{aligned} \tag{79}$$

We have seen from the Hartree solution that if the φ's are plane waves, the second and third terms on the right-hand side of (79) add up to zero. Let us now see if we can solve (79) with plane wave eigenfunctions. It turns out that we can.

The exchange term is, for unit volume,

$$(80)\qquad -\sum_l{}'\left(\int d^3y\, e^{i(\mathbf{k}_j-\mathbf{k}_l)\cdot(\mathbf{y}-\mathbf{x})}V(\mathbf{x}-\mathbf{y})\right)e^{i(\mathbf{k}_j-\mathbf{k}_l)\cdot\mathbf{x}}\varphi_l(\mathbf{x}) = -\sum_l{}' G(\mathbf{k}_j-\mathbf{k}_l)\varphi_j(\mathbf{x}),$$

where, with $\boldsymbol{\xi} = \mathbf{x} - \mathbf{y}$,

$$(81)\qquad G(\mathbf{k}) = \int d^3\xi\, e^{-i\mathbf{k}\cdot\boldsymbol{\xi}}V(\boldsymbol{\xi})$$

is the fourier transform of $V(\mathbf{x}-\mathbf{y})$. Thus the Hartree-Fock equation has plane wave eigenfunctions with the eigenvalues

$$(82)\qquad \varepsilon_j = \frac{k_j^2}{2m} - \sum_l{}' G(\mathbf{k}_j - \mathbf{k}_l);$$

the sum is over all occupied states l excluding j. The next problem is to calculate the $\sum_l G(\mathbf{k}_j - \mathbf{k}_l)$, the exchange energy.

Now

$$(83)\qquad V(\mathbf{x}-\mathbf{y}) = \frac{e^2}{|\mathbf{x}-\mathbf{y}|} = \sum_{\mathbf{K}} \frac{4\pi e^2}{K^2}e^{i\mathbf{K}\cdot(\mathbf{x}-\mathbf{y})},$$

by (1.24). Then

$$(84)\qquad G(\mathbf{k}_j-\mathbf{k}_l) = \int d^3x\, \frac{e^2}{|\mathbf{x}|}e^{i(\mathbf{k}_j-\mathbf{k}_l)\cdot\mathbf{x}} = \frac{4\pi e^2}{(\mathbf{k}_j-\mathbf{k}_l)^2}.$$

We now calculate the quantity $\sum_l{}' G(\mathbf{k}_j - \mathbf{k}_l)$ which appears in the one-electron energy. The sum is taken only over states $|ls\rangle$ of spin parallel to $|js\rangle$; antiparallel pairs do not appear in the Hartree-Fock exchange integrals. Then, for the ground state,

$$(85)\qquad \sum_l{}' G(\mathbf{k}_j-\mathbf{k}_l) = 4\pi e^2\sum_l{}'\frac{1}{(\mathbf{k}_j-\mathbf{k}_l)^2}$$

$$= \frac{4\pi e^2}{\Omega}\cdot\frac{\Omega}{(2\pi)^3}\int_{k<k_F} d^3k\,\frac{1}{(\mathbf{k}_j-\mathbf{k})^2}$$

$$= \frac{e^2}{2\pi^2}2\pi\int_0^{k_F}k^2\,dk\int_{-1}^{1}d\mu\,\frac{1}{k_j^2+k^2-2kk_j\mu}$$

$$= \frac{e^2}{\pi k_j}\int_0^{k_F}k\,dk\log\frac{k+k_j}{|k-k_j|}$$

$$= \frac{e^2}{\pi}\left(\frac{k_F^2-k_j^2}{2k_j}\log\left|\frac{k_F+k_j}{k_F-k_j}\right| + k_F\right),$$

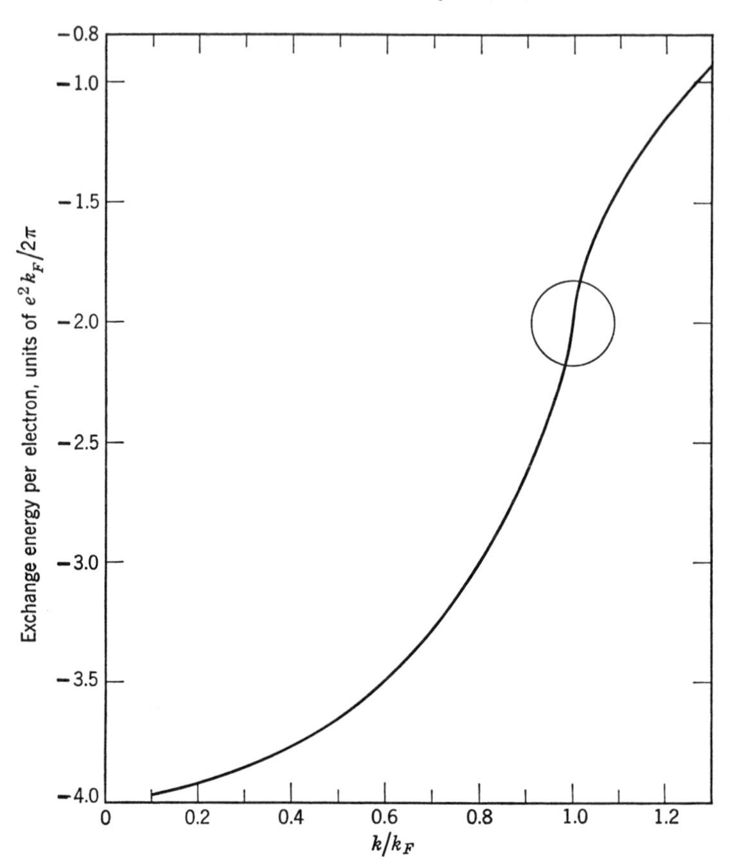

FIG. 2. Plot of exchange energy versus wavevector for free electron gas. The region inside the circle is not accurate—the slope has been exaggerated for emphasis. The derivative near $k = k_F$ is given by the expansion $-0.614 = 2 \log |x - 1|$, where $x = k/k_F$; thus the slope is reduced to 20:1 for $|x - 1| = 3.3 \times 10^{-5}$.

by elementary integration. Therefore from (82) the Hartree-Fock energy parameter is

$$\boxed{\varepsilon_j = \frac{k_j^2}{2m} - \frac{e^2}{2\pi}\left(\frac{k_F^2 - k_j^2}{k_j} \log \left|\frac{k_F + k_j}{k_F - k_j}\right| + 2k_F\right).} \tag{86}$$

The average exchange energy per particle is most easily obtained by the more direct attack below, although one can sum (86) over all occupied states. The exchange contribution to (86) is plotted in Fig. 2 as a function of k_j/k_F.

EVALUATION OF THE EXCHANGE INTEGRAL FOR THE ELECTRON GAS

We require the value of the $G(\mathbf{k}_j - \mathbf{k}_l)$ summed over all occupied states j, l. In effect we need the value of

$$\mathcal{G} = \iint_{k_1, k_2 < k_F} d^3k_1\, d^3k_2 \frac{1}{|\mathbf{k}_1 - \mathbf{k}_2|^2}. \tag{87}$$

Now, with $\mu = \cos\theta$ and $s = k_2/k_1$,

$$\frac{1}{|\mathbf{k}_1 - \mathbf{k}_2|^2} = \frac{1}{k_1^2 + k_2^2 - 2k_1k_2\mu} = \frac{1}{k_1^2} \cdot \frac{1}{1 + s^2 - 2s\mu}. \tag{88}$$

But for the half space $s < 1$ we have a well-known series expansion:

$$\frac{1}{1 + s^2 - 2s\mu} = \left[\sum_L s^L P_L(\mu)\right]^2 = \sum_{L,\lambda} s^{L+\lambda} P_L(\mu) P_\lambda(\mu), \tag{89}$$

where $P_L(\mu)$ is a legendre polynomial. Thus

$$\mathcal{G} = 2 \int_{k_1 < k_F} d^3k_1 \int_{k_2 < k_1} 2\pi k_2^2\, dk_2\, d\mu \sum_{L,\lambda} \left(\frac{k_2}{k_1}\right)^{L+\lambda} \cdot \frac{1}{k_1^2} P_L(\mu) P_\lambda(\mu); \tag{90}$$

using

$$\int_{-1}^{1} P_L(\mu) P_\lambda(\mu)\, d\mu = \frac{2}{2L+1} \delta_{L\lambda}, \tag{91}$$

we have, counting both the space $k_2 < k_1$ and the space $k_1 < k_2$,

$$\begin{aligned} \mathcal{G} &= 8\pi \int_{k_1 < k_F} d^3k_1 \int_0^{k_1} dk_2 \sum \left(\frac{k_2}{k_1}\right)^{2L+2} \frac{1}{2L+1} \\ &= 8\pi \int_{k_1 < k_F} d^3k_1\, k_1 \sum_L \frac{1}{(2L+1)(2L+3)} \\ &= 8\pi^2 k_F^4 \sum_0^\infty \frac{1}{(2L+1)(2L+3)}. \end{aligned} \tag{92}$$

The sum is easily evaluated by writing it as

$$\tfrac{1}{2} \sum_0^\infty \left(\frac{1}{2L+1} - \frac{1}{2L+3}\right) = \tfrac{1}{2} + \tfrac{1}{2} \sum_0^\infty \left(\frac{1}{2L+3} - \frac{1}{2L+3}\right) = \tfrac{1}{2}, \tag{93}$$

so that

$$\mathcal{G} = 4\pi^2 k_F^4. \tag{94}$$

Now the average exchange energy is

$$\varepsilon_{ex} = -\tfrac{1}{2}\cdot\frac{2}{N}\sum_{jl}{}' G(\mathbf{k}_j - \mathbf{k}_l) = -\frac{4\pi e^2}{n}\sum_{jl}\frac{1}{(\mathbf{k}_j - \mathbf{k}_l)^2} \tag{95}$$
$$= -\frac{4\pi e^2}{(2\pi)^6 n}\mathcal{J} = -\frac{2e^2 k_F{}^4}{(2\pi)^3 n} = -\frac{3e^2}{4\pi\alpha r_0},$$

with $\alpha = (4/9\pi)^{1/3}$ as before. We have further

$$\varepsilon_{ex} = -\frac{3}{2\pi}\frac{1}{\alpha r_s}\,\text{ry} = -\frac{0.916}{r_s}\,\text{ry}. \tag{96}$$

The Hartree-Fock energy $\varepsilon_{HF} = \langle\varepsilon_F\rangle + \varepsilon_{ex}$ is then

$$\boxed{\varepsilon_{HF} = \frac{2.21}{r_s{}^2} - \frac{0.916}{r_s}\,\text{ry}} \tag{97}$$

per particle. This is a better value of the energy than the Hartree energy, but the binding is still too weak. The defect is the neglect of the correlations in the positions of the electrons introduced by their coulomb interaction. Correlation is particularly important for pairs of electrons of antiparallel spin for which the antisymmetrization does not act to keep the pair apart. The effect of antisymmetrization on a pair of parallel spin is treated in the next section.

The correlation energy ε_c *is defined as* $\varepsilon_{exact} - \varepsilon_{HF}$, *the difference between the exact energy and the energy calculated in the Hartree-Fock approximation.* We might expect to be able to treat the coulomb interaction as a small perturbation, calculating the energy by perturbation theory. This can be done at high electron density ($r_s < 1$), but involves a careful treatment of divergent terms, as discussed in the following chapter.

One major difficulty with the Hartree-Fock result (86) is that the density of states at the fermi surface goes to zero because $d\varepsilon/dk_j \to \infty$ as $k \to k_F$; the desired derivative is just the next to the last step of (85), without the integral. The density of states involves $(dn/dk)(dk/d\varepsilon)$. The low density of states near the fermi surface on the Hartree-Fock model has important consequences for the thermal and magnetic properties of the electron gas, but none of these special consequences of the HF model is in agreement with experiment. Many-body corrections to the HF model lead to a screened coulomb potential, and for this the density of states does not vanish at the fermi surface.

We have just calculated in the HF approximation the ground-state energy of a system of electrons with a uniform rigid background of positive charge. As calculated, the ground-state coulomb energy includes:

(*a*) The self-energy of the electron gas.
(*b*) The self-energy of the positive background.
(*c*) The interaction energy of the electrons with the uniform positive background.

The sum of these three contributions is zero: both self-energies are positive and enter each with a factor $\frac{1}{2}$, thereby canceling the negative interaction energy (*c*). It will require a moment's reflection by the reader to appreciate that this cancelation is implied when we say that the energy of the system is given by the number of electrons times the average HF energy (97) per electron, corrected for the correlation energy, if we wish.

How do we go from this energy to the cohesive energy of an actual metal? Even for the simplest metal, sodium, for which the free electron model works best, there are four major things we must do. In an actual metal the positive charge is collected in a discrete ionic lattice. To take care of this in the bookkeeping we must:

(1) Subtract the self-energy E_1 of the positive background.
(2) Subtract the interaction energy E_2 of the electrons with the uniform positive background, observing that $E_2 = -2E_1$.
(3) Add the coulomb energy E_3 of the discrete lattice.
(4) Add the interaction energy of the electrons E_4 with the discrete lattice.

In the method of Wigner and Seitz developed in Chapter 13 the metal is divided up into polyhedra centered about each lattice point. The polyhedra are electrically neutral and the coulomb interaction between different polyhedra is very small. Thus the sum of the energies of the four steps we have enumerated is, for n atoms, just the energy of n Wigner-Seitz polyhedra. The terms (1) and (2) together contribute $0.6e^2/r_0$ or $1.2/r_s$ to the energy of the metal, as was calculated in Eq. (77). The sum of the terms (3) and (4) is just the energy eigenvalue for $\mathbf{k} = 0$ of the Wigner-Seitz boundary value problem discussed in Chapter 13. The cohesive energy contains the sum of

$$(\text{HF energy}) + (\text{WS energy at } \mathbf{k} = 0) + \frac{1.2}{r_s} + (\text{correlation energy}).$$

This sum will be negative for a reasonable metal, but for the metal to be stable with respect to separated neutral atoms the sum must be more negative than the first ionization energy of the neutral atom, where we express the ionization energy as a negative number. The difference between the ionization energy and the sum we have just written is the *cohesive energy*. Note that the term $1.2/r_s$ may be combined with the exchange contribution $-0.916/r_s$ to the HF energy to give $+0.284/r_s$ for the total term in r_s^{-1}. In the following chapter we discuss for sodium the magnitudes of the several terms.

Two-Electron Correlation Function. We want to calculate for a fermi gas in the Hartree-Fock approximation the probability $g(\mathbf{x},\mathbf{y})\, d^3x\, d^3y$ that there is one particle in volume element d^3x at $\mathbf{x}$ and a second particle in d^3y at $\mathbf{y}$. Even for plane wave states this probability may be nonuniform. If the spins of the two particles are antiparallel, the probability is uniform:

$$g(\mathbf{x},\mathbf{y}) = \varphi^*(\mathbf{x},\mathbf{y})\varphi(\mathbf{x},\mathbf{y}) = e^{-i(\mathbf{k}_1\cdot\mathbf{x}+\mathbf{k}_2\cdot\mathbf{y})}e^{i(\mathbf{k}_1\cdot\mathbf{x}+\mathbf{k}_2\cdot\mathbf{y})} = 1, \tag{98}$$

but for parallel spins the exclusion principle leads to a nonuniform correlation:

$$\begin{aligned} g(\mathbf{x},\mathbf{y}) &= \tfrac{1}{2}(e^{-i(\mathbf{k}_1\cdot\mathbf{x}+\mathbf{k}_2\cdot\mathbf{y})} - e^{-i(\mathbf{k}_1\cdot\mathbf{y}+\mathbf{k}_2\cdot\mathbf{x})})(e^{i(\mathbf{k}_1\cdot\mathbf{x}+\mathbf{k}_2\cdot\mathbf{y})} - e^{i(\mathbf{k}_1\cdot\mathbf{y}+\mathbf{k}_2\cdot\mathbf{x})}) \\ &= \tfrac{1}{2}(2 - e^{i(\mathbf{k}_1-\mathbf{k}_2)\cdot(\mathbf{y}-\mathbf{x})} - e^{-i(\mathbf{k}_1-\mathbf{k}_2)\cdot(\mathbf{y}-\mathbf{x})}) \\ &= 1 - \cos(\mathbf{k}_1 - \mathbf{k}_2)\cdot(\mathbf{y}-\mathbf{x}). \end{aligned} \tag{99}$$

We average (99) over the ground-state fermi sea: if there are N distinct states $\mathbf{k}$ occupied, the average correlation of parallel pairs is, with $\mathbf{r} = \mathbf{y} - \mathbf{x}$,

$$\begin{aligned} \langle g(\mathbf{r})\rangle &= \frac{1}{N^2}\sum_{ij}(1 - e^{i(\mathbf{k}_i-\mathbf{k}_j)\cdot\mathbf{r}}) \\ &= \frac{1}{N^2(2\pi)^6}\int d^3k_i \int d^3k_j\,(1 - e^{i(\mathbf{k}_i-\mathbf{k}_j)\cdot\mathbf{r}}). \end{aligned} \tag{100}$$

This is the probability of finding at $\mathbf{r}$ an electron with spin parallel to that of an electron at the origin. We may write

$$\langle g(\mathbf{r})\rangle = [1 - F^2(k_Fr)], \tag{101}$$

where

$$F(k_Fr) = \frac{1}{N(2\pi)^3}\int d^3k\, e^{i\mathbf{k}\cdot\mathbf{x}}. \tag{102}$$

The integral in (102) was evaluated in (1.68), whence

$$F(k_Fr) = 3\left(\frac{\sin k_Fr - k_Fr\cos k_Fr}{k_F{}^3r^3}\right)\delta_{ss'}, \tag{103}$$

using

$$\frac{4\pi}{3(2\pi)^3}k_F{}^3 = N. \tag{104}$$

We note that $F(k_Fr) \to 1$ and $\langle g(\mathbf{r})\rangle \to 0$ as $r \to 0$; as $r \to \infty$, $F(k_Fr) \to 0$ and $\langle g(\mathbf{r})\rangle \to 1$. We have put the δ function in the spin projections s, s' on the z axis into (103) to emphasize that $F(k_Fr)$ is zero for antiparallel spins. In (104), N is $n/2$, where n is the electron concentration counting both spin orientations. The electron deficiency near the origin is known as the *fermi hole.*

COULOMB INTERACTIONS AND THE FORMALISM OF SECOND QUANTIZATION

With the fourier transform (83) of the coulomb interaction, the hamiltonian is, for unit volume,

$$\begin{aligned} H &= \frac{1}{2m}\sum_j p_j{}^2 + \tfrac{1}{2}\sum_{ij}{}' \frac{e^2}{|\mathbf{x}_i - \mathbf{x}_j|} + v(\mathbf{x}) \\ &= \frac{1}{2m}\sum_j p_j{}^2 + \tfrac{1}{2}\sum_{ij}{}'\sum_{\mathbf{K}\neq 0}\frac{4\pi e^2}{K^2}e^{i\mathbf{K}\cdot(\mathbf{x}_i-\mathbf{x}_j)}, \end{aligned} \tag{105}$$

where the positive background removes the term in $\mathbf{K} = 0$; the term $i = j$ is excluded from the sum.

We first rewrite the potential energy in terms of the particle density fluctuation operators $\rho_\mathbf{K}$ defined as

$$\rho(\mathbf{x}) = \sum_\mathbf{K} \rho_\mathbf{K} e^{i\mathbf{K}\cdot\mathbf{x}}. \tag{106}$$

$$\begin{aligned} \int d^3x\, \rho(\mathbf{x})e^{-i\mathbf{K}'\cdot\mathbf{x}} &= \int d^3x \sum_\mathbf{K} \rho_\mathbf{K} e^{i(\mathbf{K}-\mathbf{K}')\cdot\mathbf{x}} \\ &= \sum_\mathbf{K} \rho_\mathbf{K}\Delta(\mathbf{K}-\mathbf{K}') = \rho_{\mathbf{K}'}, \end{aligned} \tag{107}$$

so that

$$\rho_\mathbf{K} = \int d^3x\, \rho(\mathbf{x})e^{-i\mathbf{K}\cdot\mathbf{x}}. \tag{108}$$

If $\rho(\mathbf{x})$ is uniform and equal to n, we have $\rho_\mathbf{K} = n\delta_{\mathbf{K}0}$. For point charges,

$$\rho(\mathbf{x}) = \sum_j \delta(\mathbf{x} - \mathbf{x}_j), \tag{109}$$

and

$$\rho_{\mathbf{K}} = \int d^3x \sum_j \delta(\mathbf{x} - \mathbf{x}_j)e^{-i\mathbf{K}\cdot\mathbf{x}} = \sum_j e^{-i\mathbf{K}\cdot\mathbf{x}_j}. \tag{110}$$

It follows that

$$\overset{+}{\rho}_{\mathbf{K}}\rho_{\mathbf{K}} = \sum_{ij} e^{i\mathbf{K}\cdot(\mathbf{x}_i - \mathbf{x}_j)}, \tag{111}$$

and

$$\sum_{ij}{}' e^{i\mathbf{K}\cdot(\mathbf{x}_i-\mathbf{x}_j)} = \sum_{ij} e^{i\mathbf{K}\cdot(\mathbf{x}_i-\mathbf{x}_j)} - n = \overset{+}{\rho}_{\mathbf{K}}\rho_{\mathbf{K}} - n, \tag{112}$$

where n is the electron concentration. Thus the hamiltonian in terms of density operators is, from (105) and (112),

$$\boxed{H = \frac{1}{2m}\sum_j p_j{}^2 + \sum_{\mathbf{K}\neq 0} \frac{2\pi e^2}{K^2}(\overset{+}{\rho}_{\mathbf{K}}\rho_{\mathbf{K}} - n).} \tag{113}$$

It is also valuable to have H expressed in terms of fermion operators. Let

$$\Psi(\mathbf{x}) = \sum_{\mathbf{k}s} c_{\mathbf{k}s}e^{i\mathbf{k}\cdot\mathbf{x}}|s\rangle \tag{114}$$

in a plane wave representation, where $|s\rangle$ is the spin part of the one-electron wavefunction. Then the kinetic energy is

$$\int d^3x\, \Psi^+(\mathbf{x})\frac{p^2}{2m}\Psi(\mathbf{x}) = \sum \varepsilon_{\mathbf{k}}\overset{+}{c}_{\mathbf{k}s}c_{\mathbf{k}s}; \qquad \varepsilon_{\mathbf{k}} = \frac{k^2}{2m}. \tag{115}$$

Now, by the definition (32),

$$\rho(\mathbf{x}) = \Psi^+(\mathbf{x})\Psi(\mathbf{x}) = \sum_{\mathbf{k}\mathbf{k}'s} \overset{+}{c}_{\mathbf{k}s}c_{\mathbf{k}'s}e^{i(\mathbf{k}'-\mathbf{k})\cdot\mathbf{x}}, \tag{116}$$

and from (108)

$$\rho_{\mathbf{K}} = \sum_{\mathbf{k}\mathbf{k}'} \overset{+}{c}_{\mathbf{k}s}c_{\mathbf{k}'s}\int d^3x\, e^{i(\mathbf{k}'-\mathbf{k}-\mathbf{K})\cdot\mathbf{x}} = \sum_{\mathbf{k}} \overset{+}{c}_{\mathbf{k}s}c_{\mathbf{k}+\mathbf{K},s}. \tag{117}$$

Further,

$$\overset{+}{\rho}_{\mathbf{K}} = \sum \overset{+}{c}_{\mathbf{k}+\mathbf{K}}c_{\mathbf{k}} = \rho_{-\mathbf{K}}. \tag{118}$$

The coulomb energy is

$$\sum_{\mathbf{K}}{}' \frac{2\pi e^2}{K^2}(\overset{+}{\rho}_{\mathbf{K}}\rho_{\mathbf{K}} - n) \rightarrow \sum_{\mathbf{K}}{}'\left(\sum_{\mathbf{k}\mathbf{k}'} \frac{2\pi e^2}{K^2}\overset{+}{c}_{\mathbf{k}+\mathbf{K}}c_{\mathbf{k}}\overset{+}{c}_{\mathbf{k}'-\mathbf{K}}c_{\mathbf{k}'} - n\right) \tag{119}$$

so that the hamiltonian becomes

$$H = \sum_{\mathbf{k}s} \varepsilon_{\mathbf{k}s} c^+_{\mathbf{k}s} c_{\mathbf{k}s} + \sum_{\mathbf{K}}{}' \left(\sum_{\substack{\mathbf{k}\mathbf{k}' \\ ss'}} \frac{2\pi e^2}{K^2} c^+_{\mathbf{k}+\mathbf{K},s} c_{\mathbf{k}s} c^+_{\mathbf{k}'-\mathbf{K},s'} c_{\mathbf{k}'s'} - n \right). \tag{120}$$

This is exact. The diagonal elements of the coulomb energy come from the terms for which $\mathbf{k} + \mathbf{K} = \mathbf{k}'$; $s = s'$; thus the factor involving the four operators has the form

$$c^+_{\mathbf{k}'} c_{\mathbf{k}} c^+_{\mathbf{k}} c_{\mathbf{k}'} = c^+_{\mathbf{k}'} c_{\mathbf{k}'} (1 - c^+_{\mathbf{k}} c_{\mathbf{k}}), \tag{121}$$

and

$$E(\text{diag}) = \sum_{\mathbf{k}s} n_{\mathbf{k}s} \varepsilon_{\mathbf{k}s} - \sum_{\mathbf{k}\mathbf{k}'s}{}' \frac{2\pi e^2}{|\mathbf{k} - \mathbf{k}'|^2} n_{\mathbf{k}s} n_{\mathbf{k}'s}, \tag{122}$$

where we have used the fact that $\Sigma\, n_{\mathbf{k}s} = n$. We have evaluated (122) previously; this is just the HF energy (97).

It is important to realize we have not shown that the plane wave solutions of the Hartree and the Hartree-Fock equations are the lowest energy solutions. Overhauser[1] has shown that both equations have solutions which give lower energies than the customary plane wave solutions. The new solutions have the form of spin density waves—there is no spatial variation of charge density, but there is a spatial variation of spin density. The spin density wave states have a lower energy because of the increase in magnitude of the exchange energy resulting from the local augmented parallelism of the spins. It is not clear what happens when further correlation effects are included in the calculation. To date all calculated improvements on the HF energy have been based on the normal HF state as the unperturbed state. The experimental situation suggests that spin density waves may occur in chromium at low temperature, but the ground state of most metals does not seem to be a spin density wave state.

PROBLEMS

1. (*a*) Verify that the anticommutation relations (6) are satisfied by (26) and (27). It will be helpful to note that

$$T^2 = 1; \qquad \{c^+,T\} = 0; \qquad \{c,T\} = 0. \tag{123}$$

(*b*) Verify that (26), (27) are equivalent to (23), (24).

[1] A. W. Overhauser, *Phys. Rev. Letters* **4,** 462 (1960) and *Phys. Rev.* **128,** 1437 (1962); W. Kohn and S. J. Nettel, *Phys. Rev. Letters* **5,** 8 (1960).

2. Find the net electron deficiency in the fermi hole; that is, evaluate

$$\int d^3x \, F^2(k_F r),$$

with $F(k_F r)$ defined by (102).

3. Find for bosons the analog to the Hartree-Fock equation (50).

4. Show that $\Psi^+(\mathbf{x}')|\text{vac}\rangle$ is a state in which there is an electron localized at $\mathbf{x}'$. *Hint:* Let the density operator $\rho(\mathbf{x}) \equiv \int d^3x'' \, \Psi^+(\mathbf{x}'')\delta(\mathbf{x} - \mathbf{x}'')\Psi(\mathbf{x}'')$ operate on the state. Show that

$$\rho(\mathbf{x})\Psi^+(\mathbf{x}')|\text{vac}\rangle = \int d^3x'' \, \delta(\mathbf{x} - \mathbf{x}'')\delta(\mathbf{x}'' - \mathbf{x}')\Psi^+(\mathbf{x}'')|\text{vac}\rangle \tag{124}$$

$$= \delta(\mathbf{x} - \mathbf{x}')\Psi^+(\mathbf{x}')|\text{vac}\rangle;$$

thus $\delta(\mathbf{x} - \mathbf{x}')$ is an eigenvalue of the density matrix and $\Psi^+(x')|\text{vac}\rangle$ is an eigenvector.

5. On the model leading to (78), calculate the ground-state energy; compare this value with the energy of metallic sodium referred to separated electrons and ion cores. The observed cohesive energy of metallic sodium is about 26 kcal/mole, referred to separated neutral sodium atoms.

6. Let N denote the operator

$$\int d^3x' \, \Psi^+(\mathbf{x}')\Psi(\mathbf{x}')$$

for the total number of particles. Show that for boson or fermion fields

$$\Psi(\mathbf{x})N = (N + 1)\Psi(\mathbf{x}). \tag{125}$$

7. In (37) let $v(\mathbf{x}') = 0$ and $V(\mathbf{x}' - \mathbf{y}) = g\delta(\mathbf{x}' - \mathbf{y})$, where g is a constant. Show that the exact equations of motion are:

$$i\dot{\Psi}_\alpha(\mathbf{x}) = \frac{1}{2m} p^2\Psi_\alpha(\mathbf{x}) + g\Psi_\beta^+(\mathbf{x})\Psi_\beta(\mathbf{x})\Psi_\alpha(\mathbf{x}); \tag{126}$$

$$i\dot{\Psi}_\alpha^+(\mathbf{x}) = -\frac{1}{2m} \Psi_\alpha^+(\mathbf{x})p^2 - g\Psi_\alpha^+(\mathbf{x})\Psi_\beta^+(\mathbf{x})\Psi_\beta(\mathbf{x}). \tag{127}$$

Here α and β are spin indices; when two β's occur in the same term, a summation is implied. We suppose that the field Ψ may be written as $\Psi_\uparrow + \Psi_\downarrow$, or at least as Ψ_1 plus its time-reversed partner (Chapter 9). In a magnetic field described by the vector potential $\mathbf{A}$, we have

$$p^2\Psi(\mathbf{x}) \rightarrow \left(\mathbf{p} - \frac{e}{c}\mathbf{A}\right)^2 \Psi \rightarrow \left(-i \text{ grad} - \frac{e}{c}\mathbf{A}\right)^2 \Psi; \tag{128}$$

$$\Psi^+(\mathbf{x})p^2 \rightarrow \Psi^+(\mathbf{x}) \left(\mathbf{p} - \frac{e}{c}\mathbf{A}\right)^2 \rightarrow \left(i \text{ grad} - \frac{e}{c}\mathbf{A}\right)\Psi^+. \tag{129}$$

The results of this problem are used in the theory of superconductivity in Chapter 21.

6 Many-body techniques and the electron gas

The major defect of the Hartree-Fock approximation to the total energy of an electron gas lies in the failure to correlate the motion of electrons of antiparallel spin. It is clear physically that the coulomb repulsion will tend to separate electrons of antiparallel spin. The neglect of correlation is not as serious for electrons of parallel spin: we have seen with the fermi hole that the exclusion principle automatically introduces a strong correlation for these.

We defined in Chapter 5 the *correlation energy* as the difference between the exact energy and the Hartree-Fock energy. This chapter is concerned with methods for the approximate calculation of the correlation energy of the degenerate electron gas, particularly at high density ($r_s < 1$). At sufficiently low density the problem is rather different: the electron gas is believed to condense into a crystalline phase of bcc structure. We refer to the low density limit as the Wigner limit; for $r_s \gtrsim 5$ it is believed[1] that the electron crystal is the stable state, whereas for smaller r_s the electron gas is stable.

There have been developed in recent years many powerful methods for calculating the properties of an electron gas. Most of the methods lead to equivalent results. The simplest of the new methods is the self-consistent field approach of Ehrenreich and Cohen, and Goldstone and Gottfried. After developing the SCF method we show the great convenience of the frequency and wavevector dependent dielectric constant $\epsilon(\omega,\mathrm{q})$ in the calculation of the properties of a many-body system. Finally, we discuss Goldstone diagrams and the linked-cluster theorem. A good general reference for this chapter is *The Many-Body Problem*, by D. Pines; this includes many relevant reprints.

The direct approach to the calculation of the correlation energy is

[1] W. J. Carr, *Phys. Rev.* **122,** 1437 (1961).

to treat the coulomb interaction as a perturbation on pairs of electrons of antiparallel spin and thus to calculate the energy correction by standard perturbation theory to second and higher order. The second-order coulomb energy $\varepsilon^{(2)}$ of two free electrons in the volume Ω in states $\mathbf{k}_1\uparrow$, $\mathbf{k}_2\uparrow$ is

$$\varepsilon_{12}^{(2)} = -\sum_{34} \frac{2m\langle 12|V|34\rangle\langle 34|V|12\rangle}{k_3^2 + k_4^2 - k_1^2 - k_2^2}, \tag{1}$$

where

$$\langle 12|V|34\rangle = \Omega^{-2} \int d^3x\, d^3y\, e^{-i(\mathbf{k}_1\cdot\mathbf{x}+\mathbf{k}_2\cdot\mathbf{y})} \left(\sum_{\mathbf{K}} \frac{4\pi e^2}{\Omega K^2} e^{i\mathbf{K}\cdot(\mathbf{x}-\mathbf{y})}\right) \cdot e^{i(\mathbf{k}_3\cdot\mathbf{x}+\mathbf{k}_4\cdot\mathbf{y})} = \frac{4\pi e^2}{\Omega q^2} \cdot \Delta(\mathbf{k}_1 + \mathbf{k}_2 - \mathbf{k}_3 - \mathbf{k}_4). \tag{2}$$

Here $\mathbf{q} = \mathbf{k}_1 - \mathbf{k}_3 = \mathbf{k}_4 - \mathbf{k}_2$ is the momentum transfer in the interaction of $\mathbf{k}_1$, $\mathbf{k}_2$ by virtual scattering to $\mathbf{k}_3$, $\mathbf{k}_4$. Thus

$$\varepsilon_{12}^{(2)} = -m\left(\frac{4\pi e^2}{\Omega}\right)^2 \sum_{\mathbf{q}} \frac{1}{q^4} \cdot \frac{1}{\mathbf{q}\cdot(\mathbf{q} + \mathbf{k}_2 - \mathbf{k}_1)}. \tag{3}$$

The summation can be converted into an integral:

$$\sum \rightarrow \frac{\Omega}{(2\pi)^3} \int_0^\infty dq\, \frac{\pi}{q^4} \int_{-1}^{1} d\mu\, \frac{1}{1 + \mu\kappa}, \tag{4}$$

where $\kappa = |\mathbf{k}_2 - \mathbf{k}_1|/q$, and $\mu = \cos\theta$; here θ is measured from the direction of $\mathbf{k}_2 - \mathbf{k}_1$. The integral over $d\theta$ in (4) has the value

$$\frac{1}{\kappa} \log \frac{1+\kappa}{1-\kappa}.$$

The integral over dq involves $1/q^3$ and is seen to diverge at the lower limit $q \rightarrow 0$.

The removal of this divergence can be accomplished by diagrammatic analysis developed initially by Brueckner, and it turns out to be possible to sum in all orders the most important terms in the perturbation expansion. We develop the method of Brueckner in an appendix. There are simpler methods of handling the correlation energy problem, but the Brueckner analysis is revealing and important.

In the Brueckner method for calculating the correlation energy of an electron gas in the high density limit there are contributions from all orders of perturbation theory. This is caused by the long range of the coulomb interaction. In the actual physical situation we expect the coulomb interaction of a pair of electrons to appear as screened (except at short distances) by the other electrons of the system. We expect the unperturbed potential e^2/r to assume for the perturbed problem a form rather like $(e^2/r)e^{-r/l_s}$, where the screening length l_s will be of the order of the fermi velocity v_F divided by the plasma frequency: $l_s \propto v_F(m/ne^2)^{1/2}$. This screened potential is an infinite series in $(e^2)^{1/2}$; such a series cannot be expected from a finite-order perturbation calculation. The calculation by Gell-Mann and Brueckner is a *tour de force,* but it also suggests that perturbation theory is not the natural way of handling the problem. The clearest method is perhaps the self-consistent field approach as described by H. Ehrenreich and M. H. Cohen, *Phys. Rev.* **115,** 786 (1959).

SELF-CONSISTENT FIELD METHOD

We consider a single particle with the one-particle hamiltonian $H = H_0 + V(\mathbf{x},t)$, where $H_0 = p^2/2m$ and $V(\mathbf{x},t)$ is the self-consistent potential arising from the interaction with all other particles of the system. We let ρ denote the statistical operator represented by the *one-particle* density matrix; thus if ψ_m is a solution of the one-particle Hartree-Fock equation, with the expansion*

$$|m\rangle \equiv \psi_m(\mathbf{x},t) = \sum_{\mathbf{k}} |\mathbf{k}\rangle\langle\mathbf{k}|m\rangle \tag{5}$$

in terms of the eigenstates $|\mathbf{k}\rangle$ of H_0, then the density matrix is defined by

$$\langle\mathbf{k}'|\rho|\mathbf{k}\rangle \equiv \sum_m \langle\mathbf{k}'|m\rangle P_m \langle m|\mathbf{k}\rangle, \tag{6}$$

where P_m is the ensemble average probability that the state m is occupied. The equilibrium statistical operator ρ_0 of the unperturbed system ($V = 0$) has the property

$$\rho_0|\mathbf{k}\rangle = f_0(\varepsilon_{\mathbf{k}})|\mathbf{k}\rangle, \tag{7}$$

where $f_0(\varepsilon)$ is the statistical distribution function.

The equation of motion of $\rho = \rho_0 + \delta\rho$ is

$$i\dot{\rho} = [H,\rho], \tag{8}$$

or, if we linearize (8) by neglecting terms of order $V\,\delta\rho$,

$$i\,\delta\dot{\rho} \cong [H_0,\delta\rho] + [V,\rho_0]. \tag{9}$$

* This ρ is not identical with the particle density operator in the second quantization form introduced in Chapter 5.

Thus, on taking matrix elements between $|\mathbf{k}\rangle$ and $|\mathbf{k}+\mathbf{q}\rangle$,

$$\begin{aligned}(10)\quad i\frac{\partial}{\partial t}\langle\mathbf{k}|\delta\rho|\mathbf{k}+\mathbf{q}\rangle &= \langle\mathbf{k}|[H_0,\delta\rho]|\mathbf{k}+\mathbf{q}\rangle + \langle\mathbf{k}|[V,\rho_0]|\mathbf{k}+\mathbf{q}\rangle \\ &= (\varepsilon_{\mathbf{k}} - \varepsilon_{\mathbf{k}+\mathbf{q}})\langle\mathbf{k}|\delta\rho|\mathbf{k}+\mathbf{q}\rangle + [f_0(\varepsilon_{\mathbf{k}+\mathbf{q}}) - f_0(\varepsilon_{\mathbf{k}})]V_{\mathbf{q}}(t),\end{aligned}$$

where

$$(11)\qquad V_{\mathbf{q}}(t) = \langle\mathbf{k}|V|\mathbf{k}+\mathbf{q}\rangle = \int d^3x\, V(\mathbf{x},t)e^{i\mathbf{q}\cdot\mathbf{x}}$$

is the $\mathbf{q}$th fourier component of $V(\mathbf{x},t)$. In this section other fourier components are defined similarly.

The potential V is composed of an external potential V^0 plus a screening potential V^s related to the induced change δn in electron density. Thus V^0 might be the potential of a charged impurity and V^s the potential of the screening charges in the electron gas as induced by V_0. Now the induced change in electron density is

$$\begin{aligned}(12)\quad \delta n(\mathbf{x}) &= \sum_m |m\rangle P_m\langle m| = \sum_m\sum_{\mathbf{k}\mathbf{k}'} |\mathbf{k}'\rangle\langle\mathbf{k}'|m\rangle P_m\langle m|\mathbf{k}\rangle\langle\mathbf{k}| \\ &= \sum_{\mathbf{k}\mathbf{k}'} |\mathbf{k}'\rangle\langle\mathbf{k}|\langle\mathbf{k}'|\rho|\mathbf{k}\rangle = \sum_{\mathbf{q}} e^{-i\mathbf{q}\cdot\mathbf{x}} \sum_{\mathbf{k}'}\langle\mathbf{k}'|\delta\rho|\mathbf{k}'+\mathbf{q}\rangle = \sum_{\mathbf{q}} e^{-i\mathbf{q}\cdot\mathbf{x}}\delta n_{\mathbf{q}}.\end{aligned}$$

The screening potential is related to $\delta n(\mathbf{x})$ by the poisson equation

$$(13)\qquad \nabla^2 V^s = -4\pi e^2\,\delta n; \qquad -q^2V^s_{\mathbf{q}}(t) = -4\pi e^2\langle\mathbf{k}|\delta n|\mathbf{k}+\mathbf{q}\rangle,$$

so that

$$(14)\qquad V^s_{\mathbf{q}}(t) = \frac{4\pi e^2}{q^2}\sum_{\mathbf{k}'}\langle\mathbf{k}'|\delta\rho|\mathbf{k}'+\mathbf{q}\rangle.$$

On combining (10) and (14) we have the equation of motion in the absence of an external perturbation:

$$\begin{aligned}(15)\quad i\frac{\partial}{\partial t}\langle\mathbf{k}|\delta\rho|\mathbf{k}+\mathbf{q}\rangle &= (\varepsilon_{\mathbf{k}} - \varepsilon_{\mathbf{k}+\mathbf{q}})\langle\mathbf{k}|\delta\rho|\mathbf{k}+\mathbf{q}\rangle \\ &\quad + \frac{4\pi e^2}{q^2}[f_0(\varepsilon_{\mathbf{k}+\mathbf{q}}) - f_0(\varepsilon_{\mathbf{k}})]\sum_{\mathbf{k}'}\langle\mathbf{k}'|\delta\rho|\mathbf{k}'+\mathbf{q}\rangle.\end{aligned}$$

This equation is essentially the same as that obtained in the random phase approximation of Bohm and Pines.[2]

It is generally helpful to express the properties of the interacting electron gas in terms of the longitudinal dielectric constant $\epsilon(\omega,\mathbf{q})$. We may define the dielectric constant in various equivalent ways. The usual definition relates the polarization component $P_{\mathbf{q}}$ to the longi-

[2] For a review see D. Pines in *Solid state physics* **1** (1955).

tudinal electric field $E_{\mathbf{q}}$ by

$$E_{\mathbf{q}} + 4\pi P_{\mathbf{q}} = \epsilon(\omega,\mathbf{q})E_{\mathbf{q}} = D_{\mathbf{q}}. \tag{16}$$

We recall that $V = V^0 + V^s$ is the net potential, where V^0 is the external potential and V^s the potential of the induced charge. The relation (16) is equivalent to

$$V_{\mathbf{q}} - V_{\mathbf{q}}^s = \epsilon(\omega,\mathbf{q})V_{\mathbf{q}}, \tag{17}$$

because the longitudinal polarization $P_{\mathbf{q}}$ gives rise to the induced electric field $-E_{\mathbf{q}}^s/4\pi$, where $E_{\mathbf{q}}^s$ is derived from the potential $V_{\mathbf{q}}^s$. Thus

$$\epsilon(\omega,\mathbf{q}) = V_{\mathbf{q}}^0/V_{\mathbf{q}}, \tag{18}$$

the ratio of the applied potential to the effective potential. Further,

$$\text{div}\,\mathbf{P} = e\,\delta n; \qquad -iqP_{\mathbf{q}} = e\,\delta n_{\mathbf{q}}; \qquad eE_{\mathbf{q}} = -iqV_{\mathbf{q}}, \tag{19}$$

so that

$$\epsilon(\omega,\mathbf{q}) = 1 + 4\pi\left(\frac{P_{\mathbf{q}}}{E_{\mathbf{q}}}\right) = 1 - 4\pi\frac{e^2\delta n_{\mathbf{q}}}{q^2V_{\mathbf{q}}}, \tag{20}$$

where ω is the frequency associated with $V_{\mathbf{q}}$. The definition (16) of $\epsilon(\omega,\mathbf{q})$ is also used in (40) below, but there it is found convenient to consider an expression (42) for the dielectric constant in terms of the density of test and induced charges.

Now if $V_{\mathbf{q}}(t)$ acts as a time-dependent driving force in (10), we have

$$\langle\mathbf{k}|\delta\rho|\mathbf{k}+\mathbf{q}\rangle = \frac{f_0(\varepsilon_{\mathbf{k}+\mathbf{q}}) - f_0(\varepsilon_{\mathbf{k}})}{\varepsilon_{\mathbf{k}+\mathbf{q}} - \varepsilon_{\mathbf{k}} + \omega + is}V_{\mathbf{q}}, \tag{21}$$

so that

$$\delta n_{\mathbf{q}} = \sum_{\mathbf{k}}\langle\mathbf{k}|\delta\rho|\mathbf{k}+\mathbf{q}\rangle = \sum_{\mathbf{k}}\frac{f_0(\varepsilon_{\mathbf{k}+\mathbf{q}}) - f_0(\varepsilon_{\mathbf{k}})}{\varepsilon_{\mathbf{k}+\mathbf{q}} - \varepsilon_{\mathbf{k}} + \omega + is}V_{\mathbf{q}}. \tag{22}$$

Finally we have the longitudinal dielectric constant (we should add to this the c.c.):

$$\boxed{\epsilon_{\mathrm{SCF}}(\omega,\mathbf{q}) = 1 - \lim_{s\to+0}\frac{4\pi e^2}{q^2}\sum_{\mathbf{k}}\frac{f_0(\varepsilon_{\mathbf{k}+\mathbf{q}}) - f_0(\varepsilon_{\mathbf{k}})}{\varepsilon_{\mathbf{k}+\mathbf{q}} - \varepsilon_{\mathbf{k}} + \omega + is}.} \tag{23}$$

This is an important result, from which we may calculate many properties of the system, including the correlation energy. The particular expression (23) is approximate because we started with a one-particle, and not a many-particle, hamiltonian. The approximation takes the electron to respond as a free particle to the average potential $V(\mathbf{x},t)$ in the system. The result is equivalent to that of the random phase approximation of Nozières and Pines. The transverse dielectric constant of an electron gas is the subject of Problem 16.3.

In the limit $\omega \gg k_F q/m$ the result (23) reduces to, with $\omega_P{}^2 = 4\pi n e^2/m$,

$$(24)\qquad \epsilon(\omega,\mathbf{q}) \cong 1 - \frac{\omega_p{}^2}{\omega^2} + i\frac{e^2}{2\pi q^2}\int d^3k\;\mathbf{q}\cdot\frac{\partial f_0}{\partial \mathbf{k}}\,\delta\left(\omega + \frac{\mathbf{k}\cdot\mathbf{q}}{m}\right),$$

where we have used the relation

$$(25)\qquad \lim_{s\to+0}\frac{1}{x+is} = \mathcal{P}\frac{1}{x} - i\pi\delta(x);$$

here $\mathcal{P}$ denotes principal part. At absolute zero the absorption given by $\mathscr{I}\{\epsilon\}$ vanishes if $\omega > k_F q/m$; we say this is the plasmon region. For $\omega < k_F q/m$ the imaginary part of ϵ is $2m^2e^2\omega/q^3$, at absolute zero. The real part of ϵ in (24) was obtained by writing

$$(26)\qquad f_0(\varepsilon_{\mathbf{k}+\mathbf{q}}) - f_0(\varepsilon_{\mathbf{k}}) \cong \mathbf{q}\cdot\frac{\partial f_0}{\partial \mathbf{k}},$$

and

$$(27)\qquad \frac{1}{\omega + \mathbf{k}\cdot\mathbf{q}/m} \cong \frac{1}{\omega}\left(1 - \frac{\mathbf{k}\cdot\mathbf{q}}{m\omega}\right).$$

The real part of ϵ agrees with the result for the dielectric constant of a plasma at $q = 0$, already familiar from Chapter 3. The equations of motion (15) of the undriven system may be solved to give the approximate eigenfrequencies as functions of $\mathbf{q}$; the eigenfrequencies are just the roots of $\epsilon(\omega,\mathbf{q})$ in (23). The equations of motion are equivalent to those considered by K. Sawada, *Phys. Rev.* **106,** 372 (1957). The eigenvalues are of two types: for one type, $\omega \cong \varepsilon_{\mathbf{k}+\mathbf{q}} - \varepsilon_{\mathbf{k}}$, which is the energy needed to create an electron-hole pair by taking an electron from $\mathbf{k}$ in the fermi sea to $\mathbf{k} + \mathbf{q}$ outside the sea. The other type of eigenvalue appears for small $\mathbf{q}$ and is $\omega^2 \cong 4\pi n e^2/m$. Thus there are collective excitations as well as quasiparticle excitations, but the total number of degrees of freedom is $3n$.

The imaginary term in (24) gives rise to a damping of plasma oscillations called Landau damping; the magnitude of the damping involves the number of particles whose velocity component $k_{\mathbf{q}}/m$ in the direction $\mathbf{q}$ of the collective excitation is equal to the phase velocity ω/q of the excitation. These special particles ride in phase with the excitation and extract energy from it. This damping is important at large values of q, and the plasmons then are not good normal modes. In a degenerate fermi gas the maximum electron velocity is v_F; for q

such that $\omega_p/q > v_F$ there are no particles in the plasma that travel with the phase velocity, and consequently the imaginary part of $\epsilon(\omega,\mathbf{q})$ vanishes for $q < q_c = \omega_p/v_F$. Using (5.71) and (5.73), we have $q_c/k_F = 0.48r_s^{1/2}$. If we consider all modes having $q > q_c$ to be individual particle modes, then the ratio of the number of plasmon modes n' to the total number of degrees of freedom $3n$ is

$$\frac{n'}{3n} = \frac{1}{2 \cdot 3}(0.48)^3 r_s^{3/2} = 0.018 r_s^{3/2}, \tag{28}$$

where the 2 in the denominator is from the spin. In sodium $r_s = 3.96$, and 14 percent of the degrees of freedom are plasmon modes.

We summarize: at low q the normal modes of the system are plasmons; at high q the normal modes are essentially individual particle excitations.

Plasmons in metals have been observed as discrete peaks in plots of energy loss versus voltage for fast electrons transmitted through thin metal films. More detailed evidence of the existence of plasmons is given by the observation[3] of photon radiation by excited plasmons. The dependence of this radiation on the angle of observation and on film thickness was predicted by R. A. Ferrell, *Phys. Rev.* **111,** 1214 (1958).

A comparison of the observed energy-loss peaks with the calculated plasma frequencies for the assumed valences are given in the accompanying table. The plasma frequencies given are corrected for the dielectric constant of the ion cores.

	Be	B	C	Mg	Al	Si	Ge
Valence	2	3	4	2	3	4	4
ω_{calc}	19	24	25	11	16	17	16 ev
ΔE_{obs}	19	19	22	10	15	17	17 ev

The comparison for the alkali metals is also striking because the inelastic-loss peaks are in close agreement with the threshold of optical transparency, as is expected.

	Li	Na	K
ω_{calc}	8.0	5.7	3.9 ev
ΔE	9.5	5.4	3.8 ev
ω_{opt}	8.0	5.9	3.9 ev

Thomas-Fermi Dielectric Constant. The Thomas-Fermi approximation to the dielectric constant of the electron gas is a quasistatic

[3] W. Steinman, *Phys. Rev. Letters* **5,** 470 (1960), *Z. Phys.* **163,** 92 (1961); R. W. Brown, P. Wessel, and E. P. Trounson, *Phys. Rev. Letters* **5,** 472 (1960).

($\omega \to 0$) approximation applicable at long wavelengths ($q/k_F \ll 1$). The basic assumption (Schiff, p. 282) is that the local electron density $n(\mathbf{x})$ satisfies

$$n(\mathbf{x}) \propto [\varepsilon_F - V(\mathbf{x})]^{3/2}, \tag{29}$$

where $V(\mathbf{x})$ is the potential energy and ε_F is the fermi energy. Thus for a weak potential

$$\delta n \cong -\frac{3n}{2\varepsilon_F} V, \tag{30}$$

or, for the fourier components,

$$\delta n_{\mathbf{q}} \cong -\frac{3n}{2\varepsilon_F} V_{\mathbf{q}} \tag{31}$$

But the poisson equation requires that a variation $\delta n_{\mathbf{q}}$ give rise to a potential $V^s_{\mathbf{q}}$:

$$\delta n_{\mathbf{q}} = \frac{q^2}{4\pi e^2} V^s_{\mathbf{q}}, \tag{32}$$

as in (13). From the definition (18) of the dielectric constant, the Thomas-Fermi dielectric constant is

$$\epsilon_{\mathrm{TF}}(\mathbf{q}) = \frac{V_{\mathbf{q}} - V^s_{\mathbf{q}}}{V_{\mathbf{q}}} = 1 - \frac{(4\pi e^2\, \delta n_{\mathbf{q}}/q^2)}{(-2\varepsilon_F\, \delta n_{\mathbf{q}}/3n)}, \tag{33}$$

or

$$\boxed{\epsilon_{\mathrm{TF}}(\mathbf{q}) = \frac{q^2 + k_s{}^2}{q^2},} \tag{34}$$

where

$$k_s{}^2 \equiv 6\pi n e^2/\varepsilon_F. \tag{35}$$

We note that $V_{\mathbf{q}} - V^s_{\mathbf{q}}$ is the external potential, and $(V_{\mathbf{q}} - V^s_{\mathbf{q}}) + V^s_{\mathbf{q}} = V_{\mathbf{q}}$ is the effective potential.

We expect the Thomas-Fermi dielectric constant (34) to be a special case of the self-consistent field dielectric constant (23) for $\omega = 0$ and for $q/k_F \ll 1$. In this limit

$$\begin{aligned} \epsilon_{\mathrm{SCF}}(0,\mathbf{q}) &\cong 1 - \left(\frac{4\pi e^2}{q^2}\right) \frac{2m}{(2\pi)^3} \int d^3k\, \frac{\mathbf{q} \cdot \partial f_0/\partial \mathbf{k}}{\mathbf{q} \cdot \mathbf{k}} \\ &= 1 + \left(\frac{4\pi e^2}{q^2}\right) \frac{8\pi n k_F}{(2\pi)^3}. \end{aligned} \tag{36}$$

This is identical with (34), because $m = k_F{}^2/2\varepsilon_F$ and $k_F{}^3 = 3n\pi^2$.

DIELECTRIC RESPONSE ANALYSIS[4]

The self-consistent field calculation of the dielectric constant is based on an independent particle model and is only approximate. We calculate now a more general expression for the dielectric constant in terms of matrix elements between *exact eigenstates* of the many-body system. Let us imagine that we introduce into the system a test charge distribution of wavevector $\mathbf{q}$ and frequency ω. We write the test charge density as

$$er_{\mathbf{q}}[e^{-i(\omega t+\mathbf{q}\cdot\mathbf{x})} + cc],$$

with $r_{\mathbf{q}}$ real. In the absence of the test charge the expectation value $\langle\rho_{\mathbf{q}}\rangle$ of all particle density fluctuation operators $\rho_{\mathbf{q}}$ will be zero. We recall from (5.118) that

$$\rho_{\mathbf{q}} = \sum_{\mathbf{k}} c^+_{\mathbf{k}-\mathbf{q}}c_{\mathbf{k}}. \tag{37}$$

In the presence of the test charge $\langle\rho_{\pm\mathbf{q}}\rangle \neq 0$; we wish to solve for this induced particle density component.

Now, by the definition of $\mathbf{D}$ and $\mathbf{E}$,

$$\operatorname{div}\mathbf{D} = 4\pi(\text{test charge density}); \tag{38}$$

$$\operatorname{div}\mathbf{E} = 4\pi \quad (\text{test charge density} + \text{induced charge density}), \tag{39}$$

so that, with $\epsilon(\omega,\mathbf{q})$ as the appropriate dielectric constant,

$$-i\mathbf{q}\cdot\mathbf{D}_{\mathbf{q}} = -i\epsilon(\omega,\mathbf{q})\mathbf{q}\cdot\mathbf{E}_{\mathbf{q}} = 4\pi er_{\mathbf{q}}e^{-i\omega t}; \tag{40}$$

$$-i\mathbf{q}\cdot\mathbf{E}_{\mathbf{q}} = 4\pi e(r_{\mathbf{q}}e^{-i\omega t} + \langle\rho_{\mathbf{q}}\rangle). \tag{41}$$

On dividing (40) by (41),

$$\frac{1}{\epsilon(\omega,\mathbf{q})} = 1 + \frac{\langle\rho_{\mathbf{q}}\rangle}{r_{\mathbf{q}}e^{-i\omega t}} = \frac{\text{total charge}}{\text{test charge}}. \tag{42}$$

We now calculate $\langle\rho_{\mathbf{q}}\rangle$, the response of the system to the test charge. The hamiltonian is $H = H_0 + H'$, where for a specimen of unit volume we have, from (5.113),

$$H_0 = \sum_i \frac{1}{2m}p_i^2 + \sum_{\mathbf{q}} \frac{2\pi e^2}{q^2}(\rho^+_{\mathbf{q}}\rho_{\mathbf{q}} - n); \tag{43}$$

and H' is the coulomb interaction between the system and the test charge:

$$H' = \frac{4\pi e^2}{q^2}\rho_{-\mathbf{q}}r_{\mathbf{q}}e^{-i\omega t+st} + cc, \tag{44}$$

[4] P. Nozières and D. Pines, *Nuovo cimento* **9**, 470 (1958).

where for adiabatic switching s is small and positive. We assume the test charge is sufficiently small that the response of the system is linear. We suppose the system is initially ($t = -\infty$) in the ground state Φ_0; in the presence of the test charge $\Phi_0 \to \Phi_0(r_{\mathbf{q}})$, where by first-order time-dependent perturbation theory in the Schrödinger picture with $\epsilon_n - \epsilon_0 = \omega_{n0}$,

$$(45)\quad \Phi_0(r_{\mathbf{q}}) = \Phi_0 - \sum_n{}' \frac{4\pi e^2}{q^2} r_{\mathbf{q}} \left(\frac{\langle n|\rho_{-\mathbf{q}}|0\rangle e^{-i\omega t+st}}{-\omega + \omega_{n0} - is} + \frac{\langle n|\rho_{\mathbf{q}}|0\rangle e^{i\omega t+st}}{\omega + \omega_{n0} - is} \right) \Phi_n.$$

Then to terms of first order in $r_{\mathbf{q}}e^{-i\omega t+st}$,

$$(46)\quad \langle \Phi_0(r_{\mathbf{q}})|\rho_{\mathbf{q}}|\Phi_0(r_{\mathbf{q}})\rangle = -\frac{4\pi e^2}{q^2} r_{\mathbf{q}} e^{-i\omega t+st} \sum_n |\langle n|\rho_{\mathbf{q}}|0\rangle^2 \left(\frac{1}{\omega + \omega_{n0} + is} + \frac{1}{-\omega + \omega_{n0} - is} \right),$$

where we have used the symmetry property $|\langle n|\rho_{\mathbf{q}}|0\rangle|^2 = |\langle n|\rho_{-\mathbf{q}}|0\rangle|^2$. From (42) we have the exact result

$$(47)\quad \boxed{\frac{1}{\epsilon(\omega,\mathbf{q})} = 1 - \frac{4\pi e^2}{q^2} \sum_n |\langle n|\rho_{\mathbf{q}}|0\rangle|^2 \left\{ \frac{1}{\omega + \omega_{n0} + is} + \frac{1}{-\omega + \omega_{n0} - is} \right\}.}$$

The eigenfrequencies of the system are given by the roots of $\epsilon(\omega,\mathbf{q}) = 0$, as for the roots the mode response is singular according to (42).

Using

$$(48)\quad \lim_{s\to+0} \frac{1}{x \pm is} = \mathcal{P}\frac{1}{x} \mp i\pi\delta(x),$$

we have for the imaginary part

$$(49)\quad \mathcal{I}\left(\frac{1}{\epsilon(\omega,\mathbf{q})}\right) = \frac{4\pi^2 e^2}{q^2} \sum_n |\langle n|\rho_{\mathbf{q}}|0\rangle|^2 [\delta(\omega + \omega_{n0}) - \delta(\omega - \omega_{n0})].$$

On integrating over all positive frequencies ω,

$$(50)\quad \int_0^\infty d\omega\, \mathcal{I}\left(\frac{1}{\epsilon(\omega,\mathbf{q})}\right) = -\frac{4\pi^2 e^2}{q^2} \sum_n |\langle n|\rho_{\mathbf{q}}|0\rangle|^2 = -\frac{4\pi^2 e^2}{q^2} \langle 0|\rho_{\mathbf{q}}^+ \rho_{\mathbf{q}}|0\rangle.$$

Recall that the states used as bases in this development are the true eigenstates of the problem including internal interactions.

The expectation value of the coulomb interaction energy in the ground state is, according to (5.113),

$$E_{\text{int}} = \langle 0| \sum_{\mathbf{q}}{}' \frac{2\pi e^2}{q^2} (\rho_{\mathbf{q}}^+ \rho_{\mathbf{q}} - n)|0\rangle, \tag{51}$$

which may now be written as

$$E_{\text{int}} = -\sum_{\mathbf{q}} \left\{ \frac{1}{2\pi} \int_0^\infty d\omega\, \mathcal{I} \left(\frac{1}{\epsilon(\omega,\mathbf{q})} \right) + \frac{2\pi n e^2}{q^2} \right\}. \tag{52}$$

This expression gives formally the coulomb energy of the exact ground state in terms of the imaginary part of the dielectric response. We do not, of course, obtain the exact ground-state energy simply by adding E_{int} to the unperturbed ground-state energy, because the kinetic energy of the ground state is modified by the coulomb interaction. That is, Φ_0 is a function of e^2. To obtain the total energy of the exact ground state, we utilize the theorem which follows.

THEOREM. Given the hamiltonian

$$H = H_0 + gH_{\text{int}}; \qquad g = \text{coupling constant}; \qquad H_0 = \text{kinetic energy}; \tag{53}$$

and the value of

$$E_{\text{int}}(g) = \langle \Phi_0(g)|gH_{\text{int}}|\Phi_0(g)\rangle; \tag{54}$$

then the exact value of the total ground-state energy

$$E_0(g) = \langle \Phi_0(g)|H_0 + gH_{\text{int}}|\Phi_0(g)\rangle \tag{55}$$

is given by

$$E_0(g) = E_0(0) + \int_0^g g^{-1} E_{\text{int}}(g)\, dg. \tag{56}$$

Proof: From (54) and (55),

$$\frac{dE_0}{dg} = g^{-1}E_{\text{int}}(g) + E_0(g) \frac{d}{dg} \langle \Phi_0(g)|\Phi_0(g)\rangle, \tag{57}$$

where the second term on the right-hand side is zero because the normalization is independent of the value of g. Here $E_0(g)$ is the exact eigenvalue and $\Phi_0(g)$ the exact eigenfunction. Thus we have a special case of Feynman's theorem:

$$\frac{dE_0}{dg} = g^{-1}E_{\text{int}}(g), \tag{58}$$

and, on integrating, we have (56).

In the electron-gas problem the ground-state energy without the coulomb interaction is, per unit volume,

$$E_0(0) = \tfrac{3}{5}n\varepsilon_F. \tag{59}$$

The coupling constant is $g = e^2$, so that given E_{int} from (52) or in some other way, we can find the total energy in the presence of the coulomb interaction from (56). If, for example, we find an approximate result for $1/\epsilon(\omega,\mathbf{q})$ by calculating (47) with matrix elements taken in a plane wave representation, then we get just the usual Hartree-Fock exchange for the interaction energy E_{int}. As Φ_0 in this approximation does not involve e^2, the ground-state energy is just the sum of the fermi energy plus E_{int}. We get a much better value of the energy using the self-consistent dielectric constant (23).

We note that with the integral representation of the delta function we may rewrite (49) as

$$\mathcal{g}\left(\frac{1}{\epsilon(\omega,\mathbf{q})}\right) = \frac{4\pi e^2}{q^2}\sum_{ij}\frac{1}{2\pi}\int dt\,(e^{i\omega t} - e^{-i\omega t})\langle e^{-i\mathbf{q}\cdot\mathbf{x}_i(0)}e^{i\mathbf{q}\cdot\mathbf{x}_j(t)}\rangle, \tag{60}$$

where the $\mathbf{x}_i$ are in the Heisenberg picture. Now Van Hove [*Phys. Rev.* **95,** 249 (1954)] discusses a function called the dynamic structure factor which may be defined as

$$\boxed{\mathcal{S}(\omega,\mathbf{q}) = \sum_{ij}\frac{1}{2\pi}\int dt\, e^{-i\omega t}\langle e^{-i\mathbf{q}\cdot\mathbf{x}_i(0)}e^{i\mathbf{q}\cdot\mathbf{x}_j(t)}\rangle;} \tag{61}$$

$\mathcal{S}$ has the property that it is the fourier transform of the pair distribution function

$$\begin{aligned} G(\mathbf{x},t) &= N^{-1}\left\langle\sum_{ij}\int d^3x'\,\delta[\mathbf{x} - \mathbf{x}_i(0) + \mathbf{x}']\,\delta[\mathbf{x}' - \mathbf{x}_j(t)]\right. \\ &= N^{-1}\left\langle\sum_{ij}\delta[\mathbf{x} - \mathbf{x}_i(0) + \mathbf{x}_j(t)]\right\rangle. \end{aligned} \tag{62}$$

That is,

$$\mathcal{S}(\omega,\mathbf{q}) = \frac{N}{2\pi}\int d^3x\,dt\,e^{i(\mathbf{q}\cdot\mathbf{x}-\omega t)}G(\mathbf{x},t). \tag{63}$$

We shall see later in connection with neutron diffraction that $\mathcal{S}(\omega,\mathbf{q})$ describes the scattering properties of the system in the first Born approximation; see also Problems (2.6), (6.9), and (6.10).

We may also write, using (5.110),

$$(64)\quad \mathcal{S}(\omega,\mathbf{q}) = \frac{1}{2\pi}\int dt\, e^{-i\omega t}\langle\rho_{\mathbf{q}}(t)\rho_{\mathbf{q}}^{+}(0)\rangle = \sum_n |\langle n|\rho_{\mathbf{q}}^{+}|0\rangle|^2\,\delta(\omega-\omega_{n0}).$$

Thus $\mathcal{S}(\omega,\mathbf{q})$ is indeed a structure factor which describes the elementary excitation spectrum of the density fluctuations of the system. It may be used equally for boson or fermion systems, with appropriate operators in the expansion of the $\rho_{\mathbf{q}}$.

From (60) and (61), we have

$$(65)\qquad \mathscr{I}\left(\frac{1}{\epsilon(\omega,\mathbf{q})}\right) = \frac{4\pi e^2}{q^2}[\mathcal{S}(-\omega,\mathbf{q}) - \mathcal{S}(\omega,\mathbf{q})].$$

This establishes the connection between the dielectric constant and the correlation function $\mathcal{S}$.

DIELECTRIC SCREENING OF A POINT CHARGED IMPURITY

An interesting application of the dielectric formulation of the many-body problem is to the problem of screening of a point charged impurity in an electron gas. The screening of the coulomb potential by the electron gas is an important effect. It is because of screening of the electron-electron interaction that the free electron or quasi-particle model works as well as it does for transport processes. The screening of charged impurities has many consequences for the theory of alloys.

Let the charge of the impurity be Z; the charge density may be written

$$(66)\qquad \rho(\mathbf{x}) = Z\,\delta(\mathbf{x}) = \frac{Z}{(2\pi)^3}\int d^3q\, e^{i\mathbf{q}\cdot\mathbf{x}}.$$

The potential of the bare charge

$$(67)\qquad V_0(\mathbf{x}) = \frac{Z}{r} = \frac{1}{(2\pi)^3}\int d^3q\,\frac{4\pi}{q^2}\,e^{i\mathbf{q}\cdot\mathbf{x}}$$

becomes, for the linear response region,

$$(68)\qquad V(\mathbf{x},\omega) = \frac{Z}{(2\pi)^3}\int d^3q\,\frac{4\pi}{q^2\epsilon(\omega,\mathbf{q})}\,e^{i\mathbf{q}\cdot\mathbf{x}}$$

in the medium. We are usually concerned with the potential $V(\mathbf{x})$ at zero frequency, because the impurity is stationary. We have also an interest in the charge $\Delta\rho(\mathbf{x})$ induced by $Z\,\delta(\mathbf{x})$. Using (68) with $\omega = 0$ and

$$\nabla^2 V = -4\pi[\Delta\rho + Z\,\delta(\mathbf{x})], \tag{69}$$

we have

$$\Delta\rho(\mathbf{x}) = \frac{Z}{(2\pi)^3}\int d^2q \left(\frac{1}{\epsilon(0,\mathbf{q})} - 1\right) e^{i\mathbf{q}\cdot\mathbf{x}}. \tag{70}$$

If $1/\epsilon(0,\mathbf{q})$ has poles, there will be oscillations in space of the screening charge density.

The total displaced charge Δ involved in screening is

$$\begin{aligned} \Delta = \int d^3x\,\Delta\rho(\mathbf{x}) &= \frac{Z}{(2\pi)^3}\int d^3q \int d^3x \left(\frac{1}{\epsilon(0,\mathbf{q})} - 1\right) e^{i\mathbf{q}\cdot\mathbf{x}} \\ &= Z\int d^3q\,\delta(\mathbf{q})\left(\frac{1}{\epsilon(0,\mathbf{q})} - 1\right) \\ &= Z\left(\frac{1}{\epsilon(0,0)} - 1\right). \end{aligned} \tag{71}$$

The value of $\epsilon(0,0)$ is sometimes not well-defined, but may depend on the order in which ω and $\mathbf{q}$ are allowed to approach zero.

We now consider the screening in several approximations.

(*a*) *Thomas-Fermi.* The dielectric constant in this approximation is given by (34) and (35):

$$\epsilon_{\mathrm{TF}}(0,\mathbf{q}) \cong 1 + \frac{k_s^{\,2}}{q^2}; \qquad k_s^{\,2} = \frac{6\pi n e^2}{\varepsilon_F} = \frac{4k_F}{\pi a_H} = 4.6/r_s; \tag{72}$$

here a_H is the bohr radius. Thus the screened potential is, from (68),

$$V(\mathbf{x}) = \frac{Z}{(2\pi)^3}\int d^3q\,\frac{4\pi}{q^2 + k_s^{\,2}}\,e^{i\mathbf{q}\cdot\mathbf{x}} = \frac{Z}{r}\,e^{-k_s r}, \tag{73}$$

corresponding to a screening length

$$l_s = 1/k_s \propto n^{-1/6}. \tag{74}$$

In copper $k_s \approx 1.8 \times 10^8$ cm^{-1}, or $l_s \approx 0.55 \times 10^{-8}$ cm. In potassium l_s is roughly double that for copper. Screening is a very important feature of the electron gas. The screening charge around an impurity is largely concentrated in the impurity sphere itself and the interaction between impurity atoms is small.

From (71) the total screening charge is $-Z$. The induced charge density is

$$\Delta\rho(\mathbf{x}) = -\frac{1}{4\pi}\nabla^2 V = -\frac{k_s{}^2 Z}{4\pi r} e^{-k_s r}. \tag{75}$$

This is singular at $r = 0$, and the magnitude decreases monotonically as r increases. The infinite electronic density on the nucleus is in disagreement with the finite lifetime of positrons in metals and with the finite Knight shift of solute atoms.

(*b*) *Hartree Approximation.* The dielectric constant is given in Problem 1. There is an infinite screening charge, because coulomb interactions between electrons are not included.

(*c*) *Self-Consistent Field or Random Phase Approximation.* The result (23) for the dielectric constant ϵ_{SCF} in the SCF or RPA approximation is seen by comparison to be simply related to the dielectric constant ε_H in the Hartree approximation given in Problem 1:

$$\epsilon_{SCF} = 2 - \frac{1}{\varepsilon_H} = 1 + \frac{k_s{}^2}{2q^2} g(q), \tag{76}$$

written for zero frequency. For a general argument connecting ϵ_{SCF} and ϵ_H, see Pines, p. 251. We see that the total screening charge is equal to $-Z$. There is a singularity in the dielectric constant at zero frequency of the type $(q - 2k_F) \log |q - 2k_F|$, so that the charge density contains oscillatory terms at large distances of the form $r^{-3} \cos 2k_F r$, as shown in Fig. 1.

What happens at $q = 2k_F$? For $q < 2k_F$ one can draw a vector $\mathbf{q}$ such that both ends lie on the fermi surface; thus the energy denominator in the expression (23) for the dielectric constant can be small and there is a corresponding large contribution to the dielectric constant. But for $q > 2k_F$ it is *not* possible to take an electron from a filled state $\mathbf{k}$ to an empty state $\mathbf{k} + \mathbf{q}$ with approximate conservation of energy. Here the energy denominator is always large and the contribution of all processes to the dielectric constant is small.

W. Kohn has made the interesting observation that the sudden drop in the dielectric constant as q increases through $2k_F$ should lead to a small sudden increase in the eigenfrequency $\omega(\mathbf{q})$ of a lattice vibration at the point $q = 2k_F$ as q is increased. The lower the dielectric constant (as a function of q), the stiffer will be the response of the electrons to the ion motion and the higher will be the lattice frequency. For a discussion of the detailed theory of fermi surface effects on phonon spectra, see E. J. Woll, Jr., and W. Kohn, *Phys. Rev.* **126,** 1693 (1962). It is not yet clear why the effect should persist in the presence of electron collisions with phonons and with impurities.

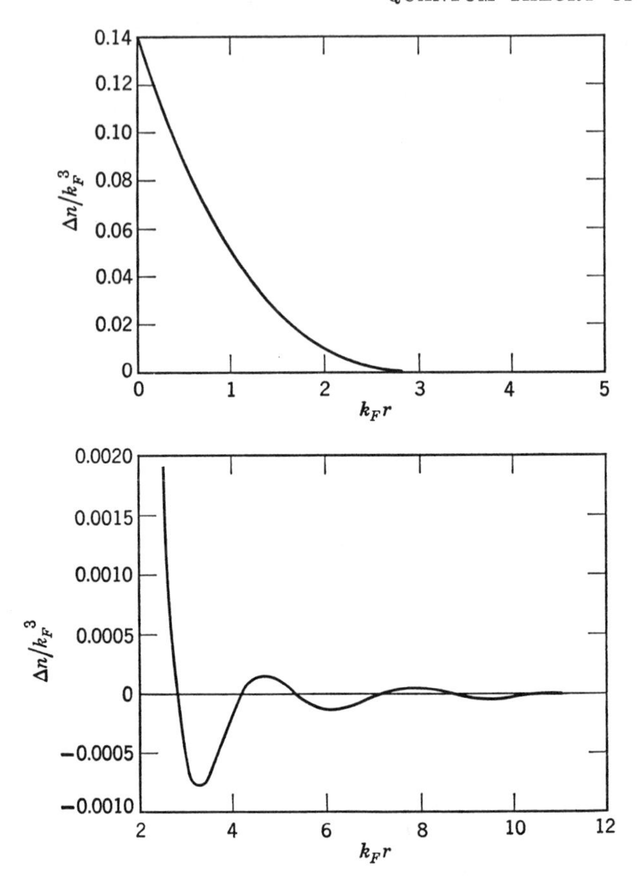

FIG. 1. Distribution of screening charge density around a point charge for an electron gas with $r_s = 3$, as calculated on many-body theory by J. S. Langer and S. H. Vosko, *J. Phys. Chem. Solids* **12,** 196 (1960).

CORRELATION ENERGY—NUMERICAL

We write the ground-state energy per electron as, using (5.97),

$$\mathcal{E}_0 = \left(\frac{2.21}{r_s^{\,2}} - \frac{0.916}{r_s} + \mathcal{E}_c\right) \text{ry}, \tag{77}$$

where the first two terms on the right-hand side give the Hartree-Fock energy, and $\mathcal{E}_c$ is the correlation energy. In the range of actual metallic densities Nozières and Pines [*Phys. Rev.* **111,** 442 (1958)] recommend the interpolation result

$$\varepsilon_c \cong (-0.115 + 0.031 \log r_s) \text{ ry}. \tag{78}$$

Their paper should be consulted for details.

It is instructive to use (77) and (78) to estimate the cohesive energy of a simple metal referred to separated neutral atoms. Sodium is a convenient example because it has an effective electronic mass close to the free electron mass: here we neglect differences between m and m^*. With $r_s = 3.96$ we have

$$\varepsilon_0 = (0.14 - 0.23 - 0.07) \text{ ry}, \tag{79}$$

where the terms are ordered as in (77). But, as we emphasized in Chapter 5, ε_0 was calculated for a uniform background of positive charge; if the positive charge is gathered together as positive ions, we should add to ε_0 the term we denoted by ε_2 in (5.77); $\varepsilon_2 = 1.2/r_s$ ry is the self-energy of a uniform distribution of electrons within the s sphere, and for sodium has the value 0.30 ry. Note that ε_2 tends to cancel the exchange and correlation contributions.

The solution for $\mathbf{k} = 0$ of the one-electron periodic potential problem for sodium gives $\varepsilon(\mathbf{k} = 0) \cong -0.60$ ry, according to calculations described in the references cited in Chapter 13. This energy is to be compared with $\varepsilon_I = -0.38$ ry, the ionization energy of a neutral sodium atom. Thus the cohesive energy of metallic sodium referred to neutral atoms is

$$\begin{aligned} \varepsilon_{\text{coh}} = -\varepsilon_I + \varepsilon(\mathbf{k} = 0) + \varepsilon_0 + \varepsilon_2 &= 0.38 - 0.60 - 0.16 \\ &+ 0.30 = -0.08 \text{ ry} = -1.1 \text{ ev}, \end{aligned} \tag{80}$$

which is rather too close to the observed value -1.13 ev.

ELECTRON-ELECTRON LIFETIME

Because of the off-diagonal parts of the electron-electron interaction, an electron put into a quasiparticle state $\mathbf{k}$ will eventually be scattered out of this initial state. The mean free path of an electron near the fermi surface is actually quite long. The cross section for scattering of an electron at the fermi level of an electron gas at temperature T is of the order of

$$\sigma \cong \sigma_0 \cdot \left(\frac{k_B T}{\varepsilon_F}\right)^2, \tag{81}$$

where σ_0 is the scattering cross section of the screened coulomb potential and $(k_B T/\varepsilon_F)^2$ is a statistical factor which expresses the requirement that the target electron must have an energy within $k_B T$ of the fermi surface if the final state of this electron is to be reasonably empty; equally, the final state of the incident electron must be available for occupancy. The fraction of states thus available is $(k_B T/\varepsilon_F)^2$.

The scattering cross section for the screened potential

$$V(r) = \frac{e^2}{r} e^{-r/l}, \tag{82}$$

has been calculated by E. Abrahams [*Phys. Rev.* **95,** 839 (1954)] by calculating phase shifts—the Born approximation is not accurate. For sodium with l_s as estimated by Pines, the result is $\sigma_0 \cong 17\pi a_H{}^2$, where a_H is the bohr radius. Numerically, in Na at 4°K the mean free path for electron-electron scattering of an electron at the fermi surface is 2.5 cm; at 300°K it is 4.5×10^{-4} cm. We see that an electron is not strongly scattered by the other electrons in a metal—this remarkable effect makes it possible to use the quasi-particle approximation to the low-lying excited states of an electron gas.

Another pertinent result is that of J. J. Quinn and R. A. Ferrell [*Phys. Rev.* **112,** 812 (1958)]. They calculate for an electron gas at absolute zero the mean free path of an electron added in a state $\mathbf{k}$ outside the fermi surface ($k > k_F$). For high electron density the mean free path Λ is

$$\Lambda k_F = \left(\frac{k - k_F}{k_F}\right)^2 \frac{3.98}{r_s{}^{1/2}}, \tag{83}$$

in the limit of high electron density. The factor $(k - k_F)^2/k_F{}^2$ is a phase space factor analogous to the factor $(k_B T/\varepsilon_F)^2$ in the thermal problem. As $k \to k_F$ the mean free path increases indefinitely—this is one reason why we may speak of a sharp fermi surface in a metal; the states at k_F are indeed well-defined.

GRAPHICAL ANALYSIS OF DIELECTRIC RESPONSE

It is instructive to carry out a graphical analysis of the most important terms in a perturbation series which contribute to the dielectric constant $\varepsilon(\omega,\mathbf{q})$ of the fermi sea at absolute zero. The diagrams we use are called Goldstone diagrams and are related to Feynman diagrams. We take the unperturbed system H_0 to be the free electron gas.

We consider the scattering caused by an external potential $v(\omega,\mathbf{q})$ through the perturbation hamiltonian

$$\begin{aligned} H'(\omega,\mathbf{q}) &= \int d^3x\, \Psi^+(\mathbf{x}) v(\omega,\mathbf{q}) e^{-i\omega t} e^{i\mathbf{q}\cdot\mathbf{x}} \Psi(\mathbf{x}) + cc \\ &= v(\omega,\mathbf{q}) e^{-i\omega t} \sum_{\mathbf{k}'\mathbf{k}} c_{\mathbf{k}'}^+ c_{\mathbf{k}} \int d^3x\, e^{i(\mathbf{k}-\mathbf{k}'+\mathbf{q})\cdot\mathbf{x}} + cc \\ &= v(\omega,\mathbf{q}) e^{-i\omega t} \sum_{\mathbf{k}} c_{\mathbf{k}+\mathbf{q}}^+ c_{\mathbf{k}} + cc. \end{aligned} \tag{84}$$

In what follows we concern ourselves for the sake of brevity with a discussion of the terms shown explicitly and not with the hermitian conjugate terms. It is as if $v(\omega,\mathbf{q})$ represented the interaction with an ultrasonic phonon, but only that part of the interaction which causes absorption of a phonon of wavevector $\mathbf{q}$ and energy ω. The creation and annihilation operators c^+, c may be written as in Chapter 5 in terms of electron and hole operators; thus combinations such as $\alpha_{\mathbf{k}+\mathbf{q}}^+\alpha_{\mathbf{k}}$; $\beta_{-\mathbf{k}-\mathbf{q}}\beta_{-\mathbf{k}}^+$; $\alpha_{\mathbf{k}+\mathbf{q}}^+\beta_{-\mathbf{k}}^+$; and $\beta_{-\mathbf{k}-\mathbf{q}}\alpha_{\mathbf{k}}$ may occur. The operation of the first pair will be spoken of as electron scattering; the second pair as hole scattering; the third as the creation of a hole-electron pair; and the fourth as the annihilation of a hole-electron pair.

The other perturbation is the coulomb interaction

$$V = \sum_{\mathbf{q}}' V(\mathbf{q})(\rho_{\mathbf{q}}^+\rho_{\mathbf{q}} - n) \tag{85}$$

between the electrons of the system; this interaction will tend to screen $v(\omega,\mathbf{q})$. In the following development we consider only terms linear in the external perturbation $v(\omega,\mathbf{q})$, but we are interested in all orders of the coulomb interaction. We recall that the term $\rho_{\mathbf{q}}^+\rho_{\mathbf{q}}$ involves a product of four operators, such as $c_{\mathbf{k}'+\mathbf{q}}^+c_{\mathbf{k}'}c_{\mathbf{k}-\mathbf{q}}^+c_{\mathbf{k}}$; in terms of the α and β there are 16 possible combinations.

We consider then the scattering caused by the external potential $v(\omega,\mathbf{q})$ with time dependence $e^{-i\omega t}$ and spatial dependence $e^{i\mathbf{q}\cdot\mathbf{x}}$. The perturbation terms in the hamiltonian are $v(\omega,\mathbf{q})$ and the coulomb interactions $V(\mathbf{q})$ among the electrons. We retain only terms linear in the external potential. In constructing the graphs which represent the terms of the perturbation series we use the hole and electron conventions shown in Fig. 2. We use the convention (5.55) and (5.56)

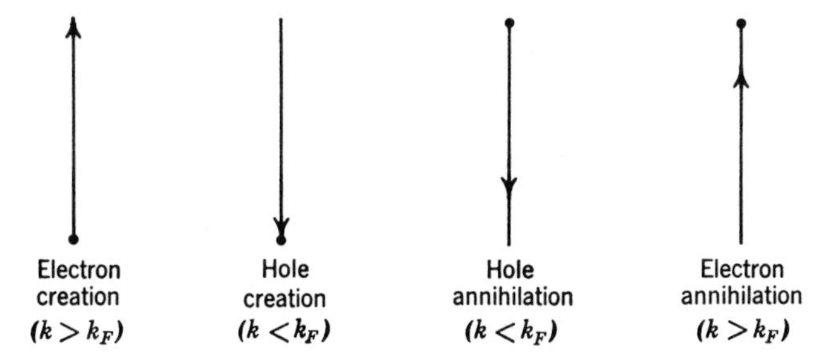

FIG. 2. Lines for Goldstone diagrams. An electron state outside the fermi sea is indicated by an arrow directed downward; a hole state (within the fermi sea) is indicated by an upward arrow. [J. Goldstone, *Proc. Roy. Soc.* (*London*) **A239**, 268 (1957); reprinted in Pines.]

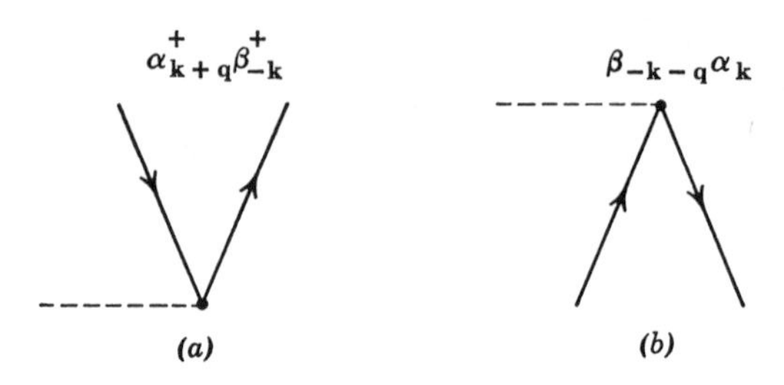

FIG. 3. Creation (*a*) and annihilation (*b*) of a hole-electron pair. The dashed line represents an interaction.

of the fermion quasiparticle discussion in Chapter 5. As in Fig. 3, the scattering of an electron with $k < k_F$ to a state outside the fermi sea is represented as the creation of a hole-electron pair. The broken line ending at the vertex represents schematically the interaction responsible for the process shown in the graph. Electron-electron and hole-hole scattering processes in lowest order are shown in Fig. 4.

In the actual electron gas problem we know that screening plays an important part in reducing the effect of $v(\omega,\mathbf{q})$. We take the screening into account by going to higher orders of a perturbation calculation in calculating the response of the electron gas to the external potential. We still stay at first order in $v(\omega,\mathbf{q})$, but consider all orders of the interelectronic coulomb potential $V(\mathbf{q})$.

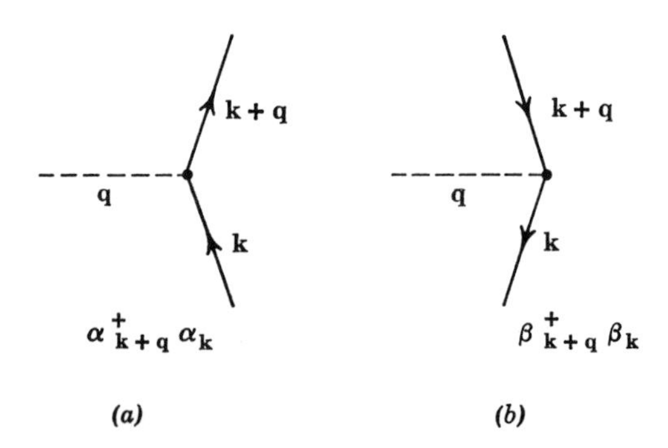

FIG. 4. Scattering in lowest order: (*a*) electron-electron; and (*b*) hole-hole, by the external potential $v(\omega,\mathbf{q})$.

A convenient question to ask is the value of the matrix element of the U operator defined by (1.55):

$$(86)\quad U(0,-\infty) = \sum_{n=0}^{\infty} (-i)^n \int_{-\infty}^{0} dt_1 \int_{-\infty}^{t_1} dt_2 \cdots \int_{-\infty}^{t_{n-1}} dt_n\, V(t_1)V(t_2) \cdots V(t_n).$$

We consider as an example the matrix element taken between a state $|i\rangle$ of the unperturbed system containing one electron in a state $\mathbf{k}_i$ outside the filled fermi sea, and a state $|f\rangle$ containing one electron in $\mathbf{k}_f$ outside the filled fermi sea. In lowest order, following (1.63), we have

$$(87)\quad \langle f|U_1(0,-\infty)|i\rangle = -\frac{\langle f|v(\omega,\mathbf{q})|i\rangle}{\varepsilon_f - \varepsilon_i - \omega - is}\Delta(\mathbf{k}_f - \mathbf{k}_i - \mathbf{q}).$$

Schematically we may write

$$(88)\quad U_{1ee}(0,-\infty) = -\sum_{\mathbf{k}} \frac{v(\omega,\mathbf{q})}{\varepsilon_f - \varepsilon_i - \omega - is}\alpha_{\mathbf{k}+\mathbf{q}}^{+}\alpha_{\mathbf{k}}$$

for the electron-electron scattering part of the process, as denoted by the subscript *ee*. This part is represented graphically by Fig. 4*a*.

We next study the second-order term $U_{2ee}(0,-\infty)$, of the form given in (1.64). The terms linear in $v(\omega,\mathbf{q})$ involve

$$(89)\quad \langle f|U_{2ee}|i\rangle = (-i)^2 \int_{-\infty}^{0} dt_1 \int_{-\infty}^{t_1} dt_2 \Big[\sum_n \langle f|v(\omega,\mathbf{q})|n\rangle e^{i(\varepsilon_f-\varepsilon_n-\omega-is)t_1}\langle n|V(\mathbf{q}')|i\rangle e^{i(\varepsilon_n-\varepsilon_i-is)t_2} + \sum_m \langle f|V(\mathbf{q}')|m\rangle e^{i(\varepsilon_f-\varepsilon_m-is)t_1}\langle m|v(\omega,\mathbf{q})|i\rangle e^{i(\varepsilon_m-\varepsilon_i-\omega-is)t_2}\Big].$$

These are the cross-product terms in $(V+v)^2$. We have not indicated in (86) the restrictions on the intermediate and final states imposed by conservation of wavevector. The restrictions are simple to discuss. We suppose, as above, that in the state $|i\rangle$ there is a single electron of wavevector $\mathbf{k}_i$ outside the filled fermi sea.

I. Consider first in the coulomb interaction the term $\mathbf{q}' = 0$: for the special case of the electron gas $V(0) = 0$ and this term does not exist, but for a more general interaction it may exist. There are then two possibilities: for $V(0)$ the state $|n\rangle$ in (89) may be identical with $|i\rangle$—this corresponds in the second quantization formalism to operators associated with $V(0)$ of the form (see 5.130):

$$\sum_{\mathbf{k}} \alpha_{\mathbf{k}_i}^{+}\alpha_{\mathbf{k}_i}\beta_{\mathbf{k}}\beta_{\mathbf{k}}^{+};$$

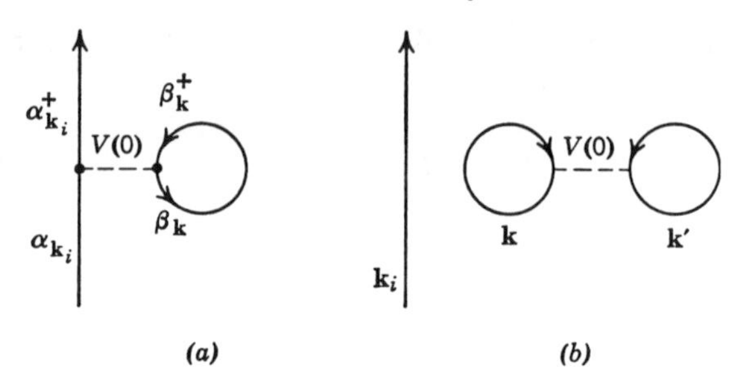

FIG. 5. Scattering processes with zero wavevector transfer; neither process affects the scattering, although (*a*) alters the energy of an added electron in $\mathbf{k}_i$ and (*b*) enters into the energy of the fermi sea.

there is no momentum exchange; a hole is created in the fermi sea at some $\mathbf{k}$, but it is annihilated in the same process. This term is represented graphically by Fig. 5*a*. Another possibility for $V(0)$ arises from operators of the form

$$\sum_{\mathbf{kk}'} \beta_{\mathbf{k}'}\beta^+_{\mathbf{k}'}\beta_{\mathbf{k}}\beta^+_{\mathbf{k}};$$

here, as shown in Fig. 5*b*, two holes $\mathbf{k}$ and $\mathbf{k}'$ are created and then annihilated.

II. For $\mathbf{q}' \neq 0$ the coulomb interaction can scatter the incident electron from $\mathbf{k}_i$ to $\mathbf{k}_f$ $(= \mathbf{k}_i + \mathbf{q}')$, accompanied by the creation of an electron-hole pair:

$$\sum_{\mathbf{k}} \alpha^+_{\mathbf{k}-\mathbf{q}'}\beta^+_{-\mathbf{k}}\alpha^+_{\mathbf{k}_i+\mathbf{q}'}\alpha_{\mathbf{k}_i}.$$

This process is represented in Fig. 6*a*. The creation of the electron-hole pair is followed by its annihilation by interaction with the external perturbation $v(\omega,\mathbf{q})$. The annihilation is represented by the operators $\alpha_{\mathbf{k}-\mathbf{q}}\beta_{-\mathbf{k}}$, where the presence of $\mathbf{q}$ is forced by the $\mathbf{q}$ dependence $e^{i\mathbf{q}\cdot\mathbf{x}}$ of the potential $v(\omega,\mathbf{q})$. But we must annihilate the same pair we create, so that $\mathbf{q}'$ above must equal $\mathbf{q}$. Thus the scattering process is $\mathbf{k}_i \rightarrow \mathbf{k}_i + \mathbf{q}$ for the indirect process of Fig. 6*a* just as for the direct process of Fig. 4*a*. The total process is

$$\sum_{\mathbf{k}} \alpha_{\mathbf{k}-\mathbf{q}}\beta_{-\mathbf{k}}\alpha^+_{\mathbf{k}-\mathbf{q}}\beta^+_{-\mathbf{k}}\alpha^+_{\mathbf{k}_i+\mathbf{q}}\alpha_{\mathbf{k}_i}.$$

If we denote this process as $v(t_1)V(t_2)$ in the time-ordered sequence, then the integral over t_2 as in (89) gives the denominator

$$\frac{1}{\varepsilon_{\mathbf{k}-\mathbf{q}} - \varepsilon_{-\mathbf{k}} + \varepsilon_{\mathbf{k}_i+\mathbf{q}} - \varepsilon_{\mathbf{k}_i} - is}. \tag{90}$$

But the subsequent integral over t_1 assures over-all energy conservation: $\omega = \varepsilon_{\mathbf{k}_i+\mathbf{q}} - \varepsilon_{\mathbf{k}_i}$, and thus (90) may be written

$$\frac{1}{\varepsilon_{\mathbf{k}-\mathbf{q}} - \varepsilon_{-\mathbf{k}} + \omega - is}. \tag{91}$$

We can indicate the requirement that $\mathbf{k} - \mathbf{q}$ be an electron and $-\mathbf{k}$ a hole by writing (91) as

$$\frac{f_o(\varepsilon_{-\mathbf{k}})[1 - f_o(\varepsilon_{\mathbf{k}-\mathbf{q}})]}{\varepsilon_{\mathbf{k}-\mathbf{q}} - \varepsilon_{-\mathbf{k}} + \omega - is}, \tag{92}$$

where f_o denotes the occupancy in the unperturbed ground state.

The process of Fig. 6*b* is

$$\sum_{\mathbf{k}} \alpha^+_{\mathbf{k}_i+\mathbf{q}}\alpha_{\mathbf{k}_i}\beta_{-\mathbf{k}}\alpha_{\mathbf{k}+\mathbf{q}}\beta^+_{-\mathbf{k}}\alpha^+_{\mathbf{k}+\mathbf{q}};$$

in the time-ordered sequence it is $V(t_1)v(t_2)$, so that the denominator from the t_2 integration is

$$\frac{1}{\varepsilon_{\mathbf{k}+\mathbf{q}} - \varepsilon_{-\mathbf{k}} - \omega - is} \rightarrow \frac{f_o(\varepsilon_{-\mathbf{k}})[1 - f_o(\varepsilon_{\mathbf{k}+\mathbf{q}})]}{\varepsilon_{\mathbf{k}+\mathbf{q}} - \varepsilon_{-\mathbf{k}} - \omega - is}. \tag{93}$$

Now $\varepsilon_{\mathbf{k}} = \varepsilon_{-\mathbf{k}}$; because $\mathbf{k}$ is a dummy index we may replace it in (92)

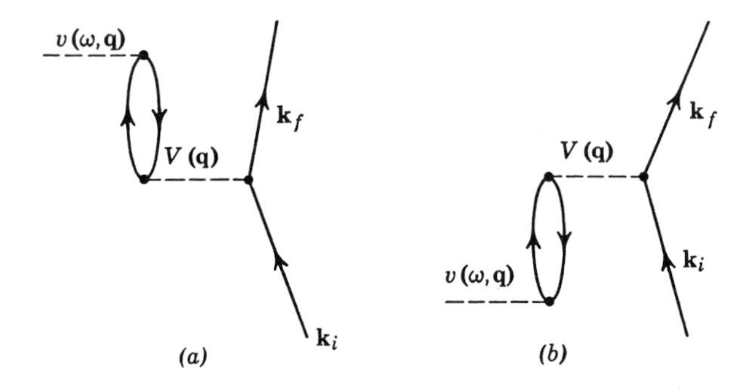

FIG. 6. Electron scattering involving an electron-hole pair in the intermediate state: (*a*) $\alpha_{\mathbf{k}-\mathbf{q}}\beta_{-\mathbf{k}}\alpha^+_{\mathbf{k}-\mathbf{q}}\beta^+_{-\mathbf{k}}\alpha^+_{\mathbf{k}_i+\mathbf{q}}\alpha_{\mathbf{k}_i}$; and (*b*) $\alpha^+_{\mathbf{k}_i+\mathbf{q}}\alpha_{\mathbf{k}_i}\beta_{-\mathbf{k}}\alpha_{\mathbf{k}+\mathbf{q}}\beta^+_{-\mathbf{k}}\alpha^+_{\mathbf{k}+\mathbf{q}}$.

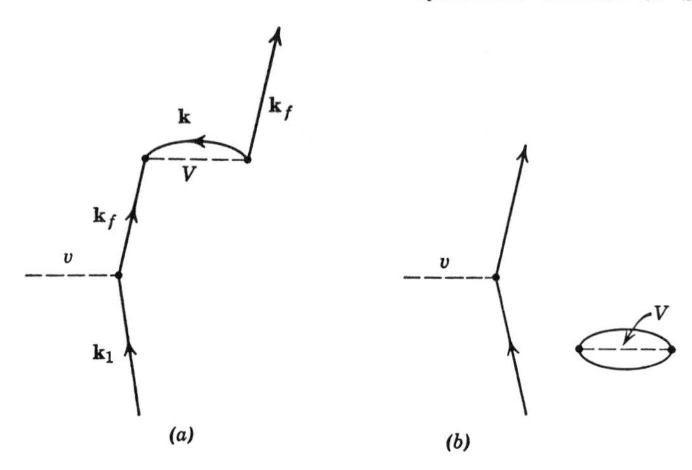

FIG. 7. Exchange graphs.

by $\mathbf{k} + \mathbf{q}$ and rewrite (92) as

$$\frac{f_o(\mathcal{E}_{\mathbf{k}+\mathbf{q}})[1 - f_o(\mathcal{E}_{\mathbf{k}})]}{\mathcal{E}_{\mathbf{k}} - \mathcal{E}_{\mathbf{k}+\mathbf{q}} + \omega - is}. \tag{94}$$

With $s = 0$ the result of the integration over t_2 for the sum of the two processes in Fig. 6 may be written, from (93) and (94), as

$$M(\omega,\mathbf{q}) = \sum_{\mathbf{k}} \frac{1}{\mathcal{E}_{\mathbf{k}+\mathbf{q}} - \mathcal{E}_{\mathbf{k}} - \omega} \{f_o(\mathcal{E}_{\mathbf{k}})[1 - f_o(\mathcal{E}_{\mathbf{k}+\mathbf{q}})] - f_o(\mathcal{E}_{\mathbf{k}+\mathbf{q}})[1 - f_o(\mathcal{E}_{\mathbf{k}})]\}. \tag{95}$$

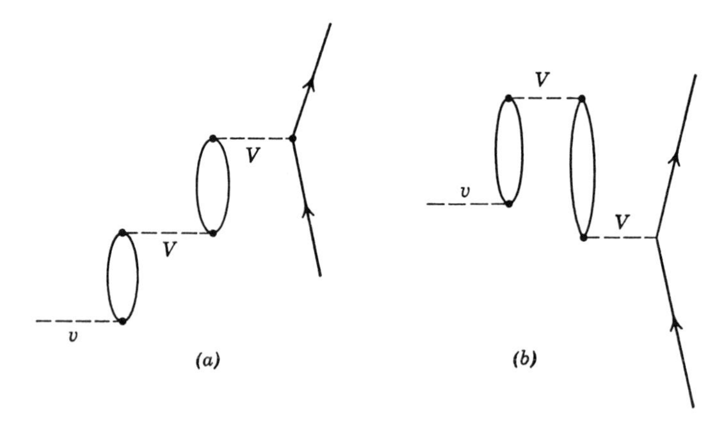

FIG. 8. Two of the 3! sequences for polarization loops in third order. At all vertices the momentum changes must be equal.

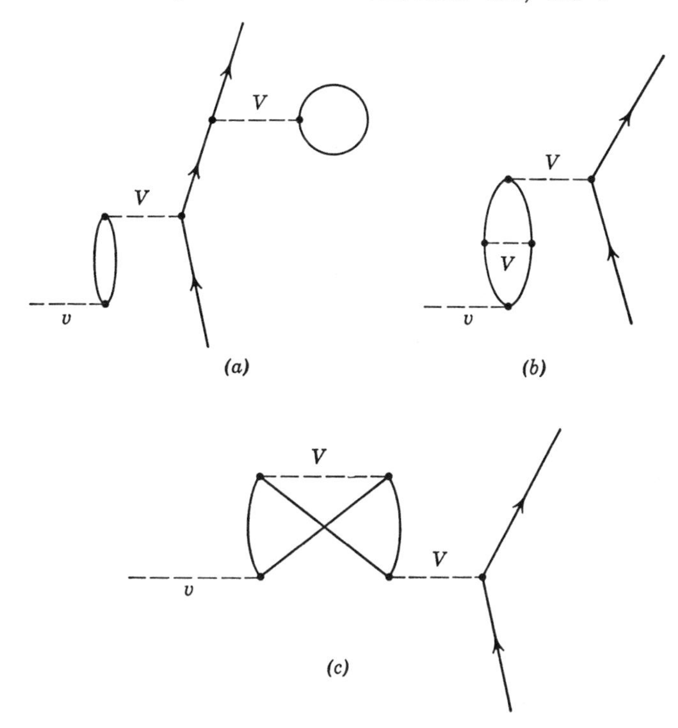

FIG. 9. Several other scattering processes in third order.

III. Two further types of scattering processes are shown in Fig. 7. Both involve electron exchange. In (*a*) the coulomb interaction contains operators

$$\alpha^{+}_{\mathbf{k}_f}\beta_{\mathbf{k}}\beta^{+}_{\mathbf{k}}\alpha_{\mathbf{k}_f},$$

representing the scattering of an electron in state $\mathbf{k}_f$ by interaction with an electron within the fermi sea, with exchange. In (*b*) the coulomb interaction creates an electron-hole pair at one vertex and annihilates it at the other vertex.

The process in Fig. 6 is of particular interest because all momentum changes in the graph are equal, and, by the same argument given in the appendix for the effect of ring diagrams on the energy, we expect such diagrams to dominate the scattering process at high electron density. In third order (first in v, second in V) the analogous graphs to Fig. 6 are given in Fig. 8. We say that such graphs contain *polarization loops*. Several other third-order scattering processes are shown in Fig. 9; these are not pure polarization loop processes.

The subset of graphs in all orders which contain only polarization loops may be written as a series. For electron-electron scattering the first term of the series is shown in Fig. 4*a*. The second term is shown in Fig. 6, and the third term is shown in Fig. 8. A careful extension of this analysis shows that the polarization loop series, including signs and numerical factors, may be written as

$$v(\omega,\mathbf{q})\alpha^{+}_{\mathbf{k}+\mathbf{q}}\alpha_{\mathbf{k}}[1 - VM + (VM)^2 - (VM)^3 + \cdots], \tag{96}$$

where $M(\omega,\mathbf{q})$ is given by (95). On summing the series we see that the effective potential is

$$v_{\text{eff}}(\omega,\mathbf{q}) = v(\omega,\mathbf{q})\,\frac{1}{1 + V(\mathbf{q})M(\omega,\mathbf{q})}. \tag{97}$$

The dielectric constant is defined as

$$\epsilon(\omega,\mathbf{q}) = \frac{v(\omega,\mathbf{q})}{v_{\text{eff}}(\omega,\mathbf{q})}, \tag{98}$$

so that

$$\epsilon(\omega,\mathbf{q}) = 1 + V(\mathbf{q})M(\omega,\mathbf{q}), \tag{99}$$

in exact agreement with the result (23) of the SCF method. We see that the SCF method is (like the random phase approximation) equivalent as far as the dielectric response is concerned to counting only those interaction terms in the U-matrix expansion which may be represented by polarization loops. The effective or screened interaction is usually represented by a double wavy line connecting two vertices; a broken line represents the bare or unscreened interaction.

LINKED-CLUSTER THEOREM

Any part of a graph which is disconnected from the rest of the graph and has no external lines going in or out is called an unlinked part. A graph containing no unlinked parts is called a linked graph. Thus linked graphs include Figs. 3, 4, 5*a*, 6, 7*a*, 8, 9; unlinked parts are found in Figs. 5*b* and 7*b*. We now derive the famous linked-cluster perturbation theorem; we use the time-dependent perturbation approach discussed at the end of Chapter 1.

The relative positions in time of the unlinked and linked parts of a graph affect the range of integration in $U\Phi_0$. Consider all those graphs containing unlinked parts which differ among themselves only in that the interactions in the unlinked part are in different positions relative to the rest of the graph. Within the linked and unlinked parts, however, the relative order of the interactions is fixed. Let the times of the interactions in an unlinked part be $t_1, t_2, \cdots, t_n$ $(0 > t_1 > t_2 > \cdots > t_n)$ and the times of the interactions in the linked part be $t_1', t_2', \cdots, t_m'$ $(0 > t_1' > t_2' > \cdots > t_m')$. The sum over all these graphs or over all the different relative positions of the linked and unlinked parts is obtained by carrying out the time integrations where the only restriction is that $0 > t_1 > t_2 \cdots > t_n$ and $0 > t_1' > t_2' > \cdots t_m'$. The sum is therefore the product of the expressions obtained separately from the two parts. This permits the terms for unlinked parts to be taken out as a factor. Then $U|0\rangle =$ (Σ terms from unlinked parts) $\times$ (Σ linked parts). But the normalizing denominator $\langle 0|U|0\rangle =$ (Σ unlinked parts), because an external line gives zero for the diagonal element. Thus, $|0) = \Sigma$ (linked parts).

We now carry out the time integration:

$$|0) = U|0\rangle = \sum_n (-i)^n \int_{-\infty}^{0} dt_1 \cdots \int_{-\infty}^{t_{n-1}} dt_n\, e^{iH_0t_n}Ve^{-iH_0t_n}e^{st_n}|0\rangle \tag{100}$$

$$= \sum_n (-i)^n \int_{-\infty}^{0} dt_1 \cdots \int_{-\infty}^{t_{n-1}} dt_n\, e^{i(H_0-E_0)t_n+st_n}V|0\rangle$$

$$= \sum_n (-i)^n \int \cdots \frac{e^{i(H_0-E_0)t_{n-1}+st_{n-1}}}{-i(E_0 - H_0 + is)} V|0\rangle$$

$$= \sum_n (-i)^{n-1} \int \cdots \int_{-\infty}^{t_{n-2}} dt_{n-1}\, e^{iH_0t_{n-1}}Ve^{-iH_0t_{n-1}} \cdot \frac{e^{i(H_0-E\ )t_{n-1}+st_{n-1}}}{(E_0 - H_0 + is)}V|0\rangle$$

$$= \sum_n (-i)^{n-1} \int \cdots \int_{-\infty}^{t_{n-2}} dt_{n-1}\, e^{i(H_0-E_0+2is)t_{n-1}}V \frac{1}{E_0 - H_0 + is} V|0\rangle;$$

hence

$$|0) = \lim_{s\to+0} \sum_L \frac{1}{E_0 - H_0 + ins} V \frac{1}{E_0 - H_0 + i(n-1)s} V \cdots \frac{1}{E_0 - H_0 + is} V|0\rangle, \tag{101}$$

where the sum L is over *linked graphs only*. Then we may write the exact result:

$$|0) = \sum_L \left(\frac{1}{E_0 - H_0} V\right)^n |0\rangle. \tag{102}$$

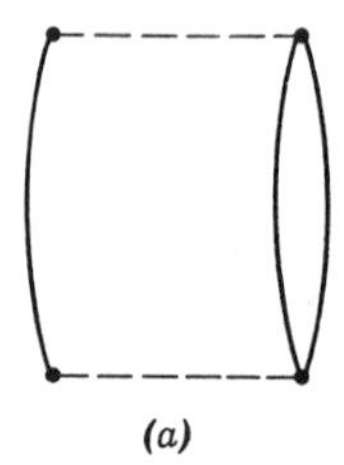

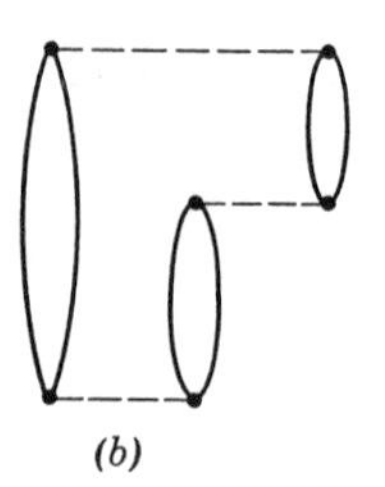

FIG. 10. Graph (*a*) and (*b*) are each connected graphs.

The exact energy shift ΔE is given by Eq. (1.45):

$$\boxed{\Delta E = \sum_{L_1} \langle 0|V\left(\frac{1}{E_0 - H_0}V\right)^n|0\rangle,} \tag{103}$$

where now contributions to the sum come only from *connected graphs* with no external lines. The normalization denominator $\langle 0|0\rangle$ is equal to unity. *A connected graph is a graph with no external lines but with the property that the entire graph may be traced out continuously*, as in Fig. 10, (*a*) or (*b*). Equations (102) and (103) are linked-cluster perturbation equations; for an alternate derivation see C. Bloch, *Nuclear Physics* **7,** 451 (1958).

As a trivial example, consider the boson hamiltonian $H = \varepsilon a^+a + \eta(a + a^+)$, where $H_0 = \varepsilon a^+a$. The only connected graph in the expansion is from

$$\langle 0|\eta a|1\rangle\langle 1|\frac{1}{-H_0}|1\rangle\langle 1|\eta a^+|0\rangle, \tag{104}$$

in which a boson is created and then destroyed. Thus

$$\Delta E = -\eta^2/\varepsilon \tag{105}$$

is the exact shift in the ground-state energy.

PROBLEMS

1. Using (47), calculate the dielectric constant of the electron gas in the Hartree approximation. Show first that

$$\frac{1}{\epsilon_H(0,\mathbf{q})} - 1 = -\frac{32\pi e^2 m}{(2\pi)^6 q^2} \iint_{\substack{k<k_F \\ k'>k_F}} d^3k\, d^3k' |\langle \mathbf{k}|e^{i\mathbf{q}\cdot\mathbf{x}}|\mathbf{k}'\rangle|^2 \mathcal{P}\frac{1}{k'^2 - k^2}, \tag{106}$$

where $\mathcal{P}$ denotes principal value. On evaluating the integrals, show that

$$\frac{1}{\epsilon_H(0,\mathbf{q})} - 1 = -\frac{k_s^2}{2q^2}\, g(\mathbf{q}), \tag{107}$$

where $k_s^2 = 6\pi ne^2/\varepsilon_F$, and

$$g(\mathbf{q}) = 1 + \frac{k_F}{q}\left(1 - \frac{q^2}{4k_F^2}\right)\log\left|\frac{q+2k_F}{q-2k_F}\right|. \tag{108}$$

Hint: In evaluating the integral show first that the related integral

$$\iint\limits_{\substack{k<k_F \\ k'<k_F}} d^3k\, d^3k'\, \delta(\mathbf{k}' + \mathbf{q} - \mathbf{k})\mathcal{P}\,\frac{1}{k'^2 - k^2} = 0. \tag{109}$$

Thus the desired integral is equal to the integral over $k < k_F$, but over *all* k'; with the use of the δ function we have

$$\int\limits_{k<k_F} d^3k\,\frac{1}{(\mathbf{k}-\mathbf{q})^2 - k^2}; \tag{110}$$

this is easily evaluated.

2. (*a*) Show that the second-order perturbation energy of the ground state of a system is always negative. Thus, if the potential energy term in the hamiltonian is of the form λV, where λ is the coupling constant, the value of $\partial^2 E_g/\partial\lambda^2$ is negative. Here E_g is the ground-state energy.

(*b*) Show further that $\partial\langle V\rangle/\partial\lambda \leqq 0$. This result has been used by R. A. Ferrell [*Phys. Rev. Letters* **1,** 444 (1958)] as a criterion for the extrapolation of expressions for the correlation energy of an electron gas.

3. For the test charge density

$$er_{\mathbf{q}}(e^{-i(\omega t + \mathbf{q}\cdot\mathbf{x})} + cc), \tag{111}$$

show, using lowest order time-dependent perturbation theory, that the rate at which the test charges do work on the electron gas is

$$\frac{dW}{dt} = 2\pi\omega\left(\frac{4\pi e^2}{q^2}\right)^2 r_{\mathbf{q}}^2|\langle n|\rho_{\mathbf{q}}|0\rangle|^2[\delta(\omega_{n0} - \omega) - \delta(\omega_{n0} + \omega)]. \tag{112}$$

On comparison with (49),

$$\frac{dW}{dt} = -\frac{8\pi e^2}{q^2}\,\omega r_{\mathbf{q}}^2 \mathcal{J}\left(\frac{1}{\epsilon(\omega,\mathbf{q})}\right). \tag{113}$$

4. Consider the equation of motion of a free electron gas having relaxation frequency η:

$$\ddot{x} + \eta\dot{x} = eE/m. \tag{114}$$

Show that the polarizability

$$\alpha(\omega,0) = \frac{nex}{E} = -\frac{ne^2}{m}\cdot\frac{1}{\omega^2 + i\eta\omega}, \tag{115}$$

where, for ω near ω_p,

(116)
$$\frac{1}{\epsilon(\omega,0)} \cong \frac{1}{2} \cdot \frac{\omega + i\eta}{\omega - \omega_p + \frac{1}{2}i\eta}.$$

Show that

(117)
$$\lim_{\eta \to +0} \mathscr{I}\left(\frac{1}{\epsilon}\right) = -\omega\pi\delta(\omega - \omega_p).$$

If we use only this pole in evaluating (52), show that

(118)
$$E_{\text{int}} = \sum_{\mathbf{q}} \left(\frac{1}{2}\omega_p - \frac{2\pi ne^2}{q^2}\right).$$

Note the contribution of the zero-point plasmon modes to the ground-state energy.

5. If in unit volume $\rho_{\mathbf{q}} = \sum_{i=1}^{n} e^{-i\mathbf{q}\cdot\mathbf{x}_i}$, and $H_0 = \sum_i \left[\frac{1}{2m}p_i^2 + V(\mathbf{x}_i)\right]$, show that

(*a*) (119)
$$[H_0,\rho_{\mathbf{q}}] = -\sum_i \frac{1}{m}\mathbf{q}\cdot(\mathbf{p} + \tfrac{1}{2}\mathbf{q})e^{-i\mathbf{q}\cdot\mathbf{x}_i};$$

(*b*) (120)
$$[[H_0,\rho_{\mathbf{q}}],\rho_{-\mathbf{q}}] = -\frac{n}{m}q^2.$$

(*c*) In the representation with H_0 diagonal,

(121)
$$\sum_m \omega_{m0}\{|\langle 0|\rho_{\mathbf{q}}|m\rangle|^2 + |\langle 0|\rho_{-\mathbf{q}}|m\rangle|^2\} = \frac{n}{m}q^2.$$

This is the Nozières-Pines longitudinal *f*-sum rule: *Phys. Rev.* **109,** 741 (1958). Usually, $|\langle 0|\rho_{\mathbf{q}}|m\rangle|^2 = |\langle 0|\rho_{-\mathbf{q}}|m\rangle|^2$; when is this true?

6. From (47) in the limit $\omega \gg \omega_{n0}$, show, using the Nozières-Pines sum rule, that

(122)
$$\epsilon(\omega,\mathbf{q}) \cong 1 - \frac{4\pi ne^2}{m\omega^2}.$$

7. Show, using the result of Problem 5, that

(123)
$$\int_0^\infty d\omega\, \omega\mathscr{I}\left(\frac{1}{\epsilon(\omega,\mathbf{q})}\right) = -\frac{\pi\omega_p^2}{2}.$$

8. Recalling that $\epsilon \equiv 4\pi i\sigma/\omega$, show that

(124)
$$\int_0^\infty d\omega\, \sigma_1(\omega,\mathbf{q}) = \tfrac{1}{8}\omega_p^2,$$

where $\sigma = \sigma_1 - i\sigma_2$. This is an important sum rule; for a discussion of the application to the superconducting transition see R. A. Ferrell and R. E. Glover, III, *Phys. Rev.* **109,** 1398 (1958); M. Tinkham and R. A. Ferrell, *Phys. Rev. Letters* **2,** 331 (1959). The proof is simple. Causality requires that ϵ be analytic with respect to ω in the upper half ω plane. Now from (47) we see that on the real axis $\omega\epsilon_1$ is an odd function of ω and $\omega\epsilon_2$ is an even function. Consider a contour integral from $-\infty$ to ∞ on the real axis and completed by a semicircle at ∞ in the upper half plane, using the result of Problem 6 for the asymptotic form of ϵ.

9. With the dynamic structure factor $\mathcal{S}(\omega,\mathbf{q})$ as given by (64), show that

$$\int_0^\infty d\omega\, \mathcal{S}(\omega,\mathbf{q}) = N\mathcal{S}(\mathbf{q}), \tag{125}$$

for N particles. Here $\mathcal{S}(\mathbf{q})$ is known as the liquid structure factor and is the fourier transform of the pair correlation function

$$p(\mathbf{x}) = N^{-1}\langle 0|\rho^+(0)\rho(\mathbf{x})|0\rangle; \tag{126}$$

$$\mathcal{S}(\mathbf{q}) = N^{-1}\langle 0|\rho_{\mathbf{q}}\rho_{\mathbf{q}}^+|0\rangle = \int d^3x\, p(\mathbf{x})e^{-i\mathbf{q}\cdot\mathbf{x}}. \tag{127}$$

10. Using the result of part c of Problem 5, show that

$$\int_0^\infty d\omega\, \omega\mathcal{S}(\omega,\mathbf{q}) = Nq^2/2m, \tag{128}$$

where m is the particle mass. We have assumed that $\mathcal{S}(\omega,\mathbf{q}) = \mathcal{S}(\omega,-\mathbf{q})$.

11. In ring graphs (see the appendix) we are concerned with coupled events described by the operators

$$A_{\mathbf{k}}^+(\mathbf{q}) = \alpha_{\mathbf{k}+\mathbf{q}}^+\beta_{-\mathbf{k}}^+; \qquad A_{\mathbf{k}}(\mathbf{q}) = \beta_{-\mathbf{k}}\alpha_{\mathbf{k}+\mathbf{q}}, \tag{129}$$

where the α, α^+ are electron operators and the β, β^+ are hole operators. Show that

$$[A_{\mathbf{k}}^+(\mathbf{q}),A_{\mathbf{k}'}^+(\mathbf{q}')] = 0, \tag{130}$$

and that

$$[A_{\mathbf{k}}(\mathbf{q}),A_{\mathbf{k}'}^+(\mathbf{q}')] = \delta_{\mathbf{k}+\mathbf{q},\mathbf{k}'+\mathbf{q}'}\delta_{\mathbf{k}\mathbf{k}'} - \delta_{\mathbf{k}\mathbf{k}'}\alpha_{\mathbf{k}'+\mathbf{q}}^+\,\alpha_{\mathbf{k}+\mathbf{q}} - \delta_{\mathbf{k}+\mathbf{q},\mathbf{k}'+\mathbf{q}'}\beta_{-\mathbf{k}'}^+\beta_{-\mathbf{k}} \approx \delta_{\mathbf{k}\mathbf{k}'}\delta_{\mathbf{q}\mathbf{q}'}, \tag{131}$$

because in the unperturbed vacuum state the electron and hole occupancies are zero. We note that in this approximation the A, A^+ have the same commutation rules as boson operators—an electron-hole pair acts as a boson.

7 Polarons and the electron-phonon interaction

Conduction electrons sense in various ways any deformation of the ideal periodic lattice of positive ion cores. Even the zero-point motion of phonons has its effect on the conduction electrons. The chief effects of the coupling of electrons and phonons are:

(*a*) To scatter electrons from one state $\mathbf{k}$ to another $\mathbf{k}'$, leading to electrical resistivity.

(*b*) To cause the absorption (or creation) of phonons: the interaction of conduction electrons and phonons is an important source of attenuation of ultrasonic waves in metals.

(*c*) To cause an attractive interaction between two electrons; this interaction is important for superconductivity and results from the virtual emission and absorption of a phonon.

(*d*) The electron will always carry with it a lattice polarization field. The composite particle, electron plus phonon field, is called a *polaron;* it has a larger effective mass than the electron in the unperturbed lattice.

In this chapter we discuss several central aspects of the electron-phonon interaction, with emphasis on features which may be presented without lengthy and detailed calculation.

THE DEFORMATION-POTENTIAL INTERACTION

In covalent crystals the electron-phonon interaction is often relatively weak, and when in semiconductors the concentration of charge carriers is low it will be valid to neglect screening effects of the carriers on each other. In this situation the deformation-potential method of Bardeen and Shockley may be applied for phonons of long wavelength.

Suppose that in an unstrained cubic covalent crystal the electron energy band of interest is nondegenerate, spherical, and given by

$$\varepsilon_0(\mathbf{k}) = \frac{k^2}{2m^*}; \tag{1}$$

here m^* is the effective mass of the conduction electron.

Let us now make a small uniform static deformation described by the strain components $e_{\mu\nu}$. The perturbed energy surface may be calculated in principle; it will be of the form

$$\varepsilon(\mathbf{k}) = \varepsilon_0(\mathbf{k}) + C_{\mu\nu}e_{\mu\nu} + C'_{\mu\nu}k_\mu k_\nu e_{\mu\nu} + \cdots, \tag{2}$$

to leading terms. In a semiconductor the k's of interest are usually low, and we set $C'_{\mu\nu}$ aside. For a spherical energy surface in the unstrained crystal, it is not possible for $\varepsilon(\mathbf{k})$ to be an odd function of the shear strain: we must therefore have $C_{\mu\nu} = 0$ for $\mu \neq \nu$. Because of the spin-orbit interaction we write for low k

$$\varepsilon(\mathbf{k}) \cong \varepsilon_0(\mathbf{k}) + C_1\Delta, \tag{3}$$

where Δ is the dilation. Here $C_1 = \partial\varepsilon(\mathbf{0})/\partial\Delta$ is a constant which may be determined in part by pressure measurements. The extension of this result to nonspherical energy surfaces has been considered by Brooks [*Adv. in Electronics* **7**, 85 (1955)] and others. It is found that if $\hat{\mathbf{k}}$ is the unit vector in the direction of $\mathbf{k}$, the shear strains add to (3) a term of the form

$$C_2(\hat{k}_\mu\hat{k}_\nu e_{\mu\nu} - \tfrac{1}{3}\Delta), \tag{4}$$

where C_2 vanishes for a spherical energy surface. Values of $|C_1|$ and $|C_2|$ are of the order of magnitude of 10 ev for the conduction band edges of Si and Ge.

It is easily shown that for a free electron gas the constant C_1 has the value $-\frac{2}{3}\varepsilon_F$, where ε_F is the fermi energy. The kinetic energy per electron is, at the fermi surface,

$$\varepsilon_F = \frac{1}{2m}\left(\frac{3\pi^2 N}{\Omega}\right)^{2/3}, \tag{5}$$

with N electrons in volume Ω. Thus

$$\frac{\delta\varepsilon}{\varepsilon_F} = -\frac{2\delta\Omega}{3\Omega} = -\tfrac{2}{3}\Delta, \tag{6}$$

or

$$\varepsilon(k_F) = \varepsilon_0(k_F) - \tfrac{2}{3}\varepsilon_0(k_F)\Delta. \tag{7}$$

This result assumes that the charges move to keep each part of the crystal electrically neutral—this is well satisfied for quasistatic perturbations of wavelength long in comparison with the screening length as defined in Chapter 6.

For acoustic phonons of long wavelength we assume that (3) may be generalized to read

$$\varepsilon(\mathbf{k},\mathbf{x}) = \varepsilon_0(\mathbf{k}) + C_1\Delta(\mathbf{x}), \tag{8}$$

with a similar generalization applying to (4). It is quite apparent that optical phonons are not covered by such a treatment; for one thing, the dilation is only related to acoustic phonons; for another, we have not included long-range electrostatic potentials which would arise from longitudinal optical phonon deformations.

In the Born approximation we are concerned with the matrix elements of $C_1\Delta(\mathbf{x})$ between the unperturbed one-electron Bloch states $|\mathbf{k}\rangle$ and $|\mathbf{k}'\rangle$, with $|\mathbf{k}\rangle = e^{i\mathbf{k}\cdot\mathbf{x}}u_{\mathbf{k}}(\mathbf{x})$, where $u_{\mathbf{k}}(\mathbf{x})$ has the periodicity of the lattice (Chapter 9). Using, from (2.84), the expansion of the dilation in phonon operators,

$$\begin{aligned} H' &= \int d^3x\, \Psi^+(\mathbf{x})C_1\Delta(\mathbf{x})\Psi(\mathbf{x}) = \sum_{\mathbf{k}'\mathbf{k}} c_{\mathbf{k}'}^+c_{\mathbf{k}}\langle\mathbf{k}'|C_1\Delta|\mathbf{k}\rangle \\ &= iC_1 \sum_{\mathbf{k}'\mathbf{k}} c_{\mathbf{k}'}^+c_{\mathbf{k}} \sum_{\mathbf{q}} (2\rho\omega_{\mathbf{q}})^{-1/2}|\mathbf{q}| \left(a_{\mathbf{q}} \int d^3x\, u_{\mathbf{k}'}^*u_{\mathbf{k}}e^{i(\mathbf{k}-\mathbf{k}'+\mathbf{q})\cdot\mathbf{x}} \right. \\ &\qquad \left. - a_{\mathbf{q}}^+ \int d^3x\, u_{\mathbf{k}'}^*u_{\mathbf{k}}e^{i(\mathbf{k}-\mathbf{k}'-\mathbf{q})\cdot\mathbf{x}}\right), \end{aligned} \tag{9}$$

where

$$\Psi(\mathbf{x}) = \sum_{\mathbf{k}} c_{\mathbf{k}}\varphi_{\mathbf{k}}(\mathbf{x}) = \sum_{\mathbf{k}} c_{\mathbf{k}}e^{i\mathbf{k}\cdot\mathbf{x}}u_{\mathbf{k}}(\mathbf{x}), \tag{10}$$

and the $a_{\mathbf{q}}^+$, $a_{\mathbf{q}}$ refer to longitudinal phonons of wavevector $\mathbf{q}$. The product $u_{\mathbf{k}'}^*(\mathbf{x})u_{\mathbf{k}}(\mathbf{x})$ involves the periodic parts of the Bloch functions and is itself periodic in the lattice; thus the integrals in (9) vanish unless

$$\mathbf{k} - \mathbf{k}' \pm \mathbf{q} = \begin{cases} 0 \\ \text{vector in the reciprocal lattice.} \end{cases} \tag{11}$$

For plane waves only the possibility zero exists, as here each $u_{\mathbf{k}}(\mathbf{x})$ is constant. In semiconductors at low temperatures the possibility zero may be the only process allowed energetically. If

$$\mathbf{k} - \mathbf{k}' \pm \mathbf{q} = 0, \tag{12}$$

the scattering process is said to be a *normal* or N process. If

$$\mathbf{k} - \mathbf{k}' \pm \mathbf{q} = \mathbf{G}, \tag{13}$$

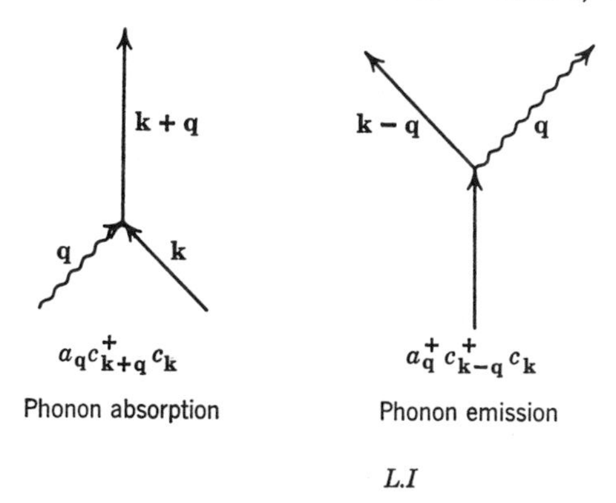

FIG. 1. Electron-phonon scattering processes in first order.

where **G** is a vector in the reciprocal lattice, the scattering process is said to be an *umklapp* or *U* process. The classification of processes as normal or umklapp depends on the choice of Brillouin zone. By "vector in the reciprocal lattice" we always mean a vector connecting two lattice points of the reciprocal lattice.

Let us limit ourselves for the present to *N* processes, and for convenience we approximate $\int d^3x\, u^*_{\mathbf{k}'} u_{\mathbf{k}}$ by unity. Then the deformation potential perturbation is

$$H' = iC_1 \sum_{\mathbf{kq}} (2\rho\omega_{\mathbf{q}})^{-\frac{1}{2}} |\mathbf{q}| (a_{\mathbf{q}} c^+_{\mathbf{k+q}} c_{\mathbf{k}} - a^+_{\mathbf{q}} c^+_{\mathbf{k-q}} c_{\mathbf{k}}); \tag{14}$$

we may equally write this as

$$\boxed{H' = iC_1 \sum_{\mathbf{kq}} (2\rho\omega_{\mathbf{q}})^{-\frac{1}{2}} |\mathbf{q}| (a_{\mathbf{q}} - a^+_{-\mathbf{q}}) c^+_{\mathbf{k+q}} c_{\mathbf{k}}.} \tag{15}$$

The field operators describe the scattering processes shown in Fig. 1.

Before going further we should see what the limitations are on the strength of the coupling parameter C_1 in order that our separation of electron and phonon energies should make sense. The existence of the electron-phonon coupling H' (14) means that an electron in a state **k** with no phonons excited cannot be an exact eigenstate of the system, but there will always be a cloud of virtual phonons accompanying the electron. The composite particle, electron plus lattice deformation, is called a *polaron*.* The phonon cloud changes the energy of the

* The term polaron is most often used for an electron plus the cloud of virtual optical phonons in ionic crystals.

electron. If the number of virtual phonons accompanying the electron is of the order of unity or larger, we can no longer trust the result of a first-order perturbation calculation. Nor can we then have much confidence in the validity of crystal wavefunctions written as a product of separate electronic and vibronic functions. This is not a trivial question: for heavy particles such as protons moving in the crystal the number of virtual phonons is very large (see Problem 1). In these circumstances the proton may become trapped locally in the crystal.

Phonon Cloud. Let us calculate by perturbation theory the number of virtual acoustic phonons accompanying a slow electron. We take as the unperturbed state of the phonon system the ground state in which no phonons are excited; the unperturbed state of the electronic system is taken as a Bloch state. Thus we write the unperturbed state of the total system as $|\mathbf{k}0\rangle$; the first-order perturbed state denoted by $|\mathbf{k}0\rangle^{(1)}$ is given by

$$|\mathbf{k}0\rangle^{(1)} = |\mathbf{k}0\rangle + \sum_{\mathbf{q}} |\mathbf{k}-\mathbf{q};1_{\mathbf{q}}\rangle \frac{\langle \mathbf{k}-\mathbf{q};1_{\mathbf{q}}|H'|\mathbf{k}0\rangle}{\varepsilon_{\mathbf{k}} - \varepsilon_{\mathbf{k}-\mathbf{q}} - \omega_{\mathbf{q}}}, \tag{16}$$

where H' is the electron-phonon interaction. The total number of phonons $\langle N\rangle$ accompanying the electron is given by taking the expectation value of $\Sigma\, a_{\mathbf{q}}^{+}a_{\mathbf{q}}$ over the state $|\mathbf{k}0\rangle^{(1)}$. On summing over the squares of the admixture coefficients we have

$$\langle N\rangle = \sum_{\mathbf{q}} \frac{|\langle \mathbf{k}-\mathbf{q};1_{\mathbf{q}}|H'|\mathbf{k}0\rangle|^2}{(\varepsilon_{\mathbf{k}} - \varepsilon_{\mathbf{k}-\mathbf{q}} - \omega_{\mathbf{q}})^2}. \tag{17}$$

For the deformation potential interaction (14),

$$|\langle \mathbf{k}-\mathbf{q};1_{\mathbf{q}}|H'|\mathbf{k}0\rangle|^2 = \frac{C_1{}^2|\mathbf{q}|}{2\rho c_s}, \tag{18}$$

where c_s is the longitudinal velocity of sound. Now, with m^* as the effective mass of the conduction electron,

$$\varepsilon_{\mathbf{k}} - \varepsilon_{\mathbf{k}-\mathbf{q}} - \omega_{\mathbf{q}} = \frac{1}{2m^*}(2\mathbf{k}\cdot\mathbf{q} - q^2) - c_s q. \tag{19}$$

For a very slow electron we neglect k in comparison with q and then write the sum in (17) as an integral,

$$\langle N\rangle = \frac{2m^{*2}C_1{}^2}{(2\pi)^3\rho c_s}\int d^3q \cdot \frac{q}{(q^2 + 2c_s m^* q)^2}, \tag{20}$$

where the integral should be carried over the first Brillouin zone of the longitudinal phonons. We shall for convenience take the integral over a sphere in $\mathbf{q}$ space out to a q_m chosen to enclose a number of modes equal to the number of atoms:

$$\langle N \rangle = \frac{1}{\pi^2} \frac{m^{*2} C_1^2}{\rho c_s} \int_0^{q_m} dq \, \frac{q}{(q + q_c)^2}, \tag{21}$$

where, with $\hbar$ restored, $q_c = 2m^* c_s/\hbar \approx 10^6 \text{ cm}^{-1}$ is essentially the electron compton wavevector in the phonon field. The numerical estimate was made using $m^* = m$ and $c_s = 5 \times 10^5$ cm/sec. The integral is standard:

$$\int_0^{q_m} dq \, \frac{q}{(q + q_c)^2} = \log \left(1 + \frac{q_m}{q_c} \right) - \frac{q_m}{q_m + q_c}. \tag{22}$$

Because $q_m \approx 10^8 \text{ cm}^{-1}$, we have $q_m/q_c \gg 1$ and the value of the integral is approximately $\log (q_m/q_c)$. Now, with $\hbar$ restored,

$$\langle N \rangle \cong \frac{1}{\pi^2} \frac{m^{*2} C_1^2}{\hbar^3 \rho c_s} \log (q_m/q_c). \tag{23}$$

Taking $C_1 \sim 5 \times 10^{-11}$ ergs; $m^* \sim 0.2 \times 10^{-27}$ gm; $\rho \sim 5$, $c_s \sim 5 \times 10^5$ cm/sec; $(q_m/q_c) \sim 10^2$, we have $\langle N \rangle \sim 0.02$. In these conditions, which are perhaps typical for covalent semiconductors, the expectation value of the number of virtual phonons around each electron is very much smaller than unity. If we do not neglect k in comparison with q, we obtain the more complete result

$$\langle N \rangle = \frac{m^{*2} C_1^2}{(2\pi)^2 \rho c_s \hbar^3 k} \left\{ (q_c - 2k) \log \left| \frac{q_c - 2k}{q_m + q_c - 2k} \right| + (q_c + 2k) \log \left| \frac{q_m + q_c + 2k}{q_c + 2k} \right| \right\}. \tag{24}$$

Relaxation Time. We see from the form of the wavefunction $|\mathbf{k}0\rangle^{(1)}$ given by (16) that in the presence of the electron-phonon interaction the wavevector $\mathbf{k}$ is not a constant of the motion for the electron alone, but the sum of the wavevectors of the electron and virtual phonon is conserved. Suppose an electron is initially in the state $|\mathbf{k}\rangle$; how long will it stay in the same state?

We calculate first the probability w per unit time that the electron in $\mathbf{k}$ will absorb a phonon $\mathbf{q}$. If $n_\mathbf{q}$ is the initial population of the phonon state,

$$\begin{aligned} w(\mathbf{k} + \mathbf{q}; n_\mathbf{q} - 1 | \mathbf{k}; n_\mathbf{q}) \\ = 2\pi |\langle \mathbf{k} + \mathbf{q}; n_\mathbf{q} - 1 | H' | \mathbf{k}; n_\mathbf{q} \rangle|^2 \delta(\varepsilon_\mathbf{k} + \omega_\mathbf{q} - \varepsilon_{\mathbf{k}+\mathbf{q}}). \end{aligned} \tag{25}$$

For the deformation potential interaction

$$|\langle \mathbf{k} + \mathbf{q}; n_\mathbf{q} - 1|H'|\mathbf{k}; n_\mathbf{q}\rangle|^2 = \frac{C_1^2 q}{2\rho c_s} n_\mathbf{q}. \tag{26}$$

The probability per unit time that an electron in **k** will emit a phonon **q** involves the matrix element through

$$|\langle \mathbf{k} - \mathbf{q}; n_\mathbf{q} + 1|H'|\mathbf{k}; n_\mathbf{q}\rangle|^2 = \frac{C_1^2 q}{2\rho c_s} (n_\mathbf{q} + 1). \tag{27}$$

The total collision rate W of an electron* in the state $|\mathbf{k}\rangle$ against a phonon system at absolute zero is, from (27) with $n_\mathbf{q} = 0$,

$$W = \frac{C_1^2}{4\pi\rho c_s} \int_{-1}^{1} d(\cos\theta_\mathbf{q}) \int_0^{q_m} dq\, q^3 \delta(\varepsilon_\mathbf{k} - \varepsilon_{\mathbf{k}-\mathbf{q}} - \omega_\mathbf{q}). \tag{28}$$

Now the argument of the delta function is

$$\frac{1}{2m^*}(2\mathbf{k}\cdot\mathbf{q} - q^2) - c_s q = \frac{1}{2m^*}(2\mathbf{k}\cdot\mathbf{q} - q^2 - qq_c), \tag{29}$$

where $q_c = 2m^*c_s$ as before. The minimum value of k for which the argument can be zero is

$$k_{\min} = \tfrac{1}{2}(q + q_c), \tag{30}$$

which for $q = 0$ reduces to

$$k_{\min} = \tfrac{1}{2}q_c = m^*c_s. \tag{31}$$

For this value of k the electron group velocity $v_g = k_{\min}/m^*$ is equal to c_s, the velocity of sound. Thus the threshold for the emission of phonons by electrons in a crystal is that the electron group velocity should exceed the acoustic velocity; this requirement resembles the Cerenkov threshold for the emission of photons in crystals by

* There is a simple connection between the first-order renormalization of the electron energy and the relaxation rate (28). The renormalized energy is

$$\varepsilon = \frac{k^2}{2m^*} + \sum_\mathbf{q} \frac{|\langle \mathbf{k} - \mathbf{q}; 1_\mathbf{q}|H'|\mathbf{k}; 0_\mathbf{q}\rangle|^2}{\varepsilon_\mathbf{k} - \varepsilon_{\mathbf{k}-\mathbf{q}} - \omega_\mathbf{q} - is},$$

where the limit $s \rightarrow +0$ is to be taken. By (1.34),

$$\mathscr{I}\{\varepsilon\} = \pi \sum_\mathbf{q} |\langle \mathbf{k} - \mathbf{q}; 1_\mathbf{q}|H'|\mathbf{k}; 0_\mathbf{q}\rangle|^2 \delta(\varepsilon_\mathbf{k} - \varepsilon_{\mathbf{k}-\mathbf{q}} - \omega_\mathbf{q}).$$

On comparison with (28),

$$W = 2\mathscr{I}\{\varepsilon\}.$$

fast electrons. The electron energy at the threshold is $\frac{1}{2}m^*c_s^2 \sim 10^{-27} \cdot 10^{11} \sim 10^{-16}$ ergs $\sim 1°$K. An electron of energy below this threshold will not be slowed down in a perfect crystal at absolute zero, even by higher order electron-phonon interactions, at least in the harmonic approximation for the phonons.

For $k \gg q_c$ we may neglect the qq_c term in (29). Then the integrals in (28) become

$$(32) \quad \int_{-1}^{1} d\mu \int dq\, q^3(2m^*/q)\delta(2k\mu - q) = 8m^* \int_0^1 d\mu\, k^2\mu^2 = 8m^*k^2/3,$$

and the emission rate is

$$(33) \qquad W(\text{emission}) = \frac{2C_1{}^2m^*k^2}{3\pi\rho c_s};$$

note that this is directly proportional to the electron energy ε_k.

The loss of the component of wavevector parallel to the original direction of the electron when a phonon is emitted at an angle θ to $\mathbf{k}$ is given by $q \cos \theta$. The fractional rate of loss of k is given by the transition rate integral with the extra factor $(q/k) \cos \theta$ in the integrand. Instead of (32), we have

$$(34) \qquad \frac{2m^*}{k} \int_0^1 d\mu\, 8k^3\mu^4 = \frac{16m^*k^2}{5},$$

so that the fractional rate of decrease of k_z is

$$(35) \qquad W(k_z) = \frac{4C_1{}^2m^*k^2}{5\pi\rho c_s}.$$

ELECTRON INTERACTION WITH LONGITUDINAL OPTICAL PHONONS

We expect electrons in ionic crystals to interact strongly with longitudinal optical phonons through the electric field of the polarization wave. This is a long-range coulomb interaction and is different from the deformation potential interaction. The interaction with transverse optical phonons will be less strong because of their smaller electric field, except at very low $\mathbf{q}$ where the electromagnetic coupling may be strong. Neglecting dispersion, the hamiltonian of the longitudinal optical phonons is approximately

$$(36) \qquad H_0 = \omega_l \sum_{\mathbf{q}} b_{\mathbf{q}}^+ b_{\mathbf{q}},$$

where b^+, b are boson operators. That is, we have N modes of different $\mathbf{q}$, but with the identical frequency ω_l. Reference to (2.83) tells us that the dielectric polarization field is proportional to the optical phonon amplitude and will have the form

$$\mathbf{P} = F \sum_{\mathbf{q}} \boldsymbol{\varepsilon}_{\mathbf{q}} (b_{\mathbf{q}} e^{i\mathbf{q}\cdot\mathbf{x}} + b_{\mathbf{q}}^+ e^{-i\mathbf{q}\cdot\mathbf{x}}), \tag{37}$$

where $\boldsymbol{\varepsilon}_{\mathbf{q}}$ is a unit vector in the direction of $\mathbf{q}$ and F is a constant to be determined. We expand the electrostatic potential in the form

$$\varphi(\mathbf{x}) = \sum (\varphi_{\mathbf{q}} e^{i\mathbf{q}\cdot\mathbf{x}} + \varphi_{\mathbf{q}}^+ e^{-i\mathbf{q}\cdot\mathbf{x}}), \tag{38}$$

whence

$$\mathbf{E} = -\operatorname{grad} \varphi = -i \sum_{\mathbf{q}} \mathbf{q} (\varphi_{\mathbf{q}} e^{i\mathbf{q}\cdot\mathbf{x}} - \varphi_{\mathbf{q}}^+ e^{-i\mathbf{q}\cdot\mathbf{x}}). \tag{39}$$

But $\operatorname{div} \mathbf{D} = 0$, so that $\mathbf{E} + 4\pi\mathbf{P} = 0$, or

$$\varphi_{\mathbf{q}} = -i4\pi F b_{\mathbf{q}}/q. \tag{40}$$

We now want to evaluate the constant F in terms of the interaction energy $e^2/\epsilon r$ between two electrons in a medium of dielectric constant ϵ. Consider electrons at $\mathbf{x}_1$ and $\mathbf{x}_2$ which interact directly through the vacuum coulomb field and indirectly through the second-order perturbation of the optical phonon field. The desired form of the effective perturbation hamiltonian in first order is obtained as the expectation value of the potential energy operator $e \int d^3x\, \rho(\mathbf{x})\varphi(\mathbf{x})$ over the state $\Psi^+(\mathbf{x}_1)\Psi^+(\mathbf{x}_2)|\text{vac}\rangle$ which represents electrons localized at $\mathbf{x}_1$ and $\mathbf{x}_2$, according to an extension of (5.124):

$$\begin{aligned} H'(\mathbf{x}_1,\mathbf{x}_2) &= e\varphi(\mathbf{x}_1) + e\varphi(\mathbf{x}_2) \\ &= -i4\pi F e \sum_{\mathbf{q}} q^{-1} (b_{\mathbf{q}} e^{i\mathbf{q}\cdot\mathbf{x}_1} - b_{\mathbf{q}}^+ e^{-i\mathbf{q}\cdot\mathbf{x}_1} + b_{\mathbf{q}} e^{i\mathbf{q}\cdot\mathbf{x}_2} \\ &\qquad - b_{\mathbf{q}}^+ e^{-i\mathbf{q}\cdot\mathbf{x}_2}). \end{aligned} \tag{41}$$

Now at absolute zero the second-order energy perturbation caused by (41) is

$$H''(\mathbf{x}_1,\mathbf{x}_2) = -2 \sum_{\mathbf{q}} \frac{\langle 0|e\varphi(\mathbf{x}_1)|\mathbf{q}\rangle\langle\mathbf{q}|e\varphi(\mathbf{x}_2)|0\rangle}{\omega_l}, \tag{42}$$

where we have dropped products in $\mathbf{x}_1$ alone or $\mathbf{x}_2$ alone, as these are self-energy terms. The factor 2 arises from the interchange of $\mathbf{x}_1$ and $\mathbf{x}_2$ in the expression for the perturbation. Here the state $|0\rangle$ denotes the vacuum phonon state; and $|\mathbf{q}\rangle$ denotes the state with one optical phonon $\mathbf{q}$ excited virtually with energy ω_l. It is supposed in using (42) that the electrons are localized and that their state does not change in the interaction process. This problem is almost identical with the interaction problem in neutral scalar meson theory, without recoil.

It is easy to evaluate H'' from (41) and (42):

$$H''(\mathbf{x}_1,\mathbf{x}_2) = -\frac{2e^2(4\pi F)^2}{\omega_l}\sum_{\mathbf{q}}\frac{1}{q^2}e^{i\mathbf{q}\cdot(\mathbf{x}_1-\mathbf{x}_2)}; \tag{43}$$

but we have seen that when summed over all $\mathbf{q}$

$$\sum_{\mathbf{q}}\frac{4\pi}{q^2}e^{i\mathbf{q}\cdot\mathbf{x}} = \frac{1}{|\mathbf{x}|}, \tag{44}$$

so that in the ground state

$$H''(\mathbf{x}_1,\mathbf{x}_2) = -\frac{8\pi F^2}{\omega_l}\frac{e^2}{|\mathbf{x}_1-\mathbf{x}_2|}. \tag{45}$$

This interaction is thus of the form of an attractive coulomb interaction between the charges e at $\mathbf{x}_1$ and $\mathbf{x}_2$: it gives exactly the ionic contribution to the interaction. It thus accounts for the difference between $e^2/\epsilon_0 r$ and $e^2/\epsilon_\infty r$, where the dielectric constant ϵ_0 includes electronic and ionic polarizabilities, and ϵ_∞ includes only the electronic contribution. The ionic term lowers the energy of the system. We have

$$\frac{1}{\epsilon_0} = \frac{1}{\epsilon_\infty} - \frac{8\pi F^2}{\omega_l}. \tag{46}$$

In step (44) the sum was carried out over all $\mathbf{q}$; actually the sum should only be over $\mathbf{q}$'s in the first Brillouin zone. The part of $\mathbf{q}$ space we should have excluded may be seen to give by itself a screened coulomb interaction. On subtracting from $1/r$ a screened interaction we are left with a potential essentially $1/r$ at long distances, but flatter at distances within $1/q_m$, where q_m is on the zone boundary.

Polaron Cloud. We now solve in the weak coupling limit for the number of optical phonons which clothe an electron. From (16)

$$\langle N\rangle = \sum_{\mathbf{q}}\frac{|\langle\mathbf{k}-\mathbf{q};1_{\mathbf{q}}|H'|\mathbf{k};0_{\mathbf{q}}\rangle|^2}{(\epsilon_{\mathbf{k}}-\epsilon_{\mathbf{k}-\mathbf{q}}-\omega_l)^2}; \tag{47}$$

but now instead of the deformation potential (14) we have

$$(48)\quad H' = \int d^3x\, e\Psi^+(\mathbf{x})\varphi(\mathbf{x})\Psi(\mathbf{x}) = -i4\pi Fe \sum_{\mathbf{kq}} q^{-1}(b_{\mathbf{q}}c^+_{\mathbf{k}+\mathbf{q}}c_{\mathbf{k}} - b^+_{\mathbf{q}}c^+_{\mathbf{k}-\mathbf{q}}c_{\mathbf{k}}),$$

using (38) and (40) for $\varphi(\mathbf{x})$. Then

$$(49)\quad |\langle \mathbf{k} - \mathbf{q};1_{\mathbf{q}}|H'|\mathbf{k};0_{\mathbf{q}}\rangle|^2 = (4\pi eF)^2/q^2,$$

and, neglecting $\mathbf{k}\cdot\mathbf{q}$ in comparison with q^2,

$$(50)\quad \langle N\rangle = 8e^2F^2(2m^*)^2 \int_0^\infty dq\, \frac{1}{(q^2 + q_p{}^2)^2},$$

where $q_p{}^2 = 2m^*\omega_l$; we have taken the upper limit as ∞ in the integration. For NaCl $q_p \approx 10^7\ \mathrm{cm}^{-1}$ if m^* is taken as the electronic mass; thus q_p times the lattice constant is of the order of $\frac{1}{2}$.

The value of the integral in (50) is $\pi/4q_p{}^3$; with $\hbar$ restored

$$(51)\quad \langle N\rangle = \frac{e^2}{4\hbar\omega_l}\left(\frac{2m^*\omega_l}{\hbar}\right)^{1/2}\left(\frac{1}{\epsilon_\infty} - \frac{1}{\epsilon_0}\right) = \frac{\alpha}{2};$$

this defines α, the dimensionless coupling constant commonly employed in polaron theory, after H. Fröhlich, H. Pelzer, and S. Zienau, *Phil. Mag.* **41,** 221 (1950). With m^* taken as the mass of the free electron, typical values of α calculated from the observed dielectric properties and infrared absorption of alkali halide crystals are:

	LiF	NaCl	NaI	KCl	KI	RbCl
α	5.25	5.5	4.8	5.9	6.1	6.4
$\langle N\rangle$	2.62	2.8	2.4	2.9	3.1	3.2

Thus for the alkali halides our estimate leads to $\langle N\rangle > 1$, so that the perturbation theory cannot be trusted to give valid quantitative results and more powerful methods are needed; we do, however, obtain an impression of the actual situation.

Polaron Effective Mass. The self-energy of a polaron for weak coupling is given in second-order perturbation theory as

$$(52)\quad \varepsilon_{\mathbf{k}} = \varepsilon^0_{\mathbf{k}} - 2m^* \sum_{\mathbf{q}} \frac{|\langle \mathbf{k} - \mathbf{q};1_{\mathbf{q}}|H'|\mathbf{k};0_{\mathbf{q}}\rangle|^2}{q^2 - 2\mathbf{k}\cdot\mathbf{q} + q_p{}^2},$$

or, using the interaction (48) appropriate to the ionic crystal problem,

$$(53)\quad \varepsilon_{\mathbf{k}} - \varepsilon^0_{\mathbf{k}} = -8e^2F^2m^* \int_{-1}^{1} d(\cos\theta) \int_0^\infty dq\, \frac{1}{q^2 - 2\mathbf{k}\cdot\mathbf{q} + q_p{}^2}.$$

Here we have used (49) and $q_p{}^2 = 2m^*\omega_l$.

The integral can be evaluated exactly, but for slow electrons ($k \ll q_p$) we might as well expand the integrand as

$$\frac{1}{1+x^2}\left(1+\frac{2\eta\mu x}{1+x^2}+\frac{4\eta^2\mu^2x^2}{(1+x^2)^2}+\cdots\right), \tag{54}$$

with $x = q/q_p$; $\mu = \cos\theta$; $\eta = k/q_p$. The integral over $d\mu$ leaves (54) as

$$\frac{1}{1+x^2}\left(2+\tfrac{8}{3}\eta^2\frac{x^2}{(1+x^2)^2}+\cdots\right); \tag{55}$$

after integrating over x from 0 to ∞ we have, using Dwight (122.3),

$$\varepsilon_k - \varepsilon_k{}^0 = -\alpha\left(\omega_l + \frac{1}{12m^*}k^2 + \cdots\right), \tag{56}$$

so that the ground-state energy is depressed by $\alpha\omega_l$ by the electron-phonon interaction, and the total polaron kinetic energy is

$$\varepsilon_{\text{kin}} = \frac{1}{2m^*}(1-\tfrac{1}{6}\alpha)k^2. \tag{57}$$

For $\alpha \ll 1$, the mass of the polaron is

$$m^*_{\text{pol}} \cong m^*(1+\tfrac{1}{6}\alpha), \tag{58}$$

in our weak coupling approximation.

For a good review of the numerous literature and of the elegant techniques which have been applied to the polaron problem when $\alpha \gg 1$, see T. D. Schultz, *Tech. Report* **9**, Solid-State and Molecular Theory Group, M.I.T., 1956. Further references include:

T. D. Lee and D. Pines, *Phys. Rev.* **92,** 883 (1953).
F. E. Low and D. Pines, *Phys. Rev.* **98,** 414 (1955).
R. P. Feynman, *Phys. Rev.* **97,** 660 (1955).

The polaron was first studied by Landau, and then extensively by Pekar and his school, and by other workers in the USSR. The Soviet literature is reviewed by Schultz; much of this literature is concerned implicitly with the limit of very strong coupling in which the lattice deformation follows the electron adiabatically.

The coupling constant α can be determined from measurements of the mobility of polarons in pure specimens of ionic crystals. Particularly thorough results are available for AgBr, as reported by D. C. Burnham, F. C. Brown, and R. S. Knox, *Phys. Rev.* **119,** 1560 (1960). Mobilities up to 50,000 cm^2/volt sec were observed; between 40 and 120°K the mobility is dominated by scattering by optical phonons, as indicated by a temperature variation of the form

$$\mu = F(\alpha)(e^{\Theta/T} - 1), \tag{59}$$

with $\Theta = 195°K$ in excellent agreement with optical data; $F(\alpha)$ is a function of the coupling constant, with a functional dependence determined by the details of the theoretical approximation employed. The form of the temperature dependence (59) can be understood simply: for $T \ll \Theta$ the relaxation rate of polarons is dominated by the absorptive collisions with optical phonons. The number of optical phonons of energy Θ is proportional to the Bose factor $(e^{\Theta/T} - 1)^{-1}$, so that the relaxation time and the mobility are proportional to $(e^{\Theta/T} - 1)$.

The electron mobility experiments on AgBr, when analyzed in terms of the calculations of Feynman, Hellwarth, Iddings, and Platzman [*Phys. Rev.* **127,** 1004 (1962)], give $\alpha = 1.60$ for the coupling constant and $m^* = 0.20m$, $m_p^* = 0.27m$ for the bare electron and the polaron effective mass, respectively. The value of m^* is related to α by the definition (51), and for this value of α we can use the approximation (57) to obtain m_p^*. Cyclotron resonance experiments by Ascarelli and Brown give $m_p^* = 0.27m$ for the electron polaron mass, thus in excellent agreement with the value inferred from the mobility experiments. Note that the polaron mass m_p^*, rather than the bare mass m^*, will be observed in a cyclotron resonance experiment as long as the cyclotron frequency is much lower than the optical phonon frequency.

ELECTRON-PHONON INTERACTION IN METALS

We consider first some aspects of a very simple model of a metal based on point positive ion cores imbedded in a uniform distribution of electrons. This model metal is sometimes called *jellium.* If we think of the electron cloud as fixed and of the positive ions as forming a fermion (or boson) gas, we see that the model is the Wigner limit (Chapter 6) for high r_s of the correlation energy problem—the positive ions do not form a gas, but in the ground state crystallize into a bcc array. The high r_s limit is applicable because the appropriate bohr radius for the unit of length is not $\hbar^2/e^2m$, but is $\hbar^2/e^2M$, where M is the mass of the ion.

In (5.78) we found an approximate value for the average energy per electron on this model in the ground state:

$$\varepsilon \cong -\frac{9}{10}\frac{e^2}{r} + \frac{3}{10\alpha^2 mr^2}, \tag{60}$$

where now r is the radius of the s sphere and $\alpha = (4/9\pi)^{1/3}$. This α must not be confused with the coupling constant defined by (51). With exchange the factor -0.90 would become -1.36; if screening were considered on the crude basis of complete screening within the s sphere the factor -0.90 would be -1.50, because one would omit the coulomb self-energy of the charge distribution.

There is little shear rigidity in jellium, but there is a substantial bulk modulus B given by, at 0°K,

$$(61)\quad B = -V\frac{\partial P}{\partial V} = V\frac{\partial^2 U}{\partial V^2} = V\left(\frac{dr}{dV}\right)^2\frac{\partial^2 U}{\partial r^2} = \frac{1}{12\pi r_0}\left(\frac{\partial^2 \varepsilon}{\partial r^2}\right)_{r_0},$$

where U is the energy and r_0 the equilibrium value of r. From (60),

$$(62)\qquad \frac{\partial \varepsilon}{\partial r} = \frac{9}{10}\frac{e^2}{r^2} - \frac{3}{5}\frac{1}{\alpha^2 m r^3};$$

at equilibrium $\partial\varepsilon/\partial r = 0$, and

$$(63)\qquad r_0 = \frac{2\hbar^2}{3\alpha^2 m e^2} = 1.30\ \text{Å},$$

with $\hbar$ restored. The observed values of r_0 at room temperature for Li, Na, K are 1.7, 2.1, 2.6 A, respectively. Further, at equilibrium,

$$(64)\qquad \left(\frac{\partial^2 \varepsilon}{\partial r^2}\right)_0 = -\frac{9e^2}{5r_0^3} + \frac{9}{5\alpha^2 m r_0^4} = \frac{9e^2}{10r_0^3},$$

using (63). Thus the bulk modulus is

$$(65)\qquad B = \frac{3e^2}{40\pi r_0^4} = \frac{1}{20\pi\alpha^2 m r_0^5},$$

and c_s, the longitudinal velocity of sound, is

$$(66)\qquad c_s^2 = \frac{B}{\rho} = \frac{m}{15M}\left(\frac{1}{\alpha r_0 m}\right)^2,$$

with $\rho = 3M/4\pi r_0^3$, where M is the atomic mass.

Now the electron velocity v_F at the fermi surface is related to r_0 by (5.71):

$$(67)\qquad v_F = k_F/m = 1/\alpha m r_0.$$

Finally, using (66) in (67),

$$(68)\qquad c_s^2 = \frac{m}{15M} v_F^2.$$

If we had counted in the compressibility *only* the contribution from the fermi gas and neglected the coulomb contribution, we would have

$$c_s^2 = \frac{1}{3}\frac{m}{M}v_F^2, \tag{69}$$

corresponding to $B = \frac{1}{3}mv_F^2/\Omega_a$, where Ω_a is the atomic volume. The same result is found in another way in (91).

We have assumed implicitly in deriving (68) that the electrons follow along with the nuclei during the compression and during the passage of a longitudinal acoustic phonon. We have also supposed that the mass which enters the fermi energy is the free-electron mass. In real metals m^* will enter, and m^* will be a function of r_0, so that the fermi energy is a more complicated function of r_0.

For Li, the experimental value of $(B/\rho)^{1/2}$ at 25°C is

$$(12.1 \times 10^{10}/0.543)^{1/2} = 4.8 \times 10^5 \text{ cm/sec};$$

the value at 0°K calculated from (69) with the fermi velocity 1.31×10^8 cm/sec as for the free-electron mass is 6.7×10^5 cm/sec, and the value calculated from (68) is 3×10^5 cm/sec.

Electron-Ion Hamiltonian in Metals. We write the hamiltonian in the form

$$H = \frac{1}{2m}\sum_i \mathbf{p}_i^2 + \sum_{ij} v(\mathbf{x}_i - \mathbf{X}_j) + H_{\text{ion-ion}} + H_{\text{coul}}, \tag{70}$$

where i is summed over the valence electrons and j over the ions; v is the electron-ion interaction; $H_{\text{ion-ion}}$ describes the ion-ion interaction; and H_{coul} is the electron-electron coulomb interaction of the valence electrons. We assume the metal is monoatomic with n ions per unit volume. The energy terms from $\mathbf{K} = 0$ charge components sum to zero, and we suppose that such terms have been eliminated.

We expand the departure of an ion from its equilibrium position $\mathbf{X}_j^0$ as, following (2.7),

$$\delta\mathbf{X}_j = \mathbf{X}_j - \mathbf{X}_j^0 = (nM)^{-1/2}\sum_{\mathbf{q}} \boldsymbol{\varepsilon}_{\mathbf{q}} Q_{\mathbf{q}} e^{i\mathbf{q}\cdot\mathbf{X}_j^0}, \tag{71}$$

where $\boldsymbol{\varepsilon}_{\mathbf{q}}$ is the longitudinal polarization vector; we do not consider here the transverse waves, as their frequencies are determined chiefly by the short-range interaction $H_{\text{ion-ion}}$. The phonon hamiltonian *exclusive* of the electron-phonon interaction may be written, following

(2.17) and (2.26),

$$H_{\text{ion-ion}} = \tfrac{1}{2}\sum_{\mathbf{q}} (P_{\mathbf{q}}P_{-\mathbf{q}} + \Omega_{\mathbf{q}}^2 Q_{\mathbf{q}}Q_{-\mathbf{q}}), \tag{72}$$

where $\Omega_{\mathbf{q}}$ is an ionic plasma frequency.

The electron-ion interaction may be expanded as

$$\sum_j v(\mathbf{x} - \mathbf{X}_j) = \sum_j v(\mathbf{x} - \mathbf{X}_j^0) - (nM)^{-1/2} \sum_{j\mathbf{q}} Q_{\mathbf{q}}[\boldsymbol{\varepsilon}_{\mathbf{q}} \cdot \boldsymbol{\nabla} v(\mathbf{x} - \mathbf{X}_j^0) e^{i\mathbf{q}\cdot\mathbf{X}_j^0}], \tag{73}$$

where the first term on the right-hand side may be combined with the electron kinetic energy to give the Bloch hamiltonian

$$H_{\text{el}} = \sum_i \left(\frac{\mathbf{p}_i^2}{2m} + \sum_j v(\mathbf{x}_i - \mathbf{X}_j^0)\right) = \sum_{\mathbf{k}} \varepsilon_{\mathbf{k}} c_{\mathbf{k}}^+ c_{\mathbf{k}}, \tag{74}$$

with the spin included in **k**. Now the **q** component of the second term of (73) may be written in second quantization as

$$-\int d^3x\, \Psi^+(\mathbf{x})\Psi(\mathbf{x})(nM)^{-1/2} Q_{\mathbf{q}}\boldsymbol{\varepsilon}_{\mathbf{q}} \cdot \boldsymbol{\nabla} \sum_j v(\mathbf{x} - \mathbf{X}_j^0) e^{i\mathbf{q}\cdot\mathbf{X}_j^0} = \sum_{\mathbf{k}} c_{\mathbf{k}+\mathbf{q}}^+ c_{\mathbf{k}} Q_{\mathbf{q}} v_{\mathbf{q}}, \tag{75}$$

where

$$v_{\mathbf{q}} = -(nM)^{-1/2} \langle \mathbf{k} + \mathbf{q} | \sum_j e^{i\mathbf{q}\cdot\mathbf{X}_j^0} \boldsymbol{\varepsilon}_{\mathbf{q}} \cdot \boldsymbol{\nabla} v(\mathbf{x} - \mathbf{X}_j^0) | \mathbf{k} \rangle \tag{76}$$

will be assumed to be independent of **k**. Note that $(v_{-\mathbf{q}})^* = v_{\mathbf{q}}$. Using the density fluctuation operator

$$\rho_{-\mathbf{q}} = \sum_{\mathbf{k}} c_{\mathbf{k}+\mathbf{q}}^+ c_{\mathbf{k}}, \tag{77}$$

the electron-phonon interaction term is

$$H_{\text{el-ph}} = \sum_{\mathbf{q}} Q_{\mathbf{q}} v_{\mathbf{q}} \rho_{-\mathbf{q}}. \tag{78}$$

The coulomb term in (70) is, for free electrons,

$$H_{\text{coul}} = \sum_{\mathbf{q}} \frac{2\pi e^2}{q^2} \rho_{-\mathbf{q}}\rho_{\mathbf{q}}. \tag{79}$$

Electron-Lattice Interaction: Self-Consistent Calculation. The phonon coordinates are contained in

$$\tfrac{1}{2}\Sigma\, \{P_{\mathbf{q}}P_{-\mathbf{q}} + \Omega_{\mathbf{q}}^2 Q_{\mathbf{q}}Q_{-\mathbf{q}} + 2v_{\mathbf{q}}Q_{\mathbf{q}}\rho_{-\mathbf{q}}\}, \tag{80}$$

where for longitudinal modes $\Omega_{\mathbf{q}}$ is essentially the plasma frequency of the ions and is nearly independent of $\mathbf{q}$; the quantity $v_{\mathbf{q}}$ is the proportionality constant connecting the interaction energy of the electron distribution and the phonon amplitude $Q_{\mathbf{q}}$. The equation of motion of $Q_{\mathbf{q}}$ is

$$\ddot{Q}_{\mathbf{q}} + \Omega_{\mathbf{q}}{}^2 Q_{\mathbf{q}} = -v_{-\mathbf{q}}\rho_{\mathbf{q}}, \tag{81}$$

on calculating $[P_{-\mathbf{q}},H_{\text{ph}} + H_{\text{el-ph}}]$.

The ions move slowly, so that we may treat the ionic motion as a quasistatic test charge density in the sense of (6.38) to (6.42), with $\omega \approx 0$. Let $\rho_{\mathbf{q}}^i$ denote the component of the ionic density. By (6.42) we have

$$\rho_{\mathbf{q}}^i = \epsilon(\mathbf{q})(\rho_{\mathbf{q}} + \rho_{\mathbf{q}}^i), \tag{82}$$

where $\rho_{\mathbf{q}}$ is the electronic density fluctuation and $\epsilon(\mathbf{q})$ is the static dielectric constant, whence

$$\rho_{\mathbf{q}} = \frac{1 - \epsilon(\mathbf{q})}{\epsilon(\mathbf{q})}\,\rho_{\mathbf{q}}^i. \tag{83}$$

The ionic density is related to the phonon coordinate $Q_{\mathbf{q}}$ in (71) by

$$\rho_{\mathbf{q}}^i = -i(n/M)^{1/2} q Q_{\mathbf{q}}. \tag{84}$$

Thus (81) becomes

$$\ddot{Q}_{\mathbf{q}} + \left[\Omega_{\mathbf{q}}^2 + i\left(\frac{n}{M}\right)^{1/2} q\,\frac{1 - \epsilon(\mathbf{q})}{\epsilon(\mathbf{q})}\,v_{-\mathbf{q}}\right] Q_{\mathbf{q}} = 0. \tag{85}$$

We now use as an approximation the Thomas-Fermi result (6.34) for the dielectric constant, whence

$$\frac{1 - \epsilon_{\text{TF}}(\mathbf{q})}{\epsilon_{\text{TF}}(\mathbf{q})} = \frac{k_s{}^2}{q^2 + k_s{}^2}; \qquad k_s{}^2 \equiv \frac{6\pi n e^2}{\varepsilon_F}. \tag{86}$$

We show in the following that for the free electron gas:

$$v_{\mathbf{q}} = \frac{4\pi e^2 i}{q}\left(\frac{n}{M}\right)^{1/2}, \tag{87}$$

so that the equation of motion becomes

$$\ddot{Q}_{\mathbf{q}} + \left[\Omega_{\mathbf{q}}^2 - \frac{4\pi e^2 n}{M}\,\frac{k_s{}^2}{q^2 + k_s{}^2}\right] Q_{\mathbf{q}} = 0. \tag{88}$$

If $\Omega_{\mathbf{q}}^2 \cong 4\pi ne^2/M$, the equation of motion becomes

$$\ddot{Q}_{\mathbf{q}} + \Omega_{\mathbf{q}}^2 \left(\frac{q^2}{q^2 + k_s^2}\right) Q_{\mathbf{q}} = 0; \tag{89}$$

thus the eigenfrequency ω_p is given by, for $q \to 0$,

$$\omega_{\mathbf{q}} \cong \Omega_{\mathbf{q}} q/k_s = c_s q, \tag{90}$$

where the longitudinal phonon velocity c_s is given by

$$c_s^2 = \frac{m}{3M} v_F^2. \tag{91}$$

The result (87) follows from the definition (76), now written explicitly for free electrons and for a phonon in the z direction, with $v(\mathbf{x}) = e^2/|\mathbf{x}|$:

$$\begin{aligned} v_{\mathbf{q}} &= -(nM)^{-\frac{1}{2}} e^2 \int d^3x \sum_j e^{-i\mathbf{q}\cdot(\mathbf{x}-\mathbf{X}_j^0)} \frac{z - Z_j^0}{|\mathbf{x} - \mathbf{X}_j^0|^3} \\ &= -\left(\frac{n}{M}\right)^{\frac{1}{2}} e^2 \int d^3x \, e^{-i\mathbf{q}\cdot\mathbf{x}} \frac{z}{|\mathbf{x}|^3}. \end{aligned} \tag{92}$$

The integral has the value

$$\begin{aligned} \int &= 2\pi \int_{-1}^{1} d\mu \int_0^{x_m} dx \cdot x^2 \cdot x\mu \cdot \frac{1}{x^3} \cdot e^{-iqx\mu} \\ &= \frac{2\pi}{(-iq)} \int_{-1}^{1} d\mu \, (e^{-iqx_m\mu} - 1) = -i\frac{4\pi}{q}\left(1 - \frac{\sin qx_m}{qx_m}\right); \end{aligned} \tag{93}$$

as $x_m \to \infty$

$$v_{\mathbf{q}}^i = \frac{4\pi e^2 i}{q}\left(\frac{n}{M}\right)^{\frac{1}{2}}. \tag{94}$$

PROBLEMS

1. Estimate the number of virtual phonons accompanying a *proton* moving in a crystal; show that $\langle N \rangle \gg 1$, which suggests that the separation of protonic and lattice modes is poor. *Suggestion:* Note that now $q_c \gg q_m$, so that the denominator of the integrand in (21) may be replaced by q_c^2. The result for $\langle N \rangle$ is

$$\langle N \rangle \approx \frac{C_1^2 q_m^2}{8\pi^2 \rho c_s^3 \hbar}. \tag{95}$$

We must remember that (16) is inadequate for strong coupling, so that (95) is only a rough estimate.

2. Examine the electron-phonon energy correction $\Delta\epsilon$ to the electron energy for an electron of wavevector $\mathbf{k}$ such that $k_m \gg |\mathbf{k}| \gg k_c$; show that

$$\Delta\epsilon \cong -\frac{C_1{}^2 m^* k_m{}^2}{4\pi^2 \rho c_s \hbar} \approx 10^{-1} \text{ ev.} \tag{96}$$

3. The results (33) and (34) apply to absolute zero. At a finite temperature T satisfying $k_B T \gg \hbar c_s k$ show that the integrated phonon emission rate is, with $\hbar$ restored,

$$W(\text{emission}) \cong \frac{C_1{}^2 m^* k k_B T}{\pi c_s{}^2 \rho \hbar^3}. \tag{97}$$

For electrons in thermal equilibrium at not too low temperatures the required inequality is easily satisfied for the rms value of k. The result is of the same form (in fact, it happens to be identical) with the Bardeen-Shockley result [*Phys. Rev.* **80,** 72 (1950)] for the relaxation rate for electrical conductivity in semiconductors.

4. The hamiltonian

$$H = \omega_l \sum b_{\mathbf{q}}^+ b_{\mathbf{q}} + e\varphi(\mathbf{x}_1) + e\varphi(\mathbf{x}_2), \tag{98}$$

with $\varphi(\mathbf{x})$ given by (41), can be solved exactly; show that the solution leads to (43) for the interaction coupling the particles at $\mathbf{x}_1$, $\mathbf{x}_2$.

5. In the weak coupling limit find an expression for the mobility of a polaron in an ionic crystal, considering only interaction processes with optical phonons of constant frequency ω_l.

6. (*a*) Consider the hamiltonian H and the canonical transformation defined by $e^{-S}He^{S}$; by expanding in a power series and collecting terms show that

$$\tilde{H} \equiv e^{-S}He^{S} = H + [H,S] + \tfrac{1}{2}[[H,S],S] + \cdots. \tag{99}$$

(*b*) Thus if $H = H_0 + \lambda H'$, show that the terms linear in λ in the transformed hamiltonian vanish if S is chosen* to make

* We can write this method in an instructive way. If H' and thus S are independent of time in the Schrödinger picture, then in the interaction picture

$$i\dot{S}_I = [S_I, H_0]; \tag{100}$$

we see this directly from $S_I \equiv e^{iH_0t}Se^{-iH_0t}$. Then the condition (104) gives

$$i\dot{S}_I = \lambda H_I'; \tag{101}$$

$$S_I(t) = -i\lambda \int_{-\infty}^{t} dt'\, H_I'(t'). \tag{102}$$

This is an explicit expression for S_I as an operator. Then (107) becomes

$$\tilde{H}_I(0) = H_0 + \frac{i}{2}\lambda^2 \int_{-\infty}^{0} dt'\, [H_I'(t'), H_I'(0)] + \cdots. \tag{103}$$

(104) $$\lambda H' + [H_0,S] = 0.$$

In a representation in which H_0 is diagonal,

(105) $$\langle n|S|m\rangle = \lambda \frac{\langle n|H'|m\rangle}{E_m - E_n},$$

provided $E_n \neq E_m$.

Using this solution for S,

(106) $$\tilde{H} = e^{-S}He^S = H_0 + [\lambda H',S] + \tfrac{1}{2}[[H_0,S],S] + \cdots,$$

so that $\tilde{H}$ has no off-diagonal terms of $0(\lambda)$. We can rewrite (106) as

(107) $$\tilde{H} = H_0 + \tfrac{1}{2}[\lambda H',S] + 0(\lambda^3).$$

(*c*) With

(108) $$H = \omega a^+a + \lambda(a^+ + a),$$

show that

(109) $$\langle n|\tilde{H}|n\rangle = n\omega - (\lambda^2/\omega),$$

where n is the expectation value of a^+a in the unperturbed boson system. Note that this result is consistent with our calculation in the text of the polaron interaction.

(*d*) Show that, at least to $0(\lambda^2)$, there are no bosons present in the ground state $\tilde{\Phi}_0$ of $\tilde{H}$ as given by (108), but that the state $\Phi_0 = e^S\tilde{\Phi}_0$ contains virtual bosons.

7. In a piezoelectric crystal the polarization P is a linear function of the elastic strain. Suppose, for example, that $P_z = C_p e_{zz}$, where C_p is a piezoelectric constant.

(*a*) Find an expression for the interaction energy of an electron with a longitudinal phonon having $\mathbf{q} \parallel \hat{z}$, noting that here $H' \propto q^{-1/2}$. For deformation potential coupling $H' \propto q^{1/2}$, and for optical phonon coupling $H' \propto q^{-1}$.

(*b*) Derive an expression for the temperature dependence of the mobility in a piezoelectric semiconductor; show that $\mu \propto T^{-1/2}$; and compare with the deformation potential result [*ISSP* (13.17)] $\mu \propto T^{-3/2}$. [See W. A. Harrison, *Phys. Rev.* **101,** 903 (1956).]

8. Discuss the form of the electron-electron interaction via virtual phonons in covalent crystals, using the deformation potential form of the electron-phonon interaction.

8 Superconductivity

In superconducting elements in which the dependence of the critical temperature T_c on the isotopic mass M has been studied it is found that

$$M^{1/2}T_c = \text{constant}, \tag{1}$$

for a given element, with the exception of several transition group elements including Ru and Os. The result (1) suggested to Fröhlich that the properties of the lattice phonons, zero point or thermal, are involved in superconductivity; it is difficult to see how else the atomic mass could enter the problem. In elements which exhibit an isotope effect it is believed that the interaction responsible for superconductivity is the attractive interaction between two electrons near the fermi surface caused by their interaction with the zero-point phonons. In the transition group elements the polarization of the d band provides an additional coupling mechanism, one which does not involve an isotope effect.

The Bardeen-Cooper-Schrieffer theory of superconductivity is a striking success of the quantum theory of solids. The BCS theory accounts in a fairly simple, although not trivial, way for the essential effects associated with superconductivity; excellent detailed agreement with experiment exists in a number of areas. The isotope effect is explained in a fairly natural way, and the discovery of flux quantization in units of one-half the natural unit ch/e is a striking confirmation of the central role of the paired electron states predicted by the theory. This chapter is devoted to a thorough explanation of the BCS theory. We shall not enter into the phenomenology of superconductors or into the technology of high critical field materials.

GENERAL REFERENCES

[1] L. Cooper, *Phys. Rev.* **104,** 1189 (1956).

[2] J. Bardeen, L. N. Cooper, and J. R. Schrieffer, *Phys. Rev.* **108,** 1175 (1957).

[3] Bogoliubov, Tolmachev, and Shirkov, *A new method in the theory of superconductivity*, Consultants Bureau, New York, 1959.

[4] J. Bardeen and J. R. Schrieffer, in *Progress in low temperature physics* **3,** 170–287 (1961). An excellent review of the theoretical and experimental position.

[5] J. Bardeen, *Encyclopedia of Physics* **15,** 274–368. A thorough review of the position before the BCS theory.

[6] M. Tinkham, *LTP*, pp. 149–230. A simple account with emphasis on electrodynamics.

INDIRECT ELECTRON-ELECTRON INTERACTION VIA PHONONS

Let us write the first-order electron-phonon interaction as, following (7.15),

$$H' = i \sum_{\mathbf{kq}} D_{\mathbf{q}} c^+_{\mathbf{k+q}} c_{\mathbf{k}} (a_{\mathbf{q}} - a^+_{-\mathbf{q}}) = \sum_{\mathbf{k}} H_{\mathbf{k}}' \tag{2}$$

here c^+, c are fermion operators and a^+, a are boson operators; $D_{\mathbf{q}}$ is a c-number, and for convenience we take it equal to D, a real constant. In first-order H' leads to electron scattering and to electrical resistivity; in second-order H' leads to a self-energy and also to coupling between two electrons. The coupling is an indirect interaction through the phonon field. One electron polarizes the lattice; the other electron interacts with the polarization.

The total hamiltonian of electrons, phonons, and their interaction is

$$H = H_0 + H' = \sum_{\mathbf{q}} \omega_{\mathbf{q}} a^+_{\mathbf{q}} a_{\mathbf{q}} + \sum \varepsilon_{\mathbf{k}} c^+_{\mathbf{k}} c_{\mathbf{k}} + iD \sum_{\mathbf{kq}} c^+_{\mathbf{k+q}} c_{\mathbf{k}} (a_{\mathbf{q}} - a^+_{-\mathbf{q}}). \tag{3}$$

We now make a canonical transformation to a new hamiltonian $\tilde{H} = e^{-S} H e^{S}$ which has no off-diagonal terms of $O(D)$. Following the result (7.102) of Problem 7.6,

$$\langle n|S|m\rangle = \frac{\langle n|H'|m\rangle}{E_m - E_n}. \tag{4}$$

To obtain the effective electron-electron coupling it is convenient to take the matrix element over the phonon operators, but to leave the fermion operators explicitly displayed. We consider the phonon system at absolute zero, so that either n or m refers to the vacuum phonon state. The final result (9) is actually independent of the phonon excitation. Then

$$\langle 1_{\mathbf{q}}|S|0\rangle = -iD \sum_{\mathbf{k}} c^+_{\mathbf{k-q}} c_{\mathbf{k}} \frac{1}{\varepsilon_{\mathbf{k}} - \varepsilon_{\mathbf{k-q}} - \omega_{\mathbf{q}}}; \tag{5}$$

$$\langle 0|S|1_{\mathbf{q}}\rangle = iD \sum_{\mathbf{k}'} c^+_{\mathbf{k'+q}} c_{\mathbf{k}'} \frac{1}{\varepsilon_{\mathbf{k}'} + \omega_{\mathbf{q}} - \epsilon_{\mathbf{k'+q}}}. \tag{6}$$

We saw in (7.107) that

$$\tilde{H} = H_0 + \tfrac{1}{2}[H',S] + 0(D^3), \tag{7}$$

whence to $0(D^2)$

$$\tilde{H} = H_0 + \tfrac{1}{2}D^2 \sum_{\mathbf{q}} \sum_{\mathbf{k}\mathbf{k}'} c^+_{\mathbf{k}'+\mathbf{q}} c_{\mathbf{k}'} c^+_{\mathbf{k}-\mathbf{q}} c_{\mathbf{k}} \times \left(\frac{1}{\varepsilon_{\mathbf{k}} - \varepsilon_{\mathbf{k}-\mathbf{q}} - \omega_{\mathbf{q}}} - \frac{1}{\varepsilon_{\mathbf{k}'} + \omega_{\mathbf{q}} - \varepsilon_{\mathbf{k}'+\mathbf{q}}} \right). \tag{8}$$

We proceed to group the terms with phonons in $\mathbf{q}$ and $-\mathbf{q}$ in the intermediate state. The terms in $-\mathbf{q}$ give a contribution of the form, with $\omega_{\mathbf{q}} = \omega_{-\mathbf{q}}$,

$$\sum_{\mathbf{k}\mathbf{k}'} c^+_{\mathbf{k}'-\mathbf{q}} c_{\mathbf{k}'} c^+_{\mathbf{k}+\mathbf{q}} c_{\mathbf{k}} \left(\frac{1}{\varepsilon_{\mathbf{k}} - \varepsilon_{\mathbf{k}+\mathbf{q}} - \omega_{\mathbf{q}}} - \frac{1}{\varepsilon_{\mathbf{k}'} + \omega_{\mathbf{q}} - \varepsilon_{\mathbf{k}'-\mathbf{q}}} \right).$$

Next interchange $\mathbf{k}$ and $\mathbf{k}'$, and reorder the operators: the sum of the terms in $\mathbf{q}$ and $-\mathbf{q}$ is

$$\tfrac{1}{4}D^2 \sum_{\mathbf{q}} \sum_{\mathbf{k}\mathbf{k}'} c^+_{\mathbf{k}'+\mathbf{q}} c_{\mathbf{k}'} c^+_{\mathbf{k}-\mathbf{q}} c_{\mathbf{k}} \cdot \frac{4\omega_{\mathbf{q}}}{(\varepsilon_{\mathbf{k}} - \varepsilon_{\mathbf{k}-\mathbf{q}})^2 - \omega_{\mathbf{q}}^2},$$

so that the electron-electron interaction may be written as

$$H'' = D^2 \sum_{\mathbf{q}} \sum_{\mathbf{k}\mathbf{k}'} \frac{\omega_{\mathbf{q}}}{(\varepsilon_{\mathbf{k}} - \varepsilon_{\mathbf{k}-\mathbf{q}})^2 - \omega_{\mathbf{q}}^2} c^+_{\mathbf{k}'+\mathbf{q}} c_{\mathbf{k}'} c^+_{\mathbf{k}-\mathbf{q}} c_{\mathbf{k}}. \tag{9}$$

The graph for H'' is shown in Fig. 1.

The electron-electron interaction (9) is attractive (negative) for excitation energies $|\epsilon_{\mathbf{k}\pm\mathbf{q}} - \epsilon_{\mathbf{k}}| < \omega_{\mathbf{q}}$; it is repulsive otherwise. Even in the attractive region the interaction is opposed by the screened coulomb repulsion, but for sufficiently large values of the interaction constant D the phonon interaction is dominant. We assume for simplicity that in superconductors the attraction is dominant when

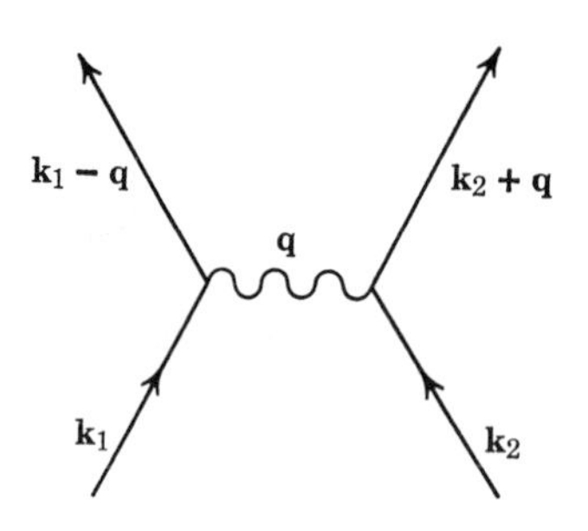

FIG. 1. Electron-electron indirect interaction through the lattice phonons.

$$\varepsilon_F - \omega_D < \varepsilon_{\mathbf{k}}, \varepsilon_{\mathbf{k}\pm\mathbf{q}} < \varepsilon_F + \omega_D, \tag{10}$$

where ω_D is the Debye energy—most of the zero-point phonons are near the Debye limit. The repulsive region of (9) is of little consequence or interest, so we drop the repulsive parts from the hamiltonian and write (9) in the simplified form

$$H'' = -V \sum_{\mathbf{q}} \sum_{\mathbf{kk'}} c^+_{\mathbf{k'+q}} c_{\mathbf{k'}} c^+_{\mathbf{k-q}} c_{\mathbf{k}}, \tag{11}$$

where we sum (Fig. 2) only over $\mathbf{q}$'s which satisfy (10); V is taken to be a positive constant. The simplified hamiltonian (11) is believed to contain the essential features of the problem. Our task now is to study the properties of a fermi gas under an attractive two-body interaction with the cutoff (10).

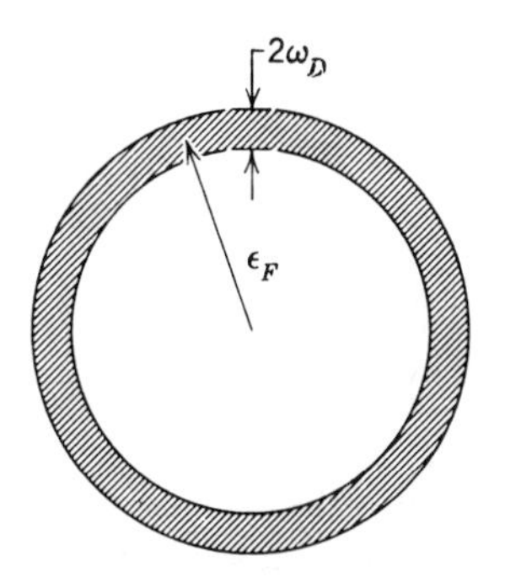

FIG. 2. Range of one-electron states (shaded area) in $\mathbf{k}$ space used in forming the BCS ground state. The thickness of the shell is twice the Debye energy. The region below $\varepsilon_F - \omega_D$ is entirely filled, but plays no active part in the superconducting properties. The states most involved actually lie in a shell of thickness $\approx 4T_c$ about the fermi surface (Fig. 4).

BOUND ELECTRON PAIRS IN A FERMI GAS

The first suggestion that unusual properties would result from attractive interactions in a fermi gas was made by Cooper[1], who proved that the fermi sea is unstable with respect to the formation of bound pairs. This important result (which we prove below) led directly to the analysis by BCS of the superconducting state. It must be emphasized that Cooper's calculation is not in itself a theory of superconductivity, but it suggests the lines such a theory might take. The BCS theory handles the many-electron problem, which is more complex than the pair problem. The BCS matrix element argument following (34) below shows why the pairs are important. The density of superconducting electrons is such that $\sim 10^3$ or more Cooper pairs would have to overlap appreciably.

We consider two free electrons, with an antisymmetric spin state. The unperturbed eigenfunction of the pair is, in unit volume,

$$\varphi(\mathbf{k}_1\mathbf{k}_2;\mathbf{x}_1\mathbf{x}_2) = e^{i(\mathbf{k}_1\cdot\mathbf{x}_1+\mathbf{k}_2\cdot\mathbf{x}_2)}. \tag{12}$$

Introduce the center-of-mass and relative-motion coordinates

$$\mathbf{X} = \tfrac{1}{2}(\mathbf{x}_1 + \mathbf{x}_2); \qquad \mathbf{x} = \mathbf{x}_1 - \mathbf{x}_2; \tag{13}$$

$$\mathbf{K} = \mathbf{k}_1 + \mathbf{k}_2; \qquad \mathbf{k} = \tfrac{1}{2}(\mathbf{k}_1 - \mathbf{k}_2); \tag{14}$$

then (12) takes the form

$$\varphi(\mathbf{Kk};\mathbf{Xx}) = e^{i(\mathbf{K}\cdot\mathbf{X}+\mathbf{k}\cdot\mathbf{x})}, \tag{15}$$

on substitution. The kinetic energy of the state (15) is $(1/m)(\frac{1}{4}K^2 + k^2)$. We now examine for convenience only states having $\mathbf{K} = 0$, so that $\mathbf{k}_1 = \mathbf{k}$; $\mathbf{k}_2 = -\mathbf{k}$. That is, the one-electron states are involved in pairs $\pm\mathbf{k}$.

Next include the electron-electron interaction (11) in the hamiltonian of the problem. We look for an eigenfunction of

$$H = \frac{1}{2m}(p_1^2 + p_2^2) + H'' = \frac{1}{m}p^2 + H'' \tag{16}$$

of the form

$$\chi(\mathbf{x}) = \sum_{\mathbf{k}} g_{\mathbf{k}} e^{i\mathbf{k}\cdot\mathbf{x}} = \sum_{\mathbf{k}} g_{\mathbf{k}} e^{i\mathbf{k}\cdot\mathbf{x}_1} e^{-i\mathbf{k}\cdot\mathbf{x}_2}, \tag{17}$$

using (13). Now if λ is the eigenvalue, $(H - \lambda)\chi(\mathbf{x}) = 0$, so that on taking a matrix element we have the secular equation

$$\int d\mathbf{x}\, e^{-i\mathbf{k}\cdot(\mathbf{x}_1-\mathbf{x}_2)}(H - \lambda)\sum_{\mathbf{k}'} g_{\mathbf{k}'} e^{i\mathbf{k}'\cdot(\mathbf{x}_1-\mathbf{x}_2)} = 0, \tag{18}$$

or, with $\varepsilon_{\mathbf{k}} = k^2/m$,

$$(\varepsilon_{\mathbf{k}} - \lambda)g_{\mathbf{k}} + \sum_{\mathbf{k}'} g_{\mathbf{k}'}\langle\mathbf{k},-\mathbf{k}|H''|\mathbf{k}',-\mathbf{k}'\rangle = 0, \tag{19}$$

where

$$\mathbf{k} = \mathbf{k}' + \mathbf{q} \text{ and } -\mathbf{k} = -\mathbf{k}' - \mathbf{q}. \tag{20}$$

If $\rho(\varepsilon)$ is the density of two-electron states $\mathbf{k}$, $-\mathbf{k}$ per unit energy range, the secular equation becomes

$$(\epsilon - \lambda)g(\varepsilon) + \int d\varepsilon'\, \rho(\varepsilon')g(\varepsilon')\langle\varepsilon|H''|\varepsilon'\rangle = 0. \tag{21}$$

In agreement with (11), we take, with V positive,

$$\langle\varepsilon|H''|\varepsilon'\rangle = -V \tag{22}$$

for an energy range $\pm\omega_D$ of one electron relative to the other; outside this range we take the interaction as zero. Let us suppose the packet (17) is made up of one-electron states above the top of the fermi sea, between ε_F and $\varepsilon_F + \omega_D$, or between k_F and k_m, where k_m is defined by

$$\frac{1}{2m}(k_m^2 - k_F^2) = \varepsilon_m - \varepsilon_F = \omega_D. \tag{23}$$

Then the secular equation (21) becomes

$$(\varepsilon - \lambda)g(\varepsilon) = V \int_{2\varepsilon_F}^{2\varepsilon_m} d\varepsilon' \, \rho(\varepsilon')g(\varepsilon') = C, \tag{24}$$

a constant independent of ε.
Thus

$$g(\varepsilon) = \frac{C}{\varepsilon - \lambda}. \tag{25}$$

Using this solution in the secular equation, (24) becomes

$$1 - V \int_{2\varepsilon_F}^{2\varepsilon_m} d\varepsilon' \, \frac{\rho(\varepsilon')}{\varepsilon' - \lambda} = 0, \tag{26}$$

where the limits refer to a pair. Over the small energy range involved one may replace $\rho(\varepsilon')$ by the constant ρ_F, the value at the fermi level, so that

$$\frac{1}{\rho_F V} = \int_{2\varepsilon_F}^{2\varepsilon_m} \frac{d\varepsilon'}{\varepsilon' - \lambda} = \log \frac{2\varepsilon_m - \lambda}{2\varepsilon_F - \lambda} = \log \frac{2\varepsilon_m - 2\varepsilon_F + \Delta}{\Delta}, \tag{27}$$

where we have written the lowest eigenvalue λ_0 as

$$\lambda_0 = 2\varepsilon_F - \Delta. \tag{28}$$

Then

$$\frac{\Delta}{2\varepsilon_m - 2\varepsilon_F + \Delta} = e^{-1/\rho_F V}, \tag{29}$$

or

$$\Delta = \frac{2\omega_D}{e^{1/\rho_F V} - 1}. \tag{30}$$

This is the binding energy of the pair with respect to the fermi level. We have thus found that for V positive (attractive interaction) we lower the energy of the system by exciting a pair of electrons above the fermi level; therefore the fermi sea is unstable. This instability modifies the fermi sea in an important way—actually a high density of pairs are formed and we must study the fermi surface carefully, taking account of the exclusion principle.

Observe that (30) may not be written as a power series in V. Therefore a perturbation calculation summed to all orders could not give the present result.

SUPERCONDUCTING GROUND STATE

We now consider the ground state of a fermi gas in the presence of the interaction (11). We write the complete hamiltonian with the one-electron Bloch energy $\varepsilon_{\mathbf{k}}$ referred to the fermi level as zero. Then on rearranging (11) we have

$$H = \sum \varepsilon_{\mathbf{k}} c_{\mathbf{k}}^{+} c_{\mathbf{k}} - V \sum c_{\mathbf{k}'+\mathbf{q}}^{+} c_{\mathbf{k}-\mathbf{q}}^{+} c_{\mathbf{k}} c_{\mathbf{k}'}, \tag{31}$$

with spin indices omitted. We recall from (5.23) and (5.24) that the fermion operators satisfy

$$c_j \Phi(n_1, \cdots, n_j, \cdots) = \theta_j n_j \Phi(n_1, \cdots, 1 - n_j, \cdots); \tag{32}$$

$$c_j^{+} \Phi(n_1, \cdots, n_j, \cdots) = \theta_j (1 - n_j) \Phi(n_1, \cdots, 1 - n_j, \cdots), \tag{33}$$

where

$$\theta_j = (-1)^{\nu_j}; \qquad \nu_j = \sum_{p=1}^{j-1} n_p. \tag{34}$$

That is, in operations with c_j, c_j^{+} there occurs multiplication by ± 1, according to the evenness or oddness of the number of occupied states which precede the state j in the ordering of states which has been adopted.

The alternation of sign in these operations is of *critical importance* to the ground state of superconductors. The interaction term in the hamiltonian connects together a large number of nearly degenerate configurations or sets of occupation numbers. If all the terms in H'' are negative, we can obtain a state of low energy, just as for the Cooper pair. But because of the sign alternation, there are for configurations picked at random about as many positive as negative matrix elements of V. This effect must be controlled by a special selection of configurations to avoid reduction of the average matrix element and reduction of the net effect of the interaction. The result for the Cooper pair suggests how this selection may be made.

We can see the alternation in sign for a simple example. Consider first $c_1^{+} c_4 \Phi(00111) = -c_1^{+} \Phi(00101) = -\Phi(10101)$; further,

$$c_1^{+} c_3 \Phi(00111) = c_1^{+} \Phi(00011) = \Phi(10011).$$

Thus the sign of the matrix element $\langle 10101 | c_1^{+} c_4 | 00111 \rangle$ is opposite to that of $\langle 10011 | c_1^{+} c_3 | 00111 \rangle$.

We can generate a coherent state of low energy by working with a subset of configurations between which the matrix elements of the interaction are always negative. This property is assured for V positive in the hamiltonian (31) if the Bloch states are always occupied only in pairs. Thus in our subspace we allow $\Phi(11;00;11)$, $\Phi(00;11;00)$, etc., but not $\Phi(10;10;11)$ or $\Phi(11;10;00)$, etc.; here the ordering of indices is arranged in pairs by the semicolons. If one member of the pair is occupied in any configuration, then the other member is also occupied. The interaction itself conserves wavevector, so that we will be most concerned with configurations for which all pairs have the same total momentum $\mathbf{k} + \mathbf{k}' = \mathbf{K}$, where $\mathbf{K}$ is usually 0. If $\mathbf{K}$ is zero, the pair is $\mathbf{k}$, $-\mathbf{k}$.

We have not discussed the spin. The exchange energy will usually be lower for an antiparallel $\downarrow\uparrow$ pair than for a pair of parallel spin. We shall work with antiparallel spins. The spin will hereafter not be considered explicitly. *We adopt below the convention that a state written explicitly as* $\mathbf{k}$ *has spin* $\uparrow$, *whereas one written as* $-\mathbf{k}$ *has spin* $\downarrow$. We suppose always that $\varepsilon_{\mathbf{k}} = \varepsilon_{-\mathbf{k}}$.

For the reasons just enumerated it is sufficient for the ground state if we work in the pair subspace using the truncated hamiltonian in which the interaction terms contain only a part of the interaction in (11):

$$H_{\text{red}} = \sum \varepsilon_{\mathbf{k}}(c_{\mathbf{k}}^{+}c_{\mathbf{k}} + c_{-\mathbf{k}}^{+}c_{-\mathbf{k}}) - V \sum c_{\mathbf{k}'}^{+}c_{-\mathbf{k}'}^{+}c_{-\mathbf{k}}c_{\mathbf{k}}. \tag{35}$$

This is known as the BCS reduced hamiltonian; it operates only within the pair subspace.

The approximate ground-state Φ_0 is shown below to be of the form

$$\Phi_0 = \prod_{\mathbf{k}} (u_{\mathbf{k}} + v_{\mathbf{k}}c_{\mathbf{k}}^{+}c_{-\mathbf{k}}^{+})\Phi_{\text{vac}}, \tag{36}$$

where Φ_{vac} is the true vacuum; $u_{\mathbf{k}}$, $v_{\mathbf{k}}$ are constants. The total number of electrons is a constant of the motion of the hamiltonian, but our state Φ_0 is not diagonal in the number of electrons. In the same way a Bloch wall in ferromagnetic domain theory is not usually described in such a way that the total spin is a constant of the motion. BCS have shown for a macroscopic system the probable number of pairs in Φ_0 is very strongly peaked about the most probable value, and so we use Φ_0 just as in the spirit of the grand canonical ensemble. In this connection see Problem 3. The ground state Φ_0 contains only pairs. The relation (36) implies $v_{\mathbf{k}} = -v_{-\mathbf{k}}$, because the c^{+} anticommute.

SOLUTION OF THE BCS EQUATION—SPIN-ANALOG METHOD

The most physical method for studying the properties of the reduced hamiltonian is due to Anderson.[7] We rearrange the hamiltonian (135) as, with $\hat{n}_{\mathbf{k}} = c_{\mathbf{k}}^{+}c_{\mathbf{k}}$,

$$H_{\text{red}} = -\sum \varepsilon_{\mathbf{k}}(1 - \hat{n}_{\mathbf{k}} - \hat{n}_{-\mathbf{k}}) - V \sum{}' c_{\mathbf{k}'}^{+}c_{-\mathbf{k}'}^{+}c_{-\mathbf{k}}c_{\mathbf{k}}, \tag{37}$$

[7] P. W. Anderson, *Phys. Rev.* **112,** 1900 (1958).

setting $\Sigma\,\varepsilon_{\mathbf{k}} = 0$ or a constant for states symmetrical about the fermi level in the range of energy $\pm\omega_D$, according to (10).

We consider first only the subspace of states defined by

$$n_{\mathbf{k}} = n_{-\mathbf{k}}; \tag{38}$$

this is the subspace in which both states $\mathbf{k}$, $-\mathbf{k}$ of a Cooper pair are occupied or both are empty. Consider the operator $(1 - \hat{n}_{\mathbf{k}} - \hat{n}_{-\mathbf{k}})$:

$$\begin{aligned} (1 - \hat{n}_{\mathbf{k}} - \hat{n}_{-\mathbf{k}})\Phi(1_{\mathbf{k}}1_{-\mathbf{k}}) &= -\Phi(1_{\mathbf{k}}1_{-\mathbf{k}}); \\ (1 - \hat{n}_{\mathbf{k}} - \hat{n}_{-\mathbf{k}})\Phi(0_{\mathbf{k}}0_{-\mathbf{k}}) &= \Phi(0_{\mathbf{k}}0_{-\mathbf{k}}). \end{aligned} \tag{39}$$

Thus this operator may be represented by the pauli matrix σ_z:

$$1 - \hat{n}_{\mathbf{k}} - \hat{n}_{-\mathbf{k}} = \begin{pmatrix} 1 & 0 \\ 0 & -1 \end{pmatrix} = \sigma_z, \tag{40}$$

if the pair state is represented in the subspace by the column matrix

$$\begin{pmatrix} 1 \\ 0 \end{pmatrix} = \text{pair empty} \leftrightarrow \Phi(0_{\mathbf{k}}0_{-\mathbf{k}}) \leftrightarrow \alpha_{\mathbf{k}} \leftrightarrow \text{“spin up”}$$

$$\begin{pmatrix} 0 \\ 1 \end{pmatrix} = \text{pair occupied} \leftrightarrow \Phi(1_{\mathbf{k}}1_{-\mathbf{k}}) \leftrightarrow \beta_{\mathbf{k}} \leftrightarrow \text{“spin down”}$$

where α, β here are the usual spin functions for spin up and spin down, respectively.

The combinations of operators in the potential energy term may be represented by other pauli matrices. We know that

$$c_{\mathbf{k}}^+ c_{-\mathbf{k}}^+ \Phi(1_{\mathbf{k}}1_{-\mathbf{k}}) = 0; \qquad c_{\mathbf{k}}^+ c_{-\mathbf{k}}^+ \Phi(0_{\mathbf{k}}0_{-\mathbf{k}}) = \Phi(1_{\mathbf{k}}1_{-\mathbf{k}}), \tag{42}$$

and

$$\sigma^- = \sigma_x - i\sigma_y = \begin{pmatrix} 0 & 0 \\ 2 & 0 \end{pmatrix}, \tag{43}$$

so that

$$c_{\mathbf{k}}^+ c_{-\mathbf{k}}^+ = \tfrac{1}{2}\sigma_{\mathbf{k}}^-. \tag{44}$$

The hermitian conjugate is

$$c_{-\mathbf{k}} c_{\mathbf{k}} = \tfrac{1}{2}\sigma_{\mathbf{k}}^+. \tag{45}$$

The hamiltonian becomes, in terms of pauli operators,

$$\begin{aligned} H_{\text{red}} &= -\sum \varepsilon_{\mathbf{k}}\sigma_{\mathbf{k}z} - \tfrac{1}{4}V \sum_{\mathbf{k}\mathbf{k}'}{}' \sigma_{\mathbf{k}'}^- \sigma_{\mathbf{k}}^+ \\ &= -\sum \varepsilon_{\mathbf{k}}\sigma_{\mathbf{k}z} - \tfrac{1}{4}V \sum{}' (\sigma_{\mathbf{k}'x}\sigma_{\mathbf{k}x} + \sigma_{\mathbf{k}'y}\sigma_{\mathbf{k}y}), \end{aligned} \tag{46}$$

on taking account of the automatic symmetrization when summing over all $\mathbf{k}'$ and $\mathbf{k}$. Within the subspace of pair states, (46) is an exact hamiltonian. It is essential to remember that we are using the σ's here not as operators on actual spins, but as operators which create or destroy pair states $\pm\mathbf{k}$. But we may use all the methods developed for the theory of ferromagnetism to find accurate approximate solutions of (46). In Problem 3 we show that the hamiltonian can be solved exactly in the strong coupling limit where all $\varepsilon_\mathbf{k} = 0$.

We define a fictitious magnetic field $\mathcal{H}_\mathbf{k}$ acting on $\sigma_\mathbf{k}$ by

$$\mathcal{H}_\mathbf{k} = \varepsilon_\mathbf{k}\hat{\mathbf{z}} + \tfrac{1}{2}V \sum_{\mathbf{k}'}{}' (\sigma_{\mathbf{k}'x}\hat{\mathbf{x}} + \sigma_{\mathbf{k}'y}\hat{\mathbf{y}}), \tag{47}$$

where $\hat{\mathbf{x}}$, $\hat{\mathbf{y}}$, $\hat{\mathbf{z}}$ are unit vectors in the directions of the coordinate axes. The form of the BCS hamiltonian suggests the application of the molecular field approximation. We rotate the spin vectors $\sigma_\mathbf{k}$ into the best possible classical arrangement, which means that each spin $\mathbf{k}$ should be parallel to the pseudofield $\mathcal{H}_\mathbf{k}$ acting on it. The molecular field approximation is very good here because the number of spins involved in $\mathcal{H}_\mathbf{k}$ is very large, so that it may be treated as a classical vector.

In the unperturbed fermi sea ($V = 0$), the effective field is $\varepsilon_\mathbf{k}\hat{\mathbf{z}}$, where $\varepsilon_\mathbf{k}$ is positive for energies above the fermi surface and negative for energies below. The stable spin state has the spins up (pairs empty) for energies above the fermi surface, and the spins down (pairs occupied) for energies below the fermi surface. The spins reverse direction precisely at the fermi energy.

Now consider an attractive interaction, that is, V positive. Exactly at the fermi surface $\varepsilon_\mathbf{k} = 0$, so that the only field acting on a spin $|\mathbf{k}| = k_F$ arises from the interaction term in the effective field:

$$\tfrac{1}{2}V \sum_{\mathbf{k}'}{}' (\sigma_{\mathbf{k}'x}\hat{\mathbf{x}} + \sigma_{\mathbf{k}'y}\hat{\mathbf{y}}). \tag{48}$$

Suppose the spin at k_F is horizontal and along $\hat{\mathbf{x}}$. Because of the interaction the spin will tend to make nearby spins line up along $\hat{\mathbf{x}}$; but as we go away from k_F the kinetic-energy terms $\epsilon_\mathbf{k}\hat{\mathbf{z}}$ will pull the spins more and more into the $\pm\hat{\mathbf{z}}$ directions. The situation is quite like the Bloch wall in ferromagnets; in the present problem we have a domain wall in $\mathbf{k}$ space, with states rotating smoothly from occupied to empty.

In the molecular field approximation to the ground state in the presence of the interaction we quantize each spin parallel to the average field which it sees; we treat the field itself as a classical vector.

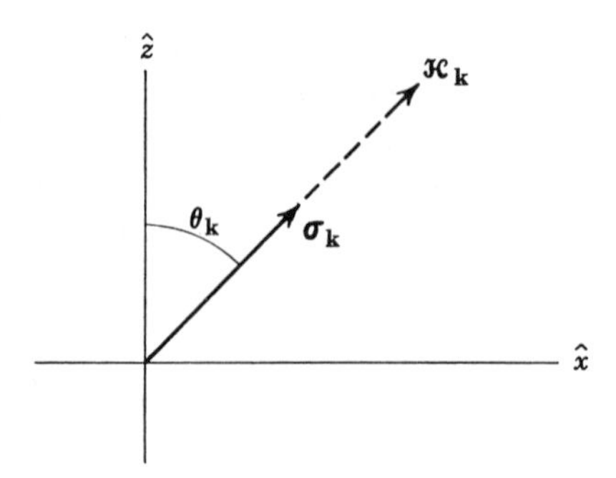

FIG. 3. Molecular field method of solution of the spin-analog BCS hamiltonian.

The molecular field is given by (47) with all $\sigma_{\mathbf{k}'y} = 0$ if, for convenience, the axes are chosen such that the spins lie in the xz plane. Then, as in Fig. 3,

$$\frac{\mathcal{H}_{\mathbf{k}x}}{\mathcal{H}_{\mathbf{k}z}} = \frac{\sigma_{\mathbf{k}x}}{\sigma_{\mathbf{k}z}} = \frac{\frac{1}{2}V \sum_{\mathbf{k}'}' \sigma_{\mathbf{k}'x}}{\epsilon_{\mathbf{k}}} = \tan \theta_{\mathbf{k}}. \tag{49}$$

Now $\sigma_{\mathbf{k}'x} = \sin \theta_{\mathbf{k}'}$, giving the BCS integral equation

$$\boxed{\tan \theta_{\mathbf{k}} = (V/2\varepsilon_{\mathbf{k}}) \sum_{\mathbf{k}'}' \sin \theta_{\mathbf{k}'}.} \tag{50}$$

To solve (50) we set

$$\Delta = \tfrac{1}{2}V \sum_{\mathbf{k}'}' \sin \theta_{\mathbf{k}'}, \tag{51}$$

so that (50) gives $\tan \theta_{\mathbf{k}} = \Delta/\varepsilon_{\mathbf{k}}$; by trigonometry

$$\sin \theta_{\mathbf{k}} = \frac{\Delta}{(\Delta^2 + \varepsilon_{\mathbf{k}}^2)^{1/2}}; \qquad \cos \theta_{\mathbf{k}} = \frac{\varepsilon_{\mathbf{k}}}{(\Delta^2 + \varepsilon_{\mathbf{k}}^2)^{1/2}}. \tag{52}$$

Using this expression for $\sin \theta_{\mathbf{k}'}$ in (51),

$$\Delta = \tfrac{1}{2}V \sum_{\mathbf{k}'}' \frac{\Delta}{(\Delta^2 + \varepsilon_{\mathbf{k}'}^2)^{1/2}}. \tag{53}$$

We replace the summation by an integral. The limits are determined by the region ω_D to $-\omega_D$ within which V is attractive; ω_D is of the order of the Debye energy. Then the fundamental equation becomes

$$1 = \tfrac{1}{2}V\rho_F \int_{-\omega_D}^{\omega_D} \frac{d\varepsilon}{(\Delta^2 + \varepsilon^2)^{1/2}} = V\rho_F \sinh^{-1} (\omega_D/\Delta). \tag{54}$$

Here ρ_F is the density of states at the fermi level, but taken for one spin direction. The evaluation of the integral is elementary. From (54)

$$\Delta = \frac{\omega_D}{\sinh(1/\rho_F V)} \cong 2\omega_D e^{-1/V\rho_F}, \tag{55}$$

if $\rho_F V \ll 1$. This is the BCS solution for the energy-gap parameter Δ; we see that Δ is positive if V is positive.

The first approximation to the excitation spectrum is obtained as the energy $E_{\mathbf{k}}$ required to reverse a fictitious spin in the field $\mathfrak{H}_{\mathbf{k}}$. We have, using (47),

$$E_{\mathbf{k}} = 2|\mathfrak{H}_{\mathbf{k}}| = 2(\varepsilon_{\mathbf{k}}^2 + \Delta^2)^{1/2}, \tag{56}$$

where we are concerned only with the positive root. The minimum excitation energy is 2Δ. Thus there is an energy gap in the excitation spectrum of a superconductor. The gap has been detected in work on heat capacity, on the transmission of far infrared radiation through thin films, and on the tunneling of electrons through barriers. The fictitious spin-reversal corresponds to the excitation of a pair $\pm\mathbf{k}$ into a state orthogonal to the ground state of the same pair. Other excitations not included in our subspace are possible for which the two electrons in the excited state have different $\mathbf{k}$'s; we study these excitations later and will see that the energy gap is also 2Δ for such excitations. Low-lying excitations like magnons do not exist for the special hamiltonian (46) because of the long range of the interactions; it is not a question of summing nearest over neighbors only. If in the magnon problem one couples *all* spins equally, then there are no low energy excitations, apart from the uniform mode.

The expectation value of the ground-state energy of the superconducting state referred to the normal state is, from (46), (50), (51), and (54),

$$\begin{aligned} E_g &= -\sum_{\mathbf{k}} \varepsilon_{\mathbf{k}} \cos\theta_{\mathbf{k}} - \tfrac{1}{4}V \sum_{\mathbf{k}\mathbf{k}'}{}' \sigma_{\mathbf{k}'x}\sigma_{\mathbf{k}x} + \sum_{\mathbf{k}} |\varepsilon_{\mathbf{k}}| \\ &= -\sum_{\mathbf{k}} \varepsilon_{\mathbf{k}}(\cos\theta_{\mathbf{k}} + \tfrac{1}{2}\sin\theta_{\mathbf{k}}\tan\theta_{\mathbf{k}}) + \sum |\varepsilon_{\mathbf{k}}|. \end{aligned} \tag{57}$$

The $\Sigma\,|\varepsilon_{\mathbf{k}}|$ term takes account of the energy $2|\varepsilon_{\mathbf{k}}|$ per electron pair up to the fermi level in the normal state. If ρ_F is the density of states at the fermi level,

$$E_g = 2\rho_F \int_0^{\omega_D} d\varepsilon \left\{ \varepsilon - \frac{\varepsilon^2}{(\varepsilon^2 + \Delta^2)^{1/2}} \right\} - \frac{\Delta^2}{V}, \tag{58}$$

where we have used the relation

$$\sum \varepsilon_{\mathbf{k}} \sin \theta_{\mathbf{k}} \tan \theta_{\mathbf{k}} = \Delta^2 \sum \frac{1}{(\varepsilon_{\mathbf{k}}^2 + \Delta^2)^{1/2}} = \frac{2\Delta^2}{V}. \tag{59}$$

The integral in (58) is elementary; on eliminating V with (54) we have

$$E_g = \rho_F \omega_D^2 \left\{ 1 - \left[1 + \left(\frac{\Delta}{\omega_D} \right)^2 \right]^{1/2} \right\} = -\frac{2\rho_F \omega_D^2}{e^{2/\rho_F V} - 1} \cong -\tfrac{1}{2}\rho_F \Delta^2. \tag{60}$$

Thus as long as V is positive the coherent state is lower in energy than the normal state: the criterion for superconductivity is simply that $V > 0$. The critical magnetic field at absolute zero is obtained by equating E_g to $H_c^2/8\pi$ for a specimen of unit volume.

The transition temperature T_c may be determined by the molecular field method, exactly as in the theory of ferromagnetism. At a finite temperature T, the ensemble average spin is directed along the effective field $\mathcal{H}_{\mathbf{k}}$ and has the magnitude, as in *ISSP* Eq. (9.19),

$$\langle \sigma_{\mathbf{k}} \rangle = \tanh (\mathcal{H}_{\mathbf{k}}/T), \tag{61}$$

in units with the boltzmann constant absorbed in the temperature. The BCS integral equation (49) is modified accordingly at a finite temperature to take account of this decrease in $\sigma_{\mathbf{k}}$; with $\varepsilon_{\mathbf{k}}$ unchanged:

$$\tan \theta_{\mathbf{k}} = (V/2\varepsilon_{\mathbf{k}}) \sum_{\mathbf{k}'}{}' \tanh (\mathcal{H}_{\mathbf{k}'}/T) \sin \theta_{\mathbf{k}'} = \Delta/\varepsilon_{\mathbf{k}}, \tag{62}$$

with, from (56),

$$\mathcal{H}_{\mathbf{k}} = \{\varepsilon_{\mathbf{k}}^2 + \Delta^2(T)\}^{1/2}. \tag{63}$$

The transition $T = T_c$ occurs on this model when $\Delta = 0$, or

$$1 = V \sum{}' \frac{1}{2\varepsilon_{\mathbf{k}'}} \tanh \frac{\varepsilon_{\mathbf{k}'}}{T_c}, \tag{64}$$

using (52), (62) and (63). This result is correct for the spin analog model which, we recall, works entirely in the pair subspace—the allowed excited states are only the real pair excitations as in (103), p. 168. If we extend the space to allow single particle excitations, as in (102), p. 168 and (110), p. 169, we double the number of possible excitations. Doubling the number of excitations doubles the entropy, which is exactly equivalent in the free energy to doubling the temperature. Hence T_c in (64) is to be replaced by $2T_c$. The energy contribution to the free energy is not affected by doubling the excitations, because two single particle excitations have the same energy as one real pair excitation of the same $|\mathbf{k}|$, according to (167) and (168). On rewriting the modified result as an integral, we have

$$\frac{2}{V\rho_F} = \int_{-\omega_D}^{\omega_D} \frac{d\varepsilon}{\varepsilon} \tanh \frac{\varepsilon}{2T_c} = 2 \int_0^{\omega_D/2T_c} dx \, \frac{\tanh x}{x}, \tag{65}$$

which is the BCS result for T_c.

If $T_c \ll \omega_D$, we may replace $\tanh x$ by unity over most of the range of integration, down to $x \approx 1$, below which it is $\approx x$; the approximate value of the integral is $1 + \log (\omega_D/2T_c)$. More closely, on graphical integration,

$$\boxed{T_c = 1.14\omega_D e^{-1/\rho_F V};} \tag{66}$$

combined with (55) for $\rho_F V \ll 1$. The energy gap is

$$2\Delta = 3.5T_c. \tag{67}$$

Experimental values of $2\Delta/T_c$ are 3.5, 3.4, 4.1, 3.3 for Sn, Al, Pb, and Cd, respectively (reference 4, p. 243).

The isotope effect—the constancy of the product $T_c M^{1/2}$ observed when the isotopic mass M is varied in a given element—follows directly from (66), because ω_D is related directly to the frequency of lattice vibrations; we suppose that V is independent of M. The frequency of an oscillator of a given force constant is proportional to $M^{-1/2}$, and thus $T_c M^{1/2}$ = constant, for isotopic variations of a given chemical element. The known exceptions to this rule include Ru and Os, both transition elements; it is likely that d-shell polarization effects are responsible. Even in simple metals V is expected to depend somewhat on M, which should spoil the agreement of observation with the simple theory; see, for example, P. Morel and P. W. Anderson, *Phys. Rev.* **125,** 1263 (1962).

The spin-analog ground state we have described may be generated by spin-rotation operations from the true vacuum state in which all pairs are empty (spin up). To generate a state with the spin in the xz plane and quantized at an angle $\theta_\mathbf{k}$ with the z axis, we operate with the spin rotation operator U (Messiah, p. 534) for a rotation $\theta_\mathbf{k}$ about the y axis:

$$U = \cos \tfrac{1}{2}\theta_\mathbf{k} - i\sigma_{\mathbf{k}y} \sin \tfrac{1}{2}\theta_\mathbf{k} = \cos \tfrac{1}{2}\theta_\mathbf{k} - \tfrac{1}{2}(\sigma_\mathbf{k}^+ - \sigma_\mathbf{k}^-) \sin \tfrac{1}{2}\theta_\mathbf{k}, \tag{68}$$

but σ^+ on the vacuum gives zero and $\frac{1}{2}\sigma_\mathbf{k}^- = c_\mathbf{k}^+ c_{-\mathbf{k}}^+$. Thus the ground state is

$$\Phi_0 = \prod_\mathbf{k} (\cos \tfrac{1}{2}\theta_\mathbf{k} + c_\mathbf{k}^+ c_{-\mathbf{k}}^+ \sin \tfrac{1}{2}\theta_\mathbf{k})\Phi_{\text{vac}}. \tag{69}$$

This is just the BCS ground state, as discussed in (88), (92), and (107). Note that for $k \ll k_F$ we have $\frac{1}{2}\theta_{\mathbf{k}} = \frac{1}{2}\pi$, and Φ_0 in this region is entirely filled with electrons.

The ground state (69) is only an approximation to the exact ground state, because we have assumed a product form for (69) and the true eigenstate is bound to be much more complicated. The result of Problem 3 suggests that (69) is excellent. Orbach[8] has made similar calculations for a ferromagnetic Bloch wall, and finds the agreement between the exact and the semiclassical energies is good and improves as the number of spins increases. There now exist a variety of proofs that the BCS solution is exact to $0(1/N)$ for the reduced hamiltonian.

SOLUTION OF THE BCS EQUATION—EQUATION-OF-THE-MOTION METHOD

It is useful to consider another approach to the solution of the BCS equation. We form the equations of motion for $c_{\mathbf{k}}$ and $c^+_{-\mathbf{k}}$ from the reduced hamiltonian (35):

$$H_{\text{red}} = \Sigma\, \varepsilon_{\mathbf{k}}(c^+_{\mathbf{k}}c_{\mathbf{k}} + c^+_{-\mathbf{k}}c_{-\mathbf{k}}) - V\,\Sigma'\, c^+_{\mathbf{k}'}c^+_{-\mathbf{k}'}c_{-\mathbf{k}}c_{\mathbf{k}}. \tag{70}$$

Then, on evaluating commutators with the use of $c_{\mathbf{k}}c_{\mathbf{k}} \equiv 0$ and $c^+_{\mathbf{k}}c^+_{\mathbf{k}} \equiv 0$, we have

$$\begin{aligned} i\dot{c}_{\mathbf{k}} &= \varepsilon_{\mathbf{k}}c_{\mathbf{k}} - c^+_{-\mathbf{k}}V\,\Sigma'\, c_{-\mathbf{k}'}c_{\mathbf{k}'}; \\ i\dot{c}^+_{-\mathbf{k}} &= -\varepsilon_{\mathbf{k}}c^+_{-\mathbf{k}} - c_{\mathbf{k}}V\,\Sigma'\, c^+_{\mathbf{k}'}c^+_{-\mathbf{k}'}. \end{aligned} \tag{71}$$

We define

$$B_{\mathbf{k}} = \langle 0|c_{-\mathbf{k}}c_{\mathbf{k}}|0\rangle = -B_{-\mathbf{k}}; \qquad B^*_{\mathbf{k}} = \langle 0|c^+_{\mathbf{k}}c^+_{-\mathbf{k}}|0\rangle. \tag{72}$$

We understand $\langle 0|c_{-\mathbf{k}}c_{\mathbf{k}}|0\rangle$ to mean $\langle 0;N|c_{-\mathbf{k}}c_{\mathbf{k}}|0;N+2\rangle$; the final wavefunction contains a mixture of states with a small spread in the numbers of particles. These matrix elements are similar to those we handled in Chapter 2 in connection with the phonon spectrum of a condensed boson gas; they are handled quite naturally in superconductivity in the Green's function method of Gorkov [*Sov. Phys. JETP* **7,** 505 (1958)] as described in Chapter 21.

We set, for $|\varepsilon_{\mathbf{k}}| < \omega_D$,

$$\Delta_{\mathbf{k}} = V\sum_{\mathbf{k}'}{}' B_{\mathbf{k}'}; \qquad \Delta^*_{\mathbf{k}} = V\sum_{\mathbf{k}'}{}' B^*_{\mathbf{k}'}, \tag{73}$$

and, for $|\varepsilon_{\mathbf{k}}| > \omega_D$,

$$\Delta_{\mathbf{k}} = \Delta^*_{\mathbf{k}} \doteq 0. \tag{74}$$

[8] R. Orbach, *Phys. Rev.* **115,** 1181 (1959).

Then the linearized equations of motion are

$$(75) \qquad i\dot{c}_{\mathbf{k}} = \varepsilon_{\mathbf{k}} c_{\mathbf{k}} - \Delta_{\mathbf{k}} c^+_{-\mathbf{k}};$$

$$(76) \qquad i\dot{c}^+_{-\mathbf{k}} = -\varepsilon_{\mathbf{k}} c^+_{-\mathbf{k}} - \Delta^*_{\mathbf{k}} c_{\mathbf{k}}.$$

This linearization is simply a generalization of the Hartree-Fock procedure to include terms such as $c^+_{\mathbf{k}} c_{\mathbf{k}'} c_{-\mathbf{k}'}$. These equations have a solution of the form $e^{-i\lambda t}$ *if*

$$(77) \qquad \begin{vmatrix} \lambda - \varepsilon_{\mathbf{k}} & \Delta_{\mathbf{k}} \\ \Delta^*_{\mathbf{k}} & \lambda + \varepsilon_{\mathbf{k}} \end{vmatrix} = 0,$$

or

$$(78) \qquad \lambda_{\mathbf{k}} = (\varepsilon_{\mathbf{k}}^2 + \Delta^2)^{1/2},$$

where $\Delta^2 = \Delta_{\mathbf{k}}\Delta^*_{\mathbf{k}}$; we now neglect the **k**-dependence of Δ.

The eigenvectors of the equations (75) and (76) are of the form

$$(79) \qquad \boxed{\begin{array}{ll} \alpha_{\mathbf{k}} = u_{\mathbf{k}} c_{\mathbf{k}} - v_{\mathbf{k}} c^+_{-\mathbf{k}}; & \alpha_{-\mathbf{k}} = u_{\mathbf{k}} c_{-\mathbf{k}} + v_{\mathbf{k}} c^+_{\mathbf{k}}. \\ \alpha^+_{\mathbf{k}} = u_{\mathbf{k}} c^+_{\mathbf{k}} - v_{\mathbf{k}} c_{-\mathbf{k}}; & \alpha^+_{-\mathbf{k}} = u_{\mathbf{k}} c^+_{-\mathbf{k}} + v_{\mathbf{k}} c_{\mathbf{k}}. \end{array}}$$

The inverse relations are

$$(80) \qquad \boxed{\begin{array}{ll} c_{\mathbf{k}} = u_{\mathbf{k}} \alpha_{\mathbf{k}} + v_{\mathbf{k}} \alpha^+_{-\mathbf{k}}; & c_{-\mathbf{k}} = u_{\mathbf{k}} \alpha_{-\mathbf{k}} - v_{\mathbf{k}} \alpha^+_{\mathbf{k}}. \\ c^+_{\mathbf{k}} = u_{\mathbf{k}} \alpha^+_{\mathbf{k}} + v_{\mathbf{k}} \alpha_{-\mathbf{k}}; & c^+_{-\mathbf{k}} = u_{\mathbf{k}} \alpha^+_{-\mathbf{k}} - v_{\mathbf{k}} \alpha_{\mathbf{k}}. \end{array}}$$

Here $u_{\mathbf{k}}$, $v_{\mathbf{k}}$ are real; $u_{\mathbf{k}}$ is even and $v_{\mathbf{k}}$ is odd: $u_{\mathbf{k}} = u_{-\mathbf{k}}$; $v_{\mathbf{k}} = -v_{-\mathbf{k}}$. We verify that the α's satisfy fermion commutation rules if $u_{\mathbf{k}}^2 + v_{\mathbf{k}}^2 = 1$:

$$(81) \qquad \{\alpha_{\mathbf{k}},\alpha^+_{\mathbf{k}'}\} = u_{\mathbf{k}} u_{\mathbf{k}'} \{c_{\mathbf{k}},c^+_{\mathbf{k}'}\} + v_{\mathbf{k}} v_{\mathbf{k}'} \{c^+_{-\mathbf{k}},c_{-\mathbf{k}'}\} = \delta_{\mathbf{k}\mathbf{k}'}(u_{\mathbf{k}}^2 + v_{\mathbf{k}}^2).$$

Further,

$$(82) \qquad \{\alpha_{\mathbf{k}},\alpha_{-\mathbf{k}}\} = u_{\mathbf{k}} v_{\mathbf{k}} \{c_{\mathbf{k}},c^+_{\mathbf{k}}\} - v_{\mathbf{k}} u_{\mathbf{k}} \{c^+_{-\mathbf{k}},c_{-\mathbf{k}}\} = u_{\mathbf{k}} v_{\mathbf{k}} - v_{\mathbf{k}} u_{\mathbf{k}} = 0.$$

If we substitute (80) in (75), we have, for $\alpha_{\mathbf{k}} \propto e^{-i\lambda t}$,

$$(83) \qquad \lambda u_{\mathbf{k}} = \varepsilon_{\mathbf{k}} u_{\mathbf{k}} + \Delta v_{\mathbf{k}};$$

on squaring

$$(84) \qquad \lambda^2 u_{\mathbf{k}}^2 = \varepsilon_{\mathbf{k}}^2 u_{\mathbf{k}}^2 + \Delta^2 v_{\mathbf{k}}^2 + 2\varepsilon_{\mathbf{k}}\,\Delta u_{\mathbf{k}} v_{\mathbf{k}} = (\varepsilon_{\mathbf{k}}^2 + \Delta^2) u_{\mathbf{k}}^2,$$

using (78). Thus

$$(85) \qquad \Delta^2(u_{\mathbf{k}}^2 - v_{\mathbf{k}}^2) = 2\varepsilon_{\mathbf{k}}\,\Delta u_{\mathbf{k}} v_{\mathbf{k}}.$$

Let us represent u, v by

$$u_{\mathbf{k}} = \cos \tfrac{1}{2}\theta_{\mathbf{k}}; \qquad v_{\mathbf{k}} = \sin \tfrac{1}{2}\theta_{\mathbf{k}}; \tag{86}$$

then (85) becomes

$$\Delta \cos \theta_{\mathbf{k}} = \varepsilon_{\mathbf{k}} \sin \theta_{\mathbf{k}}; \qquad \tan \theta_{\mathbf{k}} = \Delta/\varepsilon_{\mathbf{k}}, \tag{87}$$

so that, from (52), $\theta_{\mathbf{k}}$ has the identical meaning as in the spin-analog method.

Ground-State Wavefunction. We show now that the ground state of the system in terms of the quasiparticle operator $\alpha_{\mathbf{k}}$ is

$$\Phi_0 = \prod_{\mathbf{k}} \alpha_{-\mathbf{k}}\alpha_{\mathbf{k}}\Phi_{\text{vac}}, \tag{88}$$

or

$$\Phi_0 = \prod_{\mathbf{k}} (-v_{\mathbf{k}})(u_{\mathbf{k}} + v_{\mathbf{k}}c_{\mathbf{k}}^+ c_{-\mathbf{k}}^+)\Phi_{\text{vac}}. \tag{89}$$

We may normalize Φ_0 by omitting the factors $(-v_{\mathbf{k}})$, for then

$$\begin{aligned} \langle \Phi_0 | \Phi_0 \rangle &= \langle \Phi_{\text{vac}} | \Pi (u_{\mathbf{k}} + v_{\mathbf{k}} c_{-\mathbf{k}} c_{\mathbf{k}})(u_{\mathbf{k}} + v_{\mathbf{k}} c_{\mathbf{k}}^+ c_{-\mathbf{k}}^+) | \Phi_{\text{vac}} \rangle \\ &= \Pi (u_{\mathbf{k}}^2 + v_{\mathbf{k}}^2) \langle \Phi_{\text{vac}} | \Phi_{\text{vac}} \rangle. \end{aligned} \tag{90}$$

We verify that (88) is the ground state: we have for the quasiparticle annihilation operator

$$\alpha_{\mathbf{k}'}\Phi_0 = \alpha_{\mathbf{k}'}\alpha_{-\mathbf{k}'}\alpha_{\mathbf{k}'} \prod_{\mathbf{k}}{}' \alpha_{-\mathbf{k}}\alpha_{\mathbf{k}}\Phi_{\text{vac}} = 0, \tag{91}$$

because $\alpha_{\mathbf{k}'}\alpha_{\mathbf{k}'} \equiv 0$ for a fermion operator. Thus the normalized ground state is

$$\boxed{\Phi_0 = \prod_{\mathbf{k}} (u_{\mathbf{k}} + v_{\mathbf{k}} c_{\mathbf{k}}^+ c_{-\mathbf{k}}^+)\Phi_{\text{vac}}} \tag{92}$$

in the quasiparticle approximation. The values of $u_{\mathbf{k}}$, $v_{\mathbf{k}}$ are given by (52) and (86):

$$u_{\mathbf{k}}^2 = \cos^2 \tfrac{1}{2}\theta_{\mathbf{k}} = \tfrac{1}{2}(1 + \cos \theta_{\mathbf{k}}) = \tfrac{1}{2}[1 + (\varepsilon_{\mathbf{k}}/\lambda_{\mathbf{k}})]; \tag{93}$$

$$v_{\mathbf{k}}^2 = \sin^2 \tfrac{1}{2}\theta_{\mathbf{k}} = \tfrac{1}{2}(1 - \cos \theta_{\mathbf{k}}) = \tfrac{1}{2}[1 - (\varepsilon_{\mathbf{k}}/\lambda_{\mathbf{k}})]. \tag{94}$$

The functional dependence is shown in Fig. 4.

It is instructive to confirm the ground-state energy (60) calculated on the spin-analog model by a direct calculation using the wavefunc-

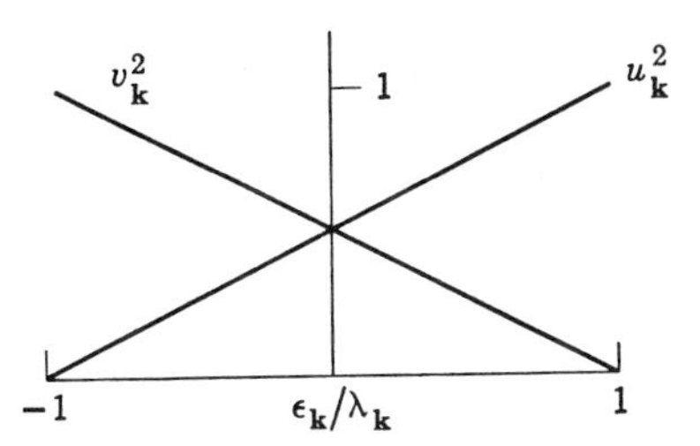

FIG. 4. Dependence of coefficients $u_{\mathbf{k}}$, $v_{\mathbf{k}}$ in the BCS state on $\varepsilon_{\mathbf{k}}/\lambda_{\mathbf{k}}$, where $\varepsilon_{\mathbf{k}}$ is the free particle kinetic energy referred to the fermi surface and $\lambda_{\mathbf{k}} = (\varepsilon_{\mathbf{k}}^2 + \Delta^2)^{1/2}$ is the quasiparticle energy parameter.

tion (88) or (92). The expectation value of the kinetic energy involves

$$(95) \qquad \langle\Phi_0|c_{\mathbf{k}'}^+c_{\mathbf{k}'}|\Phi_0\rangle = \langle\Phi_0|v_{\mathbf{k}'}{}^2\alpha_{-\mathbf{k}'}\alpha_{-\mathbf{k}'}^+|\Phi_0\rangle = v_{\mathbf{k}'}{}^2,$$

plus terms whose value is zero. The expectation value of the potential energy involves

$$(96) \qquad -\langle\Phi_0|c_{\mathbf{k}'}^+c_{-\mathbf{k}'}^+c_{-\mathbf{k}''}c_{\mathbf{k}''}|\Phi_0\rangle = \langle\Phi_0|u_{\mathbf{k}'}v_{\mathbf{k}'}u_{\mathbf{k}''}v_{\mathbf{k}''}\alpha_{-\mathbf{k}'}\alpha_{-\mathbf{k}'}^+\alpha_{-\mathbf{k}''}\alpha_{-\mathbf{k}''}^+|\Phi_0\rangle$$
$$= u_{\mathbf{k}'}v_{\mathbf{k}'}u_{\mathbf{k}''}v_{\mathbf{k}''},$$

plus terms whose value is zero. Then, with account taken in the kinetic energy of the two spin orientations,

$$(97) \qquad \langle\Phi_0|H_{\text{red}}|\Phi_0\rangle = 2\sum \varepsilon_{\mathbf{k}}v_{\mathbf{k}}{}^2 - V\sum{}' u_{\mathbf{k}}v_{\mathbf{k}}u_{\mathbf{k}'}v_{\mathbf{k}'}$$
$$= \sum \varepsilon_{\mathbf{k}}(1 - \cos\theta_{\mathbf{k}}) - \tfrac{1}{4}V\sum{}' \sin\theta_{\mathbf{k}}\sin\theta_{\mathbf{k}'}.$$

The last term on the right-hand side can be summed by applying (51) twice, giving $-\Delta^2/V$. The term $\sum \varepsilon_{\mathbf{k}}$ is zero. Thus

$$(98) \qquad \langle\Phi_0|H_{\text{red}}|\Phi_0\rangle = -\sum \varepsilon_{\mathbf{k}}\cos\theta_{\mathbf{k}} - (\Delta^2/V),$$

and the change of the ground-state energy in the superconducting state with respect to the vacuum is

$$(99) \qquad E_g = -\sum \varepsilon_{\mathbf{k}}\cos\theta_{\mathbf{k}} - (\Delta^2/V) + \sum |\varepsilon_{\mathbf{k}}|,$$

exactly as found earlier in (57) and (58); the expectation value of the energy in the state Φ_0 is identical with that given by the spin-analog method.

EXCITED STATES

With

$$(100) \qquad \Phi_0 = \prod \alpha_{-\mathbf{k}}\alpha_{\mathbf{k}}\Phi_{\text{vac}},$$

the products

(101) $$\Phi_{\mathbf{k}_1 \cdots \mathbf{k}_j} = \alpha_1^+ \cdots \alpha_j^+ \Phi_0$$

are easily seen to form a complete orthogonal set. The operation $\alpha_{\mathbf{k}}^+$ on Φ_0 creates an elementary excitation or quasiparticle with the properties of a fermion:

(102) $$\begin{aligned} \alpha_{\mathbf{k}'}^+ \Phi_0 &= (u_{\mathbf{k}'} c_{\mathbf{k}'}^+ - v_{\mathbf{k}'} c_{-\mathbf{k}'})(u_{\mathbf{k}'} + v_{\mathbf{k}'} c_{\mathbf{k}'}^+ c_{-\mathbf{k}'}^+) \prod' \cdots \\ &= c_{\mathbf{k}'}^+ \prod' \cdots . \end{aligned}$$

Here we have picked out of the product the term in $\mathbf{k}'$. The state $\alpha_{\mathbf{k}}^+ \Phi_0$ is known as a state with a *single particle excited.* The state has one particle in the state $\mathbf{k}$ and is missing a virtual pair from Φ_0.

The general two-particle excitation is $\alpha_{\mathbf{k}'}^+ \alpha_{\mathbf{k}''}^+$; when $\mathbf{k}'' = -\mathbf{k}'$ we have the special state

(103) $$\alpha_{\mathbf{k}'}^+ \alpha_{-\mathbf{k}'}^+ \Phi_0 = (u_{\mathbf{k}'} c_{\mathbf{k}'}^+ c_{-\mathbf{k}'}^+ - v_{\mathbf{k}'}) \prod' \cdots ,$$

which is said to have a *real pair excited,* with a virtual pair missing from Φ_0.

The number operator in the state $\mathbf{k}$ is, from (79),

(104) $$\hat{n}_{\mathbf{k}} = c_{\mathbf{k}}^+ c_{\mathbf{k}} = (u_{\mathbf{k}} \alpha_{\mathbf{k}}^+ + v_{\mathbf{k}} \alpha_{-\mathbf{k}})(u_{\mathbf{k}} \alpha_{\mathbf{k}} + v_{\mathbf{k}} \alpha_{-\mathbf{k}}^+).$$

The expectation value of $\hat{n}_{\mathbf{k}}$ in the ground state Φ_0 results only from the ordered term in $\alpha_{-\mathbf{k}} \alpha_{-\mathbf{k}}^+$, so that

(105) $$\langle \hat{n}_{\mathbf{k}} \rangle_0 = v_{\mathbf{k}}^2 = h_{\mathbf{k}} = \sin^2 \tfrac{1}{2}\theta_{\mathbf{k}}.$$

Here $h_{\mathbf{k}}$ is the symbol originally used by BCS to denote the expectation value of finding a pair $\mathbf{k}, -\mathbf{k}$ in the ground state Φ_0; the relation (105) establishes the connection with the expressions of BCS:

(106) $$h_{\mathbf{k}}^{1/2} = \sin \tfrac{1}{2}\theta_{\mathbf{k}} = v_{\mathbf{k}}; \qquad (1 - h_{\mathbf{k}})^{1/2} = \cos \tfrac{1}{2}\theta_{\mathbf{k}} = u_{\mathbf{k}};$$

(107) $$\Phi_0 = \prod_{\mathbf{k}} [(1 - h_{\mathbf{k}})^{1/2} + h_{\mathbf{k}}^{1/2} c_{\mathbf{k}}^+ c_{-\mathbf{k}}^+] \Phi_{\text{vac}}.$$

In our approximation, the quasiparticle hamiltonian is

(108) $$H = \sum \lambda_{\mathbf{k}} \alpha_{\mathbf{k}}^+ \alpha_{\mathbf{k}}.$$

The energies of the excited states are given directly as the sum of the roots $\lambda_{\mathbf{k}}$ of the equations of motion for the quasiparticle creation operators, (75) to (79):

(109) $$E_{1 \ldots j} = \sum_{i=1}^{j} \lambda_{\mathbf{k}_i}, \qquad \lambda_{\mathbf{k}} = (\varepsilon_{\mathbf{k}}^2 + \Delta^2)^{1/2},$$

for j excited quasiparticles. In the actual problem we must be careful to compare energies only for excited states having the same average total number of particles as the ground state. If the ground state has N pairs, we can work with an excited state having $2p$ excitations, leaving $(N - p)$ pairs. Thus the lowest allowed excitation of the type called single-particle excitation must have two particles excited:

$$\Phi_s = \alpha^+_{\mathbf{k}'}\alpha^+_{\mathbf{k}''}\Phi_0 = c^+_{\mathbf{k}'}c^+_{\mathbf{k}''}\prod_{\mathbf{k}}{}'\,(u_{\mathbf{k}} + v_{\mathbf{k}}c^+_{\mathbf{k}}c^+_{-\mathbf{k}})\Phi_{\mathrm{vac}}, \tag{110}$$

where the product over $\mathbf{k}$ is to be understood to include $(N - 1)$ pairs —without using our discretionary power it would include only $(N - 2)$ pairs. The real pair excitation

$$\Phi_p = \alpha^+_{\mathbf{k}'}\alpha^+_{-\mathbf{k}'}\Phi_0 \tag{111}$$

automatically has $(N - 1)$ unexcited pairs. For this excitation $E = 2\lambda_{k'}$. The calculation of the energy for the states (110) and (111) is left to Problem 6. It is found that the contribution of the potential energy vanishes for the excited particles; the increase of potential energy on excitation comes about from the reduction of the number of bonds among the unexcited pairs, because of the reduction of the number of unexcited pairs.

ELECTRODYNAMICS OF SUPERCONDUCTORS

The first objective of a theory of superconductivity is to explain the Meissner effect, the exclusion of magnetic flux from a superconductor. The Meissner effect follows directly [*ISSP*, Eq. (16.22)] if the London equation

$$\mathbf{j}(\mathbf{x}) = -\frac{1}{c\Lambda}\mathbf{A}(\mathbf{x}), \qquad \Lambda = \frac{m}{ne^2} = \frac{4\pi\lambda_L{}^2}{c^2}, \tag{112}$$

is satisfied. Here $\mathbf{j}$ is the current density; $\mathbf{A}$ is the vector potential; λ_L is the London penetration depth. We assume that $\mathbf{A}(\mathbf{x})$ varies only slowly in space; otherwise the London equation is replaced by an integral form due to Pippard. The integral form in effect determines $\mathbf{j}(\mathbf{x})$ by a weighted average of $\mathbf{A}$ over a distance of the order of a correlation length ξ_0 defined below. We treat the electrodynamics only in the ground state ($T = 0°$K).

We consider

$$\mathbf{A}(\mathbf{x}) = \mathbf{a}_{\mathbf{q}}e^{i\mathbf{q}\cdot\mathbf{x}}; \tag{113}$$

it is easier to treat an oscillatory potential than one linear in the coordinates. For the London equation we are concerned with the limit

$q \rightarrow 0$. The complex conjugate part of (113) need not be treated explicitly, because the following argument will apply equally to it.

The current density operator in the second-quantized form is

$$(114) \qquad \mathcal{J}(\mathbf{x}) = \frac{e}{2mi}(\Psi^+ \operatorname{grad} \Psi - \Psi \operatorname{grad} \Psi^+) - \frac{e^2}{mc}\Psi^+\mathbf{A}\Psi$$
$$= \mathcal{J}_P(\mathbf{x}) + \mathcal{J}_D(\mathbf{x}),$$

in terms of the paramagnetic and diamagnetic parts. The expression for $\mathcal{J}(\mathbf{x})$ follows as the symmetrized form of the velocity operator $[\mathbf{p} - (e/c)\mathbf{A}]/m$. We expand Ψ as, in unit volume,

$$(115) \qquad \Psi(\mathbf{x}) = \sum_{\mathbf{k}} c_{\mathbf{k}} e^{i\mathbf{k}\cdot\mathbf{x}},$$

whence

$$(116) \qquad \mathcal{J}_P(\mathbf{x}) = \frac{e}{2m}\sum_{\mathbf{kq}} c^+_{\mathbf{k}+\mathbf{q}} c_{\mathbf{k}} e^{-i\mathbf{q}\cdot\mathbf{x}}(2\mathbf{k} + \mathbf{q});$$

$$(117) \qquad \mathcal{J}_D(\mathbf{x}) = -\frac{e^2}{mc}\sum_{\mathbf{kq}} c^+_{\mathbf{k}+\mathbf{q}} c_{\mathbf{k}} e^{-i\mathbf{q}\cdot\mathbf{x}}\mathbf{A}(\mathbf{x}).$$

We suppose that the current vanishes in the absence of the field. We write the many-particle state Φ of the system as

$$(118) \qquad \Phi = \Phi_0 + \Phi_1(\mathbf{A}) + \cdots,$$

where Φ_0 is independent of $\mathbf{A}$ and Φ_1 is linear in $\mathbf{A}$:

$$(119) \qquad \Phi_1 = \sum_l{}' |l\rangle \frac{\langle l|H_1|0\rangle}{E_0 - E_l}.$$

Here to $O(\mathbf{A})$, in the coulomb gauge with div $\mathbf{A} = 0$,

$$(120) \qquad H_1 = -\int d^3x\, \Psi^+(\mathbf{x}) \frac{e}{mc}\mathbf{A}\cdot\mathbf{p}\Psi(\mathbf{x})$$
$$= -\frac{e}{mc}\sum_{\mathbf{kq}}\int d^3x\, c^+_{\mathbf{k}+\mathbf{q}} c_{\mathbf{k}} \mathbf{A}(\mathbf{x})\cdot\mathbf{k} e^{-i\mathbf{q}\cdot\mathbf{x}} = -\frac{e}{mc}\sum_{\mathbf{k}} c^+_{\mathbf{k}+\mathbf{q}} c_{\mathbf{k}}(\mathbf{k}\cdot\mathbf{a}_{\mathbf{q}}).$$

In the coulomb gauge we have

$$(121) \qquad \mathbf{q}\cdot\mathbf{a}_{\mathbf{q}} = 0.$$

The theory can be shown to be gauge-invariant, as discussed below.

To $O(\mathbf{A})$ the expectation value of the diamagnetic current operator is, letting $q \rightarrow 0$,

$$(122) \qquad \mathbf{j}_D(\mathbf{x}) = \langle 0|\mathcal{J}_D|0\rangle = -\frac{e^2}{mc}\mathbf{A}(\mathbf{x})\sum_{\mathbf{k}} c^+_{\mathbf{k}} c_{\mathbf{k}} = -\frac{ne^2}{mc}\mathbf{A}(\mathbf{x}),$$

which would be the London equation if $\mathbf{j}_P = 0$. The quantity n is the total electron concentration. The expectation value $\mathbf{j}_P(\mathbf{x})$ of the paramagnetic current operator over the state Φ is, to $O(\mathbf{A})$,

$$\mathbf{j}_P(\mathbf{x}) = \langle 0|\mathcal{J}_P|1\rangle + \langle 1|\mathcal{J}_P|0\rangle, \tag{123}$$

where $|1\rangle$ is defined by (119), so that

$$\langle 0|\mathcal{J}_P|1\rangle = \sum_l{}' \frac{1}{E_0 - E_l} \langle 0|\mathcal{J}_P|l\rangle\langle l|H_1|0\rangle. \tag{124}$$

We consider the matrix element $\langle l|H_1|0\rangle$; this has an unusual structure for the BCS ground and excited states. The structure is of central importance for the Meissner effect and for various other processes in superconductors. We have

$$\langle l|H_1|0\rangle = -(e/mc)\langle l| \sum_{\mathbf{k}} (\mathbf{k}\cdot\mathbf{a_q})c^+_{\mathbf{k+q}}c_{\mathbf{k}}|0\rangle. \tag{125}$$

Examine the contributions from a particular excited state l defined by

$$\Phi_l = \alpha^+_{\mathbf{k'+q}}\alpha^+_{-\mathbf{k'}}\Phi_0 = c^+_{\mathbf{k'+q}}c^+_{-\mathbf{k'}} \prod_{\mathbf{k}}{}' (u_{\mathbf{k}} + v_{\mathbf{k}}c^+_{\mathbf{k}}c^+_{-\mathbf{k}})\Phi_{\text{vac}}. \tag{126}$$

This state is connected with the ground state by the entry

$$(\mathbf{k'}\cdot\mathbf{a_q})c^+_{\mathbf{k'+q}}c_{\mathbf{k'}}$$

in the sum in (125), for, using (80),

$$c^+_{\mathbf{k'+q}}c_{\mathbf{k'}}\Phi_0 = u_{\mathbf{k'+q}}v_{\mathbf{k'}}\Phi_l \tag{127}$$

but there is a second entry which contributes,

$$[(-\mathbf{k'} - \mathbf{q})\cdot\mathbf{a_q}]c^+_{-\mathbf{k'}}c_{-\mathbf{k'}-\mathbf{q}},$$

for

$$c^+_{-\mathbf{k'}}c_{-\mathbf{k'}-\mathbf{q}}\Phi_0 = u_{\mathbf{k'}}v_{\mathbf{k'+q}}\Phi_l. \tag{128}$$

Noting that $(\mathbf{q}\cdot\mathbf{a_q}) = 0$, the total matrix element to a particular state l is

$$\begin{aligned}\langle l|H_1|0\rangle &= -(e/mc)(\mathbf{k'}\cdot\mathbf{a_q})(u_{\mathbf{k'+q}}v_{\mathbf{k'}} - u_{\mathbf{k'}}v_{\mathbf{k'+q}}) \\ &\propto \sin\tfrac{1}{2}(\theta_{\mathbf{k}} - \theta_{\mathbf{k+q}}),\end{aligned} \tag{129}$$

which $\rightarrow 0$ as $\mathbf{q}\rightarrow 0$, using (93) and (94). In this limit the energy denominator

$$E_0 - E_l \rightarrow -2(\varepsilon_k{}^2 + \Delta^2)^{1/2}, \tag{130}$$

and thus

$$\langle 0|\mathcal{J}_P|1\rangle \rightarrow 0 \tag{131}$$

as $\mathbf{q} \to 0$. We are left with the London equation

$$\mathbf{j}(\mathbf{x}) = -\frac{ne^2}{mc}\mathbf{A}(\mathbf{x}), \tag{132}$$

in this limit.

In the normal state the paramagnetic current approximately cancels the diamagnetic, as discussed by Bardeen.[9] The energy gap is zero in the normal state. For a normal insulator the virtual excited state is reached by a one-electron transition, and no cancellation as in (129) is found.

The gauge invariance of the theory has been demonstrated from several approaches. Schrieffer[10] gives a summary of the argument based on plasmon properties. A gauge transformation from the coulomb gauge means that one adds to our vector potential a longitudinal part $i\mathbf{q}\varphi(\mathbf{q})$. Such a term in the potential is coupled strongly to plasmon excitations and in fact shifts the plasmon coordinate. This shift does not change the plasmon frequency and thus does not change the physical properties of the system. Another approach to the gauge question is to note that a gauge transformation is equivalent to a transformation

$$\Psi \to \Psi e^{i\mathbf{K}\cdot\mathbf{x}} \tag{133}$$

on the state operator. A quantity such as $B_{\mathbf{k}}$ in (72) must be redefined, for

$$\begin{aligned}\langle 0|\Psi(\mathbf{y})\Psi(\mathbf{x})|0\rangle &\to \langle 0|\Psi(\mathbf{y})\Psi(\mathbf{x})|0\rangle e^{i\mathbf{K}\cdot(\mathbf{x}+\mathbf{y})} \\ &= \langle 0|\sum_{kk'} c_{\mathbf{k}}e^{i(\mathbf{k}+\mathbf{K})\cdot\mathbf{y}}c_{\mathbf{k}'}e^{i(\mathbf{k}'+\mathbf{K})\cdot\mathbf{x}}|0\rangle.\end{aligned} \tag{134}$$

The pair states are now such that

$$\mathbf{k} = -\mathbf{k}' - 2\mathbf{K}, \tag{135}$$

so that the pair part of (134) which we want to keep in the hamiltonian, by analogy with (72), is

$$\langle 0|\Psi(\mathbf{y})\Psi(\mathbf{x})|0\rangle = \sum_{\mathbf{k}} e^{-i\mathbf{k}\cdot(\mathbf{y}-\mathbf{x})}\langle 0|c_{-\mathbf{k}-\mathbf{K}}c_{\mathbf{k}-\mathbf{K}}0\rangle. \tag{136}$$

With these new pairs the calculation goes through unchanged. This is brought out below in the discussion of flux quantization.

[9] J. Bardeen, *Hand. Phys.* **15**, 274 (1956), p. 303 *et seq.*
[10] J. R. Schrieffer, *The many-body problem*, Wiley, 1959, pp. 573–575.

COHERENCE LENGTH

Pippard proposed on empirical grounds a modification of the London equation in which the current density at a point is given by an integral of the vector potential over a region surrounding the point:

$$\mathbf{j}(\mathbf{x}) = -\frac{3}{4\pi c\Lambda\xi_0}\int d^3y\,\frac{\mathbf{r}(\mathbf{r}\cdot\mathbf{A}(\mathbf{y}))e^{-r/\xi_0}}{r^4}, \tag{137}$$

where $\mathbf{r} = \mathbf{x} - \mathbf{y}$. The coherence distance ξ_0 is a fundamental property of the material and is of the order of 10^{-4} cm in a pure metal. For slowly varying $\mathbf{A}$ the Pippard expression (137) reduces to the London form (112), with Λ as defined there. In impure material the ξ_0 in the argument of the exponential is to be replaced by ξ, where

$$\frac{1}{\xi} = \frac{1}{\xi_0} + \frac{1}{\alpha l}. \tag{138}$$

Here α is an empirical constant of the order of unity and l is the conductivity mean free path in the normal state. The ξ_0 outside the integral remains unchanged. If $\xi < \lambda_L$, as in certain alloy superconductors, the superconductor is said to be a *hard* or Type II super, conductor; its behavior in strong magnetic fields is changed drastically, see in particular A. A. Abrikosov, *Soviet Phys. JETP* **5,** 1174 (1957); which is based on the work of V. L. Ginzburg and L. D. Landau- *J. Exptl. Theoret. Phys.* (*USSR*) **20,** 1064 (1950).

A form similar to the Pippard equation is obtained on the BCS theory; the derivation is given in the original paper. BCS make the identification

$$\boxed{\xi_0 = \frac{v_F}{\pi\Delta}.} \tag{139}$$

To understand this, recall that we obtained the London relation for $q \to 0$ from (129); this equation for small q and for k near k_F involves, from (93) and (94),

$$u_{\mathbf{k}+\mathbf{q}}v_{\mathbf{k}} - u_{\mathbf{k}}v_{\mathbf{k}+\mathbf{q}} \cong \frac{1}{4\Delta}(\varepsilon_{\mathbf{k}+\mathbf{q}} - \varepsilon_{\mathbf{k}}) \cong \frac{k_Fq}{4m\Delta}. \tag{140}$$

This term is small if

$$q \ll q_0 \equiv \frac{4m\Delta}{k_F} = \frac{4\Delta}{v_F}; \tag{141}$$

we see that q_0 defined in this way is equal to $1/\xi_0$ apart from a factor

$4/\pi$. The result (139) is quantitatively in quite good agreement with experiment. Our sketchy argument (140) to (141) is essentially an argument about the minimum extent $1/q_0$ of a wavepacket if the excess energy of the packet is to be of the order of Δ. Values of ξ_0 are of the order of 10^{-4} cm; in hard superconducting alloys ξ may be of the order of 10^{-7} cm. The penetration depth λ_L at low temperatures is of the order of 10^{-6} to 10^{-5} cm, but increases as $T \rightarrow T_c$.

MATRIX ELEMENT COHERENCE EFFECTS

In treating the matrix elements in the Meissner effect we saw that two terms contributed to the excitation of a single virtual state. In a second-order calculation the square of the matrix element enters; the contribution of the cross-product of the two terms will depend on their relative sign. This is called a *coherence effect.* For some processes the terms add to the result and for others terms subtract. Coherence effects are a striking feature of superconductivity; their explanation is a remarkable and cogent argument for the physical reality of the BCS wavefunction.

The coupling with a photon field of wavevector $\mathbf{q}$ involves the square of the matrix element worked out in (125) to (129). For photon energies less than the gap the only inelastic photon process which can occur is the scattering of a quasiparticle in an excited state. We now want to pick out of

$$c^+_{\mathbf{k+q}}c_{\mathbf{k}} - c^+_{-\mathbf{k}}c_{-\mathbf{k-q}}, \tag{142}$$

which is the sum of the terms involved in (127) and (128), not the previous term in $\alpha^+_{\mathbf{k+q}}\alpha^+_{-\mathbf{k}}$ which connected the ground state with an excited state, but the term in $\alpha^+_{\mathbf{k+q}}\alpha_{\mathbf{k}}$ which corresponds to the scattering of a quasiparticle from $\mathbf{k}$ to $\mathbf{k} + \mathbf{q}$. Thus the effective part of

$$\begin{aligned} c^+_{\mathbf{k+q}}c_{\mathbf{k}} - c^+_{-\mathbf{k}}c_{-\mathbf{k-q}} &= u_{\mathbf{k+q}}u_{\mathbf{k}}\alpha^+_{\mathbf{k+q}}\alpha_{\mathbf{k}} - v_{\mathbf{k}}v_{\mathbf{k+q}}\alpha_{\mathbf{k}}\alpha^+_{\mathbf{k+q}} + \cdots \\ &= (u_{\mathbf{k+q}}u_{\mathbf{k}} + v_{\mathbf{k}}v_{\mathbf{k+q}})\alpha^+_{\mathbf{k+q}}\alpha_{\mathbf{k}} + \cdots . \end{aligned} \tag{143}$$

The transition rate for the scattering process $\mathbf{k} \rightarrow \mathbf{k} + \mathbf{q}$ involves

$$(u_{\mathbf{k+q}}u_{\mathbf{k}} + v_{\mathbf{k}}v_{\mathbf{k+q}})^2 = \tfrac{1}{2}\left\{1 + \frac{\varepsilon_{\mathbf{k}}\varepsilon_{\mathbf{k+q}} + \Delta^2}{\lambda_{\mathbf{k}}\lambda_{\mathbf{k+q}}}\right\}. \tag{144}$$

This is the result for photon absorption and for nuclear spin relaxation in a superconductor. The process described by (143) vanishes at absolute zero because $\alpha_{\mathbf{k}}$ on the ground state gives zero.

There are other types of scattering processes which lead to a different type of result. We consider the absorption of ultrasonic waves by quasiparticles, with the deformation potential interaction hamiltonian (2):

(145) $$H' = iD \sum_{\mathbf{kq}} c^+_{\mathbf{k+q}} c_{\mathbf{k}} (a_{\mathbf{q}} - a^+_{-\mathbf{q}}).$$

The terms in H' which cause the scattering of a quasiparticle from **k** to **k** + **q**, with the absorption of a phonon **q**, are proportional to

(146) $$c^+_{\mathbf{k+q}} c_{\mathbf{k}} + c^+_{-\mathbf{k}} c_{-\mathbf{k-q}},$$

where we have a plus sign, unlike the photon problem, because D is independent of **k** for the deformation potential coupling which leads to H'. More generally, the symmetry of the phonon interaction differs from that of the photon interaction. Thus the scattering term is, from (143) with the appropriate change of sign,

(147) $$(u_{\mathbf{k+q}} u_{\mathbf{k}} - v_{\mathbf{k}} v_{\mathbf{k+q}})^2 = \tfrac{1}{2} \left\{ 1 + \frac{\varepsilon_{\mathbf{k}} \varepsilon_{\mathbf{k+q}} - \Delta^2}{\lambda_{\mathbf{k}} \lambda_{\mathbf{k+q}}} \right\}.$$

A discussion of experiments involving coherence effects of the type of (144) and (147) is given in reference 4.

QUANTIZED MAGNETIC FLUX IN SUPERCONDUCTORS

It has been observed[11, 12] that the magnetic flux through a superconducting ring or toroid is quantized in units of

(148) $$\frac{1}{2} \frac{ch}{e} = 2.07 \times 10^{-7} \text{ gauss/cm}^2.$$

We note that the unit may be written as

(149) $$\frac{1}{2} \frac{2\pi\hbar c}{e} = 2\pi \frac{e\hbar}{2mc} \frac{mc^2}{e^2} = \frac{2\pi\mu_B}{a_e},$$

where a_e is the classical radius of the electron.

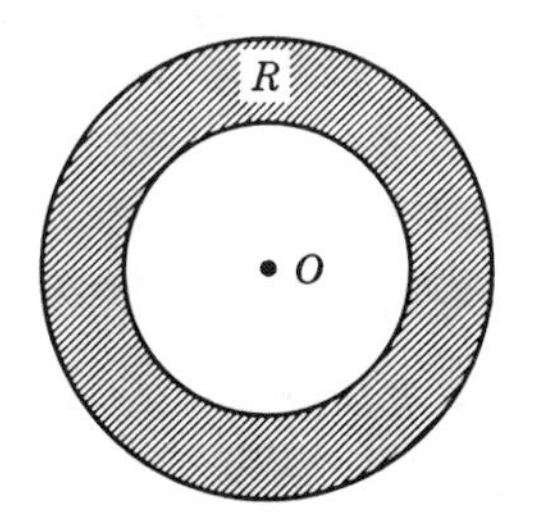

FIG. 5. Superconducting ring.

We consider a circular ring R with a tunnel O, as in Fig. 5. Let Φ denote the net magnetic flux through the ring; the flux is produced by external sources of field and by surface currents on the ring. The wave equation of the particles in the ring is

(150) $$\left\{ \sum_j \frac{1}{2m} [\mathbf{p}_j + ec^{-1} A(\mathbf{x}_j)]^2 + V \right\} \psi = E\psi,$$

where we may let V include the electron-phonon interaction and the phonon energy. Inside R we have $\mathbf{H} = 0$, if we neglect a microscopic region near the surfaces. That is, we assume the Meissner effect.

[11] B. S. Deaver, Jr., and W. M. Fairbank, *Phys. Rev. Letters* **7**, 43 (1961).
[12] R. Doll and M. Näbauer, *Phys. Rev. Letters* **7**, 51 (1961).

Then curl $\mathbf{A} = 0$ and $\mathbf{A} = \text{grad}\,\chi$, where χ cannot be a constant because it must ensure that on going once around

$$\Delta\chi = \oint \text{grad}\,\chi \cdot d\mathbf{l} = \oint \mathbf{A} \cdot d\mathbf{l} = \iint \mathbf{H} \cdot d\mathbf{S} = \Phi, \tag{151}$$

even if the path of the line integral is taken only through points where $\mathbf{H} = 0$. In cylindrical polar coordinates we take $\chi = (\varphi/2\pi)\Phi$, where φ is the angle.

We now make the transformation, noting that in this section Ψ is a one-electron wavefunction and not a field operator,

$$\Psi(\mathbf{x}) = \psi(\mathbf{x}) \exp \sum_j i(e/c)\chi(\mathbf{x}_j), \tag{152}$$

with the property that

$$\mathbf{p}\Psi = \left\{\exp \sum_j i(e/c)\chi(\mathbf{x}_j)\right\} \left\{\sum_j [-i\nabla_j + ec^{-1}\mathbf{A}(\mathbf{x}_j)]\right\} \psi. \tag{153}$$

The wavefunction Ψ satisfies

$$\left\{\frac{1}{2m} \sum_j p_j^2 + V\right\} \Psi = E\Psi, \tag{154}$$

which is identical with the equation (150) for $\mathbf{A} \equiv 0$. However, ψ must always be single-valued, whereas the change in Ψ on fixing all coordinates except one, and carrying that one around the ring, is determined by the flux Φ. Thus, from (151) and (152),

$$\Psi(\mathbf{x}) \rightarrow \Psi(\mathbf{x}) \exp(ie\Phi/c), \tag{155}$$

on carrying one electron once around the ring.

The angular dependence of ψ is

$$\psi \sim \exp\left(i \sum_j n_j\varphi_j\right), \tag{156}$$

by the cylindrical symmetry; here n_j is an integer, positive or negative. Thus the angular dependence of Ψ is

$$\Psi \sim \exp\left[i \prod_z \{n_j + (2\pi e\Phi/c)\}\varphi_j\right]. \tag{157}$$

We now consider a two-electron state:

$$\Psi = e^{in_1\varphi_1}e^{in_2\varphi_2}e^{i(2\pi e/c)\Phi(\varphi_1+\varphi_2)}, \tag{158}$$

where n_1, n_2 are integers. Transform this to center-of-mass and relative coordinates

$$\theta = \tfrac{1}{2}(\varphi_1 + \varphi_2); \qquad \varphi = \varphi_1 - \varphi_2. \tag{159}$$

Then

(160) $$\Psi = e^{i\frac{1}{2}(n_1-n_2)\varphi}e^{i\{(n_1+n_2)+(4\pi e\Phi/c)\}\theta}.$$

If the factor multiplying θ is zero, we have just the Cooper pair

(161) $$\Psi = e^{i\frac{1}{2}(n_1-n_2)\varphi_1}e^{-i\frac{1}{2}(n_1-n_2)\varphi_2}.$$

The term in θ is zero when

(162) $$4\pi e\Phi/c = -(n_1 + n_2) = \text{integer},$$

or, with $\hbar$ restored,

(163) $$\Phi = \text{integer} \times (hc/2e).$$

Thus when the flux satisfies (163) we may form pairs and deal with the equation for Ψ, which does not contain $\mathbf{A}$. The experimental verification of this, particularly of the factor $\frac{1}{2}$, is strong evidence of the importance of BCS pairing in the ground state.

The point is that we definitely need pairs to carry out the BCS procedure with the wave equation on Ψ, but the pairing condition on Ψ is only compatible with the periodic boundary condition on ψ if the flux is quantized in the units $hc/2e$. For a normal conductor we do not need to restrict ourselves to pair states, and so this special quantum condition on the flux is not energetically advantageous in the normal state.

We can do the arithmetic another way. If we take the states

(164) $$\psi_1 \sim e^{in_1\varphi_1}; \qquad \psi_2 \sim e^{-i(n_1+2e\Phi/hc)\varphi_2};$$

these go over into

(165) $$\Psi_1 \sim e^{i(n_1+e\Phi/hc)\varphi_1}; \qquad \Psi_2 \sim e^{-i(n_1+e\Phi/hc)\varphi_2}.$$

Now Ψ_1, Ψ_2 are a Cooper pair, but Ψ_2 satisfies periodic boundary conditions only if $2e\Phi/hc$ is integral.

PROBLEMS

1. Estimate in a one-dimensional metal the range of the interaction (11), subject to (10), for a Debye energy of 10^{-14} ergs.

2. (*a*) Discuss the form in relative coordinate space of the ground-state Cooper-pair wavefunction. (*b*) Estimate the range of the function for a representative superconductor, taking $\Delta \approx k_BT_c$, where T_c is the critical temperature for a superconductor.

3. In the *strong coupling* limit all $\varepsilon_\mathbf{k}$ are set equal to 0, which is the fermi surface. The equivalent spinor hamiltonian (if V is constant for N' states and zero otherwise) is

$$H = -\tfrac{1}{4}V \sum_{\mathbf{kk}'}{}' (\sigma_{\mathbf{k}x}\sigma_{\mathbf{k}'x} + \sigma_{\mathbf{k}y}\sigma_{\mathbf{k}'y}), \tag{166}$$

where $\mathbf{k}, \mathbf{k}'$ run over the N' states. (*a*) If S refers to the total spin $\mathbf{S} = \frac{1}{2}\sum_{\mathbf{k}} \boldsymbol{\sigma}_{\mathbf{k}}$ of these states, show that H may be written as

$$H = -V\mathbf{S}^2 + V\mathbf{S}_z{}^2 + \frac{N'}{2}V, \tag{167}$$

which has the exact eigenvalues

$$E = -V\{S(S+1) - M^2\} + \frac{N'}{2}V, \tag{168}$$

where S is the total spin quantum number and M is the quantum number of $\mathbf{S}_z$. The allowed values of S are $\frac{1}{2}N'$; $\frac{1}{2}N' - 1$; $\frac{1}{2}N' - 2$; $\cdots$, according to the number of reversed spins n. (*b*) Show for the state $M = 0$ that

$$E(n) = E_0 + nVN' - Vn(n-1), \tag{169}$$

so that here the energy gap $E(1) - E(0)$ is exactly $N'V$. (*c*) What are the states with different M values? (*d*) Show that the ground-state and first-pair excited-state energies calculated in the molecular field (BCS) approximation agree with the strong coupling result to $0(1/N')$. This furnishes an excellent and important check on the accuracy of the usual methods of finding the eigenvalues of the BCS reduced hamiltonian.

4. Confirm that the equations of motion (71) conserve the total number of particles.

5. The BCS creation and annihilation operators for pairs are defined as

$$b_{\mathbf{k}}^{+} = c_{\mathbf{k}\uparrow}^{+}c_{-\mathbf{k}\downarrow}^{+}; \qquad b_{\mathbf{k}} = c_{-\mathbf{k}\downarrow}c_{\mathbf{k}\uparrow}, \tag{170}$$

where the c, c^+ are one-particle fermion operators. Show that the b, b^+ satisfy the mixed commutation relations

$$\begin{aligned} [b_{\mathbf{k}}, b_{\mathbf{k}'}^{+}] &= (1 - n_{\mathbf{k}\uparrow} - n_{-\mathbf{k}\downarrow})\delta_{\mathbf{kk}'}; \\ [b_{\mathbf{k}}, b_{\mathbf{k}'}] &= 0; \\ \{b_{\mathbf{k}}, b_{\mathbf{k}'}\} &= 2b_{\mathbf{k}}b_{\mathbf{k}'}(1 - \delta_{\mathbf{kk}'}). \end{aligned} \tag{171}$$

6. Verify by direct calculation that the expectation value of the excitation energy in the state (110) is given by

$$E = \lambda_{\mathbf{k}'} + \lambda_{\mathbf{k}''}, \tag{172}$$

using the hamiltonian (70); show that

$$E = 2\lambda_{\mathbf{k}'} \tag{173}$$

for the real pair excitation (111).

7. Discuss the theory of tunneling with reference to experiments on superconductors.

8. Discuss the density of states in a superconductor near the energy gap.

9 Bloch functions—general properties

In this chapter we discuss a number of general properties of eigenfunctions in infinite periodic lattices and in applied electric and magnetic fields. The discussion is presented largely in the form of theorems.

BLOCH THEOREM

THEOREM 1. The Bloch theorem states that if $V(\mathbf{x})$ is periodic with the periodicity of the lattice, then the solutions $\varphi(\mathbf{x})$ of the wave equation

$$H\varphi(\mathbf{x}) = \left(\frac{1}{2m}\,p^2 + V(\mathbf{x})\right)\varphi(\mathbf{x}) = E\varphi(\mathbf{x}) \tag{1}$$

are of the form

$$\varphi_{\mathbf{k}}(\mathbf{x}) = e^{i\mathbf{k}\cdot\mathbf{x}}u_{\mathbf{k}}(\mathbf{x}), \tag{2}$$

where $u_{\mathbf{k}}(\mathbf{x})$ is periodic with the periodicity of the direct lattice.

Analytical proofs of this central theorem are found in the standard elementary texts on solid state theory. The most direct and elegant proof utilizes a little group theory.

Proof: With periodic boundary conditions over a volume of N^3 lattice points, the translation group is abelian. All operations of an abelian group commute. If all operations of a group commute, then all the irreducible representations of the group are one-dimensional.

Consider the lattice translation operator $\mathbf{T}$ defined by

$$\mathbf{Tx} = \mathbf{x} + \mathbf{t}_{mnp} = \mathbf{x} + m\mathbf{a} + n\mathbf{b} + p\mathbf{c}, \tag{3}$$

where m, n, p are integers; then

$$T_{mnp}\varphi_{\mathbf{k}}(\mathbf{x}) = \varphi_{\mathbf{k}}(\mathbf{x} + m\mathbf{a} + n\mathbf{b} + p\mathbf{c}). \tag{4}$$

The operations $\mathbf{T}$ form a cyclic group; because the representations are only one-dimensional,

$$T_{mnp}\varphi_{\mathbf{k}}(\mathbf{x}) = c_{mnp}\varphi_{\mathbf{k}}(\mathbf{x}), \tag{5}$$

where c_{mnp} is a constant. Because

$$T_{100}\varphi_{\mathbf{k}}(\mathbf{x}) = \varphi_{\mathbf{k}}(\mathbf{x} + \mathbf{a}) = c_{100}\varphi_{\mathbf{k}}(\mathbf{x}), \tag{6}$$

we must have in particular for a lattice having N lattice points on a side

$$T_{N00}\varphi_{\mathbf{k}}(\mathbf{x}) = \varphi_{\mathbf{k}}(\mathbf{x} + N\mathbf{a}) = (c_{100})^N\varphi_{\mathbf{k}}(\mathbf{x}). \tag{7}$$

But with periodic boundary conditions

$$\varphi_{\mathbf{k}}(\mathbf{x} + N\mathbf{a}) = \varphi_{\mathbf{k}}(\mathbf{x}), \tag{8}$$

so that

$$(c_{100})^N = 1; \tag{9}$$

thus c_{100} must be one of the N roots of unity:

$$c_{100} = e^{2\pi i\xi/N}; \qquad \xi = 1, 2, 3, \cdots N. \tag{10}$$

This condition is satisfied generally by the function

$$\varphi_{\mathbf{k}}(\mathbf{x}) = e^{i\mathbf{k}\cdot\mathbf{x}}u_{\mathbf{k}}(\mathbf{x}), \tag{11}$$

if $u_{\mathbf{k}}(\mathbf{x})$ has the period of the lattice and

$$N\mathbf{k} = \xi\mathbf{a}^* + \eta\mathbf{b}^* + \zeta\mathbf{c}^*, \qquad (\xi, \eta, \zeta \text{ integral}) \tag{12}$$

is a vector of the reciprocal lattice. For a lattice translation $\mathbf{t}$,

$$\begin{aligned}\varphi_{\mathbf{k}}(\mathbf{x} + \mathbf{t}) &= e^{i\mathbf{k}\cdot(\mathbf{x}+\mathbf{t})}u_{\mathbf{k}}(\mathbf{x} + \mathbf{t}) = e^{i\mathbf{k}\cdot\mathbf{t}}e^{i\mathbf{k}\cdot\mathbf{x}}u_{\mathbf{k}}(\mathbf{x}) = e^{i\mathbf{k}\cdot\mathbf{t}}\varphi_{\mathbf{k}}(\mathbf{x}) \\ &= e^{i2\pi(m\xi+n\eta+p\zeta)/N}\,\varphi_{\mathbf{k}}(\mathbf{x}),\end{aligned} \tag{13}$$

as required by (5) and (10).

In other language, $e^{i\mathbf{k}\cdot\mathbf{t}_{\mathbf{n}}}$ is the eigenvalue of the lattice translation operator $T_{\mathbf{n}}$:

$$\mathrm{T}_{\mathbf{n}}\varphi_{\mathbf{k}}(\mathbf{x}) = e^{i\mathbf{k}\cdot\mathbf{t}_{\mathbf{n}}}\varphi_{\mathbf{k}}(\mathbf{x}), \tag{14}$$

where $\mathbf{t}_{\mathbf{n}}$ is a lattice translation vector; $\varphi_{\mathbf{k}}(\mathbf{x})$ is an eigenvector of $T_{\mathbf{n}}$.

THEOREM 2. The function $u_{\mathbf{k}}(\mathbf{x})$ of the Bloch function $\varphi_{\mathbf{k}}(\mathbf{x}) = e^{i\mathbf{k}\cdot\mathbf{x}}u_{\mathbf{k}}(\mathbf{x})$ satisfies the equation

$$\left(\frac{1}{2m}(\mathbf{p} + \mathbf{k})^2 + V(\mathbf{x})\right)u_{\mathbf{k}}(\mathbf{x}) = \epsilon_{\mathbf{k}}u_{\mathbf{k}}(\mathbf{x}). \tag{15}$$

This is equivalent to a gauge transformation.

Proof: Note that, using $\mathbf{p} = -i\nabla$, we have the operator equation

$$\mathbf{p}e^{i\mathbf{k}\cdot\mathbf{x}} = e^{i\mathbf{k}\cdot\mathbf{x}}(\mathbf{p} + \mathbf{k}). \tag{16}$$

Thus

$$\mathbf{p}\varphi_{\mathbf{k}}(\mathbf{x}) = e^{i\mathbf{k}\cdot\mathbf{x}}(\mathbf{p} + \mathbf{k})u_{\mathbf{k}}(\mathbf{x}), \tag{17}$$

and

$$p^2\varphi_{\mathbf{k}}(\mathbf{x}) = e^{i\mathbf{k}\cdot\mathbf{x}}(\mathbf{p} + \mathbf{k})^2u_{\mathbf{k}}(\mathbf{x}), \tag{18}$$

from which (15) follows directly.

We may rewrite (15) as

$$\left(-\frac{1}{2m}(\nabla^2 + 2i\mathbf{k}\cdot\nabla) + V(\mathbf{x})\right)u_{\mathbf{k}}(\mathbf{x}) = \lambda_{\mathbf{k}}u_{\mathbf{k}}(\mathbf{x}), \tag{19}$$

with

$$\lambda_{\mathbf{k}} = \varepsilon_{\mathbf{k}} - \frac{1}{2m}k^2, \tag{20}$$

where $\varepsilon_{\mathbf{k}}$ is the eigenvalue of (15). If $V(\mathbf{x}) \equiv 0$, a solution of (19) is

$$u_{\mathbf{k}}(\mathbf{x}) = \text{constant}; \qquad \lambda_{\mathbf{k}} = 0, \tag{21}$$

and

$$\varepsilon_{\mathbf{k}} = \frac{1}{2m}k^2; \qquad \varphi_{\mathbf{k}}(\mathbf{x}) = e^{i\mathbf{k}\cdot\mathbf{x}}, \tag{22}$$

the usual plane wave. At the point $\mathbf{k} = 0$ the equation for $u_0(\mathbf{x})$ is simply

$$\left(-\frac{1}{2m}\nabla^2 + V(\mathbf{x})\right)u_0(\mathbf{x}) = \varepsilon_0u_0(\mathbf{x}); \tag{23}$$

thus the equation for $u_0(\mathbf{x})$ has the symmetry of $V(\mathbf{x})$, which is the symmetry of the crystal space group.

Spin-Orbit Interaction. The hamiltonian with spin-orbit interaction has the form (Schiff, p. 333)

$$H = \frac{1}{2m}p^2 + V(\mathbf{x}) + \frac{1}{4m^2c^2}\boldsymbol{\sigma}\times\operatorname{grad}V(\mathbf{x})\cdot\mathbf{p}, \tag{24}$$

where $\boldsymbol{\sigma}$ is the pauli spin operator, with the components

$$\sigma_x = \begin{pmatrix}0 & 1\\ 1 & 0\end{pmatrix}; \qquad \sigma_y = \begin{pmatrix}0 & -i\\ i & 0\end{pmatrix}; \qquad \sigma_z = \begin{pmatrix}1 & 0\\ 0 & -1\end{pmatrix}. \tag{25}$$

The hamiltonian (24) is invariant under lattice translations $\mathbf{T}$ if $V(\mathbf{x})$

is invariant under **T**. The eigenfunctions of (24) will be of the Bloch form, but they will not in general correspond to the pure spin states α or β for which $\sigma_z\alpha = \alpha$; $\sigma_z\beta = -\beta$. In general

$$\varphi_{\mathbf{k}\uparrow}(\mathbf{x}) = \chi_{\mathbf{k}\uparrow}(\mathbf{x})\alpha + \gamma_{\mathbf{k}\uparrow}(\mathbf{x})\beta = e^{i\mathbf{k}\cdot\mathbf{x}}u_{\mathbf{k}\uparrow}(\mathbf{x}), \tag{26}$$

where the arrow $\uparrow$ on the Bloch function $\varphi_{\mathbf{k}\uparrow}(\mathbf{x})$ denotes a state with the spin generally up in the sense that $(\varphi_{\mathbf{k}\uparrow},\sigma_z\varphi_{\mathbf{k}\uparrow})$ is positive. In the absence of spin-orbit interaction $\varphi_\uparrow$ involves only α; and $\varphi_\downarrow$ only β. The arrows on $\chi_{\mathbf{k}\uparrow}$ and $\gamma_{\mathbf{k}\uparrow}$ are labels to indicate their association with $\varphi_{\mathbf{k}\uparrow}$.

THEOREM 3. With spin-orbit interaction the function $u_{\mathbf{k}}(\mathbf{x})$ satisfies

$$\left[\frac{1}{2m}(\mathbf{p}+\mathbf{k})^2 + V(\mathbf{x}) + \frac{1}{4m^2c^2}\boldsymbol{\sigma}\times\operatorname{grad}V(\mathbf{x})\cdot(\mathbf{p}+\mathbf{k})\right]u_{\mathbf{k}}(\mathbf{x}) = \epsilon_{\mathbf{k}}u_{\mathbf{k}}(\mathbf{x}). \tag{27}$$

This follows directly from (16) and (18). The terms

$$\frac{1}{m}\left(\mathbf{p} + \frac{1}{4mc^2}\boldsymbol{\sigma}\times\operatorname{grad}V\right)\cdot\mathbf{k} = H' \tag{28}$$

are often treated as perturbations for small **k** or small changes in **k** from a special wavevector $\mathbf{k}_0$.

The quantity

$$\boldsymbol{\pi} \equiv \mathbf{p} + \frac{1}{4mc^2}\sigma\times\operatorname{grad}V \tag{29}$$

has many of the properties for the problem with spin-orbit interaction which **p** has for the problem without spin-orbit interaction.

TIME REVERSAL SYMMETRY

The time reversal transformation K takes **x** into **x**; **p** into $-\mathbf{p}$; $\boldsymbol{\sigma}$ into $-\boldsymbol{\sigma}$. The hamiltonian (24) is invariant under time reversal; thus $[H,K] = 0$. For a system of a single electron the result of Kramers (Messiah, Chapter 15, Section 18) for the time reversal operator is

$$K = -i\sigma_y K_0, \tag{30}$$

where K_0 in the Schrödinger representation is the operation of taking the complex conjugate. Thus K_0 has the property for any two states φ and ψ that

$$(\varphi,\psi) = (K_0\psi,K_0\varphi). \tag{31}$$

Further, with $\sigma_y^2 = 1$,

$$(K\psi,K\varphi) = (K_0\psi,\sigma_y^2K_0\varphi) = (\varphi,\psi); \tag{32}$$

also,

$$K^2\varphi = (-i\sigma_y)(-i\sigma_y)\varphi = -\varphi. \tag{33}$$

An important application of time-reversed pairs of states has been made by P. W. Anderson [*J. Phys. Chem. Solids* **11**, 26 (1959)]. He shows that in very impure superconductors we must consider pairs defined by the time reversal operation, rather than Bloch function pairs.

THEOREM 4. If φ is a one-electron eigenstate of H, then $K\varphi$ is also an eigenstate with the same energy eigenvalue in the absence of external magnetic fields. Further, $K\varphi$ is orthogonal to φ. This is the Kramers theorem.

Proof: The hamiltonian commutes with K; therefore $K\varphi$ must be an eigenstate if φ is an eigenstate, and the eigenvalues are the same. Now by (32) and (33)

$$(\varphi,K\varphi) = -(K^2\varphi,K\varphi) = -(\varphi,K\varphi) = 0, \tag{34}$$

so that $\varphi_{\mathbf{k}}$ and $K\varphi_{\mathbf{k}}$ are linearly independent. Q.E.D.

THEOREM 5. The states $K\varphi_{\mathbf{k}\uparrow}$ and $K\varphi_{\mathbf{k}\downarrow}$ belong to wavevector $-\mathbf{k}$, so that $\varepsilon_{\mathbf{k}\uparrow} = \varepsilon_{-\mathbf{k}\downarrow}$ and $\varepsilon_{\mathbf{k}\downarrow} = \varepsilon_{-\mathbf{k}\uparrow}$.

Proof: $K\varphi_{\mathbf{k}\uparrow} = -i\sigma_y K_0\varphi_{\mathbf{k}\uparrow} = e^{-i\mathbf{k}\cdot\mathbf{x}} \times (\text{periodic function of } \mathbf{x})_\downarrow$, so that

$$K\varphi_{\mathbf{k}\uparrow} = \varphi_{-\mathbf{k}\downarrow}, \tag{35}$$

apart from a phase factor. We recall that σ_y reverses the spin direction. We assign opposite spin arrows to φ and $K\varphi$ because

$$(\varphi_{\mathbf{k}\uparrow},\sigma_z\varphi_{\mathbf{k}\uparrow}) = (K\sigma_z\varphi_{\mathbf{k}\uparrow},K\varphi_{\mathbf{k}\uparrow}) = -(\varphi_{-\mathbf{k}\downarrow},\sigma_z\varphi_{-\mathbf{k}\downarrow}),$$

using $\sigma_y\sigma_z = -\sigma_z\sigma_y$. From (35) and Theorem 4 we have

$$\varepsilon_{\mathbf{k}\uparrow} = \varepsilon_{-\mathbf{k}\downarrow}; \qquad \varepsilon_{\mathbf{k}\downarrow} = \varepsilon_{-\mathbf{k}\uparrow}. \tag{36}$$

Q.E.D.

The bands have a twofold degeneracy in the sense that each energy occurs twice, but not at the same $\mathbf{k}$. A double degeneracy at the same energy and $\mathbf{k}$ occurs only if other symmetry elements are present; with the inversion operation J the energy surface will be double at every point in $\mathbf{k}$ space.

THEOREM 6. If the hamiltonian is invariant under space inversion, then

$$\varphi_{\mathbf{k}\uparrow}(\mathbf{x}) = \varphi_{-\mathbf{k}\uparrow}(-\mathbf{x}), \tag{37}$$

apart from a phase factor, and

$$\varepsilon_{\mathbf{k}\uparrow} = \varepsilon_{-\mathbf{k}\uparrow}. \tag{38}$$

Proof: The space inversion operator J sends $\mathbf{x}$ into $-\mathbf{x}$; $\mathbf{p}$ into $-\mathbf{p}$; and $\boldsymbol{\sigma}$ into $\boldsymbol{\sigma}$. The reason $\boldsymbol{\sigma}$ does not change sign is that it is an angular momentum and transforms as an axial vector. Thus if $JV(\mathbf{x}) = V(\mathbf{x})$, then the hamiltonian including spin-orbit interaction is invariant under J. Then $J\varphi_{\mathbf{k}\uparrow}(\mathbf{x})$ is degenerate with $\varphi_{\mathbf{k}\uparrow}(\mathbf{x})$. But

$$J\varphi_{\mathbf{k}\uparrow}(\mathbf{x}) \equiv e^{-i\mathbf{k}\cdot\mathbf{x}}u_{\mathbf{k}\uparrow}(-\mathbf{x}) \tag{39}$$

is a Bloch function belonging to $-\mathbf{k}$, because the eigenvalue of $J\varphi_{\mathbf{k}\uparrow}$ under a lattice translation operator T is $e^{-i\mathbf{k}\cdot\mathbf{t}_n}$. We may call $u_{-\mathbf{k}\uparrow}(\mathbf{x}) = u_{\mathbf{k}\uparrow}(-\mathbf{x})$, whence

$$\varphi_{-\mathbf{k}\uparrow}(\mathbf{x}) = J\varphi_{\mathbf{k}\uparrow}(\mathbf{x}), \tag{40}$$

and

$$\varepsilon_{\mathbf{k}\uparrow} = \varepsilon_{-\mathbf{k}\uparrow}. \tag{41}$$

Q.E.D.

It is simple to show directly, if one wishes, that $u_{-\mathbf{k}\uparrow}(\mathbf{x})$ satisfies the same differential equation as $u_{\mathbf{k}\uparrow}(-\mathbf{x})$.

We recall that J commutes with σ_z, so that the expectation value of σ_z over $\varphi_{\mathbf{k}\uparrow}$ and $\varphi_{-\mathbf{k}\uparrow}$ is the same. Therefore, using (36), the combined symmetry elements K and J have the consequence that

$$\varepsilon_{\mathbf{k}\uparrow} = \varepsilon_{\mathbf{k}\downarrow}, \tag{42}$$

where

$$\varphi_{\mathbf{k}\downarrow} = KJ\varphi_{\mathbf{k}\uparrow}, \tag{43}$$

apart from a phase factor.

The product operation

$$C \equiv KJ = -i\sigma_y K_0 J = JK \tag{44}$$

will be called *conjugation*. Conjugation reverses the spin of a Bloch state, but does not reverse its wavevector:

$$C\varphi_{\mathbf{k}\uparrow} = \varphi_{\mathbf{k}\downarrow}, \tag{45}$$

apart from a phase factor. A number of theorems involving the operations K and C are given as exercises at the end of the chapter.

THEOREM 7. In the momentum representation, with $\mathbf{G}$ a reciprocal lattice vector,

$$\varphi_{\mathbf{k}}(\mathbf{x}) = e^{i\mathbf{k}\cdot\mathbf{x}} \sum_{\mathbf{G}} f_{\mathbf{G}}(\mathbf{k}) e^{i\mathbf{G}\cdot\mathbf{x}}, \tag{46}$$

where the $f_{\mathbf{G}}(\mathbf{k})$ are c-numbers, the wave equation without spin-orbit interaction is

$$\frac{1}{2m} (\mathbf{k} + \mathbf{G})^2 f_{\mathbf{G}}(\mathbf{k}) + \sum_{\mathbf{g}} V(\mathbf{G} - \mathbf{g}) f_{\mathbf{g}}(\mathbf{k}) = \varepsilon_{\mathbf{k}} f_{\mathbf{G}}(\mathbf{k}), \tag{47}$$

where $\mathbf{g}$ is a reciprocal lattice vector and $V(\mathbf{G})$ is the fourier transform of $V(\mathbf{x})$ between plane wave states:

$$V(\mathbf{G}) = \int d^3x \, e^{i\mathbf{G}\cdot\mathbf{x}} V(\mathbf{x}). \tag{48}$$

The result (47) follows on operating on (46) with $H = T + V$ and taking the scalar product with $e^{i\mathbf{k}\cdot\mathbf{x}} e^{i\mathbf{G}\cdot\mathbf{x}}$.

It follows from the representation (46) that the expectation value of the velocity $\mathbf{v}$ satisfies

$$\begin{aligned} \langle \mathbf{k}|\mathbf{v}|\mathbf{k}\rangle = \langle \mathbf{k}|\mathbf{p}/m|\mathbf{k}\rangle &= m^{-1} \sum_{\mathbf{G}} (\mathbf{k} + \mathbf{G}) |f_{\mathbf{G}}(\mathbf{k})|^2 \\ &= \operatorname{grad}_{\mathbf{k}} \varepsilon(\mathbf{k}). \end{aligned} \tag{49}$$

The proof of the last step is left to the reader. An alternate derivation is given as Theorem 11.

It also follows that the effective mass tensor defined by

$$\left(\frac{1}{m^*}\right)_{\mu\nu} = \frac{\partial}{\partial k_\mu} \frac{\partial}{\partial k_\nu} \varepsilon(\mathbf{k}) \tag{50}$$

is equal to (Problem 8):

$$\left(\frac{1}{m^*}\right)_{ij} = \frac{1}{m} \left(\delta_{ij} + \sum_{\mathbf{G}} G_i \frac{\partial}{\partial k_j} |f_{\mathbf{G}}(\mathbf{k})|^2 \right). \tag{51}$$

THEOREM 8. The energy $\varepsilon_{\mathbf{k}}$ is periodic in the reciprocal lattice; that is,

$$\varepsilon_{\mathbf{k}} = \varepsilon_{\mathbf{k}+\mathbf{G}}. \tag{52}$$

Proof: Consider a state $\varphi_{\mathbf{k}}$ of energy $\varepsilon_{\mathbf{k}}$; we may write

$$\varphi_{\mathbf{k}} = e^{i\mathbf{k}\cdot\mathbf{x}} u_{\mathbf{k}}(\mathbf{x}) = e^{i(\mathbf{k}+\mathbf{G})\cdot\mathbf{x}} u_{\mathbf{k}+\mathbf{G}}(\mathbf{x}) = \varphi_{\mathbf{k}+\mathbf{G}}, \tag{53}$$

where

$$u_{\mathbf{k}+\mathbf{G}}(\mathbf{x}) \equiv e^{-i\mathbf{G}\cdot\mathbf{x}} u_{\mathbf{k}}(\mathbf{x}) \tag{54}$$

has the periodicity of the lattice. Thus $\varphi_{\mathbf{k}+\mathbf{G}}$ may be constructed from $\varphi_{\mathbf{k}}$; it follows that $\varepsilon_k = \varepsilon_{\mathbf{k}+\mathbf{G}}$.

We now develop an important theorem related to the effective mass tensor $(1/m)_{\mu\nu}$ defined by (50), which is equivalent to

$$\varepsilon_{\mathbf{k}} = \frac{1}{2m}\left(\frac{m}{m^*}\right)_{\mu\nu} k_\mu k_\nu + \cdots . \tag{55}$$

k · p PERTURBATION THEORY

THEOREM 9. If the state $\varphi_{\mathbf{k}}$ at $\mathbf{k} = 0$ in the band γ is nondegenerate, except for the time reversal degeneracy, the effective mass tensor at this point is given by

$$\boxed{\left(\frac{m}{m^*}\right)_{\mu\nu} = \delta_{\mu\nu} + \frac{2}{m}\sum_{\delta}{}' \frac{\langle\gamma 0|\pi_\mu|0\delta\rangle\langle\delta 0|\pi_\nu|0\gamma\rangle}{\varepsilon_{\gamma 0} - \varepsilon_{\delta 0}},} \tag{56}$$

where δ, γ are band indices and the zeros stand for $\mathbf{k} = 0$. Without spin-orbit interaction $\mathbf{p}$ replaces $\boldsymbol{\pi}$, and usually it is sufficiently accurate to write $\mathbf{p}$ for $\boldsymbol{\pi}$. The result (56) is referred to also as the f-sum rule for $\mathbf{k} = 0$.

Proof: In the equation (27) for $u_{\mathbf{k}}(\mathbf{x})$ we treat

$$H' = \frac{1}{m}\left(\mathbf{p} + \frac{1}{4mc^2}\,\boldsymbol{\sigma} \times \operatorname{grad} V\right)\cdot \mathbf{k} = \frac{1}{m}\,\boldsymbol{\pi}\cdot\mathbf{k} \tag{57}$$

as a perturbation, with the hamiltonian for $\mathbf{k} = 0$ treated as the unperturbed hamiltonian. We could equally expand about any other wavevector, say $\mathbf{k}_0$.

Let us consider first the diagonal matrix elements of H'. If the crystal has a center of symmetry, then

$$\langle\gamma 0|\boldsymbol{\pi}|0\gamma\rangle = 0 \tag{58}$$

by parity; further

$$\langle\gamma 0|\boldsymbol{\pi}|C\gamma\rangle = 0, \tag{59}$$

by Exercise 5, with $|C\gamma\rangle$ denoting the conjugate state to $|0\gamma\rangle$. The spin indices are not shown in our present notation. If the crystal does not have a center of symmetry, we must consider the matrix elements for the particular symmetry involved. Thus at the point Γ in the zinc-blende structure the twofold representations Γ_6 and Γ_7 of the double group satisfy the selection rules (see Chapter 10)

$$\Gamma_6 \times \Gamma_V = \Gamma_7 + \Gamma_8; \qquad \Gamma_7 \times \Gamma_V = \Gamma_6 + \Gamma_8, \tag{60}$$

where Γ_V is the vector representation. These rules are given by G. Dresselhaus, *Phys. Rev.* **100,** 580 (1955), along with the character table. Because $\boldsymbol{\pi}$ transforms as a vector, the rules tell us that $\boldsymbol{\pi}$ does not have diagonal matrix elements within the twofold representations. Thus the first-order energy correction from H' vanishes. In Problem (14.4), we have a situation in which the first-order energy does not vanish.

The energy to second order is

$$\varepsilon_\gamma(\mathbf{k}) = \varepsilon_\gamma(0) + \frac{k^2}{2m} + \frac{1}{m^2} \sum_\delta{}' \frac{\langle\gamma 0|\pi_\mu k_\mu|0\delta\rangle\langle\delta 0|\pi_\nu k_\nu|0\gamma\rangle}{\varepsilon_{\gamma 0} - \varepsilon_{\delta 0}}, \tag{61}$$

where on the right-hand side we have included the kinetic energy associated with the $e^{i\mathbf{k}\cdot\mathbf{x}}$ modulation. The result (56) is obtained if we write (61) in the form

$$\varepsilon_\gamma(\mathbf{k}) = \varepsilon_\gamma(0) + \frac{1}{2m}\left(\frac{m}{m^*}\right)_{\mu\nu} k_\mu k_\nu + \cdots. \tag{62}$$

By going to higher orders in the perturbation theory we may construct the entire energy surface. The method is referred to as $\mathbf{k}\cdot\mathbf{p}$ perturbation theory.

The eigenfunction to first order in $\mathbf{k}$ is

$$|\mathbf{k}\gamma\rangle = e^{i\mathbf{k}\cdot\mathbf{x}}\left(|0\gamma\rangle + \frac{1}{m}\sum_\delta{}' |0\delta\rangle \frac{\langle\delta 0|\mathbf{k}\cdot\boldsymbol{\pi}|0\gamma\rangle}{\varepsilon_{\gamma 0} - \varepsilon_{\delta 0}}\right). \tag{63}$$

If further degeneracy exists at the point $\mathbf{k} = 0$, we must apply degenerate perturbation theory, as in Schiff, pp. 156–158. The valence band edge in important semiconductor crystals is degenerate; the form of the energy surfaces is considered in a later chapter on semiconductor bands, but an example will be given below.

We can draw some immediate conclusions from the form of (61). If one $\varepsilon_{\gamma 0} - \varepsilon_{\delta 0}$ is very small, the form of the band γ near $\mathbf{k} = 0$ will be determined largely by the matrix elements connecting it with the band δ; and, vice versa, δ will be determined by γ. Further, if the energy denominator is very small, the effective mass ratio m^*/m will be very small. An extreme example may be cited: It is believed that the energy gap in the semiconductor crystal $Cd_xHg_{1-x}Te$ ($x = 0.136$) is less than 0.006 ev, and the experiments suggest also that $m^*/m \leqq 4 \times 10^{-4}$ at the bottom of the conduction band.

According to calculations by F. S. Ham, *Phys. Rev.* **128,** 82 (1962), the effective masses at $\mathbf{k} = 0$ in the conduction bands of the alkali metals have the following values:

Metal	Li	Na	K	Rb	Cs
Band index	2s	3s	4s	5s	6s
m^*/m	1.33	0.965	0.86	0.78	0.73

Suppose that the order of the bands near $\mathbf{k} = 0$ in an alkali metal is the same as the order of the states in a free atom. Then in Li all the perturbations on the $2s$ conduction band will come from p levels higher in energy than $2s$, as there is no $1p$ level; for Li $\varepsilon_{s0} - \varepsilon_{p0} < 0$, so that $m < m^*$. For Na the $3s$ conduction band is perturbed about equally, but in opposite directions, by the $2p$ levels below and the $3p$ levels above $3s$ in energy, and thus $m^* \cong m$. As we go further along in the alkali series, the perturbations from below increase in effect relative to those from above, and $m^* < m$.

Degenerate $\mathbf{k} \cdot \mathbf{p}$ *Perturbation Theory.* The simplest example of $\mathbf{k} \cdot \mathbf{p}$ perturbation theory for degenerate bands occurs in uniaxial crystals with a center of symmetry. Suppose we have a band of s-like symmetry at $\mathbf{k} = 0$ lying above by an energy E_g a pair of bands degenerate at $\mathbf{k} = 0$ and transforming at this point like x and y. The symmetry axis is along the z direction. The state which is z-like at $\mathbf{k} = 0$ will be neglected implicitly: we assume that the crystal potential splits z off from the other states by an energy large in comparison with E_g. We neglect spin-orbit interaction in this example.

We note that the first-order energy correction vanishes from the perturbation $(1/m)\mathbf{k} \cdot \mathbf{p}$, by parity. The second-order energy involves the matrix elements

$$(64) \qquad \langle s|H''|x\rangle = \frac{1}{m^2E_g} \sum_j \langle s|\mathbf{k} \cdot \mathbf{p}|j\rangle\langle j|\mathbf{k} \cdot \mathbf{p}|x\rangle = 0,$$

also by parity; here $j = x, y$. Further,

$$(65) \quad \langle x|H''|x\rangle = \frac{1}{m^2E_g} \langle x|\mathbf{k} \cdot \mathbf{p}|s\rangle\langle s|\mathbf{k} \cdot \mathbf{p}|x\rangle = -\frac{k_x{}^2}{m^2E_g} |\langle x|p_x|s\rangle|^2;$$

$$(66) \quad \langle x|H''|y\rangle = \frac{1}{m^2E_g} \langle x|\mathbf{k} \cdot \mathbf{p}|s\rangle\langle s|\mathbf{k} \cdot \mathbf{p}|y\rangle = -\frac{k_xk_y}{m^2E_g} \langle x|p_x|s\rangle\langle s|p_y|y\rangle.$$

By symmetry $\langle s|p_y|y\rangle = \langle s|p_x|x\rangle$; thus we may write, for $i, j = x$ or y,

$$(67) \qquad \langle i|H''|j\rangle = -Ak_ik_j; \qquad A = \frac{1}{m^2E_g} |\langle x|p_x|s\rangle|^2.$$

The secular equation for the three states is

$$(68) \quad \begin{vmatrix} E_g + A(k_x{}^2 + k_y{}^2) - \lambda & 0 & 0 \\ 0 & -Ak_x{}^2 - \lambda & -Ak_xk_y \\ 0 & -Ak_yk_x & -Ak_y{}^2 - \lambda \end{vmatrix} = 0.$$

One solution is

$$\varepsilon_s(\mathbf{k}) = \frac{k^2}{2m} + \lambda = E_g + \frac{1}{2m}k^2 + A(k_x^2 + k_y^2); \tag{69}$$

this shows that to second order in k the energy band structure of the s state near the band edge is spheroidal, with the free electron mass m in the z direction and with the effective mass in the xy plane given by

$$\frac{1}{m^*} = \frac{1}{m} + \frac{2}{m^2 E_g}|\langle x|p_x|s\rangle|^2; \tag{70}$$

here $m^* < m$.

The energies of the degenerate bands involve the solutions of

$$\begin{vmatrix} Ak_x^2 + \lambda & Ak_xk_y \\ Ak_yk_x & Ak_y^2 + \lambda \end{vmatrix} = 0, \tag{71}$$

or

$$\lambda = 0, \qquad \text{and} \qquad -(k_x^2 + k_y^2). \tag{72}$$

Therefore the energies of the two bands degenerate at $\mathbf{k} = 0$ are, to second order in $\mathbf{k}$,

$$\varepsilon_\alpha(\mathbf{k}) = \frac{1}{2m}k^2; \qquad \varepsilon_\beta(\mathbf{k}) = \frac{1}{2m}k^2 - A(k_x^2 + k_y^2). \tag{73}$$

The form of the secular equation (68) is not the most general form: the coefficients of the off-diagonal elements are usually not equal to the coefficients of the diagonal elements. Suppose that somewhere above the s state there lie two degenerate d states which transform as xz and yz. Then the diagonal elements of the secular equation will also involve

$$\langle x|\mathbf{k}\cdot\mathbf{p}|xz\rangle\langle xz|\mathbf{k}\cdot\mathbf{p}|x\rangle = k_z^2|\langle x|p_z|xz\rangle|^2, \tag{74}$$

which is also the value of $\langle y|\mathbf{k}\cdot\mathbf{p}|yz\rangle\langle yz|\mathbf{k}\cdot\mathbf{p}|y\rangle$. The contributions of the d states to the off-diagonal elements vanish. Thus (71) becomes, in general,

$$\begin{vmatrix} Ak_x^2 + Bk_z^2 + \lambda & Ak_xk_y \\ Ak_yk_x & Ak_y^2 + Bk_z^2 + \lambda \end{vmatrix} = 0, \tag{75}$$

which has the eigenvalues

$$\begin{aligned} \lambda_{\mathbf{k}} &= -Bk_z^2; \\ \lambda_{\mathbf{k}} &= -Bk_z^2 - A(k_x^2 + k_y^2). \end{aligned} \tag{76}$$

The surfaces of constant energy are figures of revolution about the z axis. One surface (that with the + sign) describes heavy holes; the other surface describes light holes.

ACCELERATION THEOREMS

THEOREM 10. In a steady applied electric field $\mathbf{E}$ the acceleration of an electron in a periodic lattice is described by

$$\dot{\mathbf{k}} = e\mathbf{E}, \tag{77}$$

and the electron remains within the same band. We suppose that the band is nondegenerate.

First proof: If the electric field is included in the hamiltonian in the usual way as a scalar potential $\varphi = -e\mathbf{E}\cdot\mathbf{x}$, the nonboundedness of $\mathbf{x}$ causes some mathematical difficulty. The simplest approach to the problem is to establish the electric field by a vector potential which increases linearly with time. We set

$$\mathbf{A} = -c\mathbf{E}t; \tag{78}$$

thus

$$\mathbf{E} \equiv -\operatorname{grad}\varphi - \frac{1}{c}\frac{\partial\mathbf{A}}{\partial t} = \mathbf{E}, \tag{79}$$

as required. The one-electron hamiltonian is

$$\begin{aligned} H &= \frac{1}{2m}\left(\mathbf{p} - \frac{e}{c}\mathbf{A}\right)^2 + V(\mathbf{x}) \\ &= \frac{1}{2m}(\mathbf{p} + e\mathbf{E}t)^2 + V(\mathbf{x}). \end{aligned} \tag{80}$$

It is useful to become familiar with the classical motion of free electrons in the vector field $\mathbf{A} = -c\mathbf{E}t$:

$$H = \frac{1}{2m}(\mathbf{p} + e\mathbf{E}t)^2; \tag{81}$$

the hamiltonian equations are

$$\dot{\mathbf{p}} = -\partial H/\partial\mathbf{x} = 0; \qquad \dot{\mathbf{x}} = \partial H/\partial\mathbf{p} = (\mathbf{p} + e\mathbf{E}t)/m. \tag{82}$$

On quantum theory for a free electron

$$i\dot{\mathbf{p}} = [\mathbf{p},H] = 0; \qquad i\dot{\mathbf{x}} = [\mathbf{x},H] = i(\mathbf{k}_0 + e\mathbf{E}t)/m, \tag{83}$$

where $\mathbf{k}_0$ is the eigenvalue of $\mathbf{p}$, which is a constant of the motion.

Observe that the hamiltonian (80) has the periodicity of the lattice, whether or not $\mathbf{E}$ is present. Therefore the solutions are precisely of the Bloch form:

$$\varphi_{\mathbf{k}}(\mathbf{x},\mathbf{E},t) = e^{i\mathbf{k}\cdot\mathbf{x}}u_{\mathbf{k}}(\mathbf{x},\mathbf{E},t), \tag{84}$$

where $u_{\mathbf{k}}(\mathbf{x},\mathbf{E},t)$ has the periodicity of the lattice; here the time t is viewed as a parameter. The functions $u_{\gamma\mathbf{k}}(\mathbf{x},\mathbf{E},t)$ for band γ can be expanded as a linear combination of $u_{\alpha\mathbf{k}}(\mathbf{x},0)$—the eigenfunctions of all bands for $\mathbf{E} = 0$. We see that bands can be defined rigorously in the electric field and $\mathbf{k}$ is a good quantum number: in this formulation $\mathbf{k}$ is not changed by the electric field!

We now treat the time t as a parameter and compare the kinetic energy term $(\mathbf{p} + e\mathbf{E}t + \mathbf{k})^2/2m$ in the effective hamiltonian for $u_{\mathbf{k}}(\mathbf{E},t)$ with the kinetic energy term $(\mathbf{p} + e\mathbf{E}t' + \mathbf{k}')^2/2m$ in the hamiltonian for $u_{\mathbf{k}'}(\mathbf{E},t')$. The two hamiltonians will be identical if

$$e\mathbf{E}t + \mathbf{k} = e\mathbf{E}t' + \mathbf{k}', \tag{85}$$

so that the state and the energy at $\mathbf{k},t$ are identical with those at $\mathbf{k}',t'$ if (85) is satisfied. Thus an electron which stays in a given state $\mathbf{k}$ will *appear* to change its properties in terms of the states classified in $\mathbf{k}$ at $t = 0$ as if

$$\boxed{\dot{\mathbf{k}} = e\mathbf{E}.} \tag{86}$$

That is, an electron in $\varphi_{\mathbf{k}}$ at $t = 0$ will at a later time t be in a state having the original $\mathbf{k}$, but with all the other properties (including the energy) of the state originally at $\mathbf{k} - e\mathbf{E}t$. The current in the state $\mathbf{k}$ is related to the expectation value of $\mathbf{p} - (e/c)\mathbf{A}$; the current will tend to increase linearly with time because $\mathbf{A} \propto t$.

Because $e\mathbf{E}(t - t')$ is invariant under spatial translation, it will not cause $\mathbf{k}$ to change. We must still show that an electron at $\mathbf{k}$ in band γ will at time t still be in the same band. That is, we need the adiabatic theorem, which states that a transition between states α and γ is unlikely to occur if the change in the hamiltonian during the period $1/\omega_{\alpha\gamma}$ is small in comparison with the energy difference $\omega_{\alpha\gamma}$:

$$\frac{\partial H}{\partial t}\frac{1}{\omega_{\alpha\gamma}} \cdot \frac{1}{\omega_{\alpha\gamma}} \ll 1. \tag{87}$$

Our states α and γ are states of the same $\mathbf{k}$, but in different bands. The condition (87) is very easily satisfied—it is difficult to violate over an extended volume of a crystal. The argument of the present theorem is due to Kohn and to Shockley. The vector potential $\mathbf{A} = -c\mathbf{E}t$ can be established in a ring-shaped crystal by changing magnetic flux at a uniform rate through an infinite solenoid running through the inside of the ring.

Second proof: We write $H = H_0 + H'$, where

$$H_0 = \frac{1}{2m} p^2 + V(\mathbf{x}); \qquad H' = -\mathbf{F} \cdot \mathbf{x}, \tag{88}$$

with $\mathbf{F} = e\mathbf{E}$ as the force on an electron in the electric field. Now note that

$$\text{grad}_{\mathbf{k}}\, \varphi_{\mathbf{k}\gamma}(\mathbf{x}) = i\mathbf{x}\varphi_{\mathbf{k}\gamma}(\mathbf{x}) + e^{i\mathbf{k}\cdot\mathbf{x}}\, \text{grad}_{\mathbf{k}}\, e^{-i\mathbf{k}\cdot\mathbf{x}} \varphi_{\mathbf{k}\gamma}(\mathbf{x}). \tag{89}$$

Then

$$H = H_{\mathbf{F}} + i\mathbf{F} \cdot \text{grad}_{\mathbf{k}}, \tag{90}$$

where

$$H_{\mathbf{F}} = H_0 - ie^{i\mathbf{k}\cdot\mathbf{x}}\mathbf{F} \cdot \text{grad}_{\mathbf{k}}\, e^{-i\mathbf{k}\cdot\mathbf{x}} \tag{91}$$

acts as invariant under a lattice translation because the term in $\mathbf{F}$ does not mix states of different $\mathbf{k}$, but only of the same $\mathbf{k}$ of different bands. If $\varphi_{\mathbf{k}\gamma}(\mathbf{x})$ are the eigenstates of H_0, then

$$\begin{aligned} -i\langle \delta\mathbf{k}' | e^{i\mathbf{k}\cdot\mathbf{x}}\mathbf{F} \cdot \text{grad}_{\mathbf{k}}\, e^{-i\mathbf{k}\cdot\mathbf{x}} | \mathbf{k}\gamma\rangle \\ = -i \int d^3x\, e^{i(\mathbf{k}-\mathbf{k}')\cdot\mathbf{x}}\, u^*_{\mathbf{k}'\delta}\mathbf{F} \cdot \text{grad}_{\mathbf{k}}\, u_{\mathbf{k}\gamma}, \end{aligned} \tag{92}$$

which vanishes except for $\mathbf{k} = \mathbf{k}'$ because the term $u^*_{\mathbf{k}'\delta}\, \text{grad}_{\mathbf{k}}\, u_{\mathbf{k}\gamma}$ is invariant under a lattice translation. It follows that $H_{\mathbf{F}}$ gives interband mixing, but only the term $i\mathbf{F} \cdot \text{grad}_{\mathbf{k}}$ in the hamiltonian (90) can cause a change of $\mathbf{k}$. Notice that in the present formulation, unlike the earlier one with a time-dependent vector potential, $\mathbf{k}$ is not a constant of the motion.

Consider the problem of a free electron

$$\varphi_{\mathbf{k}} = e^{i\mathbf{k}\cdot\mathbf{x}} e^{-i\alpha(t)} \tag{93}$$

in an electric field. The time-dependent Schrödinger equation is

$$i\frac{d\varphi}{dt} = \left(\frac{1}{2m} p^2 - \mathbf{F} \cdot \mathbf{x}\right)\varphi. \tag{94}$$

But, from (93),

$$\frac{d\varphi}{dt} = \frac{\partial\varphi}{\partial t} + \frac{d\mathbf{k}}{dt} \cdot \text{grad}_{\mathbf{k}}\, \varphi = \left(-i\frac{d\alpha}{dt} + i\frac{d\mathbf{k}}{dt} \cdot \mathbf{x}\right)\varphi, \tag{95}$$

so that

$$\frac{d\alpha}{dt} - \frac{d\mathbf{k}}{dt} \cdot \mathbf{x} = \frac{1}{2m} k^2 - \mathbf{F} \cdot \mathbf{x}, \tag{96}$$

or

$$\frac{d\mathbf{k}}{dt} = \mathbf{F}. \tag{97}$$

The same argument applies in a crystal. We define a set of functions $\chi_{\mathbf{k}\gamma}(\mathbf{x})$ as the eigenfunctions of

$$H_{\mathbf{F}}\chi_{\mathbf{k}\gamma} = \varepsilon_{\mathbf{k}\gamma}\chi_{\mathbf{k}\gamma}. \tag{98}$$

The time-dependent equation is

$$i\,\frac{d\chi_{\mathbf{k}}}{dt} = (H_{\mathbf{F}} + i\mathbf{F} \cdot \mathrm{grad}_{\mathbf{k}})\chi_{\mathbf{k}}. \tag{99}$$

We try a solution with $\chi_{\mathbf{k}}$ confined to one band:

$$\chi_{\mathbf{k}} = e^{i\mathbf{k}\cdot\mathbf{x}}e^{-i\alpha(t)}u_{\mathbf{k}\gamma}(\mathbf{x}); \tag{100}$$

the derivative is

$$i\,\frac{d\chi_{\mathbf{k}}}{dt} = \left(\frac{d\alpha}{dt} + i\,\frac{d\mathbf{k}}{dt} \cdot \mathrm{grad}_{\mathbf{k}}\right)\chi_{\mathbf{k}}, \tag{101}$$

or, on comparing (99) with (101),

$$\frac{d\mathbf{k}}{dt} = \mathbf{F}. \tag{102}$$

Thus the acceleration theorem is valid in the basis $\chi_{\mathbf{k}\gamma}$ of Bloch states for which the polarization effect of the electric field has been taken into account by the hamiltonian $H_{\mathbf{F}}$.

For very short time intervals it can be shown that the motion of an electron in a crystal is governed by the free electron mass and not by the effective mass; see, for example, E. N. Adams and P. N. Argyres, *Phys. Rev.* **102,** 605 (1956).

We give now a theorem which connects the expectation value of the velocity with the wavevector, thereby enabling us to use the acceleration theorem to connect the change of velocity and the applied force; see also (49).

THEOREM 11. If $\langle\mathbf{v}\rangle$ is the expectation value of the velocity in a state $|\mathbf{k}\gamma\rangle$, then

$$\langle\mathbf{v}\rangle = i\langle[H,\mathbf{x}]\rangle = \mathrm{grad}_{\mathbf{k}}\, \varepsilon_{\mathbf{k}\gamma}, \tag{103}$$

in the absence of magnetic fields.

Proof: We consider the matrix element in the band γ:

$$\langle \mathbf{k}|[H,\mathbf{x}]|\mathbf{k}\rangle = \int d^3x\, u_{\mathbf{k}}^*(\mathbf{x})e^{-i\mathbf{k}\cdot\mathbf{x}}[H,\mathbf{x}]e^{i\mathbf{k}\cdot\mathbf{x}}u_{\mathbf{k}}(\mathbf{x}). \tag{104}$$

Now

$$\begin{aligned}\operatorname{grad}_{\mathbf{k}} (e^{-i\mathbf{k}\cdot\mathbf{x}}He^{i\mathbf{k}\cdot\mathbf{x}}) &= -ie^{-i\mathbf{k}\cdot\mathbf{x}}\mathbf{x}He^{i\mathbf{k}\cdot\mathbf{x}} \\ &\quad + ie^{-i\mathbf{k}\cdot\mathbf{x}}H\mathbf{x}e^{i\mathbf{k}\cdot\mathbf{x}} = ie^{-i\mathbf{k}\cdot\mathbf{x}}[H,\mathbf{x}]e^{i\mathbf{k}\cdot\mathbf{x}};\end{aligned} \tag{105}$$

further, we have seen in (15) that

$$H(\mathbf{p},\mathbf{x})e^{i\mathbf{k}\cdot\mathbf{x}} = e^{i\mathbf{k}\cdot\mathbf{x}}H(\mathbf{p}+\mathbf{k},\mathbf{x}). \tag{106}$$

Thus

$$\begin{aligned}\langle \mathbf{k}|[H,\mathbf{x}]|\mathbf{k}\rangle &= -i\int d^3x\, u_{\mathbf{k}}^*(\mathbf{x})(\operatorname{grad}_{\mathbf{k}} e^{-i\mathbf{k}\cdot\mathbf{x}}He^{i\mathbf{k}\cdot\mathbf{x}})u_{\mathbf{k}}(\mathbf{x}) \\ &= -i\int d^3x\, u_{\mathbf{k}}^*(\mathbf{x})(\operatorname{grad}_{\mathbf{k}} H(\mathbf{p}+\mathbf{k},\mathbf{x}))u_{\mathbf{k}}(\mathbf{x}).\end{aligned} \tag{107}$$

Now use the Feynman theorem, namely

$$\frac{\partial}{\partial\lambda}\langle \mathbf{k}|H|\mathbf{k}\rangle = \left\langle \mathbf{k}\right| \frac{\partial H}{\partial\lambda}\left|\mathbf{k}\right\rangle, \tag{108}$$

where λ is a parameter in the hamiltonian. Thus (107) becomes $-i \operatorname{grad}_{\mathbf{k}} \varepsilon_{\mathbf{k}}$, and

$$\langle \dot{\mathbf{x}}\rangle = \operatorname{grad}_{\mathbf{k}} \varepsilon_{\mathbf{k}}. \tag{109}$$

Q.E.D.

Further, as $\langle \dot{\mathbf{x}}\rangle$ is a function of $\mathbf{k}$ alone,

$$\frac{d}{dt}\langle \dot{\mathbf{x}}\rangle = \frac{d\mathbf{k}}{dt} : \operatorname{grad}_{\mathbf{k}} \operatorname{grad}_{\mathbf{k}} \varepsilon_{\mathbf{k}}, \tag{110}$$

or, by (55),

$$\frac{d}{dt}\langle \dot{x}_\mu\rangle = \frac{dk_\nu}{dt}\left(\frac{1}{m^*}\right)_{\nu\mu} = F_\nu\left(\frac{1}{m^*}\right)_{\nu\mu}. \tag{111}$$

If $\varepsilon_{\mathbf{k}} = k^2/2m^*$, then

$$m^*\frac{d}{dt}\langle \dot{\mathbf{x}}\rangle = \mathbf{F}. \tag{112}$$

It is more difficult to treat rigorously the motion of a lattice electron in a magnetic field. Particular problems are treated at several points in the text. For a general discussion and further references, see G. H. Wannier, *Rev. Mod. Phys.* **34,** 645 (1962) and E. J. Blount in *Solid state physics* **13,** 306. For electrons in nondegenerate bands and not-too-strong magnetic fields the result of the detailed calculations is that the equation of motion (111) may be generalized to

$$\mathbf{F} = e\left(\mathbf{E} + \frac{1}{c}\mathbf{v} \times \mathbf{H}\right). \tag{113}$$

We now give several theorems concerning special functions—Wannier functions—which are sometimes used in discussions of the motion of lattice electrons in perturbed potentials and in electric and magnetic fields.

WANNIER FUNCTIONS

Let $\varphi_{\mathbf{k}\gamma}(\mathbf{x})$ be a Bloch function in the band γ; the Wannier functions are defined by

$$w_\gamma(\mathbf{x} - \mathbf{x}_n) = N^{-1/2} \sum_{\mathbf{k}} e^{-i\mathbf{k}\cdot\mathbf{x}_n}\varphi_{\mathbf{k}\gamma}(\mathbf{x}), \tag{114}$$

where N is the number of atoms and $\mathbf{x}_n$ is a lattice point.

THEOREM 12. The Bloch functions may be expanded in terms of Wannier functions as

$$\varphi_{\mathbf{k}}(\mathbf{x}) = N^{-1/2} \sum_{n} e^{i\mathbf{k}\cdot\mathbf{x}_n} w(\mathbf{x} - \mathbf{x}_n). \tag{115}$$

Proof: From the definition of w,

$$\begin{aligned}\varphi_{\mathbf{k}}(\mathbf{x}) &= N^{-1/2} \sum_{n} e^{i\mathbf{k}\cdot\mathbf{x}_n} N^{-1/2} \sum_{\mathbf{k}'} e^{-i\mathbf{k}'\cdot\mathbf{x}_n}\varphi_{\mathbf{k}'}(\mathbf{x}) \\ &= N^{-1} \sum_{\mathbf{k}',n} e^{i(\mathbf{k}-\mathbf{k}')\cdot\mathbf{x}_n}\varphi_{\mathbf{k}'}(\mathbf{x}) = \varphi_{\mathbf{k}}(\mathbf{x}).\end{aligned} \tag{116}$$

THEOREM 13. Wannier functions about different lattice points are orthogonal, that is,

$$\int d^3x\, w^*(\mathbf{x})w(\mathbf{x} - \mathbf{x}_n) = 0, \qquad \mathbf{x}_n \neq 0. \tag{117}$$

Proof:

$$\begin{aligned}\int d^3x\, w^*(\mathbf{x})w(\mathbf{x} - \mathbf{x}_n) &= N^{-1} \sum_{\mathbf{k}\mathbf{k}'} \int d^3x\, e^{-i\mathbf{k}\cdot\mathbf{x}_n}\varphi_{\mathbf{k}}^*(\mathbf{x})\varphi_{\mathbf{k}'}(\mathbf{x}) \\ &= N^{-1} \sum_{\mathbf{k}} e^{-i\mathbf{k}\cdot\mathbf{x}_n} = \delta_{0n}.\end{aligned} \tag{118}$$

The Wannier functions tend to be peaked around the individual lattice sites $\mathbf{x}_n$. We examine this under the special assumption that

$$\varphi_{\mathbf{k}} = e^{i\mathbf{k}\cdot\mathbf{x}}u_0(\mathbf{x}), \tag{119}$$

where $u_0(\mathbf{x})$ is independent of $\mathbf{k}$. Then

$$w(\mathbf{x} - \mathbf{x}_n) = N^{-\frac{1}{2}} u_0(\mathbf{x}) \sum_{\mathbf{k}} e^{i\mathbf{k}\cdot(\mathbf{x}-\mathbf{x}_n)}. \tag{120}$$

In one dimension with lattice constant a,

$$k = m\frac{2\pi}{Na}, \tag{121}$$

where m is an integer between $\pm\frac{1}{2}N$. Then

$$\sum_k e^{ik\xi} = \sum_m e^{i(2\pi m\xi/Na)} \cong \frac{\sin(\pi\xi/a)}{(\pi\xi/Na)}, \tag{122}$$

for $N \gg 1$, and

$$w(x - x_n) = N^{\frac{1}{2}} u_0(x) \frac{\sin\{\pi(x - x_n)/a\}}{\{\pi(x - x_n)/a\}}. \tag{123}$$

In three dimensions we have the product of three similar functions. Thus the Wannier function assumes its largest value within the lattice cell about $\mathbf{x}_n$, and it tails off as we go out from the central cell.

THEOREM 14. If $\varepsilon(\mathbf{k})$ is the solution of the unperturbed one-particle periodic potential problem for a nondegenerate energy band, then the eigenvalues with a slowly varying perturbation $H'(\mathbf{x})$ are given by the eigenvalues λ of the equation

$$[\varepsilon(\mathbf{p}) + H'(\mathbf{x})]U(\mathbf{x}) = \lambda U(\mathbf{x}), \tag{124}$$

where $\varepsilon(\mathbf{p})$ is the operator obtained on substituting $\mathbf{p}$ or $-i$ grad for $\mathbf{k}$ in $\varepsilon(\mathbf{k})$ in the band γ; $U(\mathbf{x})$ has the property that

$$\chi(\mathbf{x}) = \sum_n U(\mathbf{x}_n)w(\mathbf{x} - \mathbf{x}_n), \tag{125}$$

where $\chi(\mathbf{x})$ is the solution of the Schrödinger equation

$$[H_0 + H'(\mathbf{x})]\chi(\mathbf{x}) = \lambda\chi(\mathbf{x}). \tag{126}$$

Proof: A clear proof is given by J. C. Slater, *Phys. Rev.* **76,** 1592 (1949). A treatment of a similar problem for weakly bound donor and acceptor states in semiconductors is given in Chapter 14; the method given there is the one most often used in practice when quantitative calculations are carried out.

In a magnetic field (124) becomes

$$\left[\varepsilon\left(\mathbf{p} - \frac{e}{c}\mathbf{A}\right) + H'(\mathbf{x})\right] U(\mathbf{x}) = \lambda U(\mathbf{x}), \tag{127}$$

as demonstrated by J. M. Luttinger, *Phys. Rev.* **84,** 814 (1951). In the expansion of $\varepsilon(\mathbf{k})$ any product of $\mathbf{k}$'s is to be written as a symmetrized product before making the substitution $\mathbf{k} \rightarrow \mathbf{p} - e/c\mathbf{A}$. An example of effects arising from the noncommutativity of the components of $\mathbf{k}$ in a magnetic field is given in Chapter 14.

PROBLEMS

1. If O_1 has the property

$$KO_1K^{-1} = O_1^+, \tag{128}$$

show that

$$\langle\varphi|O_1|K\varphi\rangle = 0. \tag{129}$$

For O_1 we may have a symmetrized product of an even number of momentum components, or any function of $\mathbf{x}$.

2. For O_1 as defined in the first problem, show that

$$\langle\varphi|O_1|\varphi\rangle = \langle K\varphi|O_1|K\varphi\rangle. \tag{130}$$

3. If O_2 has the property

$$KO_2K^{-1} = -O_2^+, \tag{131}$$

show that

$$\langle\varphi|O_2|\varphi\rangle = -\langle K\varphi|O_2|K\varphi\rangle. \tag{132}$$

4. Show that the results of 1, 2, 3 hold if everywhere $C \equiv KJ$ is written for K; the states are now assumed to be eigenstates of a hamiltonian invariant under C.

5. If $COC^{-1} = O^+$, show that

$$\langle\uparrow\mathbf{k}|O|\mathbf{k}\downarrow\rangle = 0; \tag{133}$$

here O might be $\mathbf{p}$, a symmetrized product of an even number of linear momenta, or the spin-orbit interaction; show further that

$$\langle\uparrow\mathbf{k}|O|\mathbf{k}\uparrow\rangle = \langle\downarrow\mathbf{k}|O|\mathbf{k}\downarrow\rangle. \tag{134}$$

6. If $COC^{-1} = -O^+$, show that

$$\langle\uparrow\mathbf{k}|O|\mathbf{k}\uparrow\rangle = -\langle\downarrow\mathbf{k}|O|\mathbf{k}\downarrow\rangle; \tag{135}$$

here O might be $\mathbf{L}$ or $\boldsymbol{\sigma}$.

7. Prove (49); use (47) and the normalization condition $\text{grad}_{\mathbf{k}} \sum_{\mathbf{G}} |f_{\mathbf{G}}(\mathbf{k})|^2 = 0$.

8. Prove (51).

9. Evaluate the effective mass tensor (56), with $\mathbf{p}$ written for $\boldsymbol{\pi}$, in the limit of separated atoms. The wavefunctions may be written in the tight binding form as

$$\varphi_{\mathbf{k}\gamma}(\mathbf{x}) = N^{-\frac{1}{2}} \sum_j e^{i\mathbf{k}\cdot\mathbf{x}_j} v_\gamma(\mathbf{x} - \mathbf{x}_j), \tag{136}$$

where v is an atomic function in the state γ. It is assumed that v's centered on different lattice sites do not overlap. We find that $\langle\gamma\mathbf{k}|\mathbf{p}|\mathbf{k}\delta\rangle = \langle v_\gamma|\mathbf{p}|v_\delta\rangle$, where v_γ and v_δ are different states of the same atom. Now

$$\frac{i}{m}\langle\gamma|\mathbf{p}|\delta\rangle = (\varepsilon_\delta - \varepsilon_\gamma)\langle\gamma|\mathbf{x}|\delta\rangle, \tag{137}$$

so that

$$\left(\frac{m}{m^*}\right)_{xx} = \left[1 - 2m \sum_\delta{}' (\varepsilon_\delta - \varepsilon_\gamma)|\langle\gamma|x|\delta\rangle|^2\right] = 0, \tag{138}$$

on application of the atomic f-sum rule. Show that (136) satisfies the translational symmetry requirement (14).

10. (*a*) Show that an electron in a crystal in an electric field $\boldsymbol{\mathcal{E}}$ will oscillate according to

$$e(\mathbf{x} - \mathbf{x}_0) \cdot \boldsymbol{\mathcal{E}} = \varepsilon(\mathbf{k}_0 + e\boldsymbol{\mathcal{E}}t) - \varepsilon(\mathbf{k}_0), \tag{139}$$

from conservation of energy. The amplitude Δx of oscillation is $\Delta x \cong \Delta\varepsilon/e|\boldsymbol{\mathcal{E}}|$, where $\Delta\varepsilon$ is the width of the band. (*b*) Estimate Δx for a reasonable electric field. (*c*) Estimate the frequency of the motion.

11. Consider a Bloch state which is nondegenerate at $\mathbf{k} = 0$. Using $\varphi_{\mathbf{k}}(\mathbf{x})$ as an expansion of $\varphi_0(\mathbf{x})$ to first order in $\mathbf{k}\cdot\mathbf{p}$, show by direct calculation that

$$\langle\mathbf{k}|p_\mu|\mathbf{k}\rangle \cong k_\alpha \left(\frac{m}{m^*}\right)_{\alpha\mu}, \tag{140}$$

to first order in $\mathbf{k}$.

10 Brillouin zones and crystal symmetry

We have seen that the energy eigenvalues of the periodic potential problem are periodic in the reciprocal lattice:

$$\varepsilon_{\mathbf{k}+\mathbf{G}} = \varepsilon_{\mathbf{k}};$$

thus to label the eigenvalues uniquely it is necessary to restrict **k** to a primitive cell of the reciprocal lattice. The primitive cell may be chosen in various ways, but the standard convention is to bound the cell by the planes which bisect the lines joining $\mathbf{k} = 0$ to the nearest points of the reciprocal lattice. This cell is called the Brillouin zone, or first Brillouin zone. Unless otherwise specified, our **k**'s are understood to be reduced to this zone. The Brillouin zone of the linear lattice is shown in Fig. 1, of the square lattice in Fig. 2, the sc lattice in Fig. 3, the bcc lattice in Fig. 5, and the fcc lattice in Fig. 7. The construction of the zones is described in *ISSP*, Chapter 12.

There are certain useful symmetry properties of the hamiltonian for the periodic crystal potential which are most easily discussed with the help of elementary group theory. The reader without benefit of group theory may acquire the needed elements from Chapter 12 in Landau and Lifshitz. The modest object of this chapter is to make it possible for a reader equipped with a knowledge of point symmetry groups and their representations to extend his knowledge to the important symmetry properties of the Brillouin zone. We also summarize in tabular form results which are frequently used. The symmetry properties are first discussed without spin, and later the spin is added.

In a crystal the group G of the hamiltonian is the space group of the crystal structure plus the operation of time reversal. We recall that the lattice and thus the hamiltonian is invariant under all translations of the form

$$Tx = x + t, \tag{1}$$

where **t** is a vector in the direct lattice:

$$\mathbf{t} = l\mathbf{a} + m\mathbf{b} + n\mathbf{c}; \qquad l, m, n = \text{integers}; \tag{2}$$

and **a**, **b**, **c** are the primitive basis vectors. Thus for a Bloch function $\varphi_{\mathbf{k}}$,

$$T\varphi_{\mathbf{k}} = e^{i\mathbf{k}\cdot\mathbf{t}}\varphi_{\mathbf{k}}; \tag{3}$$

the $\varphi_{\mathbf{k}}$ belong to one-dimensional representations of the translation group T and have the eigenvalue $e^{i\mathbf{k}\cdot\mathbf{t}}$. We restrict our discussion at the beginning to crystal structures which are themselves bravais lattices. That is, we defer discussion of space groups which contain screw axes or glide planes.[1] Crystallographic nomenclature is summarized in *ISSP*, Chapter 1.

We now study the effect of the operations of the point group R. Let P_R be an operator of the point group. The result of operating on a function $f(\mathbf{x})$ with R is defined to be

$$P_R f(\mathbf{x}) \equiv f(R^{-1}\mathbf{x}), \tag{4}$$

where R is a real orthogonal transformation.

The rotation R transforms a Bloch function $\varphi_{\mathbf{k}}(\mathbf{x}) = e^{i\mathbf{k}\cdot\mathbf{x}}u_{\mathbf{k}}(\mathbf{x})$ into a new function $\varphi_{\mathbf{k}'}(\mathbf{x})$, where $\mathbf{k}'$ is derived from $\mathbf{k}$ by a rotation R applied in $\mathbf{k}$ space. This result is intuitively obvious on observing that $\mathbf{k} \cdot R^{-1}\mathbf{x} = \mathbf{x} \cdot R\mathbf{k}$.

THEOREM. If $\varphi_{\mathbf{k}}(R^{-1}\mathbf{x})$ is a solution of the wave equation, then $\varphi_{R[\mathbf{k}]}(\mathbf{x})$ is a solution with the same energy, where R is an element of the group of the Schrödinger equation.

Proof: We have

$$\varphi_{\mathbf{k}}(R^{-1}\mathbf{x}) = e^{i\mathbf{k}\cdot R^{-1}\mathbf{x}}u_{\mathbf{k}}(R^{-1}\mathbf{x}) = e^{iR[\mathbf{k}]\cdot\mathbf{x}}u_{\mathbf{k}}(R^{-1}\mathbf{x}). \tag{5}$$

Now $u_{\mathbf{k}}(\mathbf{x})$ is a solution of

$$\left\{\frac{1}{2m}(p^2 + 2\mathbf{k}\cdot\mathbf{p} + k^2) + V(\mathbf{x})\right\} u_{\mathbf{k}}(\mathbf{x}) = \lambda_{\mathbf{k}}u_{\mathbf{k}}(\mathbf{x}), \tag{6}$$

[1] For full details of these space groups, see H. Jones, *Theory of Brillouin zones and electronic states in crystals*, North-Holland, Amsterdam, 1960; V. Heine, *Group theory in quantum mechanics*, Pergamon, London, 1960; G. F. Koster, *Solid state physics* **5**, 174–256.

and $u_{\mathbf{k}}(R^{-1}\mathbf{x})$ is a solution of

$$\left\{\frac{1}{2m}(p^2 + 2\mathbf{k}\cdot R^{-1}p + k^2) + V(\mathbf{x})\right\} u_{\mathbf{k}}(R^{-1}\mathbf{x}) = \lambda_{\mathbf{k}} u_{\mathbf{k}}(R^{-1}\mathbf{x}), \tag{7}$$

where we have used the relations

$$V(R^{-1}\mathbf{x}) = V(\mathbf{x}), \tag{8}$$

and

$$R^{-1}\mathbf{p}\cdot R^{-1}\mathbf{p} = p^2. \tag{9}$$

But observe that

$$R[\mathbf{k}]\cdot\mathbf{p} = \mathbf{k}\cdot R^{-1}[\mathbf{p}], \tag{10}$$

and

$$R[\mathbf{k}]\cdot R[\mathbf{k}] = k^2, \tag{11}$$

so that $\varphi_{R[\mathbf{k}]}(\mathbf{x})$ is a solution of the same equation as $\varphi_{\mathbf{k}}(R^{-1}\mathbf{x})$ and has the same energy. Note that $\varphi_{R[\mathbf{k}]}(\mathbf{x})$ is an eigenfunction of the lattice translation operator T, with the eigenvalue $e^{iR[\mathbf{k}]\cdot\mathbf{t}}$.

We can therefore generate a representation of R by letting R operate on $\mathbf{k}$ in $\mathbf{k}$ space or by letting R^{-1} operate on $\mathbf{x}$ in real space. If there is only a single $\varphi_{\mathbf{k}}$ for each $\mathbf{k}$, we may replace (4) by

$$P_R\varphi_{\mathbf{k}}(\mathbf{x}) = \varphi_{R[\mathbf{k}]}(\mathbf{x}). \tag{12}$$

If the point group has n elements, the degenerate $\varphi_{\mathbf{k}}$ form (for a non-special $\mathbf{k}$) an n-dimensional representation of the group of the Schrödinger equation.

If a certain $\mathbf{k}_0$ is invariant with $\mathbf{k}_0 = R'\mathbf{k}_0$ under certain operations R' forming a subgroup of R, these operations form the group of $\mathbf{k}_0$. That is, if there are symmetry elements which leave special wavevectors invariant, these symmetry elements form a group which is called the *group of the wavevector.* Because of the periodicity of the reciprocal lattice we treat $\mathbf{k}$ and $\mathbf{k} + \mathbf{G}$ as *identical* (not merely equivalent) wavevectors, where $\mathbf{G}$ is a vector in the reciprocal lattice. This statement is consistent with the correct enumeration of states (Chapter 1). Suppose that the states $\varphi_{\mathbf{k}\mu}$ of given $\mathbf{k}$ are degenerate in energy: the operations of the group of $\mathbf{k}$ transform $\varphi_{\mathbf{k}\mu}$ into a $\varphi_{\mathbf{k}\lambda}$ with the same $\mathbf{k}$, and the φ's are said to form a representation of the group of $\mathbf{k}$. The representation is known as the *small representation.*

We consider first the trivial example of a linear lattice, of lattice constant a; the Brillouin zone is shown in Fig. 1. The first zone is

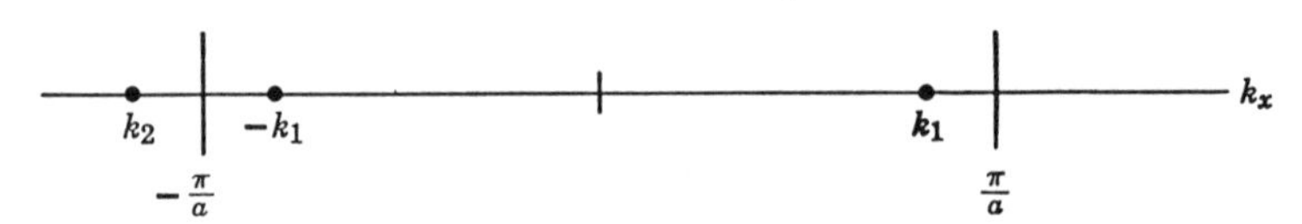

FIG. 1. Brillouin zone of linear lattice.

bounded by $-\pi/a$ and π/a. If the potential $V(x)$ is even with $V(-x) = V(x)$, then the group of the hamiltonian includes the reflection operation in a plane through the origin, and k_1 and $-k_1$ are degenerate in energy. We denote the reflection operation normal to the x axis by m_x.

The special points in **k** space in this example are $-\pi/a$ and π/a; they differ by the reciprocal lattice vector $2\pi/a$ and are therefore *identical* points in every respect. It follows that

$$m_x\left[\frac{\pi}{a}\right] = -\frac{\pi}{a} \equiv \frac{\pi}{a}; \tag{13}$$

thus the point π/a is invariant under m_x. The operations E and m_x, where E is the identity, are the group of the wavevector π/a. The representations of this group are one dimensional and are trivial; they are either even or odd under m_x, so that $\varphi_{\pi/a} = \pm\varphi_{-\pi/a}$, and either

$$\varphi_{\pi/a} = \sin(\pi x/a)u_{\pi/a}(x), \tag{14}$$

or

$$\varphi_{\pi/a} = \cos(\pi x/a)u_{\pi/a}(x). \tag{15}$$

We see that the Bloch functions at the boundaries of this zone are standing waves. The u's in (14) and (15) need not be identical, because the φ's belong to different representations and thus to different energies.

We notice another feature of the zone boundary. The point $k_2 = k_1 - (2\pi/a)$ is identical with k_1 because the points differ only by a vector in the reciprocal lattice. Recall that $\pm k_1$ are degenerate in energy. Thus the energies satisfy

$$\varepsilon(k_1) = \varepsilon(k_2) = \varepsilon(-k_1); \tag{16}$$

if we let k_1 approach π/a, we see that k_2 approaches $-\pi/a$, so that

$$\lim_{\delta\to+0} \varepsilon\left(\frac{\pi}{a} - \delta\right) = \lim_{\delta\to+0} \varepsilon\left(-\frac{\pi}{a} - \delta\right) = \lim_{\delta\to+0} \varepsilon\left(-\frac{\pi}{a} + \delta\right). \tag{17}$$

This implies that the energy is even about $\pm\pi/a$, whence

$$\frac{\partial}{\partial k}\,\varepsilon(k) = 0 \tag{18}$$

at the points $\pm\pi/a$.

SQUARE LATTICE

The Brillouin zone of the square lattice is shown in Fig. 2. The point group symmetry of the lattice is denoted by the symbol $4mm$; there is a fourfold axis containing two sets of mirror planes, one set made up of m_x and m_y, and the other made up of two diagonal planes designated m_d, $m_{d'}$. For a discussion of crystal symmetry elements, see *ISSP*, Chapter 1.

There are six special types of points or lines in the Brillouin zone of the square lattice—the points Γ, M, X, and the lines Δ, Z, Σ. The point Γ which lies at $\mathbf{k} = 0$ transforms into itself under all the operations of the point group. Under the same operations M transforms directly into itself or into the other corners of the square. The corners are connected to each other by vectors in the reciprocal lattice, and therefore the four corners represent only a *single* point. Transforming one corner into another is equivalent to taking a corner M into itself. The point X is invariant under the operations 2_z, m_x, m_y, where the reflection m_x and the twofold rotation 2_z carries π/a into the identical point $-\pi/a$.

The special lines Σ, Δ, and Z are invariant respectively under the mirror operations m_d, m_y, and m_x; the invariance of Z under m_x follows

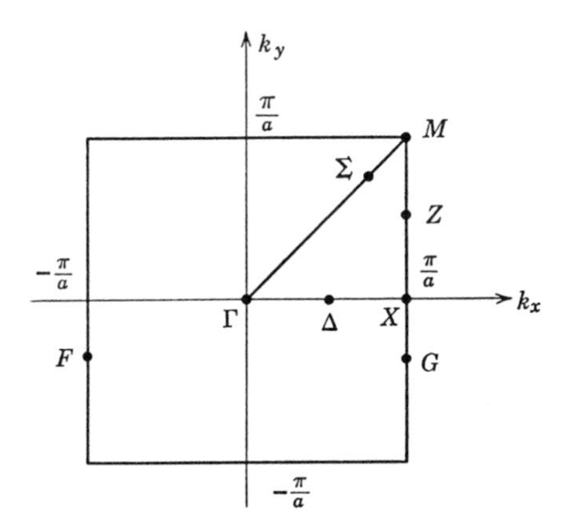

FIG. 2. Brillouin zone of the square lattice; the point group symmetry is $4mm$.

because the operation takes Z into a point connected to Z by a reciprocal lattice vector. Thus the points labeled G and F are related by m_x and differ by the reciprocal lattice vector $(2\pi/a,0,0)$. For the square lattice we see that the argument of (17) applies to every point on the zone boundary, so that $\text{grad}_{\mathbf{k}}\, \varepsilon = 0$ on the zone boundary. This property of the energy is inseparable from the presence of the mirror plane; the property does not hold, for example, at the (111) faces in the fcc problem (Fig. 7) because there is no mirror plane normal to the [111] direction.

TABLE 1

CHARACTER TABLES OF THE SMALL REPRESENTATIONS OF THE SPECIAL POINTS AND LINES OF THE SQUARE LATTICE

Γ, M	E	2_z	$4_z, 4_z^3$	m_x, m_y	$m_d, m_{d'}$
Γ_1, M_1	1	1	1	1	1
Γ_2, M_2	1	1	1	−1	−1
Γ_3, M_3	1	1	−1	1	−1
Γ_4, M_4	1	1	−1	−1	1
Γ_5, M_5	2	−2	0	0	0

X	E	2_z	m_x	m_y
X_1	1	1	1	1
X_2	1	1	−1	−1
X_3	1	−1	1	−1
X_4	1	−1	−1	1

Δ Σ Z	E E E	m_y m_d m_x
Δ_1, Σ_1, Z_1	1	1
Δ_2, Σ_2, Z_2	1	−1

The character tables for the special points and lines of the square lattice are given in Table 1. We can usefully label a band by the set of labels of its irreducible representations at special points; thus a band might be labeled as $\Gamma_5\Delta_1X_4Z_2M_5\,\Sigma_1$.

COMPATIBILITY RELATIONS

Within a single energy band the representations at the special points and lines are not entirely independent. The representations must be compatible. Suppose that the band has the representation Z_2 at Z, so that the state is odd under the mirror operation m_x. This representation is not compatible at the point X on the line Z with the representations X_1 and X_3, which are even under m_x; further, Z_2 is not compatible with M_1 and M_3, because these representations are even under m_x. The question of M_5 requires attention, but M_5 is reducible under the group E, m_x into Z_1 and Z_2, so that M_5 is compatible with either Z_1 or Z_2. The complete compatibility relations for the square lattice are simple to work out and are given in Table 2.

TABLE 2

COMPATIBILITY RELATIONS FOR THE SQUARE LATTICE

Representation	Compatible with
Δ_1	Γ_1, Γ_3, Γ_5; X_1, X_4
Δ_2	Γ_2, Γ_4, Γ_5; X_2, X_3
Σ_1	Γ_1, Γ_4, Γ_5; M_1, M_4, M_5
Σ_2	Γ_2, Γ_3, Γ_5; M_2, M_3, M_5
Z_1	X_1, X_3; M_1, M_3, M_5
Z_2	X_2, X_4; M_2, M_4, M_5

SIMPLE CUBIC LATTICE

The full cubic point group is $4/m\,\overline{3}\,2/m$. There are four special points R, M, X, Γ and five special lines Δ, S, T, Σ, Z, as shown in Fig. 3.

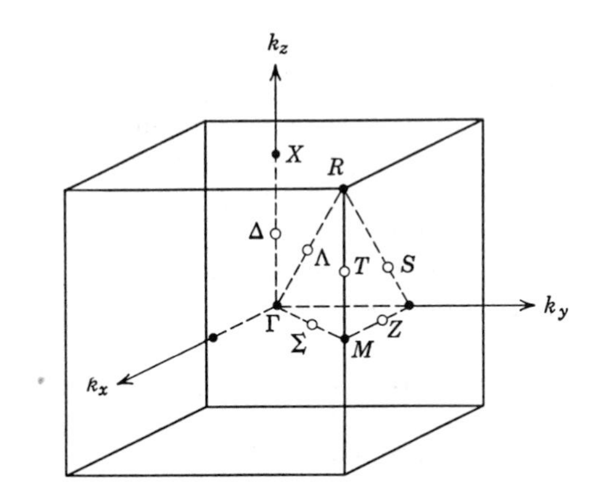

FIG. 3. Brillouin zone of the simple cubic lattice, with special points labeled.

Γ, *R*. The point Γ at the center of the zone obviously transforms into itself under all the operations of the cubic group. The point *R* at the corner is connected to the other corners by reciprocal lattice vectors, so that all eight corners are a single point. The eight corners transform into each other under the cubic group; thus *R* and Γ have the same representations, as given in Table 3, which is given in every textbook on group theory. The point *H* is included for use with the bcc lattice.

TABLE 3

CHARACTER TABLE OF THE SMALL REPRESENTATIONS OF Γ, *R*, *H*

	E	4^2	4	2	3	J	$J4^2$	$J4$	$J2$	$J3$
Γ_1	1	1	1	1	1	1	1	1	1	1
Γ_2	1	1	−1	−1	1	1	1	−1	−1	1
Γ_{12}	2	2	0	0	−1	2	2	0	0	−1
Γ_{15}'	3	−1	1	−1	0	3	−1	1	−1	0
Γ_{25}'	3	−1	−1	1	0	3	−1	−1	1	0
Γ_1'	1	1	1	1	1	−1	−1	−1	−1	−1
Γ_2'	1	1	−1	−1	1	−1	−1	1	1	−1
Γ_{12}'	2	2	0	0	−1	−2	−2	0	0	1
Γ_{15}	3	−1	1	−1	0	−3	1	−1	1	0
Γ_{25}	3	−1	−1	1	0	−3	1	1	−1	0

In Table 4 we compare three of the common notations used for the representations of the cubic group, and give the lowest-order basis functions which transform according to these representations. For x, y, z we could equally well write k_x, k_y, k_z. Basis functions of higher orders are given in Table II of the paper by Von der Lage and Bethe.

X, *M*. The point equivalent to *X* lies at the intersection of the k_z axis with the lower face of the cube. There are three points equivalent to *M*, at the intersections of the k_xk_y plane with the vertical edges; the points *X* and *M* have the same symmetry elements, $4/mmm$. The representations and symmetry types are given in Table 5. Character tables may be found in Jones, pp. 99 and 104; Bouckaert, Smoluchowski and Wigner, p. 64.

Δ, *T*. The point *T* is equivalent to three points on the other vertical edges. The point group is $4mm$; the point Δ has the same point group. The basis functions below are referred to the z axis:

Δ_1	Δ_2	Δ_2'	Δ_1'	Δ_5
1	$x^2 - y^2$	xy	$xy(x^2 - y^2)$	$\{x,y\}$

TABLE 4

SYMMETRY TYPES AT POINTS Γ, R, H OF CUBIC LATTICES

Notation			
BSW	LB	Chem	Basis Functions
Γ_1	α	A_{1g}	1
Γ_2	β'	A_{2g}	$x^4(y^2 - z^2) + y^4(z^2 - x^2) + z^4(x^2 - y^2)$
Γ_{12}	γ	E_g	$\{z^2 - \frac{1}{2}(x^2 + y^2), (x^2 - y^2)\}$
Γ_{15}'	δ'	T_{1g}	$\{xy(x^2 - y^2), yz(y^2 - z^2), zx(z^2 - x^2)\}$
Γ_{25}'	ε	T_{2g}	$\{xy,yz,zx\}$
Γ_1'	α'	A_{1u}	$xyz[x^4(y^2 - z^2) + y^4(z^2 - x^2) + z^4(x^2 - y^2)]$
Γ_2'	β	A_{2u}	xyz
Γ_{12}'	γ'	E_u	$\{xyz[z^2 - \frac{1}{2}(x^2 + y^2)], xyz(x^2 - y^2)\}$
Γ_{15}	δ	T_{1u}	$\{x,y,z\}$
Γ_{25}	ε'	T_{2u}	$\{z(x^2 - y^2), x(y^2 - z^2), y(z^2 - x^2)\}$

Note: BSW ≡ Bouckaert, Smoluchowski, and Wigner, *Phys. Rev.* **50,** 58(1936).
LB ≡ Von der Lage and Bethe, *Phys. Rev.* **71,** 612(1947).
Chem ≡ used by most chemists; also in the text by Heine, reference 1.

TABLE 5

SYMMETRY TYPES OF POINTS X, M OF CUBIC LATTICES (REFERRED TO THE z AXIS)

Representation	Basis Functions
X_1, M_1	1
X_2, M_2	$x^2 - y^2$
X_3, M_3	xy
X_4, M_4	$xy(x^2 - y^2)$
X_5, M_5	$\{yz,zx\}$
X_1', M_1'	$xyz(x^2 - y^2)$
X_2', M_2'	xyz
X_3', M_3'	$z(x^2 - y^2)$
X_4', M_4'	z
X_5', M_5'	$\{x,y\}$

Λ. The point group is $3m$. The basis functions are referred to the [111] axis:

Λ_1	Λ_2	Λ_3
1	$xy(x - y) + yz(y - z) + zx(z - x)$	$\{(x - z), (y - z)\}$

The representations of F are identical with that of Λ.

Σ, S. The groups are holomorphic to $2mm$. For operations referred to $k_x = k_y$ and $k_z = 0$, the basis functions are:

Σ_1	Σ_2	Σ_3	Σ_4
1	$z(x - y)$	z	$x - y$

Z. The point Z has two mirror planes and a twofold axis; with basis functions referred to the z axis:

Z_1	Z_2	Z_3	Z_4
1	yz	y	z

The representations of G, K, U, D are identical with those of Z.

The compatibility relations for the sc lattice are given in Table 6.

TABLE 6

COMPATIBILITY RELATIONS FOR THE SIMPLE CUBIC LATTICE

Γ_1	Γ_2	Γ_{12}	Γ_{15}'	Γ_{25}'	Γ_1'	Γ_2'	Γ_{12}'	Γ_{15}	Γ_{25}
Δ_1	Δ_2	$\Delta_1\Delta_2$	$\Delta_1'\Delta_5$	$\Delta_2'\Delta_5$	Δ_1'	Δ_2'	$\Delta_1'\Delta_2'$	$\Delta_1\Delta_5$	$\Delta_2\Delta_5$
Λ_1	Λ_2	Λ_3	$\Lambda_2\Lambda_3$	$\Lambda_1\Lambda_3$	Λ_2	Λ_1	Λ_3	$\Lambda_1\Lambda_3$	$\Lambda_2\Lambda_3$
Σ_1	Σ_4	$\Sigma_1\Sigma_4$	$\Sigma_2\Sigma_3\Sigma_4$	$\Sigma_1\Sigma_2\Sigma_3$	Σ_2	Σ_3	$\Sigma_2\Sigma_3$	$\Sigma_1\Sigma_3\Sigma_4$	$\Sigma_1\Sigma_2\Sigma_4$

X_1	X_2	X_3	X_4	X_5	X_1'	X_2'	X_3'	X_4'	X_5'
Δ_1	Δ_2	Δ_2'	Δ_1'	Δ_5	Δ_1'	Δ_2'	Δ_2	Δ_1	Δ_5
Z_1	Z_1	Z_4	Z_4	Z_2Z_3	Z_2	Z_2	Z_3	Z_3	Z_1Z_4
S_1	S_4	S_1	S_4	S_2S_3	S_2	S_3	S_2	S_3	S_1S_4

M_1	M_2	M_3	M_4	M_5	M_1'	M_2'	M_3'	M_4'	M_5'
Σ_1	Σ_4	Σ_1	Σ_4	$\Sigma_2\Sigma_3$	Σ_2	Σ_3	Σ_2	Σ_3	$\Sigma_1\Sigma_4$
Z_1	Z_1	Z_3	Z_3	Z_2Z_4	Z_2	Z_2	Z_4	Z_4	Z_1Z_3
T_1	T_2	T_2'	T_1'	T_5	T_1'	T_2'	T_2	T_1	T_5

CLASSIFICATION OF PLANE WAVE STATES IN THE EMPTY LATTICE

In Chapter 13 on the calculation of energy bands we shall discover reasons why the sequence of bands in a crystal often has a strong resemblance to the sequence of plane wave states, with the crystal field considered as a potential which lifts certain accidental degeneracies which occur for plane waves. One gains a very powerful insight into band structure by considering the perturbed plane waves.

We wish to rewrite a plane wave of general wavevector $\mathbf{k}'$:

$$\varphi_{\mathbf{k}'} = e^{i\mathbf{k}'\cdot\mathbf{x}}, \qquad \varepsilon_{\mathbf{k}'} = \frac{1}{2m} k'^2, \tag{19}$$

so that it appears in the reduced zone scheme. We can always find a reciprocal lattice vector $\mathbf{G}$ such that

$$\mathbf{k} = \mathbf{k}' - \mathbf{G} \tag{20}$$

lies in the first Brillouin zone. Then we define

$$\varphi_{\mathbf{k}} \equiv \varphi_{\mathbf{k}'} = e^{i\mathbf{k}\cdot\mathbf{x}} u_{\mathbf{k}}(\mathbf{x}), \tag{21}$$

with

$$u_{\mathbf{k}}(\mathbf{x}) = e^{i\mathbf{G}\cdot\mathbf{x}}, \qquad \varepsilon_{\mathbf{k}} = \frac{1}{2m} (\mathbf{k} + \mathbf{G})^2 = \varepsilon_{\mathbf{k}'}. \tag{22}$$

Here $e^{i\mathbf{G}\cdot\mathbf{x}}$ has the periodicity of the direct lattice, as required.

Energy versus $\mathbf{k}$ *in Reduced Zone, SC Lattice.* We consider now the behavior and degeneracies of the energy bands in the reduced zone, for an empty sc lattice of unit lattice constant ($a = 1$) and $V(\mathbf{x}) \equiv 0$. The lowest energy occurs for $\mathbf{G} = 0$ and gives us the band A sketched in Fig. 4:

$$\varepsilon_{A\mathbf{k}} = \frac{1}{2m} k^2. \tag{23}$$

Define band B for $\mathbf{G} = 2\pi(\bar{1}00)$; according to (22)

$$\varepsilon_{B\mathbf{k}} = \frac{1}{2m} \{(k_x - 2\pi)^2 + k_y{}^2 + k_z{}^2\}. \tag{24}$$

At $\mathbf{k} = 0$, $\varepsilon_{B0} = (1/2m)(2\pi)^2$, and $\varepsilon_{B\mathbf{k}}$ drops as we go out in the [100] direction to make contact with band A at the point X ($\mathbf{k} = \pi 00$). The band C is defined for $\mathbf{G} = 2\pi(100)$. The bands D, E, F, G are defined for $\mathbf{G}/2\pi = (010), (0\bar{1}0), (001), (00\bar{1})$. The next set contains 12 bands, for $\mathbf{G} = 2\pi(110)$ and equivalent $\mathbf{G}$'s.

We consider the effect of a weak cubic crystal potential in lifting the accidental parts of the degeneracies evident in the band scheme of Fig. 4. There is in the empty lattice a sixfold degeneracy at Γ for $\mathbf{G} = 2\pi(100)$ and equivalent $\mathbf{G}$'s. The unperturbed wavefunctions at Γ may be written as

$$\begin{aligned} &\varphi_1 = e^{2\pi ix}; \qquad \varphi_2 = e^{-2\pi ix}; \qquad \varphi_3 = e^{2\pi iy}; \\ &\varphi_4 = e^{-2\pi iy}; \qquad \varphi_5 = e^{2\pi iz}; \qquad \varphi_6 = e^{-2\pi iz}. \end{aligned} \tag{25}$$

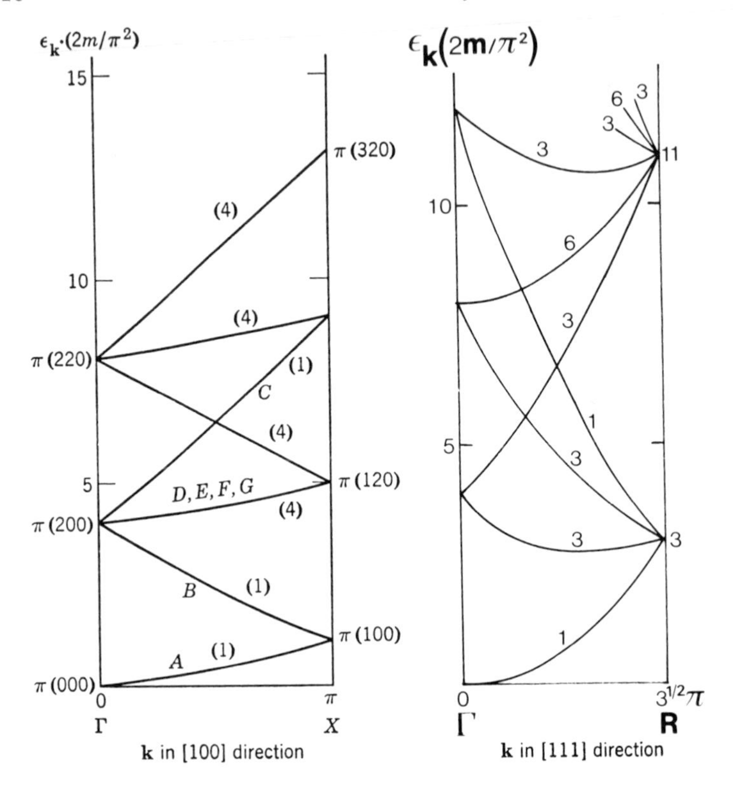

FIG. 4. Reduced zone scheme for free electrons in empty simple cubic lattice for $a = 1$. Degeneracies are indicated in parentheses. Representative values of the extended wavevectors are given for several boundary points.

At Γ the group of the wavevector is that of the full cubic point group. We can construct irreducible representations of this group from the φ_i by determining the characters of the φ_i and then reducing the representation or, perhaps more easily, by expanding the φ_i in series for small arguments and then using ingenuity, guided by Table 4. Thus, to quadratic terms in the coordinates,

$$\begin{aligned} &\varphi_1 \cong 1 + 2\pi ix - (2\pi)^2x^2; \qquad &&\varphi_2 \cong 1 - 2\pi ix - (2\pi)^2x^2; \\ (26)\quad &\varphi_3 \cong 1 + 2\pi iy - (2\pi)^2y^2; \qquad &&\varphi_4 \cong 1 - 2\pi iy - (2\pi)^2y^2; \\ &\varphi_5 \cong 1 + 2\pi iz - (2\pi)^2z^2; \qquad &&\varphi_6 \cong 1 - 2\pi iz - (2\pi)^2z^2. \end{aligned}$$

We emphasize that the elements R' of the group of the wavevector are elements which operate on the coordinates. We may form several representations from (26):

(27) $\Gamma_1\colon \varphi_1 + \varphi_2 + \varphi_3 + \varphi_4 + \varphi_5 + \varphi_6$

(28) $\Gamma_{15}\colon \varphi_1 - \varphi_2 \sim x; \qquad \varphi_3 - \varphi_4 \sim y; \qquad \varphi_5 - \varphi_6 \sim z$

(29) $\Gamma_{12}\colon \varphi_1 + \varphi_2 - \varphi_3 - \varphi_4 \sim x^2 - y^2;$
$$\varphi_1 + \varphi_2 + \varphi_3 + \varphi_4 - 2\varphi_5 - 2\varphi_6 \sim x^2 + y^2 - 2z^2.$$

Thus for $\mathbf{G} = 2\pi(100)$ we reduce the sixfold degenerate Γ to

$$\Gamma = \Gamma_1 + \Gamma_{15} + \Gamma_{12} \sim s + p + d_\gamma, \tag{30}$$

by analogy with atomic orbitals. The sixfold degeneracy splits into one, two, and threefold states.

At the lowest point X the states are

$$\varphi_1 = e^{\pi i x}; \qquad \varphi_2 = e^{-\pi i x}. \tag{31}$$

From these we may form the combinations

$$X_1 \sim \cos \pi x; \qquad X_4' \sim \sin \pi x. \tag{32}$$

If the ion-core potential is attractive, it is likely that X_1 will lie lower in energy than X_4', because the cosine piles up more charge on an ion core centered at $\mathbf{x} = 0$ than does the sine.

BODY-CENTERED CUBIC LATTICE

The Brillouin zone of the bcc lattice is the rhombic dodecahedron shown in Fig. 5; the form of the zone is derived in *ISSP*, Chapter 12. The symmetry operations for Γ, Δ, Λ, Σ are identical with those of the

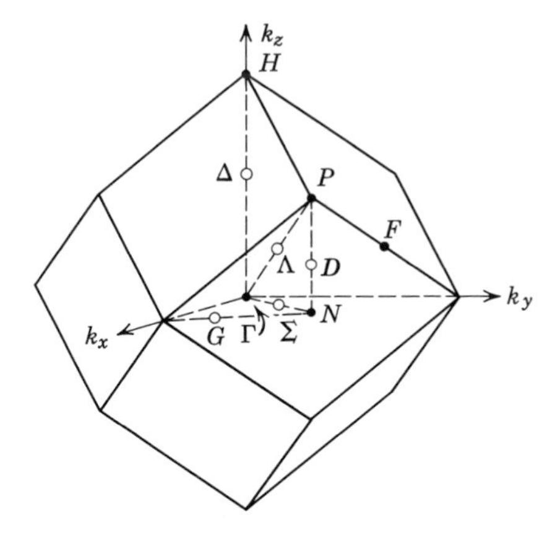

FIG. 5. The Brillouin zone of the body-centered cubic lattice showing the symmetry points and axes.

same points in the sc lattice. The point H has the full cubic symmetry, as for Γ. The characters and symmetry types of N and P are given by Jones. The classification of the representations of the energy bands in the empty lattice is given in Fig. 6.

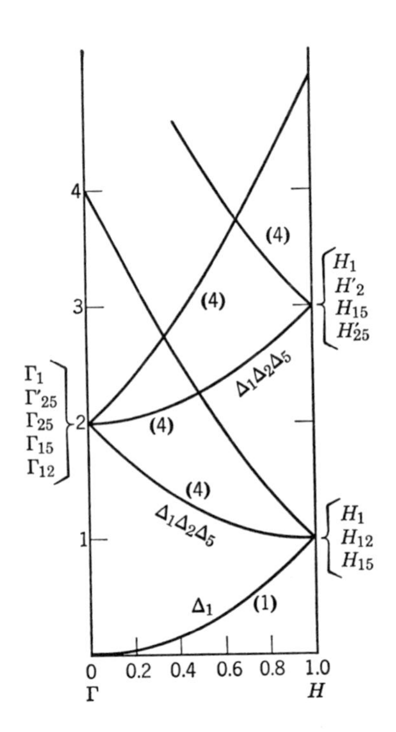

FIG. 6. Free particle energy bands for a body-centered cubic lattice.

FACE-CENTERED CUBIC LATTICE

The form of the Brillouin zone is given in Fig. 7; the zone is a truncated octahedron. Special points of unusual interest are L at the center of each hexagonal face; X at the center of each square face; and W at each corner formed by two hexagons and one square. Their coordinates in terms of the side a of the unit cube of the direct lattice are

$$X = \frac{2\pi}{a}(001); \tag{33}$$

$$L = \frac{2\pi}{a}(\tfrac{1}{2}\,\tfrac{1}{2}\,\tfrac{1}{2});$$

$$W = \frac{2\pi}{a}(\tfrac{1}{2}\,0\,1);$$

$$K = \frac{2\pi}{a}(\tfrac{3}{4}\,\tfrac{3}{4}\,0).$$

The classification of the representations of the energy bands is given in Fig. 8. We note that there is no mirror plane normal to the [111] direction in the fcc lattice; thus there is no need for $\text{grad}_{\mathbf{k}}\ \varepsilon$ to be zero across the hexagonal faces. For details of the behavior on the hexagonal faces, see Jones, p. 47.

HEXAGONAL CLOSE-PACKED AND DIAMOND STRUCTURES

The space groups of these structures contain glide planes or screw axes which are not inherent in the primitive translational lattice. The irreducible representations of wavevectors lying within the zone are not radically changed by the new operations, but there is a serious change at the surface points of the zone. Because of the new operations it can happen that at a special point, along a whole line, or on a

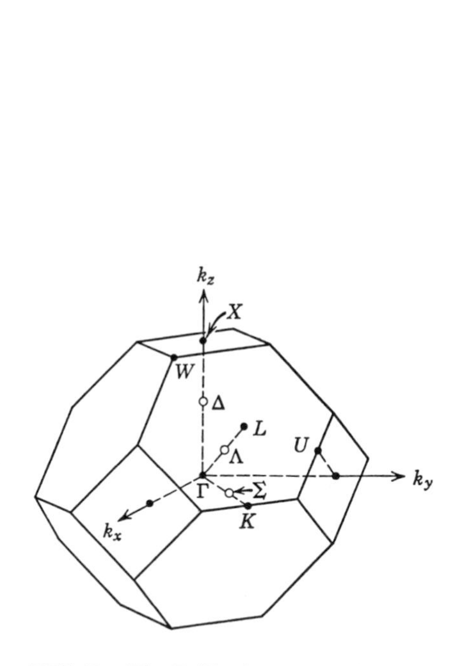

FIG. 7. The Brillouin zone of the face-centered cubic lattice showing the symmetry points and axes.

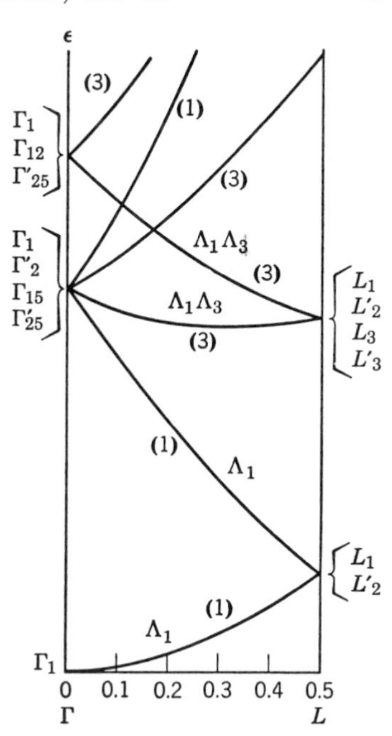

FIG. 8. Free particle energy bands for a face-centered cubic lattice.

whole zone face the irreducible representations are *only* two dimensional. One says that the glide planes and screw axes cause the bands to *stick together* on the special surface lines and planes. We emphasize that our discussion of zone symmetry is still worded as if electron spin did not exist.

We illustrate the effect by a simple example in two dimensions. Consider the rectangular Brillouin zone shown in Fig. 9; let the space group of the direct crystal have mirror planes m normal to the **a** axis at $x = \frac{1}{4}a$ and $\frac{3}{4}a$, and a glide plane g parallel to the **a** axis. The space group is $p2mg$. Suppose that $X(x,y)$ is a solution of the wave equation at $\mathbf{k} = \pi/a(10)$, which we denote by X. The glide operation implies

$$gX(x,y) = X(x + \tfrac{1}{2}a;-y). \tag{34}$$

This space group necessarily contains the inversion J, so that

$$JX(x,y) = X(-x;-y). \tag{35}$$

The mirror plane at $x = \frac{1}{4}a$ implies

$$mX(x,y) = X(-x + \tfrac{1}{2}a;y). \tag{36}$$

Then we see that

$$gX(x,y) = mJX(x,y). \tag{37}$$

Suppose that the representation were one-dimensional: then because $J^2X(x,y) = X(x,y)$, we would have $JX(x,y) = \pm X(x,y)$. Because $m^2X(x,y) = X(x,y)$, it would follow that $mX(x,y) = \pm X(x,y)$. From (37)

$$g^2X(x,y) = mJmJX(x,y) = (\pm 1)^2(\pm 1)^2X(x,y) = X(x,y). \tag{38}$$

However,

$$\begin{aligned} g^2X(x,y) &= gX(x + \tfrac{1}{2}a, -y) = X(x + a,y) \\ &= e^{ik_xa}X(x,y) = e^{i\pi}X(x,y) = -X(x,y), \end{aligned} \tag{39}$$

which contradicts (38). Therefore the representation in this space group at X cannot be one-dimensional. The bands must stick at X.

By using time reversal invariance we show that the bands stick on the boundary line through XZ. Let K be the time reversal operator; we know from the preceding chapter that in the absence of spin

$$K\varphi_{\mathbf{k}}(x,y) = \varphi_{-\mathbf{k}}(x,y). \tag{40}$$

But on the boundary $\mathbf{k} = (\pi/a,k_y)$ we have

$$-\mathbf{k} = \left(-\frac{\pi}{a}, -k_y\right) \equiv \left(\frac{\pi}{a}, -k_y\right). \tag{41}$$

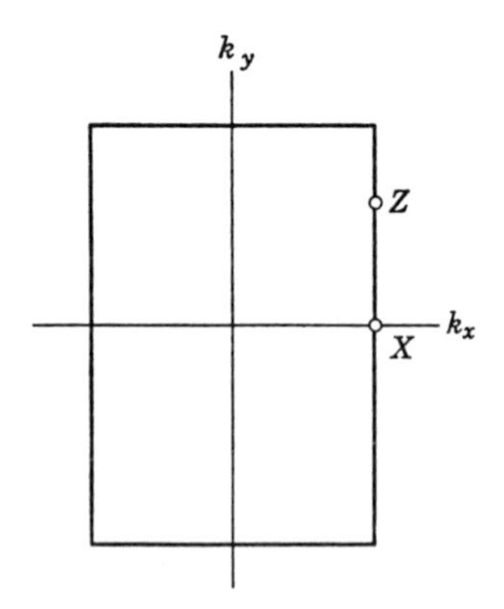

FIG. 9. Brillouin zone for a simple rectangular lattice.

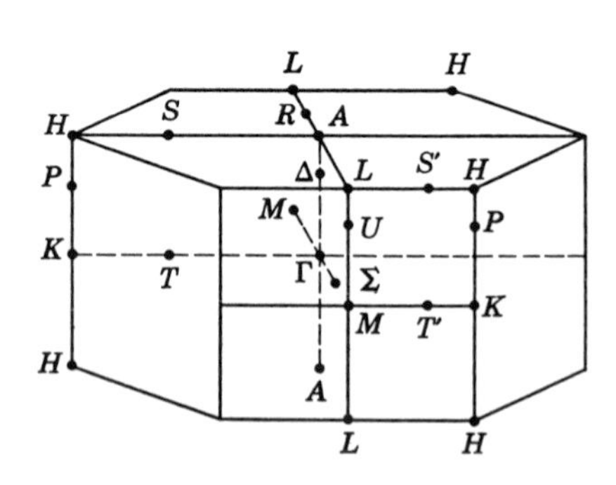

FIG. 10. Brillouin zone of the hexagonal close-packed structure.

Thus, from (37) and assuming the state $\mathbf{k}$ to be nondegenerate,

$$gK\varphi_{\mathbf{k}}(x,y) = mJ\varphi_{-\mathbf{k}}(x,y) = m\varphi_{\mathbf{k}}(x,y) = \varphi_{\mathbf{k}}(x,y), \tag{42}$$

where in the last step we have used the relation $m_x^{-1}\left[\left(\frac{\pi}{a}, k_y\right)\right] = \left(\frac{\pi}{a}, k_y\right)$, instead of letting m_x act on the coordinates. However,

$$gKgK\varphi_{\mathbf{k}}(x,y) = g^2\varphi_{\mathbf{k}}(x,y);$$

and by the argument of (39)

$$g^2\varphi_{\mathbf{k}}(x,y) = \varphi_{\mathbf{k}}(x + a,y) = -\varphi_{\mathbf{k}}(x,y), \tag{43}$$

which is inconsistent with (42). Thus the $\varphi_{\mathbf{k}}$ and $gK\varphi_{\mathbf{k}}$ must be independent functions; their energy must be the same because the hamiltonian is invariant under the operations g and K. The bands stick on the boundary line through XZ; there can be no energy gap between these two bands.

A similar argument applied to the hexagonal close-packed structure [C. Herring, *Phys. Rev.* **52,** 361 (1937); *J. Franklin Institute* **233,** 525 (1942)] shows that only doubly-degenerate states exist on the hexagonal faces of the Brillouin zone, Fig. 10. This result is important because it implies that no energy gap exists across the hexagonal face. It has been shown, however, [M. H. Cohen and L. M. Falicov, *Phys. Rev. Letters* **5,** 544 (1960)] that the degeneracy is lifted by the spin-orbit interaction. For the diamond structure, see the paper by Elliott cited in footnote 2; for the zinc blende structure, the paper by Dresselhaus.

SPIN-ORBIT COUPLING[2]

We now add the electron spin to the Brillouin zone problem. Adding the spin without turning on the spin-orbit coupling simply doubles the degeneracy of every state. But the spin-orbit interaction will lift some of the degeneracy—not, however, at $\mathbf{k}$ values for which the states are only doubly-degenerate in the presence of spin, because the Kramer theorem on time reversal symmetry requires at least twofold degeneracy. A p-like state (such as Γ_{15} in a cubic crystal, according to Table 4) is threefold degenerate without spin; sixfold degenerate with spin; and with spin-orbit interaction the state behaves as a $p_{3/2}$, $p_{1/2}$ set in atomic spectroscopy: the sixfold level splits into one fourfold level like $p_{3/2}$ and one twofold level like $p_{1/2}$.

[2] R. J. Elliott, *Phys. Rev.* **96,** 280 (1954); G. Dresselhaus, *Phys. Rev.* **100,** 580 (1955).

The small representations are changed by spin-orbit coupling. For the group Γ of Table 4, the representations with spin are given in terms of the representations without spin.

Γ_i	Γ_1	Γ_2	Γ_{12}	Γ_{15}'	Γ_{25}'	Γ_1'	$\Gamma_2' \cdots$
$\Gamma_i \times D_{1/2}$	Γ_6	Γ_7	Γ_8	$\Gamma_6 + \Gamma_8$	$\Gamma_7 + \Gamma_8$	Γ_6'	$\Gamma_7' \cdots$

The representations Γ_6, Γ_7, Γ_8 are two-, two-, and four-dimensional, respectively. Character tables for special points are given in the paper by Elliott; results for the point Γ are given in Table 7.

TABLE 7
CHARACTER TABLE OF THE EXTRA REPRESENTATIONS IN THE DOUBLE GROUP OF Γ

(The operations with bars overhead are isomorphous to those without them)

	E	$\bar{E}$	4^2	$\bar{4}^2$	4	$\bar{4}$	2	$\bar{2}$	3	$\bar{3}$	$J \times Z$
Γ_6, Γ_6'	2	-2	0	0	$2^{1/2}$	$-2^{1/2}$	0	0	1	-1	$\pm\chi(Z)$
Γ_7, Γ_7'	2	-2	0	0	$-2^{1/2}$	$2^{1/2}$	0	0	1	1	$\pm\chi(Z)$
Γ_8, Γ_8'	4	-4	0	0	0	0	0	0	-1	1	$\pm\chi(Z)$

Further effects of spin-orbit coupling on band structure are best discussed by important specific examples. Several of these are given in the chapter on the band structure of semiconductors.

PHONONS

The symmetry properties of phonons in crystals may be described in the same way as we have described the symmetry properties of electrons. For a discussion of selection rules in processes involving both phonons and electrons, see R. J. Elliott and R. Loudon, *Phys. Chem. Solids* **15,** 146 (1960); M. Lax and J. J. Hopfield, *Phys. Rev.* **124,** 115 (1961).

PROBLEMS

1. Show for the square lattice that $\Gamma_5 = \Delta_1 + \Delta_2$; $\Gamma_5 = \Sigma_1 + \Sigma_2$; $M_5 = \Sigma_1 + \Sigma_2$; $M_5 = Z_1 + Z_2$.

2. Show that the bands D, E, F, G in Fig. 4 on a line Δ reduce into the representations Δ_1, Δ_2, Δ_5.

3. Show that without spin only doubly-degenerate states exist at general points of the hexagonal faces of the Brillouin zone of the hcp structure.

4. Confirm the labeling of all the bands shown in Fig. 8 for the fcc lattice.

11 Dynamics of electrons in a magnetic field: de Haas-van Alphen effect and cyclotron resonance

In a crystal the behavior of electrons in a magnetic field is vastly more interesting than their behavior in an electric field. In a magnetic field the orbits are usually closed and quantized; occasionally the orbits are open, with unique consequences. Observations on the orbits provide quite direct information on the fermi surface. The more interesting and revealing observations include cyclotron resonance, the de Haas-van Alphen effect, magnetoacoustic attenuation, and magnetoresistance. It is unfortunate, but probably not generally serious, that the theoretical basis of the motion of electrons in crystals in a magnetic field has not yet been developed in a clean and closed form as concise as the earlier theorem on the motion of electrons in an electric field. One or two magnetic problems have been solved exactly, and a wide family of problems have been solved in a semiclassical approximation. Other solutions have been obtained in the form of a series expansion in the magnetic field, with several terms in the series calculated. The semiclassical analysis of the motion of an electron on a fermi surface in a magnetic field appears to be sufficiently accurate for most purposes.

FREE ELECTRON IN A MAGNETIC FIELD

We first consider the motion of a free particle of mass m and charge e in a uniform magnetic field H directed parallel to the z axis. The vector potential in the Landau gauge is $\mathbf{A} = H(0,x,0)$. The hamiltonian is

$$(1) \qquad H = \frac{1}{2m}\left(\mathbf{p} - \frac{e}{c}\mathbf{A}\right)^2 = \frac{1}{2m}\left[p_x{}^2 + p_z{}^2 + (p_y + m\omega_c x)^2\right],$$

where the cyclotron frequency is defined for an electron as

$$\omega_c \equiv -eH/mc. \tag{2}$$

We first examine the classical equations of motion in order to gain familiarity with their form in this gauge. We have

$$\dot{x} = \partial H/\partial p_x = p_x/m; \qquad \dot{y} = \partial H/\partial p_y = (p_y + m\omega_c x)/m; \qquad \dot{z} = p_z/m; \tag{3}$$

$$\dot{p}_x = -\partial H/\partial x = -(p_y + m\omega_c x)\omega_c; \qquad \dot{p}_y = 0; \qquad \dot{p}_z = 0. \tag{4}$$

Thus p_y, p_z are constants of the motion, which we may denote by k_y, k_z, respectively. From the equations for $\dot{x}$ and $\dot{p}_x$, we have

$$m\ddot{x} = -k_y\omega_c - m\omega_c^2 x = -m\omega_c^2\left(x + \frac{1}{m\omega_c}k_y\right), \tag{5}$$

which is the equation of a linear harmonic oscillator of frequency ω_c with origin at

$$x_0 = -\frac{1}{m\omega_c}k_y. \tag{6}$$

Note too that $\dot{y}$ is not a constant of the motion, although p_y is constant. A typical solution is

$$x = -\frac{1}{m\omega_c}\dot{k}_y + \rho\cos\omega_c t; \qquad y = y_0 + \rho\sin\omega_c t, \tag{7}$$

where y_0 is disposable, as is k_y.

The quantum equations of motion are:

$$i\dot{x} = [x,H] = ip_x/m; \qquad i\dot{y} = [y,H] = i(p_y + m\omega_c x)/m; \qquad i\dot{z} = ip_z/m; \tag{8}$$

$$i\dot{p}_x = [p_x,H] = -i(p_y + m\omega_c x)\omega_c; \qquad \dot{p}_y = 0; \qquad \dot{p}_z = 0, \tag{9}$$

in agreement with the classical equations (3) and (4). If we make the substitutions

$$p_y = k_y; \qquad p_z = k_z; \qquad x = -\frac{1}{m\omega_c}k_y + q = x_0 + q, \tag{10}$$

the hamiltonian (1) takes the form

$$H = \frac{1}{2m}(p_x^2 + m^2\omega_c^2 q^2) + \frac{1}{2m}k_z^2, \tag{11}$$

which has the eigenvalues

$$\varepsilon = (\lambda + \tfrac{1}{2})\omega_c + \frac{1}{2m} k_z^2, \tag{12}$$

where λ is a positive integer. The eigenfunctions are of the form

$$\varphi(\mathbf{x}) = e^{i(k_y y + k_z z)} \times \text{harmonic oscillator function of } (x - x_0). \tag{13}$$

The maximum value of k_y is determined through (6) by the requirement that x_0 fall within the specimen. We suppose that the electron gas is in a rectangular parallelepiped of sides L_x, L_y, L_z. If

$$-\tfrac{1}{2}L_x < x_0 < \tfrac{1}{2}L_x, \tag{14}$$

then

$$-\tfrac{1}{2}m\omega_c L_x < k_y < \tfrac{1}{2}m\omega_c L_x. \tag{15}$$

The number of allowed values of k_y in this range in k_y space is

$$\frac{L_y}{2\pi} m\omega_c L_x = L_x L_y \frac{eH}{2\pi c}, \tag{16}$$

neglecting spin. This is the degeneracy of a state of fixed k_z and energy quantum number λ. The separation in energy of states $\Delta\lambda = 1$ is ω_c, so that the number of states per unit energy range is $mL_xL_y/2\pi$, for fixed k_z.

The density of states of a two-dimensional gas in zero magnetic field is

$$\frac{L_x L_y}{(2\pi)^2} 2\pi k \frac{dk}{d\varepsilon} = \frac{L_x L_y m}{2\pi}, \tag{17}$$

with $\varepsilon = k^2/2m$. Thus the average density of states is unaffected by the magnetic field; what the field does is to pull together a large number of states into a single level.

At absolute zero all states are filled up to the fermi level ε_F; above, all states are empty. Consider a plane slab in **k** space, of thickness δk_z at k_z, with the z axis parallel to the magnetic field. The number of allowed values of k_z in the range δk_z is $(L_z/2\pi)\delta k_z$, so that the total degeneracy of the state λ in the slice δk_z is, from (16),

$$\frac{eHL_xL_y}{2\pi c} \frac{L_z}{2\pi} \delta k_z = L_x L_y L_z \frac{e\delta k_z}{4\pi^2 c} H \equiv L_x L_y L_z \xi H. \tag{18}$$

We define the degeneracy parameter ξ as the degeneracy per unit magnetic field per unit volume of the specimen:

$$\xi = \frac{e\delta k_z}{4\pi^2 c}, \tag{19}$$

apart from spin.

DE HAAS-VAN ALPHEN EFFECT[1, 2, 3]

Many of the electronic properties of pure metals at low temperatures are periodic functions of $1/H$. The Schubnikow-de Haas effect is the periodic variation of electric resistivity with $1/H$. The de Haas-van Alphen effect is the periodic variation of the magnetic susceptibility; it is a spectacular and important effect. It provides one of the best tools for the investigation of the fermi surfaces of metals.

We treat first the de Haas-van Alphen effect in a free-electron gas at absolute zero. The fermi level ε_F can be shown to be approximately constant as H is varied; for the periodic effects in $1/H$ in which we are interested the variation in the fermi level may be neglected entirely. At absolute zero all levels λ in the slice δk_z defined above will be completely filled (Fig. 1) for which

$$(\lambda + \tfrac{1}{2})\omega_c < \varepsilon_F - \frac{1}{2m} k_z^2 \equiv \varepsilon'_F, \tag{20}$$

and above this all orbits are empty. If the highest filled level is λ', then the total number of electrons n in the slice of thickness δk_z is

$$n = (\lambda' + 1)\xi H, \tag{21}$$

using (18) and recalling that $\lambda = 0$ is a filled level.

The number n increases linearly with H until λ' coincides with the fermi level ε_F. An infinitesimal further increase of H will raise λ' above ε_F, and all the electrons in λ' will empty out into orbits in other slices of the fermi surface; that is, slices with other values of k_z and ε'_F. The discontinuous evacuation occurs whenever an integer λ' satisfies

$$(\lambda' + \tfrac{1}{2}) = \frac{\varepsilon'_F}{\omega_c} = \frac{mc\varepsilon'_F}{e}\frac{1}{H}, \tag{22}$$

so that the population δn is approximately a periodic function of $1/H$, with period $e/mc\varepsilon'_F$. The population of the slice oscillates with ampli-

[1] A. B. Pippard, LTP.

[2] D. Shoenberg, in *Progress in low temperature physics* **2,** 226 (1957); this contains a full review of the experimental situation up to 1957.

[3] A. H. Kahn and H. P. R. Frederikse, *Solid state physics* **9,** 257 (1959).

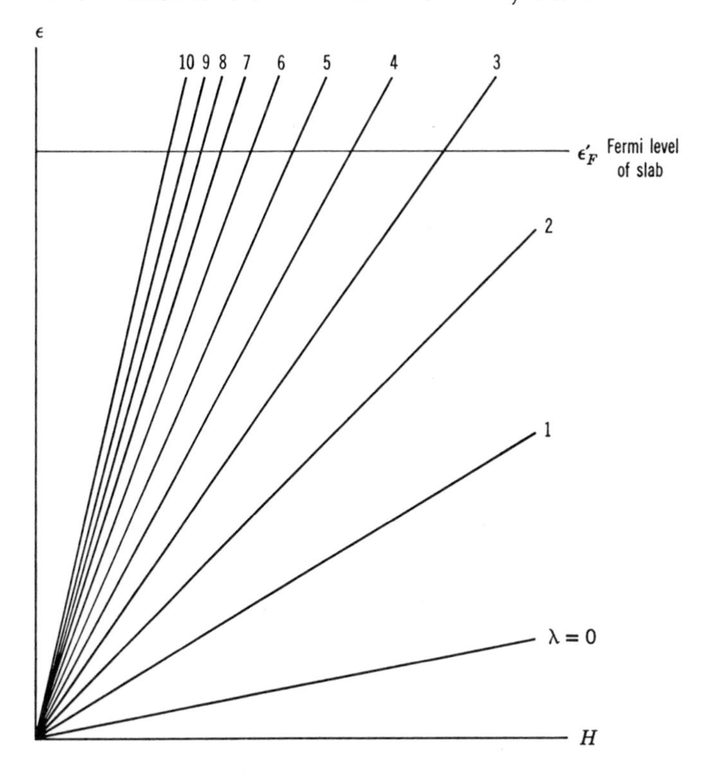

FIG. 1. Spectrum of Landau levels versus H; as H is increased, states of lower and lower λ burst up through the fermi level $\epsilon_{F'}$.

tude $\frac{1}{2}\xi H$ about the value n_0, which is the total number of electrons in the slice δk_z in zero magnetic field. The variation is shown in Fig. 2.

The energy of the electrons in the slice δk_z in a magnetic field such that the population is n_0 is

$$(23) \quad U_0 = \xi H \omega_c \sum_0^{\lambda'} (\lambda + \tfrac{1}{2}) + n_0 \frac{k_z^2}{2m} = \tfrac{1}{2}\xi H \omega_c (\lambda' + 1)^2 + n_0 \frac{k_z^2}{2m}$$

$$= \frac{1}{2} \frac{\omega_c}{\xi H} n_0^2 + n_0 \frac{k_z^2}{2m},$$

using $n_0 = (\lambda' + 1)\xi H$ from (21) for this value of H. For nearby H with the same value of λ' for the topmost filled state,

$$(24) \qquad U = \frac{1}{2} \frac{\omega_c}{\xi H} n^2 + n \frac{k_z^2}{2m} + (n_0 - n)\varepsilon_F.$$

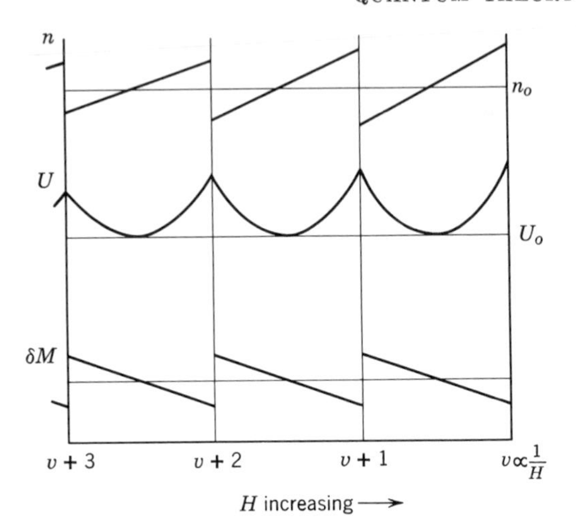

FIG. 2. Variation of population n, energy U, and magnetic moment δM of slice δk_z, as the magnetic field is increased. The successive values of λ' are denoted by $\nu + 3$, $\nu + 2$, $\nu + 1$, ν, $\cdot \cdot \cdot$. The horizontal scale is linear in $1/H$, which decreases to the right.

The first two terms on the right-hand side are the energy of the electrons in the slice δk_z; the term $(n_0 - n)\varepsilon_F$ is the change in energy of the rest of the fermi sea arising from the transfer of $n - n_0$ electrons at the fermi level. By including the transfer, we have in $U - U_0$ the entire energy change of the whole fermi surface, provided ε_F remains exactly constant. We may write $\omega_c/2\xi H = \mu/\xi$, with

$$\mu \equiv e/2mc. \tag{25}$$

Thus, with $\varepsilon'_F \equiv \varepsilon_F - k_z{}^2/2m$,

$$\delta U = U - U_0 = (\mu/\xi)(n^2 - n_0{}^2) + (n_0 - n)\varepsilon'_F. \tag{26}$$

Now $\varepsilon'_F = (n_0/\xi H)\omega_c = 2\mu n_0/\xi$, so that

$$\delta U = (\mu/\xi)(n - n_0)^2. \tag{27}$$

This is always positive.

The magnetization of the slice at absolute zero is

$$\delta M = -\frac{\partial U}{\partial H} = -(2\mu/\xi)(n - n_0)\frac{dn}{dH}; \tag{28}$$

by (21) and (22),

$$\frac{dn}{dH} = (\lambda' + 1)\xi \cong \varepsilon'_F(\xi/\omega_c) = \varepsilon'_F(\xi/2\mu H), \tag{29}$$

so that

$$\boxed{\delta M \cong -\frac{\varepsilon'_F}{H}(n - n_0).} \tag{30}$$

We have seen that as H is increased the quantity $n - n_0$ varies cyclically with extrema $\pm\frac{1}{2}\xi H$ as the population of the level λ' varies between ξH and 0. The magnetization varies between $\mp\frac{1}{2}\xi\varepsilon'_F$. This cyclic variation of magnetization as a periodic function of $1/H$ is the de Haas-van Alphen effect. The period in $1/H$ is, as we have seen, $e/mc\varepsilon'_F$.

To determine the variation of the total magnetization with H, we have to sum the contributions from slices at all k_z; for each slice δn and ε'_F are different. We analyze δM in a fourier series:

$$\delta M = \delta k_z \sum_{p=1}^{\infty} A_p \sin px, \qquad x = \pi\varepsilon'_F/\mu H. \tag{31}$$

Now, for $-\pi < x < \pi$,

$$\delta M = -\frac{1}{2\pi}\xi\varepsilon'_F x = \frac{1}{\pi}\xi\varepsilon'_F \sum_{p=1}^{\infty}(-1)^p \frac{\sin px}{p}, \tag{32}$$

using the *Smithsonian Mathematical Formulae*, 6.810. Thus

$$A_p = \frac{1}{p\pi}\varepsilon'_F(-1)^p\left(\frac{\xi}{\delta k_z}\right) = (-1)^p \frac{e\varepsilon'_F}{4p\pi^3 c}. \tag{33}$$

We sum over all k_z:

$$M = \frac{e}{4\pi^3 c}\sum \frac{(-1)^p}{p}\int_{-k_F}^{k_F} dk_z \cdot \varepsilon'_F \sin\left[\frac{p\pi}{\mu H}\left(\varepsilon_F - \frac{1}{2m}k_z^2\right)\right]. \tag{34}$$

Now in metals and semimetals $\mu H \ll \varepsilon_F$, and the integrand oscillates very rapidly as a function of k_z, except for $k_z \approx 0$. Thus we may replace ε'_F by ε_F in the integrand; after using the trigonometric formula for the sine of the difference of two angles, we are left with Fresnel integrals. The value of the integral in (34) to high accuracy is

$$\varepsilon_F\left(\frac{m\mu H}{p}\right)^{1/2}\sin\left(\frac{p\pi\varepsilon_F}{\mu H} - \frac{\pi}{4}\right); \tag{35}$$

here we have used

$$\int_0^{\infty} dx \cos\frac{\pi}{2}x^2 = \int_0^{\infty} dx \sin\frac{\pi}{2}x^2 = \tfrac{1}{2}. \tag{36}$$

Then

$$M = \frac{\varepsilon_F e m^{1/2} (\mu H)^{1/2}}{4\pi^3 c} \sum_p \frac{(-1)^p}{p^{3/2}} \sin\left(\frac{p\pi\varepsilon_F}{\mu H} - \frac{\pi}{4}\right). \tag{37}$$

It is important to note that this result refers only to the properties of a stationary section of the fermi surface, in this instance the section with $k_z = 0$. As in many other fermi-surface problems, we are involved with the properties of that section of the surface for which the integrand is stationary, that is, surfaces for which $\partial S/\partial k_{\mathrm{H}} = 0$. Here S is the area of the section of the surface at $k_H =$ constant, where k_H is the projection of $\mathbf{k}$ on the direction of the magnetic field. Thus *measurements of the de Haas-van Alphen effect usually relate to the characteristics only of the stationary sections of the fermi surface.* For a given orientation of the magnetic field relative to the crystal axes there may be several stationary sections. At absolute zero a section may enclose either all filled states or all empty states.

As the temperature is increased from absolute zero, the orbital states near the fermi level will be partly populated, instead of being either filled or empty. The distribution of population tends to average out the oscillations in the magnetic moment. The relevant parameter is the ratio of k_BT to the magnetic splitting ω_c; more precisely, it is found from the full analysis that the pth term in the sum in (37) is multiplied by the factor

$$L_p = \frac{x_p}{\sinh x_p}, \tag{38}$$

where $x_p = 2\pi^2 p k_B T/\omega_c$. With $H = 10^5$ oersteds and ω_c as for the free-electron mass, the factor $L_1 \approx 0.71$ at 1°K; further, $L_2 \approx 0.30$; $L_3 \approx 0.10$; $L_4 \approx 0.03$. Thus at a finite temperature the oscillations are more sinusoidal than sawtooth in form. In any event, the de Haas-van Alphen effect is limited to low temperatures: at 4°K we have $L_1 \approx 0.03$ in the foregoing conditions.

It is also necessary that the specimen be quite pure in order that collisions do not blur the quantization. On a simple model the effect of a relaxation frequency $1/\tau$ is to replace k_BT by $k_BT + 1/\pi\tau$ in (38).

For a general fermi surface the period of the de Haas-van Alphen effect may be expressed in terms of the area S in $\mathbf{k}$ space of the stationary section of the fermi surface, taken normal to the direction of the magnetic field. With the general quantization condition given in (61) below, it is apparent that in (37) we should make the replacement

$$\frac{p\pi\varepsilon_F}{\mu H} \to \frac{pcS}{eH}, \tag{39}$$

apart from phase. This result is valid even for nonquadratic fermi surfaces. For free electrons

$$S = \pi k_F^2 = 2\pi m \varepsilon_F, \tag{40}$$

at the extremal section.

The de Haas-van Alphen effect is a powerful tool for the investigation of fermi surfaces. In recent years the range of experimental work has been extended to include a number of alkali and noble metals, as well as transition-group metals. The earlier work was largely concerned with semimetals or with small pockets of the Brillouin zone, because low values of ε_F or of S make for less difficult experiments in terms of the values of T and H required.

Landau Diamagnetism. The change of energy δU of the slice δk_z in a magnetic field is given by (27). When δU is averaged over a cycle between the extrema $n - n_0 = \pm\frac{1}{2}\xi H$, we have, from (27),

$$\langle \delta U \rangle = \tfrac{1}{3}(\mu/\xi)(\tfrac{1}{2}\xi H)^2 = \tfrac{1}{12}\xi\mu H^2. \tag{41}$$

Now

$$\langle \delta M \rangle = -\frac{d\langle \delta U \rangle}{dH} = -\tfrac{1}{6}\xi\mu H, \tag{42}$$

and

$$\delta\chi = \langle \delta M \rangle / H = -\tfrac{1}{6}\xi\mu = -\frac{e\mu}{24\pi^2 c}\,\delta k_z. \tag{43}$$

For the whole fermi surface the average diamagnetic susceptibility is

$$\chi = -\frac{e\mu}{24\pi^2 c}(2k_F) = -\frac{\mu^2 m}{6\pi^2}k_F = -\frac{\mu^2 m}{6\pi^2}(2m\varepsilon_F)^{1/2}. \tag{44}$$

The total electron concentration, counting *both* spin orientations, is given by $3\pi^2 n = (2m\varepsilon_F)^{3/2}$, so that after multiplying (44) by 2 we have

$$\boxed{\chi = -\frac{n\mu^2}{2\varepsilon_F}.} \tag{45}$$

This is the Landau diamagnetic susceptibility of an electron gas. Note that n now refers to the whole fermi sea; earlier it referred to the slice δk_z.

SEMICLASSICAL DYNAMICS OF AN ELECTRON IN A MAGNETIC FIELD[4]

If the fermi surface is itself independent of the magnetic field $\mathbf{H}$, we believe following Chapter 9 that we are justified except at special

[4] L. Onsager, *Phil. Mag.* **43,** 1006 (1952); A. Pippard, *LTP*.

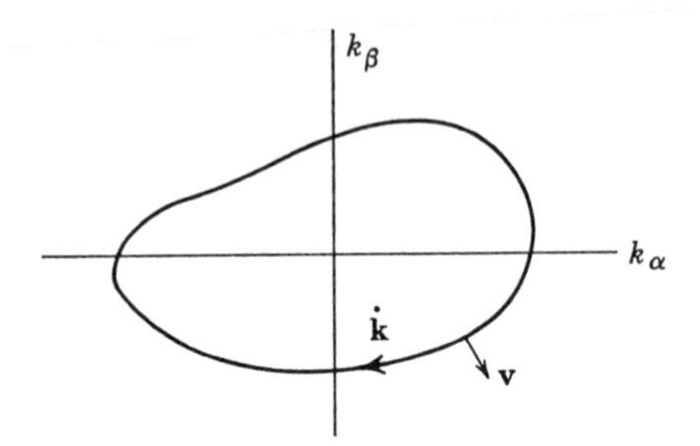

FIG. 3. Motion of an electron in **k** space around an orbit of constant energy. The magnetic field is normal to the k_α, k_β plane.

points in **k** space in describing the motion of an electron in the magnetic field by

$$\dot{\mathbf{k}} = (e/c)\mathbf{v} \times \mathbf{H}, \tag{46}$$

where the right-hand side is just the lorentz force. Thus in **k** space the electron moves in a plane normal to **H**. We have seen that

$$\mathbf{v_k} = \text{grad}_\mathbf{k}\, \varepsilon(\mathbf{k}), \tag{47}$$

where $\varepsilon(\mathbf{k})$ is the energy as calculated in the absence of the magnetic field; therefore the velocity is always normal to the constant-energy surface (Fig. 3). The lorentz force causes **k** to change only along the curve of constant energy formed by the intersection of the constant-energy surface with a plane $\perp \mathbf{H}$; the value of $k_\mathbf{H}$, the wavevector component parallel to **H**, is constant and is determined by the initial conditions. Let **K** be the two-dimensional wavevector in the section and let $\boldsymbol{\rho}$ be the components in real space of **x** in a plane normal to **H**. Then (46) may be written

$$\dot{\mathbf{K}} = (e/c)\dot{\boldsymbol{\rho}} \times \mathbf{H}; \tag{48}$$

as the electron describes an orbit in **k** space, it also describes a similar orbit in real space. From (48) we see that $\boldsymbol{\rho}(t)$ is derived from $\mathbf{K}(t)$ by multiplying $\mathbf{K}(t)$ by c/eH and rotating by $\pi/2$. The transformation $\mathbf{K} \leftrightarrow \boldsymbol{\rho}$ is independent of the shape of the energy surface. We rewrite (48) as

$$\dot{K} = (e/c)v_\perp H, \tag{49}$$

where $v_\perp$ is the velocity in the plane $\perp \mathbf{H}$.

Cyclotron Resonance Frequency—Geometrical Interpretation. Consider two orbits in the same plane section in **K** space, one orbit of energy ε and the other of energy $\varepsilon + \delta\varepsilon$. The separation in **K** space of these orbits is

$$\delta K = \frac{\delta\varepsilon}{|\nabla_{\mathbf{K}}\varepsilon|} = \frac{\delta\varepsilon}{v_{\mathbf{K}}} = \frac{\delta\varepsilon}{\dot{\rho}}. \tag{50}$$

Imagine an electron moving on one orbit because of a magnetic field; the rate at which it may be said to sweep out the area (Fig. 4) between the two orbits is

$$\dot{K}\,\delta K = (e/c)\dot{\rho}H\,\delta\epsilon/\dot{\rho} = (eH/c)\,\delta\epsilon, \tag{51}$$

which is constant at constant $\delta\varepsilon$. Let T denote the period of the orbit; then $T(eH/c)\,\delta\varepsilon$ is equal to the area of the annulus $(dS/d\varepsilon)\,\delta\varepsilon$, where $S(\varepsilon)$ is the area in **K** space of the orbit of energy ε. Thus

$$T = \frac{c}{eH}\frac{dS}{d\varepsilon}, \tag{52}$$

or, for the cyclotron frequency,

$$\omega_c = \frac{2\pi}{T} = \frac{2\pi eH}{c}\frac{1}{dS/d\varepsilon} = \frac{eH}{m_c c}. \tag{53}$$

This equation defines m_c, the effective mass of the orbit for cyclotron resonance. Thus

$$\boxed{m_c = \frac{1}{2\pi}\frac{dS}{d\varepsilon},} \tag{54}$$

which relates the cyclotron frequency and the effective mass to the

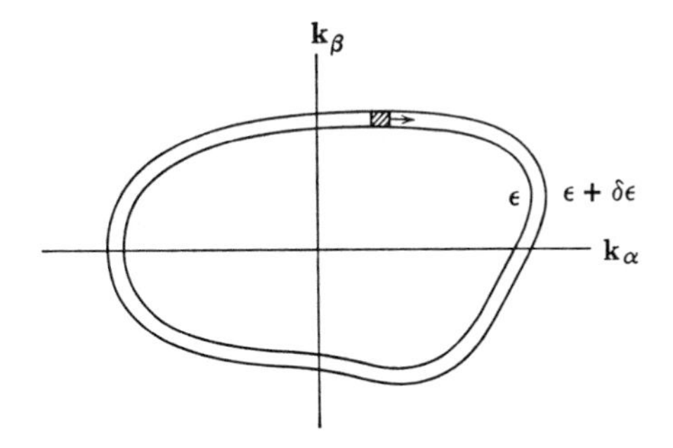

FIG. 4. Geometrical representation of the cyclotron frequency; the area in **k** space of the orbit at energy ε is denoted by $S(\varepsilon)$.

geometry of the fermi surface. Another form for m_c is given in Eq. (72).

Quantization of the Orbits. The periodic character of the motion leads us to expect that the energy levels are quantized, with orbits in **k** space separated in energy by ω_c. We may show this by the Bohr-Sommerfeld method. We have assumed that the effective hamiltonian in a magnetic field is obtained by the substitution

$$\mathbf{k} \leftrightarrow \mathbf{p} - ec^{-1}\mathbf{A}$$

in the energy $\varepsilon(\mathbf{k})$.

$$\oint \mathbf{p} \cdot d\mathbf{q} = \oint (\mathbf{k} + ec^{-1}\mathbf{A}) \cdot d\mathbf{q} = \oint ec^{-1}(\boldsymbol{\varrho} \times \mathbf{H} + \mathbf{A}) \cdot d\boldsymbol{\varrho}, \tag{55}$$

where we have integrated (3) to give $\mathbf{k} = ec^{-1}(\boldsymbol{\varrho} \times \mathbf{H})$ apart from an additive constant which vanishes on integration. The vector potential is **A**. Now

$$\oint \boldsymbol{\varrho} \times \mathbf{H} \cdot d\boldsymbol{\varrho} = -\mathbf{H} \cdot \oint \boldsymbol{\varrho} \times d\boldsymbol{\varrho} = -2\Phi, \tag{56}$$

where Φ is the flux contained within the orbit in real space; further

$$\oint \mathbf{A} \cdot d\boldsymbol{\varrho} = \int \operatorname{curl} \mathbf{A} \cdot d\boldsymbol{\sigma} = \int \mathbf{H} \cdot d\boldsymbol{\sigma} = \Phi, \tag{57}$$

by the stokes theorem; here $d\boldsymbol{\sigma}$ is the area element in real space. Thus

$$\oint \mathbf{p} \cdot d\mathbf{q} = -ec^{-1}\Phi. \tag{58}$$

The quantum condition is

$$\oint \mathbf{p} \cdot d\mathbf{q} = 2\pi(\lambda + \gamma), \tag{59}$$

where λ is an integer and γ is a phase factor having the value $\frac{1}{2}$ for a free electron in a magnetic field. Thus, apart from sign,

$$\Phi = (2\pi c/e)(\lambda + \gamma); \tag{60}$$

the different orbits differ by integral multiples of the unit of flux $2\pi c/e$; or, in usual units, hc/e. This result, due to I. M. Lifshitz and to Onsager, may be contrasted with the quantum of flux $hc/2e$ observed in superconductors, where the pairing condition makes the effective charge $2e$.

The result (60) is translated into **k** space by multiplying by $(eH/c)^2$; we have directly for the area of allowed orbits

$$\boxed{S = \frac{2\pi eH}{c}(\lambda + \gamma).} \tag{61}$$

Thus in the discussion of the de Haas-van Alphen effect we should replace the quantum condition (20) on the frequency by the condition (61) on the area of the orbit in **k** space. Apart from phase, the argument of the pth oscillatory term in the expression (34) for the magnetization is replaced as follows:

$$\frac{p\pi}{\mu H}\,\varepsilon_F \rightarrow \frac{pcS}{eH} = 2\pi(\lambda + \gamma), \tag{62}$$

where S is the area of a stationary section of the fermi surface, taken normal to **H**. With $\hbar$, the right-hand side is $pcS/e\hbar H$. At a stationary section $\partial S/\partial k_H = 0$. The de Haas-van Alphen effect thus gives us the area of the fermi surface at a stationary plane $\perp\mathbf{H}$. The cyclotron resonance frequency gives us $dS/d\epsilon$, which is related to the velocity at the fermi surface.

In strong enough magnetic fields the semiclassical quantization procedure fails—as the field is increased the energy levels are broadened and eventually reform into a different set of levels. A simple analysis of the situation is given in a paper by A. B. Pippard, *Proc. Roy. Soc. (London)* **A270,** 1 (1962); the magnetic breakthrough failure is likely to occur when $\omega_c \gg E_g^2/\varepsilon_F$, where E_g is the energy gap between bands.

TOPOLOGICAL PROPERTIES OF ORBITS IN A MAGNETIC FIELD[5,6]

We have seen from (48) that as an electron describes an orbit in **k** space it also describes an orbit in real space. If the section of the constant-energy surface in **k** space is a closed curve, the electron will describe a helix in real space. The helix is periodic in the sense that every turn is a repetition of the last turn.

Some orbits are not closed curves in **k** space. The existence of open orbits comes as a surprise at first sight. It is essential to recall from the chapter on Bloch functions that the energy is periodic in the reciprocal lattice; thus the constant-energy surfaces in each band extend periodically through the whole of **k** space. The surfaces are not confined within one Brillouin zone. When an electron meets the boundary of a zone it simply passes into the next zone.

We consider several different situations:

(*a*) If the fermi surface lies entirely within the zone boundaries, the surface will be closed and all orbits in a magnetic field will also be closed.

[5] I. M. Lifshitz and M. I. Kaganov, *Soviet Physics—Uspekhi* **2,** 831 (1959).
[6] R. G. Chambers, in *The fermi surface,* Wiley, 1960.

(*b*) If the fermi surface consists of sections attached to separate zone faces or around separate corners, these sections may be combined in the periodic zone scheme to form simple closed surfaces. Such a closed surface will consist of pieces from more than one cell in the extended zone scheme. When the lorentz force brings an electron to the zone boundary, it will continue into the adjacent cell of the extended zone scheme.

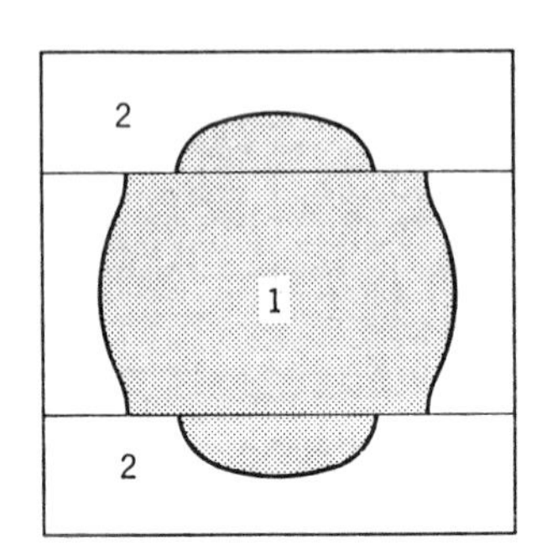

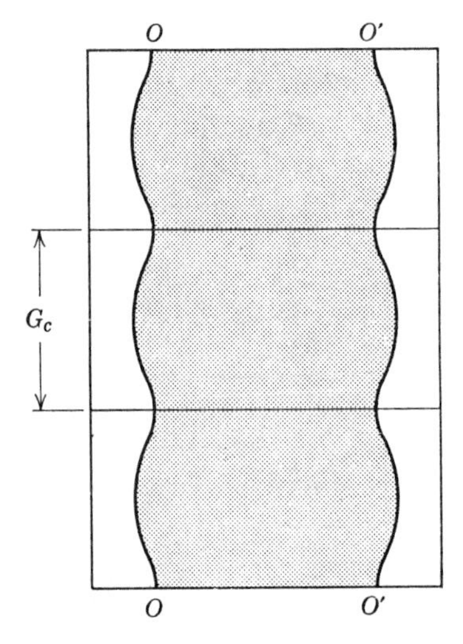

FIG. 5. Fermi surface in a simple rectangular lattice in two dimensions, on the almost-free electron model. (*a*) First and second zones. (*b*) First zone on extended zone scheme. The open orbits are marked 00 and 0′0′.

(*c*) The closed orbits may be either *electron* orbits or *hole* orbits. An electron orbit encloses states of lower energy; the velocity vector $\mathbf{v} = \text{grad}_{\mathbf{k}}\, \varepsilon$ points outward from an electron orbit. A hole orbit is defined as enclosing states of high energy; $\mathbf{v}$ points inward. An electron in a magnetic field traverses a hole orbit in the opposite sense to an electron orbit, and thus an electron in a hole orbit acts as if positively charged. Sometimes it is possible for a given point on the fermi surface to be part of either an electron or hole orbit, according to the direction of **H**.

(*d*) If the fermi surface extends across one cell, from one face to another or from one corner to another, then in the extended zone scheme the fermi surface will be a multiply-connected surface extending continuously throughout **k** space. The simplest example of an *open orbit* occurs in the rectangular lattice in two dimensions on the almost free electron model when the fermi surface cuts into the second zone, as in Fig. 5. For a sc lattice a possible fermi surface containing open orbits is shown in Figs. 6 and 7. In Fig. 8 are shown schematically two types of sections at constant k_z of such an open surface. One

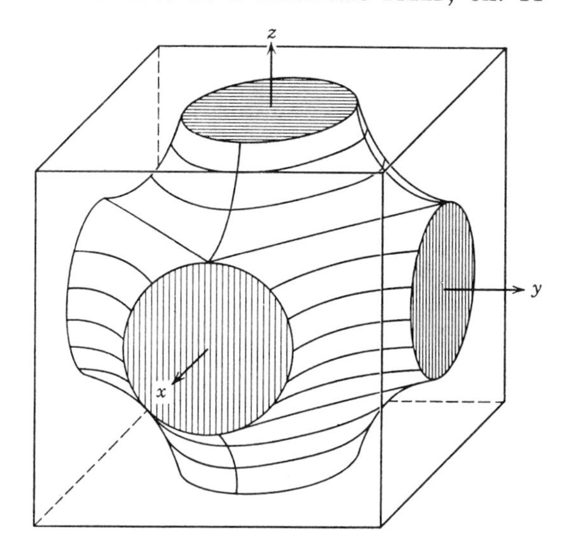

FIG. 6. Possible fermi surface in a simple cubic lattice. (After Sommerfeld and Bethe.)

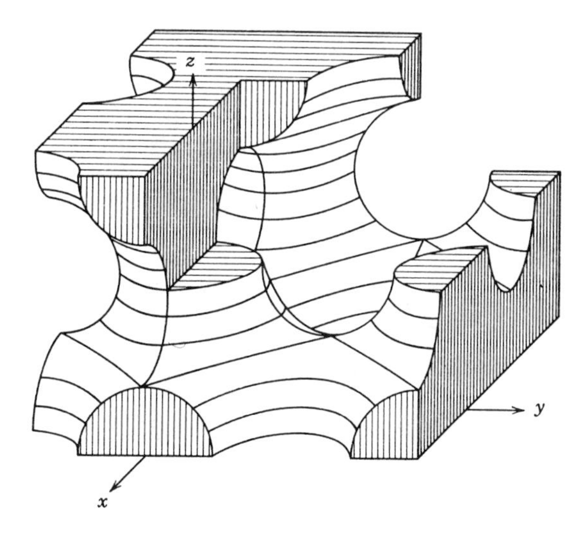

FIG. 7. The fermi surface of Fig. 6 continued in **k** space. This is an open fermi surface. (After Sommerfeld and Bethe.)

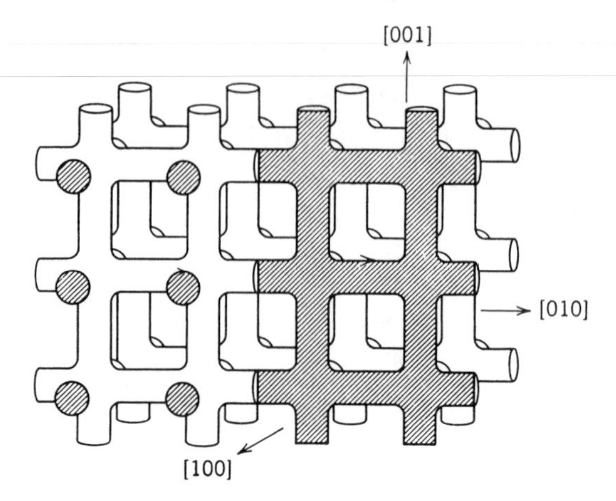

FIG. 8. One possible type of open fermi surface for a cubic metal, showing two sections of constant k_μ, for **H** along [100]. *Left:* electron orbits; *right:* hole orbits. (After R. G. Chambers.)

cut contains electron orbits, the other contains hole orbits. The shaded regions denote filled states; the boundary of the shaded regions is the fermi surface. For a particular value of k_z (intermediate between the two shown) there is a transition between electron and hole orbits; here open orbits occur which run throughout **k** space.

(*e*) The open orbits become clearer if we tilt **H** slightly in the k_xk_z plane, as in Fig. 9. Between the closed electron orbits there are open orbits unbounded in **k** space. For such an orbit $dS/d\varepsilon \rightarrow \infty$,

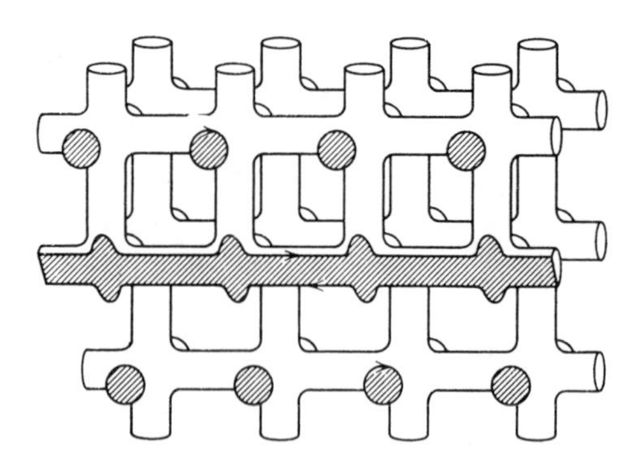

FIG. 9. Section of fermi surface for **H** in (010) plane, showing the periodic open orbits bounding the central shaded strip. Electron orbits above and below. (After R. G. Chambers.)

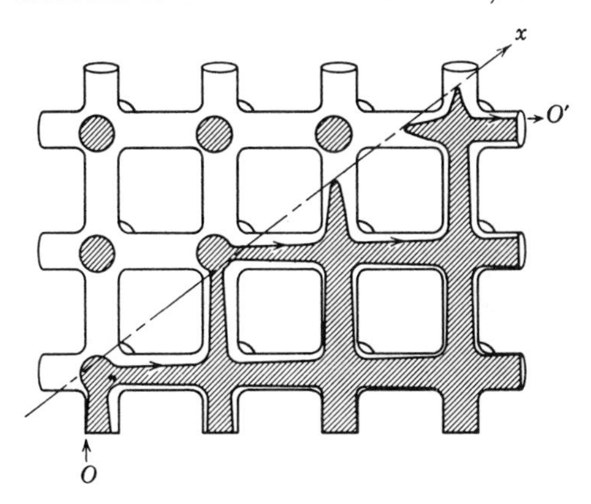

FIG. 10. As in Fig. 9, but with **H** tilted slightly away from [100] in an arbitrary direction. Regions of electron orbits (top left) and hole orbits (bottom right), separated by an aperiodic open orbit OO'. Direction of open orbit taken as x axis. (After R. G. Chambers.)

where S is the area of the orbit; from (53) and (54) the cyclotron frequency $\omega_c \rightarrow 0$ and the cyclotron mass $m_c \rightarrow \infty$. The open orbits shown exist for a range of values of k_H, in front of and behind the value for the section shown. Open orbits also exist when **H** is tilted slightly away in a general direction from the normal to a principal plane, as shown in Fig. 10.

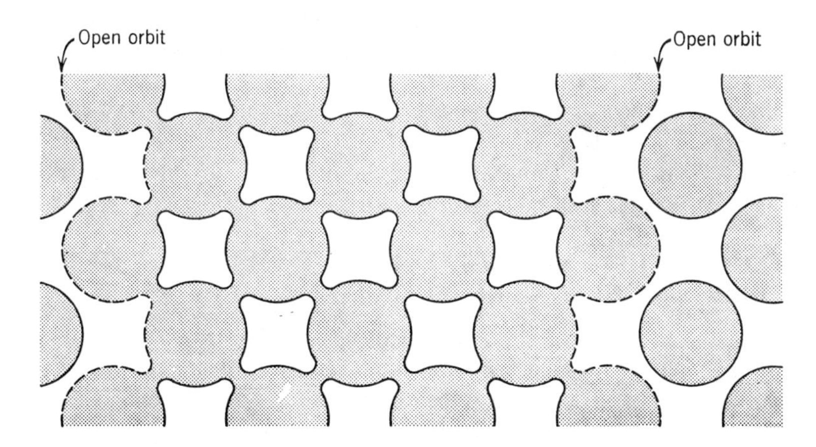

FIG. 11. Periodic open orbits. Slice through extended BZ scheme of fcc crystal, with **H** tilted in yz plane slightly from z direction. (After Pippard.)

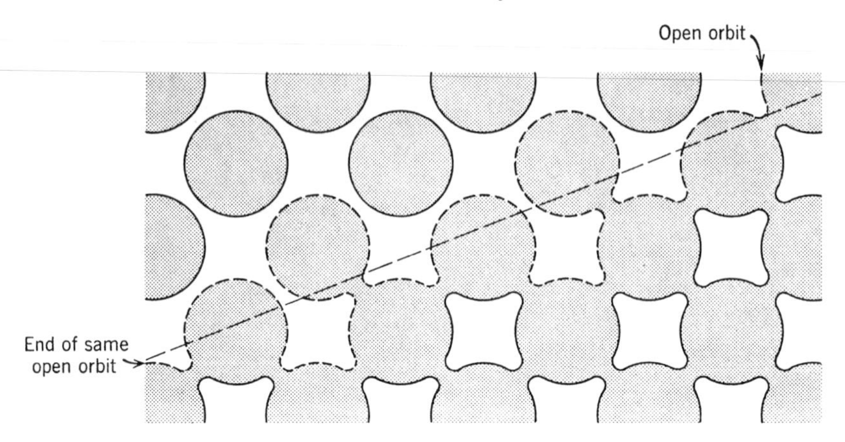

FIG. 12. Aperiodic open orbit. (After Pippard.)

(*f*) The situation of Fig. 9 is shown again in Fig. 11, but for a fcc crystal. Because **H** lies in a plane of high symmetry, successive elements of the open orbit are identical. This is called a *periodic open orbit.* Alternate periodic open orbits are traversed in opposite senses.

(*g*) The situation of Fig. 10 is shown again in Fig. 12, but for a fcc lattice in real space. Such open orbits are not in general replicated exactly as we move along the orbit, and they are called *aperiodic open orbits.*

(*h*) An *extended orbit* is a closed orbit which cannot be contained within a single cell, however the origin of the Brillouin zone is chosen.

The importance of the open orbits is that they act as two-dimensional conductors; they carry current only in the plane containing the magnetic field and the general line of the orbit. They have a striking effect on magnetoresistance, as we shall see in the following chapter.

CYCLOTRON RESONANCE ON SPHEROIDAL ENERGY SURFACES

The neighborhood of the conduction-band edge in germanium and silicon consists of a set of spheroidal energy surfaces located in equivalent positions in **k** space. We now discuss the semiclassical theory of cyclotron resonance for surfaces of this character, neglecting spin and the spin-orbit interaction. The effect of these is considered in Chapter 14.

We make the replacement

$$\varepsilon(\mathbf{k}) \rightarrow H(\mathbf{p} - ec^{-1}\mathbf{A}), \tag{63}$$

where $\varepsilon(\mathbf{k})$ is the energy in the absence of the magnetic field:

$$\varepsilon(\mathbf{k}) = \frac{1}{2m_t}(k_x^2 + k_y^2) + \frac{1}{2m_l}k_z^2, \tag{64}$$

where m_t is the transverse effective mass of the spheroid, that is, the mass for motion in the xy plane; m_l is the longitudinal effective mass—the mass for motion parallel to the z or figure axis. The vector potential

$$A_x = A_z = 0; \qquad A_y = H(x\cos\theta + z\sin\theta) \tag{65}$$

describes a uniform magnetic field H in the xz plane at an angle θ with the z axis. Then the transcription (63) gives

$$H = \frac{1}{2m_t}p_x^2 + \frac{1}{2m_t}[p_y - ec^{-1}H(x\cos\theta + z\sin\theta)]^2 + \frac{1}{2m_l}p_z^2. \tag{66}$$

The equations of motion of the momenta are, with $\omega_t = -eH/m_tc$ and $\omega_l = -eH/m_lc$,

$$\begin{aligned} i\dot{p}_x &= [p_x,H] = -i\omega_t(x\cos\theta + z\sin\theta)\cos\theta; \\ i\dot{p}_y &= 0; \qquad i\dot{p}_z = -i\omega_l(x\cos\theta + z\sin\theta)\sin\theta. \end{aligned} \tag{67}$$

We have further

$$\begin{aligned} -\ddot{p}_x &= \omega_t^2\cos^2\theta p_x + \omega_l\omega_t\sin\theta\cos\theta p_z; \\ -\ddot{p}_z &= \omega_t^2\sin\theta\cos\theta p_x + \omega_l\omega_t\sin^2\theta p_z. \end{aligned} \tag{68}$$

With time dependence $e^{-i\omega t}$, this set of equations has a solution if

$$\begin{vmatrix} \omega^2 - \omega_t^2\cos^2\theta & -\omega_t\omega_l\sin\theta\cos\theta \\ -\omega_t^2\sin\theta\cos\theta & \omega^2 - \omega_t\omega_l\sin^2\theta \end{vmatrix} = 0, \tag{69}$$

or

$$\omega^2 = \omega_t^2\cos^2\theta + \omega_t\omega_l\sin^2\theta; \tag{70}$$

in addition, there are roots at $\omega = 0$ for the analogue of motion parallel to the field. From (70) we see that the effective mass m_c determining the cyclotron frequency is

$$\left(\frac{1}{m_c}\right)^2 = \frac{\cos^2\theta}{m_t^2} + \frac{\sin^2\theta}{m_lm_t}. \tag{71}$$

Cyclotron resonance in metals is considered in Chapter 16.

PROBLEMS

1. Find and solve the quantum equations of motion of a free particle in a uniform magnetic field, using the gauge $\mathbf{A} = (-\frac{1}{2}Hy;\frac{1}{2}Hx;0)$.

2. For the magnetic field H parallel to the x axis of the spheroidal energy surface (64), evaluate the area S of the orbit in **k** space and calculate m_c; compare this result with that deduced from (71).

3. Show that the period of an electron on an energy surface in a magnetic field is

$$T = \frac{c}{eH} \oint \frac{dl}{v_\perp}; \tag{72}$$

here dl is the element of length in **k** space. Show that the area of the surface may be written as

$$S(\varepsilon,p_z) = \iint dp_x\, dp_y = \int d\varepsilon \oint \frac{dl}{v_\perp}, \tag{73}$$

so that

$$T = \frac{c}{eH} \frac{\partial S}{\partial \varepsilon}, \tag{74}$$

in agreement with (52).

4. Consider the hamiltonian of a many-body system:

$$H = \frac{1}{2m} \sum_i \mathbf{P}_i^2 + \sum_{ij} V(\mathbf{x}_i - \mathbf{x}_j), \tag{75}$$

where $\mathbf{P}_i = \mathbf{p}_i - ec^{-1}\mathbf{A}$, with $\mathbf{A}$ for a uniform magnetic field $\mathbf{H}$. Form $\mathbf{P} = \Sigma\, \mathbf{P}_i$; show that

$$\dot{\mathbf{P}} = (e/mc)\mathbf{P} \times \mathbf{H}. \tag{76}$$

Show further that if Ψ_0 is an eigenstate of H having energy E_0, then $(P_x + iP_y)\Psi_0 = \Psi_1$ is an exact eigenstate of the many-body system with energy $E_1 = E_0 + \omega_c$, where ω_c is the cyclotron energy.

5. Show that

$$\mathbf{k} \times \mathbf{k} = \frac{ie\mathbf{H}}{c}. \tag{77}$$

12 Magnetoresistance

In this chapter we consider an important transport problem—the electrical conductivity of metals in a magnetic field. A large effort of theoretical physicists in recent years has gone into the derivation of improved solutions to transport problems, in gases, plasmas, and metals. Pioneer papers dealing with the quantum theory of charge transport in metals include those by J. M. Luttinger and W. Kohn, *Phys. Rev.* **109,** 1892 (1958) and by I. M. Lifshitz, *Soviet Phys.—JETP* **5,** 1227 (1957). The classical theory of magnetoresistance is developed rather fully in the books by Wilson and by Ziman. In Chapters 16 and 17 we treat several interesting, but somewhat complicated, problems by classical methods. But the startling highlights of the observed magnetoresistive phenomena in solids can be elucidated qualitatively by relatively elementary methods. The analysis of the experimental results bears directly on the shape and connectivity of the fermi surface.

By magnetoresistance we mean the increase in the electrical resistance of a metal or semiconductor when placed in a magnetic field. The effect of greatest interest is the transverse magnetoresistance, which is usually studied in the following geometrical arrangement: a long thin wire is directed along the x axis, and a d-c electric field E_x is established in the wire by means of an external power supply. A uniform magnetic field H_z is applied along the z axis, thus normal to the axis of the wire. The most interesting experiments are those carried out at low temperatures on very pure specimens in strong magnetic fields, as here the product $|\omega_c|\tau$ of the cyclotron frequency and the relaxation time may be $\gg 1$. In these conditions the details of the collision processes are suppressed and the details of the fermi surface enhanced.

In the geometry described, which we shall refer to as the *standard geometry*, the effect of a weak magnetic field ($|\omega_c|\tau \ll 1$) is to increase the resistance by an additive term proportional to H^2. The additive term may be of the order of magnitude of $(\omega_c\tau)^2$:

$$\frac{R(H) - R(0)}{R(0)} \approx (\omega_c\tau)^2. \tag{1}$$

On dimensional grounds we could not expect much else, bearing in mind that a term linear in H is inconsistent with the obvious symmetry of the problem with respect to the sign of the magnetic field. We note (*ISSP*, p. 238) that in copper at room temperature the relaxation time is $\approx 2 \times 10^{-14}$ sec; for $m^* = m$ and $H = 30$ koe we have $|\omega_c| \approx 8 \times 10^{11}$ sec^{-1}, so that $|\omega_c|\tau \approx 0.02$. At 4°K the conductivity of a fairly pure crystal of copper may be higher than at room temperature by 10^3 or more; thus τ is lengthened by 10^3 and in the same magnetic field $|\omega_c|\tau \approx 20$.

In very strong fields, that is, for $|\omega_c|\tau \gg 1$, the transverse magnetoresistance of a crystal may generally do one of three quite different things.

(*a*) The resistance may saturate, that is, may become independent of H, perhaps at a resistance of several times the zero field value. Saturation occurs for all orientations of the crystal axes relative to the measurement axes.

(*b*) The resistance may continue to increase up to the highest fields studied for all crystal orientations.

(*c*) The resistance may saturate in some crystal directions, but not saturate in other, often very nearby, crystal directions. This behavior is exhibited as an extraordinary anisotropy of the resistance in a magnetic field, as illustrated by Fig. 1.

Crystals are known in all three categories. We shall see that the first category comprises crystals with *closed* fermi surfaces, such as In, Al, Na, and Li. The second category comprises crystals with equal numbers of electrons and holes, such as Bi, Sb, W, and Mo. The third category comprises crystals with fermi surfaces having *open orbits* for some directions of the magnetic field; this category is known to include Cu, Ag, Au, Mg, Zn, Cd, Ga, Tl, Sn, Pb, and Pt. The value of magnetoresistance as a tool is that it tells us whether the fermi surface is closed or contains open orbits, and in which directions the open orbits lie. There are geometrical situations possible for which open fermi surfaces do not contain open orbits.

Many interesting features can be explained by an elementary drift velocity treatment or by simple extensions thereof. We give this

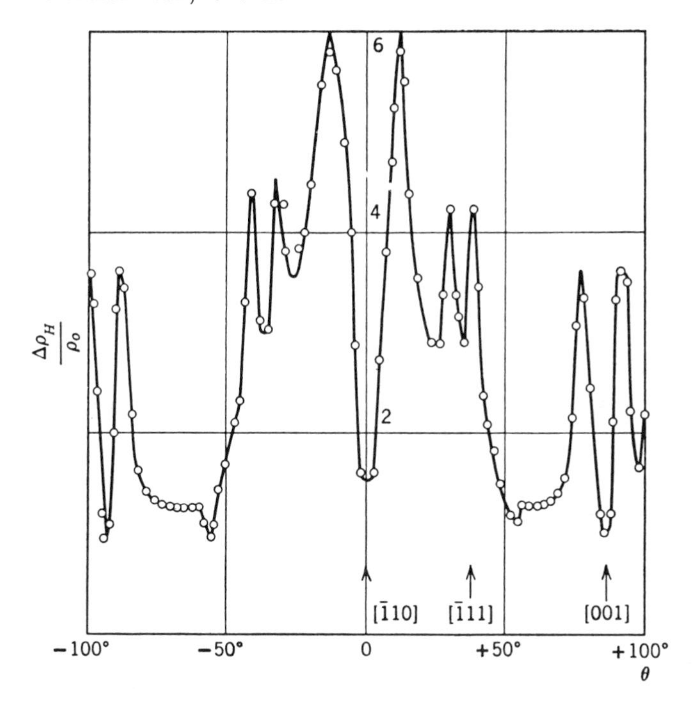

FIG. 1. Variation of transverse magnetoresistance with field direction in a field of 23.5 kG, for a single crystal Au specimen with current ∥ [110]. (From Gaidukov, 1959.)

treatment first as a preliminary to the application of more detailed transport theory. The drift velocity $\mathbf{v}$ is defined as the average carrier velocity:

$$\mathbf{v} = \frac{1}{N} \sum_i \mathbf{v}_i. \tag{2}$$

Single Carrier-Type Isotropic Effective Mass and Constant Relaxation Time. The equation of motion of the drift velocity of a gas of carriers having isotropic mass m^* is, according to *ISSP*, Chapter 10,

$$m^*\left(\dot{\mathbf{v}} + \frac{1}{\tau}\mathbf{v}\right) = e\left(\mathbf{E} + \frac{1}{c}\mathbf{v} \times \mathbf{H}\right), \tag{3}$$

where τ is the relaxation time of the charge carriers. The relaxation time is approximately related to the mean free path Λ by $\Lambda \cong \overline{|\mathbf{v}_i|}\tau$, where $\overline{|\mathbf{v}_i|}$ is the mean magnitude of the particle velocity. We take $\mathbf{H}$

in the z direction. In the steady state $\dot{\mathbf{v}} = 0$, so that

$$\mathbf{v} = \frac{e\tau}{m^*}\left(\mathbf{E} + \frac{1}{c}\,\mathbf{v} \times \mathbf{H}\right). \tag{4}$$

If we set

$$\mu \equiv e\tau/m^*; \qquad \xi \equiv \mu H = eH\tau/m^*c = -\omega_c\tau, \tag{5}$$

then (4) becomes

$$v_x = \mu E_x + \xi v_y; \qquad v_y = \mu E_y - \xi v_x; \qquad v_z = \mu E_z. \tag{6}$$

Thus, on solving for v_x and v_y,

$$v_x = \mu E_x + \mu\xi E_y - \xi^2 v_x; \qquad v_y = \mu E_y - \mu\xi E_x - \xi^2 v_y, \tag{7}$$

or

$$v_x = \frac{\mu}{1+\xi^2}\,(E_x + \xi E_y); \qquad v_y = \frac{\mu}{1+\xi^2}\,(E_y - \xi E_x). \tag{8}$$

The current density component j_λ is obtained from the velocity component v_λ by multiplying by ne, where n is the carrier concentration. The conductivity tensor component $\sigma_{\lambda\nu}$ is defined by

$$j_\lambda = \sigma_{\lambda\nu} E_\nu. \tag{9}$$

From (8) we have, for $H \parallel \hat{\mathbf{z}}$,

$$\bar{\bar{\sigma}} = \frac{ne\mu}{1+\xi^2}\begin{pmatrix} 1 & \xi & 0 \\ -\xi & 1 & 0 \\ 0 & 0 & 1+\xi^2 \end{pmatrix}. \tag{10}$$

The components satisfy the condition

$$\sigma_{\lambda\nu}(H) = \sigma_{\nu\lambda}(-H), \tag{11}$$

as a general consequence of the theory of the thermodynamics of irreversible processes.

In our standard geometry the boundary conditions permit current flow only in the x direction, thus

$$j_y = j_z = 0. \tag{12}$$

From (8) we see that the boundary conditions can be satisfied only if

$$E_y = \xi E_x; \qquad E_z = 0. \tag{13}$$

The field E_y is known as the hall field. From (10) and (13),

$$j_x = \frac{ne\mu}{1+\xi^2}\,(E_x + \xi E_y) = ne\mu E_x; \tag{14}$$

thus in this geometry the *effective* conductivity in the x direction as calculated on our model is independent of the magnetic field in the z direction, even though the conductivity tensor components (10) do involve the magnetic field. That is, our model gives zero for the transverse magnetoresistance.

The resistivity tensor $\bar{\bar{\rho}}$ is the inverse of the conductivity tensor, so that $E_\lambda = \rho_{\lambda\nu} j_\nu$. The components are given by

$$\rho_{\lambda\nu} = \Delta_{\lambda\nu}/\Delta, \tag{15}$$

where Δ is the determinant of $\bar{\bar{\sigma}}$; $\Delta_{\lambda\nu}$ is the $\lambda\nu$-th minor; and

$$\Delta = \frac{(ne\mu)^3}{1 + \xi^2}. \tag{16}$$

Thus for $\bar{\bar{\sigma}}$ given by (10) the resistivity tensor is

$$\bar{\bar{\rho}} = \frac{1}{ne\mu}\begin{pmatrix} 1 & -\xi & 0 \\ \xi & 1 & 0 \\ 0 & 0 & 1 \end{pmatrix}. \tag{17}$$

This is consistent with (6), from which $\bar{\bar{\rho}}$ is most easily found. For the standard geometry with $j_y = 0$ we have from (17) that

$$E_x = \frac{1}{ne\mu} j_x; \qquad E_y = \frac{\xi}{ne\mu} j_x = \frac{H_z}{ne} j_x = \xi E_x, \tag{18}$$

in agreement with (13) and (14).

The absence of magnetoresistance on this model in the standard geometry is the result of the hall electric field E_y, which just balances the lorentz force of the magnetic field. The balance can be maintained only for the one kinematical quantity $\mathbf{v}$ included in the equations of motion. But usually the relaxation time depends on the speed v_i of an individual carrier, so that we cannot expect to describe the motion of the carriers in terms of a single drift velocity. Then the cancellation will not take place. The experimental situation is that a transverse magnetoresistance is always, or nearly always, observed. A simple and important alteration of the drift velocity model is to introduce a second carrier type. With two carrier types the identical hall electric field cannot rectify the orbits of both carrier types at once. This is an important practical situation—the two carrier types may be electrons and holes; s electrons and d electrons; open orbits and closed orbits; etc.

Two Carrier Types—High Field Limit. There is a special value in treating the problem of two carrier types in the high field limit. In any field the steady state drift velocity equations are, by analogy with (4),

$$\mathbf{v}_1 = (e\tau_1/m_1^*)\mathbf{E} + (e\tau_1/m_1^*c)\mathbf{v}_1 \times \mathbf{H}; \tag{19}$$

$$\mathbf{v}_2 = -(e\tau_2/m_2^*)\mathbf{E} - (e\tau_2/m_2^*c)\mathbf{v}_2 \times \mathbf{H}; \tag{20}$$

where the carriers of type 1 are taken to be electrons of effective mass m_1^*, relaxation time τ_1, and concentration n_1. The carriers of type 2 are taken to be holes. We now consider fields such that $|\omega_{c1}|\tau_1 \gg 1$ and $|\omega_{c2}|\tau_2 \gg 1$. Then we can neglect $\mathbf{v}_1$, $\mathbf{v}_2$ when they appear on the left-hand side of (19) and (20), whence for the x components of these equations we have

$$E_x + \frac{H}{c} v_{1y} = 0; \qquad E_x + \frac{H}{c} v_{2y} = 0. \tag{21}$$

Thus

$$j_y \equiv n_1 e v_{1y} - n_2 e v_{2y} = \frac{(n_2 - n_1)ec}{H} E_x, \tag{22}$$

whence

$$\boxed{\sigma_{yx} = (n_2 - n_1)ec/H.} \tag{23}$$

This is a crucial result, because it shows that for equal numbers of holes and electrons $\sigma_{yx} = 0$. But if $\sigma_{yx} = 0$, there is no hall voltage E_y, as $j_y = 0$ without benefit of an E_y. Without E_y the effective resistivity becomes simply $1/\sigma_{xx}$, where σ_{xx} is given by (10) and in this limit

$$\sigma_{xx} \cong \frac{n|e|}{H^2}\left(\frac{1}{|\mu_1|} + \frac{1}{|\mu_2|}\right), \tag{24}$$

where $n = n_1 = n_2$. Thus the *transverse magnetoresistance does not saturate if there are equal numbers of holes and electrons.*

Divalent metals having one atom (and two valence electrons) per primitive cell will necessarily have equal numbers of holes and electrons $(n_- = n_+)$, provided there are no open orbits. There is one point in $\mathbf{k}$ space in each Brillouin zone for each primitive cell in the crystal. Equality of electron and hole concentrations can also occur in metals of odd valence if the primitive cell contains an even number of atoms. Under these conditions it is observed that the transverse magnetoresistance does not saturate. The topology of the equality of electrons and holes is easily understood in two dimensions; see, for example, Fig. 2, where the fermi surface has been constructed with parts in two zones, but with the total filled area just equal to the area of one zone.

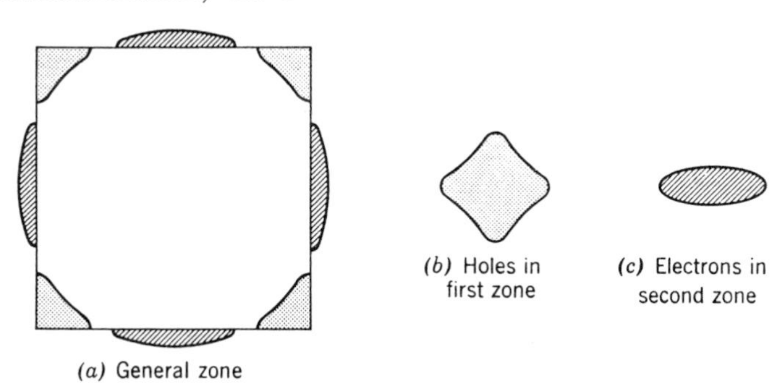

FIG. 2. (a) Fermi surface in two dimensions enclosing an area equal to the area of one Brillouin zone. (b) Hole orbit as connected. (c) One electron orbit as connected.

This result (23) holds also for a general fermi surface, at least in the semiclassical approximation developed in the preceding chapter for the dynamics of an electron in a magnetic field. We consider a thin section of an electron fermi surface, bounded by planes normal to H, with α states per unit area of the section. In constant fields H_z and E_x the energy ε of an electron in the section changes according to

$$\dot{\varepsilon} = ev_xE_x = -c\dot{k}_yE_x/H, \tag{25}$$

because $\dot{k}_y = -(e/c)v_xH$, if we may neglect collisions. Thus the shift of the fermi surface from equilibrium is given by

$$\Delta\varepsilon = -ck_yE_x/H, \tag{26}$$

within an additive constant. The resultant current in the y direction is, integrating over the surface of the section,

$$J_y = \alpha e \int dk_x\, dk_y \frac{\partial\epsilon}{\partial k_y} = -\alpha ec(E_x/H_z)\int dk_x\, dk_y. \tag{27}$$

The integral on the right-hand side is just the area of the section, so that J_y is the number of states in the section times ecE_x/H. For the whole fermi surface we integrate over dk_z, recalling that $\alpha \int dk_x\, dk_y\, dk_z$ gives the number of states in the volume. Thus the total current density is

$$j_y = \frac{ec}{H}(n_+ - n_-)E_x, \tag{28}$$

where n_+ is the hole concentration and n_- the electron concentration. For $n_- = n_+$ we have $\sigma_{yx} = 0$, in agreement with (23), and by the above argument the magnetoresistance does not saturate, in agreement with experiment. This result is independent of crystal orientation and explains the second variety of magnetoresistive behavior, as enumerated earlier.

INFLUENCE OF OPEN ORBITS

It is a remarkable experimental fact that in some crystals the magnetoresistance saturates *except* for certain special crystal orientations. The absence of saturation in certain directions may be explained in terms of open orbits. In strong magnetic fields an open orbit carries current essentially only in a single direction in the plane normal to the magnetic field; thus the open orbit cannot be saturated by the field. Suppose that for a given crystal orientation there are open orbits parallel to k_x; in real space these orbits carry current parallel to the y axis. We can associate a conductivity σ_{yy} with the open orbits; let us write the open-orbit conductivity as equal to $sne\mu$; this defines s. The high field limit of the conductivity tensor (10) is

$$\bar{\bar{\sigma}} \approx ne\mu \begin{pmatrix} \xi^{-2} & \xi^{-1} & 0 \\ -\xi^{-1} & \xi^{-2} & 0 \\ 0 & 0 & 1 \end{pmatrix}, \tag{29}$$

not considering the contribution of open orbits. We recall that $\xi \propto H$. With the contribution of the open orbits we have

$$\bar{\bar{\sigma}} \approx ne\mu \begin{pmatrix} \xi^{-2} & \xi^{-1} & 0 \\ -\xi^{-1} & s & 0 \\ 0 & 0 & 1 \end{pmatrix}. \tag{30}$$

Here we have dropped ξ^{-2} in comparison with s in the term σ_{yy}. We have assumed for convenience that $\sigma_{xz} = 0 = \sigma_{zx}$; this is not the most general situation. The applicability of (29) and (30) as high field limits is demonstrated by Pippard, *LTP*, pp. 93–95. We note that an open orbit parallel to k_x has an average velocity component only in the y direction and does not contribute to σ_{xy}, σ_{xx}, etc. The strength of the magnetic field does not affect the average carrier velocity v_y on the orbit; the field strength only affects the rate $\dot{k}_x$ at which the open orbit is traversed in **k** space.

With (30) we obtain $j_y = 0$ when

$$-\frac{E_x}{\xi} + sE_y = 0, \qquad \text{or} \qquad E_y = \frac{E_x}{s\xi}; \tag{31}$$

thus

$$j_x \approx ne\mu(\xi^{-2}E_x + \xi^{-1}E_y) = ne\mu\left(1 + \frac{1}{s}\right)\xi^{-2}E_x; \tag{32}$$

whence the effective resistivity is

$$\rho \approx \frac{\xi^2}{ne\mu}\frac{s}{s+1}. \tag{33}$$

The resistivity does not saturate, but increases as H^2. Crystal distributions tend to reduce the exponent towards 1. We have thus accounted for the third variety of magnetoresistive behavior, namely nonsaturation only in special crystal orientations.

Suppose the crystal is oriented so that the open orbit carries current in the x direction. Then

$$\bar{\sigma} \cong ne\mu\begin{pmatrix} s & \xi^{-1} & 0 \\ -\xi^{-1} & \xi^{-2} & 0 \\ 0 & 0 & 1 \end{pmatrix}, \tag{34}$$

and $j_y = 0$ if $E_y = \xi E_x$, so that

$$j_x \approx (s+1)ne\mu E_x. \tag{35}$$

For this orientation the magnetoresistance saturates.

If the open orbit runs in a general direction in the xy plane, the conductivity tensor has the form, again in the limit $\xi \gg 1$,

$$\bar{\bar{\sigma}} \approx ne\mu\begin{pmatrix} s\sin^2\theta + \xi^{-2} & -s\sin\theta\cos\theta + \xi^{-1} & 0 \\ s\sin\theta\cos\theta - \xi^{-1} & s\cos^2\theta + \xi^{-2} & 0 \\ 0 & 0 & 1 \end{pmatrix} \tag{36}$$

This gives $j_y = 0$ when $E_y = (\xi^{-1} - s\sin\theta\cos\theta)E_x/(s\cos^2\theta + \xi^{-2})$ for $\theta \neq 0$; we have

$$j_x = ne\mu\left\{s\sin^2\theta + \xi^{-2} + \frac{(s\sin\theta\cos\theta - \xi^{-1})^2}{s\cos^2\theta + \xi^{-2}}\right\}E_x \tag{37}$$

$$\rightarrow 2ne\mu s\sin^2\theta E_x.$$

Thus the *magnetoresistance saturates except when the open orbit carries current almost precisely parallel to the y direction.* By the geometrical rules discussed in Chapter 11 this requirement is that the orbit in **k** space must run in the k_x direction.

The circumstance that the magnetoresistance saturates in sufficiently strong magnetic fields, except when there are open orbits in the k_x direction, explains the extraordinary anisotropy of the transverse magnetoresistance observed in single crystals. The anisotropy is a striking feature of the experimental results, as illustrated for gold in

Fig. 1. Thus high-field studies of the angular dependence of the transverse magnetoresistance in single crystals can provide information on the presence of open orbits and on the connectivity of the fermi surface. In directions in which open orbits exist the resistance does not saturate; in other directions it does, except in special directions in which the metal may simulate a two-band metal with equal numbers of electrons and holes.

TRANSPORT EQUATIONS FOR MAGNETORESISTANCE

The Chambers or kinetic formulation of the transport equation is rather more revealing for the magnetoresistance problem with a general fermi surface than is the usual iterative formulation of the boltzmann method, in which magnetic effects appear only in second order. We first work with a version of the theory linearized in **E**. If the distribution is $f = f_0 + f_1$, where f_0 is the equilibrium distribution, then the electric current density is

$$\mathbf{j} = \frac{2e}{(2\pi)^3} \int d^3k\, \mathbf{v} f_1 = \frac{2e}{(2\pi)^3} \int d^3k\, \mathbf{v}\, \frac{df_0}{d\varepsilon}\, \Delta\varepsilon, \tag{38}$$

where $\Delta\varepsilon$ is the mean energy gained by the electron from the electric field **E** in the time between collisions. It is assumed that immediately after a collision the electron is in the equilibrium distribution. Then

$$\Delta\varepsilon = e \int_{-\infty}^{0} dt\, \mathbf{E} \cdot \mathbf{v}(t) e^{t/\tau}, \tag{39}$$

if the relaxation time τ is a constant. Here $e^{-|t|/\tau}$ is the probability that the last collision took place at least a time $|t|$ from the next collision, taken at $t = 0$. It is a simple matter to generalize (39) to problems in which τ is known as a function of **k**.

The general nonlinearized result of Chambers [*Proc. Phys. Soc.* (*London*) **A65,** 458 (1952)] for the distribution function is

$$f = \int_{-\infty}^{t} \frac{dt'}{\tau(\mathbf{k}(t'))} f_0(\varepsilon - \Delta\varepsilon(t')) \exp\left(-\int_{t'}^{t} \frac{ds}{\tau(\mathbf{k}(s))}\right), \tag{40}$$

where

$$\Delta\varepsilon = \int_{t'}^{t} dt''\, \mathbf{F} \cdot \mathbf{v}(t'') \tag{41}$$

is the energy gain from the force **F** between times t' and t in the absence of collisions; τ is the relaxation time; and f_0 is the equilibrium distribution function. A proof is given by H. Budd, *Phys. Rev.* **127,** 4 (1962), that (40) satisfies the boltzmann equation.

An electron in a high magnetic field will traverse a closed orbit many times between collisions and the integral (39) will approach zero for the velocity components in the plane normal to the magnetic field $\mathbf{H}$; the component parallel to $\mathbf{H}$ $(= H_z)$ may be replaced by the average $\langle v_{\mathbf{H}}(k_{\mathbf{H}})\rangle$ taken over the orbit at constant $k_{\mathbf{H}}$. Thus if the magnetic field is in the z direction,

$$\sigma_{xx} = \sigma_{xy} = \sigma_{yy} = \sigma_{yx} = 0; \tag{42}$$

$$\sigma_{zz} = \frac{2e^2}{(2\pi)^3\tau} \int d^3k\, v_z\langle v_z\rangle \frac{df_0}{d\varepsilon}. \tag{43}$$

This value of σ_{zz} for $H = \infty$ is in general lower than the value

$$\sigma_{zz}(0) \frac{2e^2}{(2\pi)^3\tau} \int d^3k\, v_z{}^2 \frac{df_0}{d\varepsilon} \tag{44}$$

for zero magnetic field by an amount depending on the anisotropy of v_z around the orbit. Thus there is a longitudinal magnetoresistance, which always saturates.

If open orbits are present in the k_x direction, then $\langle v_x\rangle = 0$, but $\langle v_y\rangle \neq 0$. Then $\sigma_{yy} \neq 0$ in the high-field limit, just as found in (30).

Let us apply the kinetic method to transverse magnetoresistance in weak magnetic fields such that $|\omega_c|\tau \ll 1$. We are concerned with

$$\int_{-\infty}^{0} dt\, v_\mu(t)e^{t/\tau},$$

for constant relaxation time τ. We expand

$$v_\mu(t) = v_\mu(0) + t\dot{v}_\mu(0) + \tfrac{1}{2}t^2\ddot{v}_\mu(0) + \cdots. \tag{45}$$

The integrals are trivial and we have

$$\int_{-\infty}^{0} dt\, v_\mu(t)e^{t/\tau} = \tau v_\mu(0) + \tau^2\dot{v}_\mu(0) + \tau^3\ddot{v}_\mu(0) + \cdots. \tag{46}$$

As an example, consider

$$\sigma_{xy} = \frac{2e^2}{(2\pi)^3} \int d^3k\, v_x \frac{df_0}{d\varepsilon} (\tau v_y + \tau^2\dot{v}_y + \tau^3\ddot{v}_y + \cdots). \tag{47}$$

The term in v_y vanishes on integration, by symmetry. The term in $\dot{v}_y$ for a free-electron gas is obtained by using

$$m\dot{v}_y = -\frac{e}{c} v_x H; \qquad \dot{v}_y = \omega_c v_x. \tag{48}$$

Thus to $O(H)$

$$\sigma_{xy} = \frac{2e^2}{(2\pi)^3} \omega_c\tau^2 \int d^3k\, v_x{}^2 \frac{df_0}{d\varepsilon} = \omega_c\tau\sigma_{xx}(0). \tag{49}$$

The term of $O(H^2)$ in σ_{xx} or σ_{yy} is obtained by evaluating $\ddot{v}_y$, for example. We have

$$\ddot{v}_y = \omega_c \dot{v}_x = -\omega_c{}^2 v_y, \tag{50}$$

so that

$$\sigma_{yy} = \sigma_{yy}(0) - (\omega_c\tau)^2\sigma_{yy}(0). \tag{51}$$

These results for the free-electron gas agree to the appropriate order with the drift-velocity treatment given earlier, but the present formulation can be applied to general energy surfaces. For example, to evaluate σ_{xy} we need

$$\dot{v}_y = \dot{k}_\nu \frac{\partial^2 \varepsilon}{\partial k_\nu\, \partial k_y} = \frac{eH}{c}\left(\frac{\partial \varepsilon}{\partial k_y}\frac{\partial^2 \varepsilon}{\partial k_x\, \partial k_y} - \frac{\partial \varepsilon}{\partial k_x}\frac{\partial^2 \varepsilon}{\partial k_y{}^2}\right), \tag{52}$$

whence to $O(H)$ we have a well-known result:

(53)

$$\sigma_{xy} = \frac{2e^2}{(2\pi)^3}\cdot\frac{eH\tau^2}{c}\int d^3k\frac{\partial f_0}{\partial \varepsilon}\left[\frac{\partial \varepsilon}{\partial k_x}\left(\frac{\partial \varepsilon}{\partial k_y}\frac{\partial^2 \varepsilon}{\partial k_x \partial k_y} - \frac{\partial \varepsilon}{\partial k_x}\frac{\partial^2 \varepsilon}{\partial k_y{}^2}\right)\right].$$

For $\omega_c > k_BT$ and $\omega_c\tau \gg 1$ an oscillatory behavior is observed in the conductivity components. The quantum oscillations in transport properties have the same origin as the susceptibility oscillations in the de Haas-van Alphen effect considered in Chapter 11.

PROBLEMS

1. Show that the transverse magnetoresistance vanishes for a conductor with an isotropic relaxation time and an ellipsoidal mass surface. An elegant derivation of this result is quoted by Chambers in *The fermi surface*, Wiley.

2. Discuss the magnetoresistive properties of a conductor with a fermi surface in the form of an infinite circular cylinder.

3. Consider the magnetoresistance of the two-carrier-type problem for all values of the magnetic field, but assuming $m_1 = m_2$; $\tau_1 = \tau_2$. Show that for the standard geometry

$$j_x = (n_1 + n_2)e\mu\frac{1}{1+\xi^2}\left[1 + \frac{(n_1 - n_2)^2}{(n_1+n_2)^2}\xi^2\right]E_x. \tag{54}$$

13 Calculation of energy bands and Fermi surfaces

The title of this chapter could more appropriately be used as the title of a substantial monograph which would derive and evaluate the principal terms and corrections which contribute to the one-electron energy band structure of a crystal. The monograph would also develop in detail computational aids and instructions, and it would contain a careful comparison of the theoretical calculations by various methods with the experimental observations. Such a monograph would be extremely useful, but if compiled at the present time even by a committee of the experts learned in the field, it would have several defects:

(*a*) The relevance of many-body corrections to the band structure and to the interpretation of the experiments is known quite incompletely.

(*b*) Our confidence in our ability to calculate the electronic properties of transition metals is low.

(*c*) The calculations on other metals are improving rapidly because of improvements in modified plane wave methods, and pertinent experimental data are being acquired at a substantial rate. In perhaps five years time the situation for the simpler metals will have stabilized somewhat.

There is now no such monograph. There are a number of useful reviews, among which are:

[1] F. S. Ham, *Solid state physics* **1,** 127 (1955).

[2] J. R. Reitz, *Solid state physics* **1,** 1 (1955).

[3] T. O. Woodruff, *Solid state physics* **4,** 367 (1957).

[4] J. Callaway, *Solid state physics* **7,** 100 (1958).

[5] H. Brooks, *Nuovo cimento supplement* **7,** 165 (1958); F. S. Ham, *Phys. Rev.* **128,** 82, 2524 (1962).

[6] W. A. Harrison, *Phys. Rev.* **118,** 1190 (1960).

[7] L. Pincherle, *Repts. Prog. Physics* **23,** 355 (1959) and the unpublished reports of the group associated with Slater at the Massachusetts Institute of Technology.

But what we can do in one chapter is to develop the principles of those methods which give a deep physical insight into the nature of the wavefunctions in the crystals to which the methods apply. We can also give a good idea of the accuracy of the methods. Those which we are going to discuss in detail are the Wigner-Seitz method applicable to the alkali metals, and various modified plane wave methods applicable over an extraordinarily wide range of crystals. This modest program does not by any means exhaust the methods which are used in practice and which give excellent results: the method of Kuhn and Van Vleck, as extended by Brooks, is discussed by Ham;[1] the variational method of Kohn has been applied widely by Ham;[5] an augmented plane wave method has been used extensively by the M.I.T. group. The first realistic band calculations were carried out by Wigner and Seitz[8] for lithium and sodium.

THE WIGNER-SEITZ METHOD

We consider first the state $\mathbf{k} = 0$ of the $3s$ conduction band of metallic sodium. At room temperature the crystal structure is body-centered cubic and the lattice constant is 4.28A. The distance between nearest-neighbor atoms is 3.71A. In the bcc and fcc lattices it is possible to fill the whole space with quasispherical polyhedra by constructing planes bisecting the lines which join each atom to its nearest neighbors and (for bcc) next-nearest neighbors. The atomic polyhedra thus obtained are shown in Fig. 1. Each polyhedron contains one lattice point and is a possible choice for the primitive cell.

An ion core is found at the center of each polyhedron. The ion core is often small in comparison with the half-distance to nearest neighbors. In sodium the half-distance is 1.85A; the ionic radius is given as 0.95A or 0.98A, but the core potential is only strong over less than 0.6A. Thus the potential energy is small over most of the volume of an atomic polyhedron. Inside the ion core where the potential is large it is approximately spherical. Thus to a good approximation the potential $V(\mathbf{x})$ may be taken to be spherical within each polyhedron.

The point $\mathbf{k} = 0$ is the special point Γ, if the state φ_0 (at Γ) is

[8] Li: F. Seitz, *Phys. Rev.* **47,** 400 (1935); J. Bardeen, *J. Chem. Phys.* **6,** 367 (1938). Na: E. Wigner and F. Seitz, *Phys. Rev.* **43,** 804 (1933); **46,** 509 (1934); E. Wigner, *Phys. Rev.* **46,** 1002 (1934).

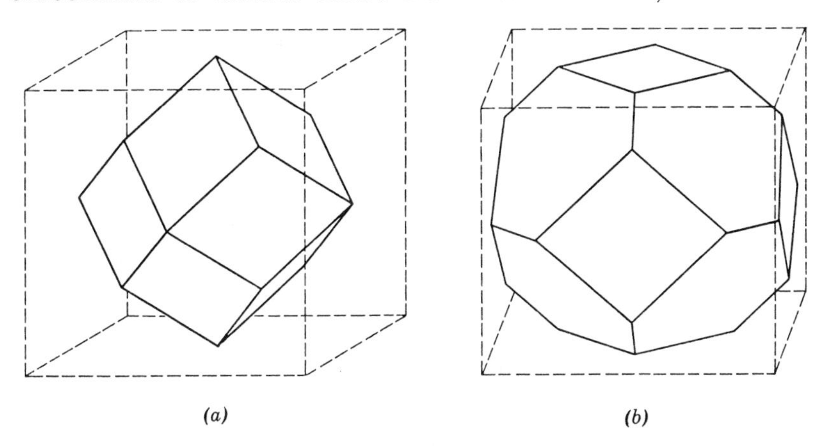

FIG. 1. Atomic polyhedra surrounding an atom for (*a*) face-centered cubic structure and (*b*) body-centered cubic structure.

nondegenerate and derived from a 3*s* atomic function, it will transform into itself under the operations of the full cubic group. The state is also periodic with the period of the lattice. The boundary planes of the polyhedral cells are related by reflection in a mirror plane through the center; consequently the normal derivative of φ_0 must vanish on the polyhedral surfaces:

$$\frac{\partial \varphi_0}{\partial \mathbf{n}} = 0. \tag{1}$$

In the actual calculation Wigner and Seitz replaced the polyhedra by spheres of equal volume. The boundary condition (1) is replaced by

$$\left(\frac{\partial \varphi_0}{\partial r}\right)_{r_s} = 0. \tag{2}$$

The spheres are known as *s* spheres; for the bcc lattice of cube edge a the radius r_s of an *s* sphere is determined by

$$\frac{4\pi}{3} r_s{}^3 = \tfrac{1}{2}a^3; \qquad \text{or} \qquad r_s \cong 0.49a; \tag{3}$$

there are two lattice points in a unit cube of the bcc lattice. Thus u_0 is the solution of

$$\left[-\frac{1}{2mr^2}\frac{d}{dr}\left(r^2\frac{d}{dr}\right) + V(r)\right] u_0(r) = \varepsilon_0 u_0(r), \tag{4}$$

subject to the boundary condition (2). Note that the boundary condition differs from that for the free atom where $u(r)$ must vanish at $r = \infty$.

We assumed in writing (4) that the lowest energy solution at $\mathbf{k} = 0$ is an s function. The next possible solution which transforms as Γ_1 is the g function ($l = 4$) with angular dependence described by

$$(x^2y^2 + y^2z^2 + z^2x^2) - \tfrac{1}{3}(x^4 + y^4 + z^4). \tag{5}$$

This function is tabulated as a cubic harmonic by F. C. Von der Lage and H. A. Bethe, *Phys. Rev.* **71,** 612 (1947). In the free Na atom the lowest possible g state is $5g$, about 5 ev higher in energy than $3s$. It is quite reasonable to neglect the possible mixture of $5g$ in $3s$ in the crystal. The $3p$ states transform as Γ_{15} and are about 2 ev higher in the free atom. Their energy is not split in first order in a cubic field, so there is no reason why in the crystal at $\mathbf{k} = 0$ the p state should be lowered relative to the s state.

The potential $V(r)$ was calculated on the assumption that screening by the fermi sea of the interaction between one conduction electron and one ion core was negligible when both were within the same cell and was complete otherwise. That is, it is assumed that only one conduction electron exists within each cell. This is a crude model of correlation effects among the conduction electrons, but the model does not work at all badly. The potential within each s sphere is then just the potential of the free ion. The actual free-ion potential contains exchange terms; the effect of these was approximated simply by constructing a potential $V(r)$ which reproduced empirically to high accuracy the spectroscopic term values of the free atom. The ground-state wavefunction is shown in Fig. 2. The function is quite constant over 90 percent of the

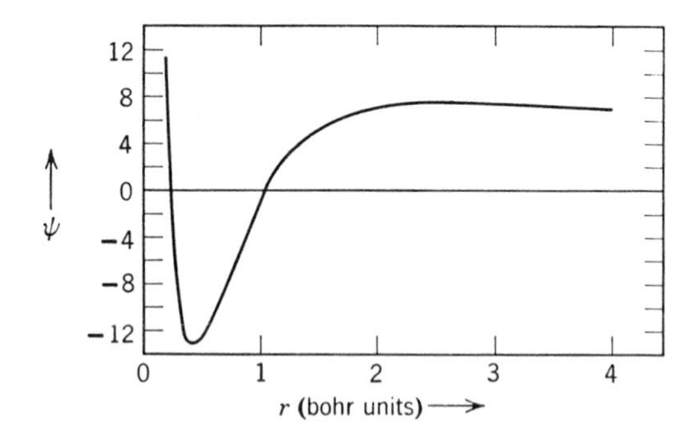

FIG. 2. The lowest wavefunction of metallic sodium.

atomic volume. The potential in the outer regions of the s sphere is itself not constant, but is $-e^2/r$; however, the boundary condition (2) forces the solution to look rather flat.

The electronic energy of the state u_0 is generally lowered with respect to the free atom because of the change in the boundary conditions. This effect can be seen easily by an explicit example. Consider the harmonic oscillator wave equation

$$\frac{d^2u}{d\xi^2} + (\lambda - \xi^2)u = 0; \tag{6}$$

in the ground state with the usual boundary condition $u(\xi) \to 0$ as $|\xi| \to \infty$ the eigenvalue is $\lambda = 1$ and $u(\xi) = e^{-\xi^2/2}$. We now look for the ground state with the Wigner-Seitz boundary condition $du/d\xi = 0$ at $\xi = \pm\xi_0$. Let

$$u(\xi) = \sum_{n=0}^{\infty} c_n \xi^n. \tag{7}$$

The differential equation is satisfied if

$$(n + 1)(n + 2)c_{n+2} + \lambda c_n - c_{n-2} = 0. \tag{8}$$

We set $c_0 = 1$ for a solution even in ξ; then from $n = 0$ and $n = 2$ we have

$$c_2 = -\tfrac{1}{2}\lambda; \qquad c_4 = \tfrac{1}{24}\lambda^2 + \tfrac{1}{12}. \tag{9}$$

Thus

$$\begin{aligned} u(\xi) &= 1 - \frac{\lambda}{2}\xi^2 + \tfrac{1}{12}(1 + \tfrac{1}{2}\lambda^2)\xi^4 + \cdots; \\ du/d\xi &= -\lambda\xi + \tfrac{1}{3}(1 + \tfrac{1}{2}\lambda^2)\xi^3 + \cdots. \end{aligned} \tag{10}$$

If we require $du/d\xi = 0$ at ξ_0 and cut off the series (10) at the terms given, we have

$$\frac{3\lambda}{1 + \frac{1}{2}\lambda^2} \cong \xi_0^2. \tag{11}$$

For $\xi_0 = 1$, for example, the eigenvalue is $\lambda \cong 0.353$, which is much lower than the eigenvalue $\lambda = 1$ for the unbounded harmonic oscillator. The term in c_6 gives a negligible correction for $\xi_0 = 1$. Thus the Wigner-Seitz boundary condition permits the ground-state energy to be lowered substantially.

The ground-state energy calculated for sodium with $\partial u_0/\partial r = 0$ at $r_s = 3.96$ atomic units is about -8.3 ev, as compared with the experimental atomic ground-state energy of -5.16 ev; thus the $3s$ energy at $\mathbf{k} = 0$ is lower by 3.1 ev in the metal than in the atom. But the conduction electrons at finite $\mathbf{k}$ in the metal have extra kinetic energy; the mean fermi energy per electron is

$$\langle \varepsilon_F \rangle = \frac{3}{5} \frac{1}{2m^*} k_F{}^2 = \frac{1.10}{r_s{}^2} \frac{e^2}{a_H} \left(\frac{m}{m^*} \right), \tag{12}$$

or 2.0 ev in Na with m^* taken as m. The net binding energy is $-8.3 + 2.0 = -6.3$ ev, as compared with -5.16 ev in the free atom. The difference, -1.1 ev, is in good agreement with the observed cohesive energy -1.13 ev. For the other alkali metals it is not a good approximation to take $m = m^*$, but a fairly good value of the effective mass may be calculated by the $\mathbf{k} \cdot \mathbf{p}$ perturbation theory discussed in the chapter on Bloch functions.

These estimates omit all coulomb, exchange, and correlation corrections. The corrections are all large, but nearly cancel among themselves, as we saw in (6.80). This is not entirely surprising; our present calculation builds in a great deal of electron correlation by taking one electron per cell. If this model corresponds fairly well to reality, as we expect it to, the sum of all other corrections should not be very large.

A review of calculations on the alkali metals is given by Ham.[5] The physical properties of the alkalis relative to one another appear to be fairly well understood, although the experimental situation is not yet as clear as in less reactive metals. The fermi surfaces are believed to be nearly spherical in Na and K, but are quite anisotropic in Li and in Cs. The theoretical calculations account moderately well for the cohesive energies and other properties.

ALMOST-FREE ELECTRON APPROXIMATION—GENERALIZED OPW METHOD

It has been realized for a long time that even in complex polyvalent metals the over-all width in energy of the filled bands is not greatly different from the width one expects from a free electron gas of the same concentration. This result must follow if the kinetic energy of the fermi gas is comparable with or larger than the differences in potential energy of the individual electrons. In 1958 A. V. Gold[9] made the remarkable observation for lead that his de Haas-van Alphen

[9] A. V. Gold, *Phil. Trans. Roy. Soc.* (*London*) **A251,** 85 (1958).

data could be explained on a nearly free electron approximation. That is, he assumed the conduction electrons behave as if free, but subject at zone boundaries to Bragg reflection from a weak periodic potential. Lead has four valence electrons outside the filled $5d^{10}$ shell; the nuclear charge is 82 and there are 78 electrons in the ion core. Lead would appear at first sight to be a complicated nontransition metal; the fermi surface is complicated; yet it appears that a simple model of nearly free electrons will account closely for the form of the fermi surface.

It is becoming increasingly evident[10] that many features of the band structure of polyvalent metals may be calculated in a simple way, using a model of free electrons weakly perturbed by the crystal lattice. This circumstance is of tremendous value to the theory of metals. Our program in this section is to describe how the fermi surfaces are determined by this method and then to explain the theoretical basis of the method. The method applies best to crystals whose ion cores are not in contact; thus it is less useful in the noble metals where the cores touch. The result does *not* imply that the actual wavefunctions resemble plane waves.

We consider first in one dimension the perturbation of nearly free electrons by the real periodic potential

$$V = V_1(e^{iG_1x} + e^{-iG_1x}), \tag{13}$$

where $G_1 = 2\pi/a$ is the first or shortest vector in the reciprocal lattice; the lattice constant in the direct lattice is a. The matrix elements of V in a plane wave representation are

$$\begin{aligned}\langle k'|V|k\rangle &= V_1 \int_0^1 dx\, e^{i(k-k')x}(e^{iG_1x} + e^{-iG_1x}) \\ &= V_1[\Delta(k - k' + G_1) + \Delta(k - k' - G_1)].\end{aligned} \tag{14}$$

Thus the perturbation mixes states differing by the reciprocal lattice vector G_1. Such states are degenerate in energy at the Brillouin zone boundary $k = \pm\pi/a$, and here the perturbation will have its maximum effect on the energy. Away from the boundary the perturbation will not change the energy seriously as long as the potential V_1 is small compared with the kinetic energy difference, as measured typically by $G_1^2/2m$.

We find the energy at the zone boundary π/a by degenerate pertur-

[10] J. C. Phillips and L. Kleinman, *Phys. Rev.* **116,** 287, 880 (1959); W. A. Harrison, *Phys. Rev.* **118,** 1190 (1960).

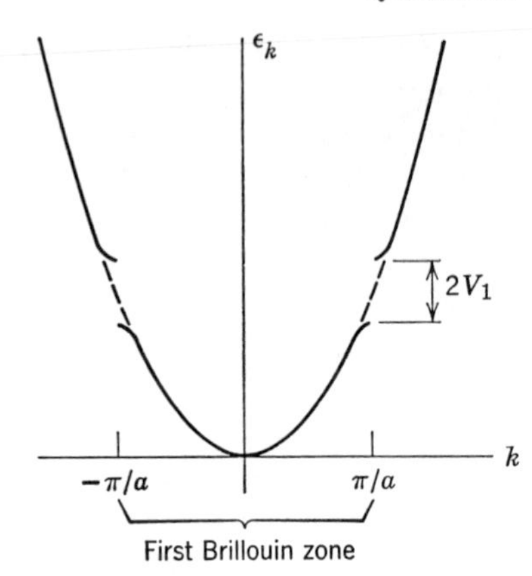

FIG. 3. Band structure of a linear lattice as calculated on the nearly free electron approximation.

bation theory on the states k and $k - G_1$. The secular determinant is

$$(15)\quad \begin{vmatrix} \langle k|H|k\rangle - \varepsilon & \langle k|H|k - G_1\rangle \\ \langle k - G_1|H|k\rangle & \langle k - G_1|H|k - G_1\rangle - \varepsilon \end{vmatrix} = \begin{vmatrix} \dfrac{k^2}{2m} - \varepsilon & V_1 \\ V_1 & \dfrac{(k - G_1)^2}{2m} - \varepsilon \end{vmatrix} = 0.$$

For k exactly at the boundary $k^2 = (k - G_1)^2$ and the equation reduces to

$$(16)\quad \left(\varepsilon - \frac{\pi^2}{2a^2m}\right)^2 = V_1{}^2; \qquad \varepsilon = \frac{\pi^2}{2a^2m} \pm V_1.$$

An energy gap of magnitude $E_g = 2V_1$ exists at the zone boundary (Fig. 3). The features of the result are: (1) the dependence of ε on k over most of the zone is close to that of a free electron with $m^* \cong m$; (2) near the zone boundaries the energy is perturbed, with effective masses $m^* = (\partial^2\varepsilon/\partial k^2)^{-1}$ markedly different from m and possibly much smaller than m; (3) the magnitude of the energy gap is simply related to the single matrix element $\left\langle \frac{\pi}{a} - G_1 \middle| V \middle| \frac{\pi}{a} \right\rangle = V_1$ of the crystal potential; (4) the eigenfunctions at the gap are either even or odd with respect to x.

In three dimensions a single plane wave, suitably orthogonalized to the core, gives good results except near a zone face, where two waves are necessary; near a zone edge, three; near a zone corner, four.

CONSTRUCTION OF FERMI SURFACES

We observe that in two and three dimensions the fermi surface can appear to be quite complicated even if the energy varies everywhere exactly as k^2. In Fig. 4*a* we show the first three Brillouin zones of a simple square lattice in two dimensions. A circle is drawn about the central reciprocal-lattice point as origin. The circle represents a constant-energy surface for a free electron gas. If this particular energy is the fermi energy, the fermi surface will have segments in the second, third, and fourth zones. The first zone lies entirely within the surface and at absolute zero will be filled entirely. In an actual crystal there may be energy gaps across all zone boundaries; an electron moving on a constant-energy surface will remain within one zone. All electrons on the fermi surface have the same energy.

It is interesting to look at the several portions of the fermi surface

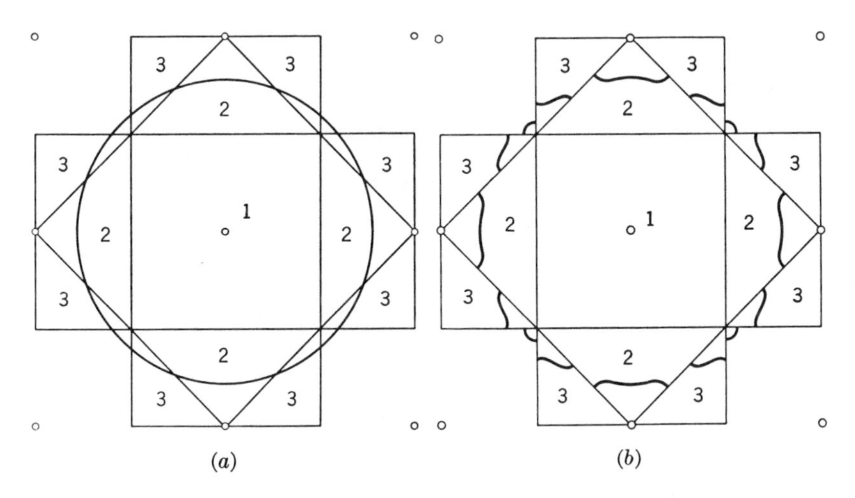

FIG. 4(*a*). First three Brillouin zones of a simple square lattice. Reciprocal lattice points are shown as dots. Zone boundaries are defined as the bisectors of lines joining reciprocal lattice points; the boundaries are drawn. The numbers denote the zone in order of increasing energy (as measured by k^2) to which the segment belongs; all segments can be moved to the first zone by a translation by a reciprocal lattice vector. The circle drawn might represent a fermi surface with area in zones 2 and 3, and also in zone 4, which is not shown. (*b*) Same as Fig. 4*a*, but showing distortion of fermi surface by a weak crystal potential. The fermi surface is made to intersect the zone boundaries at right angles.

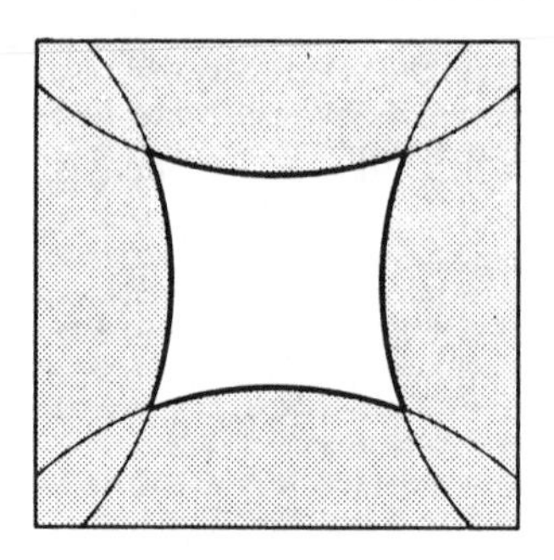

FIG. 5. Fermi surface in the second Brillouin zone of Fig. 4, shown in the reduced zone. The orbit is a hole orbit.

of Fig. 4 in the reduced zone scheme as in Figs. 5 and 6. Looking at any one of the surfaces by itself, we might not guess that they result from free electrons. The surfaces give the appearance of being profoundly modified by the crystal structure.

The free-electron surfaces in Figs. 5 and 6 were constructed by translating the several portions of Fig. 4 by the appropriate reciprocal-lattice vectors. Harrison has given a simpler construction for fermi surfaces on the free-electron model. The reciprocal lattice is constructed and a free-electron sphere is drawn around *each* lattice point, as in Fig. 7. Any point in **k** space which lies within at least one of the spheres corresponds to an occupied state in the first zone. Points

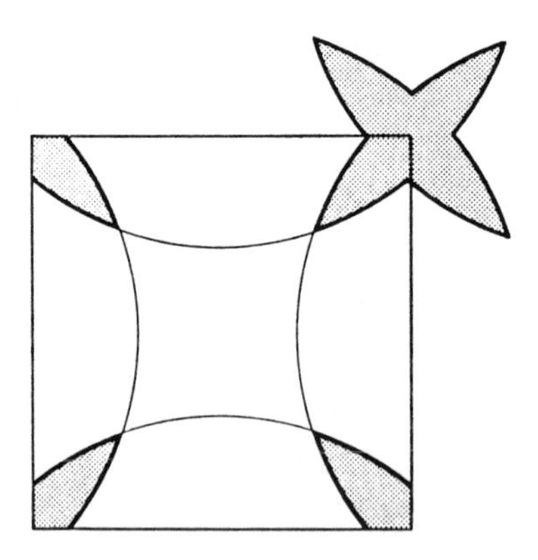

FIG. 6. Fermi surface in the third zone. The orbits form rosettes when combined with adjacent third zones; the orbits are electron orbits.

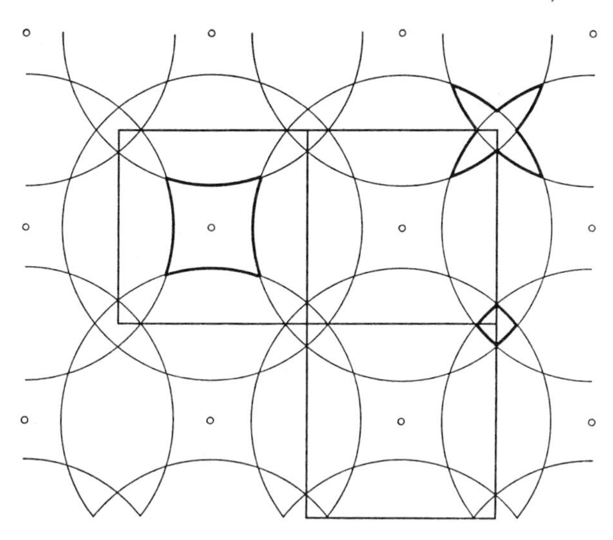

FIG. 7. Harrison construction of the fermi surfaces in the second, third, and fourth zones for the problem of Fig. 4.

lying within at least two spheres correspond to occupied states in the second zone, etc.

In Fig. 7 we have drawn the circles equivalent to the circle of Fig. 4. In one square, we have enclosed within a heavy line the regions which lie within two or more circles; in another square, three or more; in another square, four or more. These are the fermi surfaces in the second, third, and fourth zones, in the free-electron approximation.

It is evident immediately that the introduction of the slightest amount of a periodic potential will change the surfaces near the boundaries; thus compare Figs. 4*a* and 4*b*. In a square lattice the

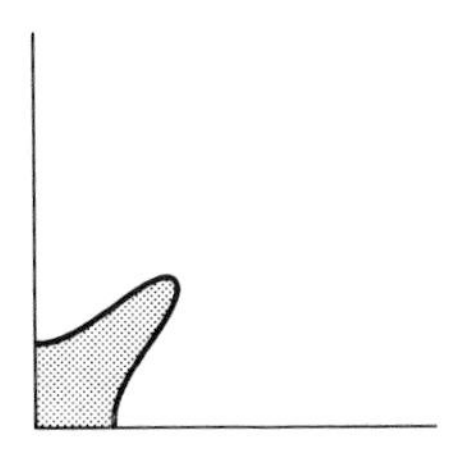

FIG. 8. One corner of the third zone of Fig. 6 as it might be affected by a weak crystal potential. The line of constant energy intersects the boundary at normal incidence.

existence of the mirror planes $m_x m_y$ forces the condition

$$\frac{\partial \varepsilon}{\partial \mathbf{n}} = 0 \tag{16a}$$

at each zone boundary, where **n** is the normal to the boundary. Therefore in an actual crystal the lines of constant energy in Figs. 4, 6, and 7 must come in at normal incidence to the boundary. A possible form of a corner of the third zone is shown in Fig. 8.

We note that the perturbation of the free-electron fermi surface

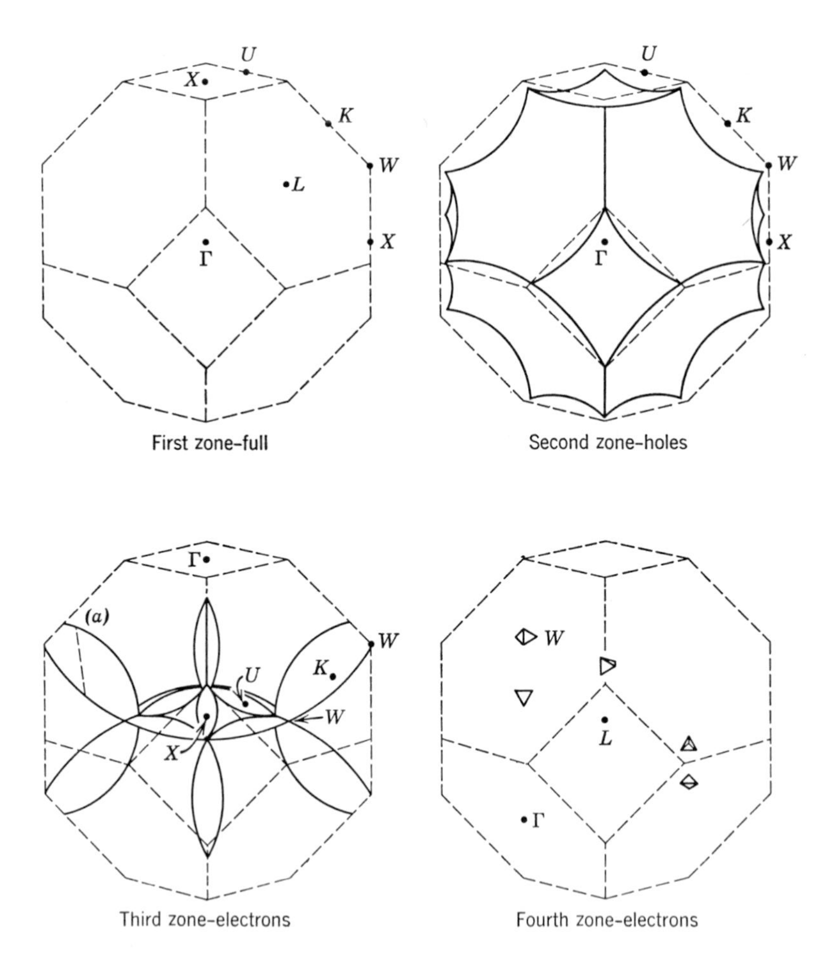

FIG. 9. Free-electron fermi surface of aluminum in the reduced zone scheme. (After Harrison.)

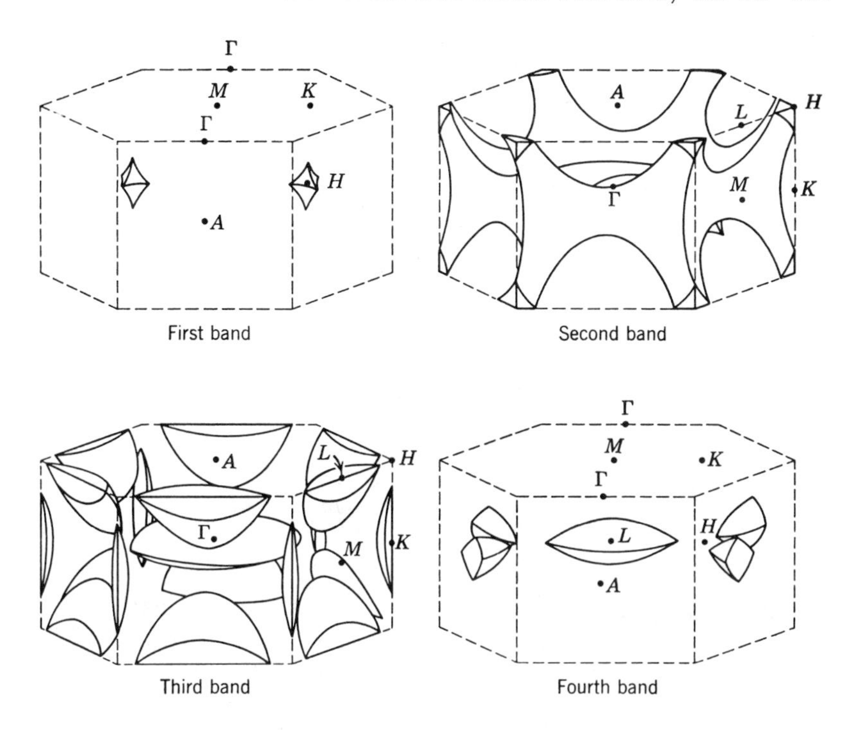

FIG. 10. The one-OPW or nearly free-electron fermi surface for a divalent, hexagonal close-packed metal of ideal c/a ratio. Spin-orbit splittings are included with their consequent reduction of the double zone. (After Harrison.)

by the crystal lattice acts to reduce the exposed *area* of the fermi surface. A relatively weak perturbation can reduce the area by one-half, with little effect on the average energy at the fermi surface.

The free-electron construction of the fermi surface of aluminum is shown in Fig. 9, after Harrison. The crystal structure of aluminum is the fcc structure. There are three valence electrons; the radius of the free-electron sphere is chosen to enclose three electrons per atomic volume. The first zone is full. The second zone contains hole states with a large area of fermi surface. The section of the fermi surface shown in the third zone is known as the *monster;* the monster has eight tentacles. There are two similar monsters in this zone related by symmetry to the one shown. The data on the de Haas-van Alphen effect are compatible with this fermi surface. In the fourth zone only little electron pockets exist. These are empty when the periodic potential is included in the calculation, because the potential acts to raise considerably the energy of the pockets. Table 1 compares the free-electron energies with the results of the band calculations for aluminum by Heine. The extent of the agreement is quite remarkable.

TABLE 1

ENERGIES IN RYDBERGS IN ALUMINUM RELATIVE TO THE BAND MINIMUM FOR SEVERAL POINTS OF HIGH SYMMETRY

(After Harrison)

		Free Electron	Heine
Fermi level		1.11	1.09
First zone:	X	0.89	0.81
	W	1.09	1.01
Second zone:	L	0.69	0.72
	X	0.89	0.93
	W	1.09	1.01
Third zone:	W	1.09	1.06
Fourth zone:	W	1.09	1.18

THEORETICAL BASIS OF THE NEARLY FREE-ELECTRON MODEL

It is not immediately obvious that the nearly free-electron model is justified theoretically. The expectation value of the potential energy of a conduction electron in a solid is usually not small in comparison with the fermi energy; if V_1 in (13) is not small, the assumptions on which the calculation (13) to (16) rest are unfounded.

However, there are two essentially different regions within the unit cell, the region outside the core and the region inside the core. Outside the core the potential energy is weak and the wavefunction is smooth; inside the core the wavefunction oscillates rapidly to gain kinetic energy to cancel approximately the strong potential of the core. The core potential has little real connection with the solution of the eigenvalue problem, but the core potential is a great complication in finding a solution.

A series of developments due to Herring,[11] to Phillips and Kleinman,[10] and to Cohen and Heine,[12] have shown that one can transform the hamiltonian into a modified hamiltonian whose solutions vary smoothly and resemble plane waves. Herring observed that the wavefunction of a conduction electron in a solid is nearly a plane wave in the region between the ion cores and that the oscillations of the wavefunction near the nuclei can be represented by subtracting

[11] C. Herring, *Phys. Rev.* **57,** 1169 (1940).

[12] M. H. Cohen and V. Heine, *Phys. Rev.* **122,** 1821 (1961); see also E. Brown, *Phys. Rev.* **126,** 421 (1962); Austin, Heine, and Sham, *Phys. Rev.* **127,** 276 (1962).

filled-core orbitals from the plane wave. The subtraction makes it possible for the amended function to be orthogonal to the core states. The method is called the orthogonalized plane wave, or OPW, method. Thus a conduction state in sodium, derived from the $3s$ level, is represented by a plane wave minus $1s$, $2s$, and $2p$ contributions. We emphasize that the true eigenfunctions do not resemble plane waves, but the eigenvalues are close to the free-electron energies.

Let φ_c denote the states in the filled core; then

$$\psi_{\mathrm{OPW}}(\mathbf{k}) = e^{i\mathbf{k}\cdot\mathbf{x}} - \sum_c \langle \varphi_c | e^{i\mathbf{k}\cdot\mathbf{x}} \rangle \varphi_c(\mathbf{x}) \tag{17}$$

is an *orthogonalized plane wave*, where the sum runs over core states φ_c. The core states are usually taken in the tight-binding approximation; thus $\varphi_c(\mathbf{x})$ is a condensed notation $\sum_j e^{i\mathbf{k}\cdot\mathbf{x}_j}\varphi_c(\mathbf{x} - \mathbf{x}_j)$. We note that ψ_{OPW} is orthogonal to any core state $\varphi_{c'}$:

$$\langle \varphi_{c'} | \psi_{\mathrm{OPW}} \rangle = \langle \varphi_{c'} | e^{i\mathbf{k}\cdot\mathbf{x}} \rangle - \langle \varphi_{c'} | e^{i\mathbf{k}\cdot\mathbf{x}} \rangle \langle \varphi_{c'} | \varphi_{c'} \rangle = 0. \tag{18}$$

The solution $\psi_{\mathrm{OPW}}(\mathbf{k})$ of (17) can be improved by adding terms in $\mathbf{k} + \mathbf{G}$, where $\mathbf{G}$ is a reciprocal lattice vector; thus

$$\psi_{\mathrm{OPW}} = \sum_{\mathbf{G}} C_{\mathbf{G}} [e^{i(\mathbf{k}+\mathbf{G})\cdot\mathbf{x}} - \sum_c \langle \varphi_c | e^{i(\mathbf{k}+\mathbf{G})\cdot\mathbf{x}} \rangle \varphi_c(\mathbf{x})], \tag{19}$$

where $C_{\mathbf{G}}$ is a constant to be determined by the solution of a secular equation.

Now, with $|\mathbf{k} + \mathbf{G}\rangle = e^{i(\mathbf{k}+\mathbf{G})\cdot\mathbf{x}}$, we have

$$H\psi_{\mathrm{OPW}} = \sum_{\mathbf{G}} C_{\mathbf{G}} \left\{ \left[\frac{(\mathbf{k} + \mathbf{G})^2}{2m} + V(\mathbf{x}) \right] |\mathbf{k} + \mathbf{G}\rangle - \sum_c \langle \varphi_c | \mathbf{k} + \mathbf{G} \rangle \varepsilon_c \varphi_c(\mathbf{x}) \right\}, \tag{20}$$

where $H\varphi_c = \varepsilon_c\varphi_c$, and we assume that ε_c is independent of $\mathbf{k}$, as for a very narrow core band. The secular equation is obtained by taking the scalar product of this equation with one plane wave, say $\mathbf{k} + \mathbf{G}'$:

$$\begin{aligned} \sum_{\mathbf{G}} C_{\mathbf{G}} \bigg[\frac{(\mathbf{k} + \mathbf{G})^2}{2m} \delta_{\mathbf{G}\mathbf{G}'} &+ \langle \mathbf{G}' | V | \mathbf{G} \rangle - \sum_c \varepsilon_c \langle \varphi_c | \mathbf{k} + \mathbf{G} \rangle \langle \mathbf{k} + \mathbf{G}' | \varphi_c \rangle \bigg] \\ &= \lambda \sum_{\mathbf{G}} C_{\mathbf{G}} \left[\delta_{\mathbf{G}\mathbf{G}'} - \sum_c \langle \varphi_c | \mathbf{k} + \mathbf{G} \rangle \langle \mathbf{k} + \mathbf{G}' | \varphi_c \rangle \right]. \end{aligned} \tag{21}$$

The eigenvalues are the roots λ of this set of equations.

It is useful to summarize the actual procedure by which a band calculation is carried out in the OPW method. We take for the potential of the ion core at each ion site a Hartree or Hartree-Fock potential. We then suppose that the exact wavefunction is expanded in a series of orthogonalized plane waves as in (19), and we form the secular equation (21). We may work with a reasonably large secular equation and for easier computation reduce it at a symmetry point, or we may often approximate the secular equations by considering only two or three $\mathbf{G}$ values appropriate for a specific region in $\mathbf{k}$ space. The core terms actually tend to subtract in (21) from the plane wave matrix elements of the crystal potential. Heine[13] has carried out a careful calculation for aluminum and finds that the cancellation is remarkably complete, within about 5 percent.

One can express the method in terms of an effective wave equation. Let $\psi_{\mathbf{k}}$ be an exact eigenstate of $H_0\psi_{\mathbf{k}} = \varepsilon_{\mathbf{k}}\psi_{\mathbf{k}}$ in the band of interest. We want to write $\psi_{\mathbf{k}}$ as the sum of some smoothly varying function $\chi_{\mathbf{k}}$ and of a linear combination of filled-core states φ_c. The exact $\psi_{\mathbf{k}}$ cannot be written as a single plane wave. We write

$$\psi_{\mathbf{k}} = \chi_{\mathbf{k}} - \sum_c \varphi_c\langle\varphi_c|\chi_{\mathbf{k}}\rangle, \tag{22}$$

which is automatically orthogonal to the core states. We look at the wave equation for $\chi_{\mathbf{k}}$:

$$H_0\psi_{\mathbf{k}} = \varepsilon_{\mathbf{k}}\psi_{\mathbf{k}} = H_0\chi_{\mathbf{k}} - \sum \varepsilon_c\varphi_c\langle\varphi_c|\chi_{\mathbf{k}}\rangle = \varepsilon_{\mathbf{k}}\chi_{\mathbf{k}} - \sum_c \varepsilon_{\mathbf{k}}\varphi_c\langle\varphi_c|\chi_{\mathbf{k}}\rangle, \tag{23}$$

with $H_0\varphi_c = \varepsilon_c\varphi_c$. Now define an operator V_R such that

$$V_R\chi_{\mathbf{k}} = \sum_c (\varepsilon_{\mathbf{k}} - \varepsilon_c)\varphi_c\langle\varphi_c|\chi_{\mathbf{k}}\rangle. \tag{24}$$

Thus V_R is an integral operator or nonlocal potential such that

$$V_R f(\mathbf{x}) = \int d^3x'\, V_R(\mathbf{x},\mathbf{x}')f(\mathbf{x}'), \tag{25}$$

where $f(\mathbf{x})$ is any function, and

$$V_R(\mathbf{x},\mathbf{x}') = \sum_c (\varepsilon_{\mathbf{k}} - \varepsilon_c)\varphi_c^*(\mathbf{x}')\varphi_c(\mathbf{x}). \tag{26}$$

Because $\varepsilon_{\mathbf{k}} > \varepsilon_c$, this potential is repulsive at $\mathbf{x}'$ near $\mathbf{x}$. We can separate V_R into the sum of $s, p, d, \cdots$ contributions.

[13] V. Heine, *Proc. Roy. Soc. (London)* **A240,** 354, 361 (1957).

With V_R defined by (24), the wave equation for $\chi_{\mathbf{k}}$ is

$$(H_0 + V_R)\chi_{\mathbf{k}} = \varepsilon_{\mathbf{k}}\chi_{\mathbf{k}}; \tag{27}$$

here the repulsive V_R cancels part of the attractive crystal potential.

We can add to $\chi_{\mathbf{k}}$ any linear combination of core states without changing $\psi_{\mathbf{k}}$: whatever we add to $\psi_{\mathbf{k}}$ from the core is cancelled exactly by the change in $\Sigma\, \varphi_c\langle\varphi_c|\chi_{\mathbf{k}}\rangle$. Thus we are at liberty to impose an additional constraint on $\chi_{\mathbf{k}}$. We could, for example, demand that $\chi_{\mathbf{k}}$ be chosen to minimize the expectation value of the kinetic energy

$$\bar{T} = \langle\chi_{\mathbf{k}}|T|\chi_{\mathbf{k}}\rangle/\langle\chi_{\mathbf{k}}|\chi_{\mathbf{k}}\rangle; \tag{28}$$

this criterion may be said to make $\chi_{\mathbf{k}}$ as smooth as possible. The variation equation is

$$\langle\delta\chi|T|\chi\rangle - \bar{T}\langle\delta\chi|\chi\rangle = 0. \tag{29}$$

If we write the variation in χ as

$$\delta\chi = \Sigma\, \alpha_c\varphi_c, \tag{30}$$

then

$$\langle\varphi_c|T|\chi\rangle - \bar{T}\langle\varphi_c|\chi\rangle = 0. \tag{31}$$

It follows from $(V + V_R)\chi = (\varepsilon - T)\chi$ that

$$\langle\varphi_c|V + V_R|\chi\rangle = (\varepsilon - \bar{T})\langle\varphi_c|\chi\rangle; \tag{32}$$

with (24), we have

$$(V + V_R)\chi_{\mathbf{k}} = [V\chi_{\mathbf{k}} - \sum_c \langle\varphi_c|V|\chi_{\mathbf{k}}\rangle\varphi_c] + [(\varepsilon_{\mathbf{k}} - \bar{T}) \sum \varphi_c\langle\varphi_c|\chi_{\mathbf{k}}\rangle]. \tag{33}$$

The terms in the second bracket on the right-hand side involve, after taking the scalar product with χ, the quantity $\Sigma\, |\langle\varphi_c|\chi_{\mathbf{k}}\rangle|^2$, which turns out to be of the order of 0.1. It appears then that

$$(V + V_R)\chi_{\mathbf{k}} \cong V\chi_{\mathbf{k}} - \sum_c \langle\varphi_c|V|\chi_{\mathbf{k}}\rangle\varphi_c. \tag{34}$$

We show below that if the φ_c formed a complete set of functions, the scalar product of any φ_c with the right-hand side would give zero; that is, a complete cancellation of V and V_R. The φ_c for the filled-core states alone are, however, quite a good set for the representation of V over the extent of the ion core; thus the cancelation in practice is quite good. This close cancellation is the basis of the applicability of the almost free-electron approximation. For actual calculations a modified treatment of (22) has been recommended by Harrison.[14]

[14] W. A. Harrison, *Phys. Rev.* **126,** 497 (1962).

Now suppose that we can write

$$V(\mathbf{x})e^{i\mathbf{k}\cdot\mathbf{x}} = \sum_c f_{c\mathbf{k}} \sum_i e^{i\mathbf{k}\cdot\mathbf{x}_i}\varphi_c(\mathbf{x} - \mathbf{x}_i), \tag{35}$$

where the $\varphi_c(\mathbf{x} - \mathbf{x}_i)$ are atomic core functions. Then we can find a solution of the wave equation of the form

$$\Psi_{\mathbf{k}}(\mathbf{x}) = e^{i\mathbf{k}\cdot\mathbf{x}} + \sum_c d_{c\mathbf{k}} \sum_j e^{i\mathbf{k}\cdot\mathbf{x}_j}\varphi_c(\mathbf{x} - \mathbf{x}_j), \tag{36}$$

with $\varepsilon_{\mathbf{k}} = k^2/2m$. For

$$\begin{aligned} H\Psi_{\mathbf{k}} &= \varepsilon_{\mathbf{k}}e^{i\mathbf{k}\cdot\mathbf{x}} + \sum_c f_{c\mathbf{k}} \sum_j e^{i\mathbf{k}\cdot\mathbf{x}_j}\varphi_c(\mathbf{x} - \mathbf{x}_j) \\ &\qquad + \sum_c \varepsilon_c d_{c\mathbf{k}} \sum_j e^{i\mathbf{k}\cdot\mathbf{x}_j}\varphi_c(\mathbf{x} - \mathbf{x}_j) \\ &= \varepsilon_{\mathbf{k}}e^{i\mathbf{k}\cdot\mathbf{x}} + \sum_c \varepsilon_{\mathbf{k}} d_{c\mathbf{k}} \sum_j e^{i\mathbf{k}\cdot\mathbf{x}_j}\varphi_c(\mathbf{x} - \mathbf{x}_j), \end{aligned} \tag{37}$$

provided that

$$d_{c\mathbf{k}} = f_{c\mathbf{k}}/(\varepsilon_{\mathbf{k}} - \varepsilon_c). \tag{38}$$

Thus if $V(\mathbf{x})$ can be expanded in the form (35), there are no energy gaps!

Actually, of course, the φ_c are not a complete set and there will always be a residue $\delta V(\mathbf{x})$ after an attempted expansion. This residue, which is not unique, acts as an effective potential for the band or scattering problem. We hope that δV can be made small in comparison with V. When this is true, the scattering cross section of impurities in insulating crystals will be much smaller than the geometric cross section. We can always choose the φ_c appropriate to the particular local environment, even if we are dealing with a mixed crystal or a liquid. The message is that imperfect condensed phases are often much less disordered, when viewed by an electron, than one might guess.

PROBLEMS

1. By crude numerical integration using a desk calculator, or else by programming a modern computer, calculate $u_0(r)$ and ε_0 for metallic sodium at $r_s = 4$ atomic units. Use the potential given by W. Prokofjew, as corrected by E. Wigner and F. Seitz, *Phys. Rev.* **43,** 807 (1933).

2. Consider the OPW solution

$$\Psi_{\mathbf{k}} = \Omega^{-1/2}e^{i\mathbf{k}\cdot\mathbf{x}} - \sum_c \varphi_c\langle c|\mathbf{k}\rangle, \tag{39}$$

where $|\mathbf{k}\rangle$ denotes $\Omega^{-\frac{1}{2}}e^{i\mathbf{k}\cdot\mathbf{x}}$ normalized in the cell volume Ω. Suppose that all core functions c vanish outside a core volume Δ. Then show that

$$\sum_c |\langle c|\mathbf{k}\rangle|^2 \leqq \Delta/\Omega. \tag{40}$$

This result shows that the core admixtures are small when the cores occupy only a small fraction of the volume of the cell.

3. Define a function $\Psi_\mathbf{k}(\mathbf{x})$ such that the Bloch function

$$\varphi_\mathbf{k}(\mathbf{x}) = u_0(\mathbf{x})\Psi_\mathbf{k}(\mathbf{x}), \tag{41}$$

where $u_0(\mathbf{x})$ is the eigenfunction for $\mathbf{k} = 0$. (*a*) Find the wave equation satisfied by $\Psi_\mathbf{k}(\mathbf{x})$. (*b*) Consider an s-state at the point N in the Brillouin zone of the bcc lattice, Fig. 10.5. Show that at N the function $\Psi_N(x)$ satisfies the equation

$$u_0(\mathbf{x})\left[\frac{1}{2m}\,p^2 + (\varepsilon_0 - \varepsilon_N)\right]\Psi_N(\mathbf{x}) = \frac{1}{m}\,(\mathbf{p}u_0)\cdot(\mathbf{p}\Psi_N). \tag{42}$$

Now $\mathbf{p}u_0(\mathbf{x})$ is small over the outer part of the cell—why? (*c*) Show that $\mathbf{p}\Psi_N$ is approximately zero over the interior of the cell, and thus that

$$\varepsilon_N \cong \varepsilon_0 + \frac{1}{2m}\,k_N{}^2. \tag{43}$$

This result is due to Cohen and Heine, *Adv. in Phys.* **7,** 395 (1958).

4. Show that the exact eigenvalue equation for the harmonic oscillator with the Wigner-Seitz boundary condition at ξ_0 is

$$F(\tfrac{1}{4} - \tfrac{1}{4}\lambda|\tfrac{1}{2}|\xi_0{}^2) - (1 - \lambda)F(\tfrac{5}{4} - \tfrac{1}{4}\lambda|\tfrac{3}{2}|\xi_0{}^2) = 0, \tag{44}$$

where the hypergeometric function

$$F(a|c|z) = 1 + \frac{a}{c}\,z + \frac{a(a+1)}{2!c(c+1)}\,z^2 + \frac{a(a+1)(a+2)}{3!c(c+1)(c+2)}\,z^3 + \cdots.$$

(P. Farber)

5. The jacket on this book is an example of artistic license. In what respects does the color scheme violate the crystal symmetry requirements?

14 Semiconductor crystals: I. Energy bands, cyclotron resonance, and impurity states

In this and the following chapter we discuss the band structures of several important semiconductors, and we then treat phenomena which involve the structure at the band edge: cyclotron resonance, spin resonance, impurity states, optical transitions, oscillatory magnetoabsorption, and excitons.

ENERGY BANDS

The most important semiconductor crystals have the diamond structure or structures closely related to diamond. The diamond structure is based on a fcc bravais lattice (*ISSP*, pp. 36–37) with a basis of two atoms at 000; $\frac{1}{4}\,\frac{1}{4}\,\frac{1}{4}$, as shown in Fig. 1. The structures of the valence bands are similar in diamond, Si, and Ge, with the point of maximum energy at $\mathbf{k} = 0$. The point of maximum energy is called the *band edge.* The valence band edge for these crystals would be threefold degenerate (p-like) in the absence of spin and of spin-orbit interaction; with the spin-orbit interaction we shall see that this 3×2 = sixfold degenerate band edge splits into fourfold ($p_{3/2}$-like) and twofold ($p_{1/2}$-like) levels.

The valence electrons in the ground state of the free atoms have the configuration ns^2np^2, with $n = 2, 3, 4$ for diamond, Si, and Ge, respectively. In the crystal the ground state is formed from the configuration $nsnp^3$. Using the language of chemistry, we say that the valence electrons form directed sp^3 tetrahedral bonding orbitals of the form

$$s + p_x + p_y + p_z; \qquad s + p_x - p_y - p_z;$$
$$s - p_x + p_y - p_z; \qquad s - p_x - p_y + p_z.$$

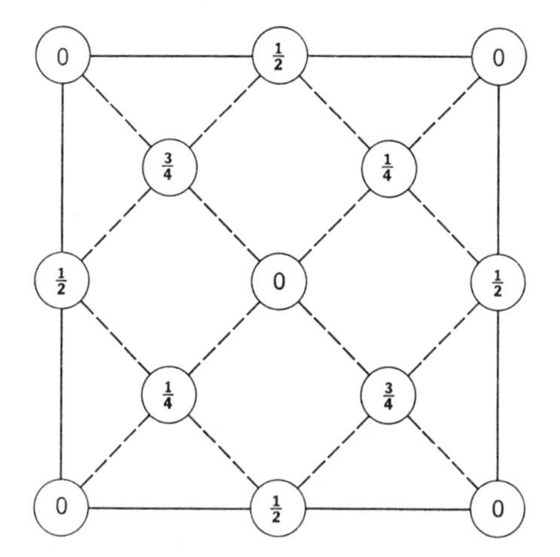

FIG. 1. Atomic positions in the unit cell of the diamond structure projected on a cube face; fractions denote height above base in units of a cube edge. The points at 0 and $\frac{1}{2}$ are on the fcc lattice; those at $\frac{1}{4}$ and $\frac{3}{4}$ are on a similar lattice displaced among the body diagonal by one-fourth of its length.

Each atom in the diamond structure is at the center of a tetrahedron with the nearest-neighbor atoms at the vertices. The four orbitals just enumerated have lobes pointing in the tetrahedral directions. These orbitals form the basis of a reducible representation of the tetrahedral point group $\bar{4}3m$; the representation may be reduced into the identical representation Γ_1 and the vector representation Γ_{15}. The representation Γ_1 is believed to occur at the bottom of the valence band (Fig. 2) at the center of the zone; Γ_1 is like an s state and is formed from the sum of the sp^3 orbitals above. Each of the two atoms in the primitive cell of the diamond structure furnishes one electron to the lowest band. This band turns out to be s-like at the points Γ, X, and L.

The valence band edge lies at the center of the zone and has the threefold representation Γ'_{25}, transforming as xy, yz, xz about the *center of the line joining the two atoms in the primitive cell.* The representation may be formed from p orbitals on the individual atoms, taken to be symmetrical with respect to inversion about the center of the line connecting the two atoms; the symmetrical combination is said to be *bonding.* The antibonding combination forms the representation Γ_{15} and in diamond lies about 5.7 ev above Γ'_{25}.

It is useful to consider the form of the wavefunctions at the points Γ in terms of a tight-binding model, also called a linear combination of atomic orbitals. The two interpenetrating fcc lattices of diamond are displaced from the other by the vector

$$\mathbf{t} = \tfrac{1}{4}a(1,1,1), \tag{1}$$

referred to the edges of the unit cube shown in Fig. 1. At $\mathbf{k} = 0$ the tight-binding functions have the form

$$\Psi_j^{\pm}(\mathbf{x}) = (2N)^{-\frac{1}{2}} \sum_n [\varphi_j(\mathbf{x} - \mathbf{x}_n) \pm \varphi_j(\mathbf{x} - \mathbf{x}_n - \mathbf{t})], \tag{2}$$

where $\mathbf{x}_n$ runs over all the lattice points of one fcc lattice; the φ_j are atomic or Wannier functions with $j = s,\ p_x,\ p_y,$ or p_z. The $\pm$ sign indicates the two independent ways in which the atomic functions may be combined on the two lattices. Tight-binding functions are not a good approximation to the actual wavefunctions, but they form an easy pictorial representation of the symmetry properties of the exact solutions. One may readily show by examining the transformation properties that Ψ_s^+ forms a representation of Γ_1; Ψ_s^- of Γ_2'; $\Psi_{x,y,z}^-$ of Γ_{25}'; and $\Psi_{x,y,z}^+$ of Γ_{15}.

We can understand qualitatively some of the features of the band structure of diamond by reference to the free-electron energy bands in an fcc bravais lattice, as illustrated in Fig. 10.8. We omit from the treatment the electrons of the $1s^2$ core, because these go into narrow Γ_1 and Γ_2' bands quite low in energy. The lowest point (Γ_1) shown in Fig. 10.8 is formed in the tight-binding approximation by taking $2s$ functions on each lattice, with a positive choice of the sign in (2):

$$\Psi(\Gamma_1) = (2N)^{-\frac{1}{2}} \sum_{\mathbf{n}} [\varphi_s(\mathbf{x} - \mathbf{x}_\mathbf{n}) + \varphi_s(\mathbf{x} - \mathbf{x}_\mathbf{n} - \mathbf{t})], \tag{3}$$

where $\mathbf{x}_\mathbf{n}$ runs over all lattice points of an fcc lattice. This combination is called bonding. There is no other plausible way of forming the low-lying state Γ_1.

Next in energy at Γ on the free-electron model are eight degenerate states having $\mathbf{G} = (2\pi/a)(\pm 1;\ \pm 1;\ \pm 1)$. The states belong to four different representations of the cubic group; in the crystal the eightfold degeneracy will be lifted and we will have the threefold levels Γ_{25}' and Γ_{15}, and the nondegenerate levels Γ_1 and Γ_2'. The Γ_1 component will require $3s$ orbitals and is therefore expected to lie quite high in energy.

We expect Γ_{25}' to lie below Γ_{15}. The arrangement of p orbitals along the shortest line joining two atoms is schematically as

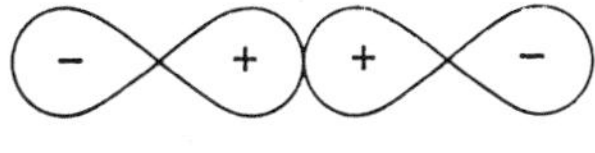

Bonding p orbitals

for Γ'_{25}; this arrangement is even under inversion at the center of the line. For Γ_{15} the arrangement must be odd under inversion:

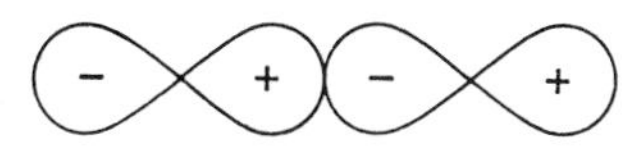

Antibonding p orbitals

but this contains fourier components at twice the wavevector as for Γ'_{25}. Our rough argument suggests that Γ'_{25} lies lower than Γ_{15}; in fact, in all crystals of the diamond and zinc blende structures that is believed to be realized.

We cannot order the Γ'_2 antibonding s level by the same argument. In diamond and silicon $\varepsilon(\Gamma_{15}) < \varepsilon(\Gamma'_2)$; the order is inverted in the heavier elements, probably because the stronger core potentials lower s relative to p.

STRUCTURE OF VALANCE BAND EDGE

The valence band edge in diamond-type crystals has a threefold orbital degeneracy; with spin the degeneracy is sixfold. The spin-orbit interaction lifts some of the degeneracy by splitting the p-like states into $p_{3/2}$ and $p_{1/2}$ states. In diamond (Table 1) the spin-orbit splitting Δ is estimated to be 0.006 ev, which is very much less than the band

TABLE 1

ENERGY-BAND DATA OF SEVERAL SEMICONDUCTOR CRYSTALS

(At helium temperature unless specified)

	Diamond	Si	Ge	InSb
E_g: minimum energy gap (ev)	5.33*	1.14	0.744	0.23
Vertical gap at $\mathbf{k} = 0$ (ev)		(2.5)	0.898	$0.23\text{–}10^{-4}$
Valence band width (ev)	(20)	17	7.0	
Δ: valence band spin-orbit splitting (ev)	0.006	0.04	0.29	(0.9)
m_l/m at conduction band edge		0.98	1.64	0.014
m_t/m at conduction band edge		0.19	0.082	0.014
Valence band edge parameters $2mA/\hbar^2$		−4.0	−13.1	
Valence band edge parameters $2m\|B\|/\hbar^2$		1.1	8.3	
Valence band edge parameters $2m\|C\|/\hbar^2$		4.1	12.5	

* Room temperature.

gap, 5.3 ev. As we advance along the periodic table, the spin-orbit splitting increases markedly and the energy gap may decrease. In InSb the spin-orbit splitting is 0.9 ev and the band gap is 0.23 ev.

Thus the spin-orbit splitting may be larger than the band gap; in heavy elements the splitting is one of the important factors determining the gap. Even in diamond the splitting is important for experiments on holes if their effective temperature is less than about 50°K. There is, however, a mathematical convenience in first setting up the $\mathbf{k} \cdot \mathbf{p}$ perturbation theory for the valence band edge with the neglect of spin-orbit interaction, and later including it.

We make an arbitrary choice of a basis for the representation Γ'_{25} at $\mathbf{k} = 0$, taking the three degenerate orbital states to transform as

$$\varepsilon'_1 \sim yz; \qquad \varepsilon'_2 \sim zx; \qquad \varepsilon'_3 \sim xy. \tag{4}$$

The second-order perturbation matrix (Schiff, pp. 156–158) has the form

$$\langle \varepsilon'_r | H'' | \varepsilon'_s \rangle = \frac{1}{m^2} \sum_{\delta}{}' \frac{\langle r | \mathbf{k} \cdot \mathbf{p} | \delta \rangle \langle \delta | \mathbf{k} \cdot \mathbf{p} | s \rangle}{\varepsilon_s - \varepsilon_\delta}, \tag{5}$$

where r, s denote 1, 2, or 3 above and the sum is over all states at $\mathbf{k} = 0$ except those in the valence band edge level under consideration. The dependence of $\langle r | H'' | s \rangle$ on the components of $\mathbf{k}$ is found by considering the form of the sum if all energy denominators were equal. Then by the completeness relation

$$\langle r | H'' | s \rangle \propto \langle r | (\mathbf{k} \cdot \mathbf{p})^2 | s \rangle; \tag{6}$$

with (4) we have, on examining the derivatives,

$$\langle 1 | H'' | 2 \rangle = 2k_x k_y \langle 1 | p_y p_x | 2 \rangle, \tag{7}$$

and similarly for the other matrix elements.

The secular equation is then of the form

$$\begin{vmatrix} Lk_x^2 + M(k_y^2 + k_z^2) - \lambda & Nk_x k_y & Nk_x k_z \\ Nk_x k_y & Lk_y^2 + M(k_x^2 + k_z^2) - \lambda & Nk_y k_z \\ Nk_x k_z & Nk_y k_z & Lk_z^2 + M(k_x^2 + k_y^2) - \lambda \end{vmatrix} = 0. \tag{8}$$

The cubic symmetry of the crystal enables us to express the coefficients in terms of the three constants L, M, N. The energy eigenvalue is given by $\varepsilon(\mathbf{k}) = \varepsilon(0) + (1/2m)k^2 + \lambda$. Expressions for L, M, N, as simplified by the use of symmetry, are given by Dresselhaus, Kip, and Kittel, *Phys. Rev.* **98,** 368 (1955).

To include spin-orbit effects, we take as the basis the six functions $\varepsilon_1'\alpha$, $\varepsilon_2'\alpha$, $\varepsilon_3'\alpha$, $\varepsilon_1'\beta$, $\varepsilon_2'\beta$, $\varepsilon_3'\beta$, where α, β are the spin functions. We include in the perturbation the spin-orbit interaction

$$H_{so} = \frac{1}{4m^2c^2}(\boldsymbol{\sigma} \times \text{grad } V) \cdot \mathbf{p}, \tag{9}$$

and neglect the corresponding term having $\mathbf{k}$ written for $\mathbf{p}$. Suppose that we transform from the basis just given to a basis in which the quantum numbers J,m_J are diagonal, where $\mathbf{J}$ is the operator for the total angular momentum. Then in the new 6×6 secular equation the spin-orbit interaction simply subtracts the splitting Δ from the two diagonal terms involving the states $|\frac{1}{2};\frac{1}{2}\rangle$ and $|\frac{1}{2};-\frac{1}{2}\rangle$.

If we restrict ourselves to energies $k^2/2m\Delta \ll 1$, the 6×6 secular equation has the approximate eigenvalues[1]

$$\varepsilon(\mathbf{k}) = Ak^2 \pm [B^2k^4 + C^2(k_x^2k_y^2 + k_y^2k_z^2 + k_x^2k_z^2)]^{1/2}; \tag{10}$$

$$\varepsilon(\mathbf{k}) = -\Delta + Ak^2. \tag{11}$$

There are three roots, given by (10) and (11); each root is double, as required by the time reversal and inversion invariance of the hamiltonian. The solutions (10) converge at $\mathbf{k} = 0$ to a fourfold degenerate state which belongs to the Γ_8 representation of the cubic group; the representation may be built up from $p_{3/2}$ atomic functions on each atom. The solutions (11) converge at $\mathbf{k} = 0$ to a twofold degenerate state which belongs to the Γ_7 representation; it may be viewed as built up from atomic $p_{1/2}$ functions. The band at Γ_7 is called the split-off band and lies lower in energy than Γ_8. The presence of Γ_7 was first deduced from the analysis of experiments on optical absorption in p-Ge. Values of A, B, C, as determined by cyclotron resonance, are given in Table 1. The form (10) is established at the end of this chapter.

Diamond. Representative theoretical calculations of the band structure of diamond are given by F. Herman, *Phys. Rev.* **93,** 1214 (1954); and J. C. Phillips and L. Kleinman, *Phys. Rev.* **116,** 287 (1959). The results are shown in Fig. 2. The valence band edge is Γ_{25}'. The conduction band edge is believed to lie along the Δ axis; the electron energy surfaces are six equivalent spheroids, one along each 100 axis. The calculated band gap of 5.4 ev agrees well with the observed 5.33 ev. It is known experimentally that the band edges are indirect; that is, the

[1] Solutions valid over a wider range of $\mathbf{k}$ have been given by E. O. Kane, *Phys. Chem. Solids* **1,** 82 (1956).

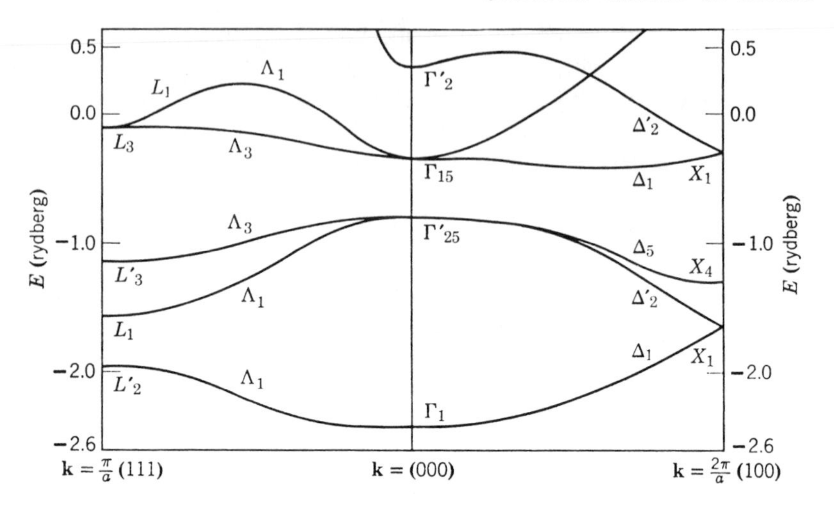

FIG. 2. Energy bands of diamond along (100) and (111) axes of the Brillouin zone. The valence band edge is at $\mathbf{k} = 0$ and has the representation Γ'_{25}; the conduction band edge should lie along Δ. (After Phillips and Kleinman.)

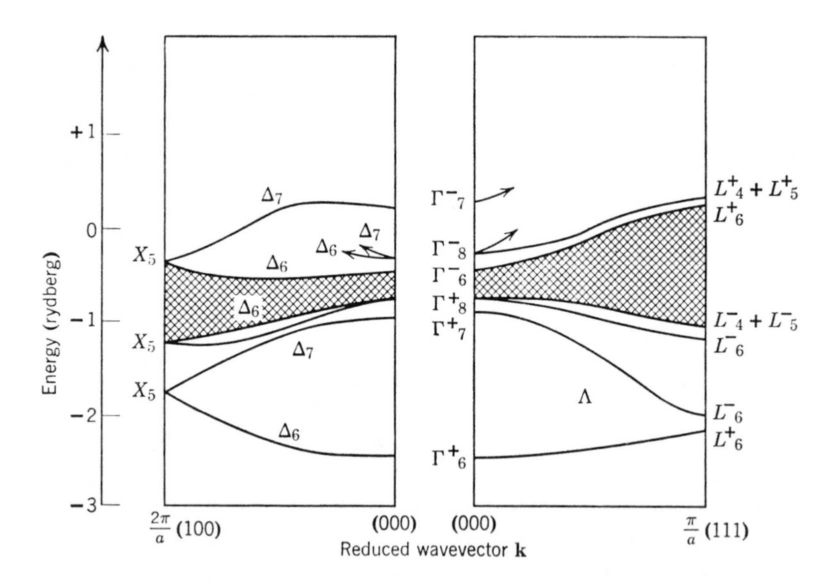

FIG. 3. Schematic representation of the energy bands of diamond including spin-orbit coupling effects. (Based on Herman's calculations, after R. J. Elliott.)

valence and conduction band edges are connected by a nonzero **k**. Cyclotron resonance experiments on p-type diamond give $m^*/m \cong 0.7$ and 2.2 for the light and heavy hole bands at the band edge and $m^*/m \cong 1.06$ for the band split off by the spin-orbit interaction. The splitting caused by spin-orbit interaction is shown in Fig. 3.

Silicon. The energy bands of silicon are shown in Fig. 4 without spin-orbit interaction. The band structure is similar to that of diamond: the valence band edge belongs to the representation Γ'_{25} and the conduction band edge belongs to Δ_1 at a general point along the axis between Γ and X. There is a good deal of evidence that one minimum is at $(2\pi/a)$ (0.86,0,0); there are six equivalent minima, one along each cube edge. The transverse and longitudinal effective masses are

$$m_t = 0.19m; \qquad m_l = 0.98m.$$

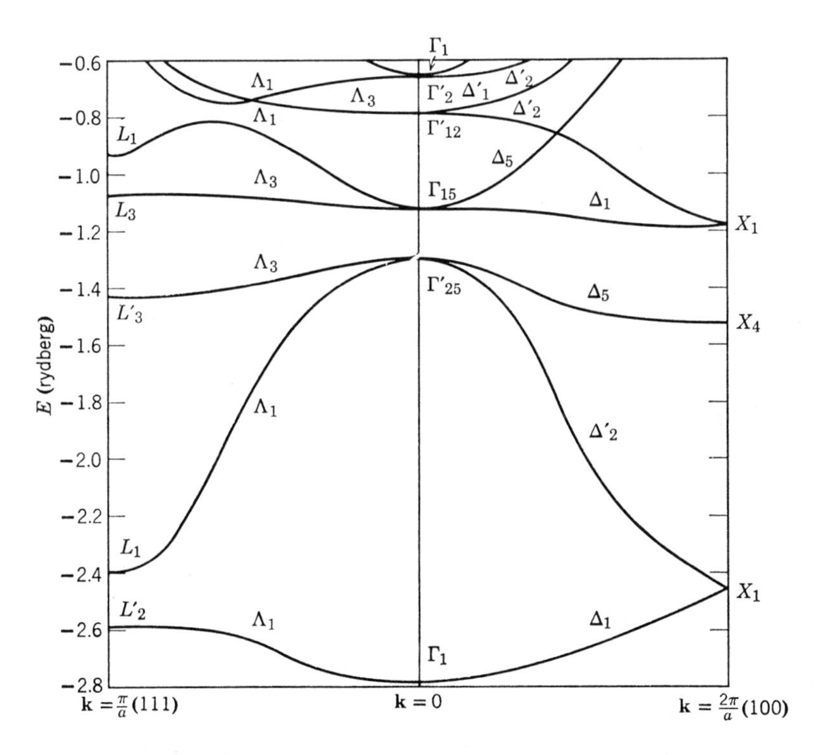

FIG. 4. Energy band structure of silicon. (After Kleinman and Phillips.) The valence band edge is at Γ'_{25}; the conduction band edge is along Δ_1. Spin-orbit interaction has not been considered in this figure.

The minimum energy gap is 1.14 ev; it does not occur vertically in an energy-band diagram in **k** space. The gap at the center of the zone between Γ_{15} and Γ'_{25} is believed to be about 2.5 ev. Values of the valence-band constants A, B, C calculated by Kleinman and Phillips [*Phys. Rev.* **118,** 1153 (1960)] are in excellent agreement with the values determined by cyclotron resonance.

Germanium. The band structure is shown in Fig. 5, without spin-orbit interaction. The valence-band edge is Γ'_{25}, with a spin-orbit splitting of 0.29 ev. The minimum of the conduction band occurs in the 111 directions at the edge of the zone; that is, the conduction-band edge occurs at the point L and is presumed to have the representation L_1. The effective masses of the prolate spheroidal surfaces are highly anisotropic: $m_l/m = 1.64$; $m_t/m = 0.082$. The effective mass in the state Γ'_2 at $\mathbf{k} = 0$ is isotropic and has the value $m^*/m = 0.036$; this state is normally vacant, but the splitting of the optical absorption

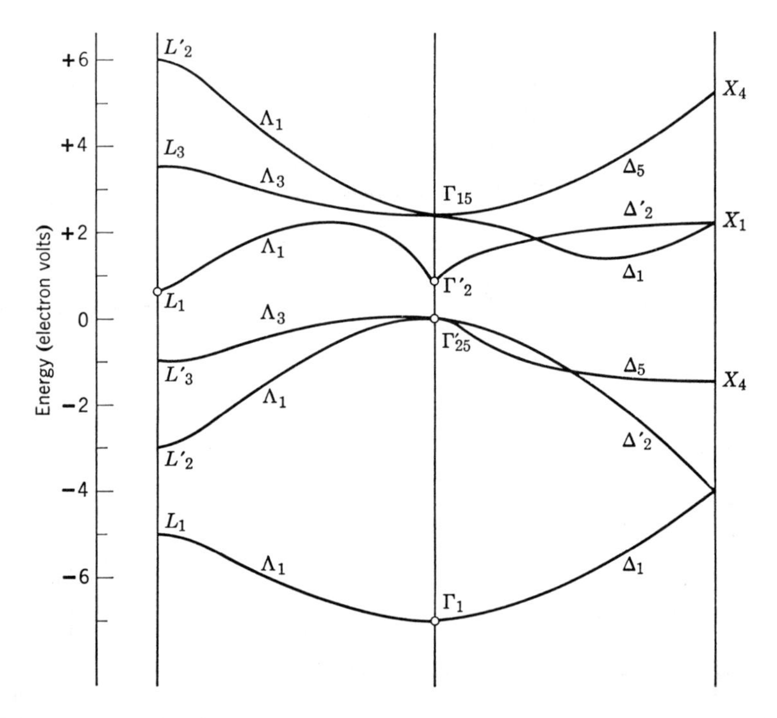

FIG. 5. Energy bands in germanium along the [100] and [111] axes according to the experimental information and the calculations of Herman. Spin-orbit coupling is neglected. Levels determined by experiment are circled. (After J. Callaway.)

in a magnetic field gives the mass. One major change which occurs in going from Si to Ge is that the lowest conduction-band state at $\mathbf{k} = 0$ changes from Γ_{15} in Si to the nondegenerate Γ_2' in Ge. In grey Sn, beyond Ge, it is believed that Γ_2' remains lower than Γ_{15}.

Indium Antimonide. The crystal InSb has the zinc blende structure, which differs from the diamond structure in an important respect. The diamond structure is composed of two identical interpenetrating fcc lattices, but in InSb one of the lattices contains In atoms and the other lattice contains Sb atoms. The chemical valences are 3 and 5; InSb is an example of 3-5 compound. The crystal symmetry resembles that of diamond, but no longer has an inversion center. Thus we can no longer say that the energy levels at fixed $\mathbf{k}$ have the twofold conjugation invariance. The time reversal operation K still commutes with the hamiltonian, so that $\varepsilon_{\mathbf{k}\uparrow}$ is degenerate with $\varepsilon_{-\mathbf{k}\downarrow}$. A number of changes in the band structure occur with respect to the corresponding 4-4 crystals because of the loss of the inversion J as a symmetry element; this introduces a component of the crystal potential antisymmetric with respect to a point midway between the two atoms in a cell.

In 3-5 crystals in which the lowest conduction-band state at $\mathbf{k} = 0$ belongs to Γ_{15} it is expected that the antisymmetric potential will mix the representation Γ_{25}' of the original valence-band edge and the representation Γ_{15}. The antisymmetric potential by itself will split these representations if, as for a plane wave model, they were degenerate; if the representations are already split, it will act to increase the splitting. The gap at $\mathbf{k} = 0$ in BN is calculated[2] to be about 10 ev, about twice as large as in diamond. We note that the gap observed in AlP is 3.0 ev, compared with 1.1 ev in Si. In InSb and grey tin the conduction band at $\mathbf{k} = 0$ is Γ_2'; the antisymmetric potential increases the gap from 0.07 ev in grey tin to 0.23 ev in InSb. The conduction-band edge is at $\mathbf{k} = 0$ in both crystals.[3]

The spin-orbit splitting Δ of the valence-band edge in InSb is estimated to be 0.9 ev, or about four times as large as the band gap. In this situation it makes no sense to treat the representations of the problem without spin. With spin, the valence-band edge belongs to the fourfold cubic representation Γ_8, with the split-off band belonging to Γ_7. The conduction-band edge may belong to Γ_6 or Γ_7, both twofold.

[2] L. Kleinman and J. C. Phillips, *Phys. Rev.* **117,** 460 (1960).

[3] For a detailed theoretical treatment of the band-edge structure of InSb, see G. Dresselhaus, *Phys. Rev.* **100,** 580 (1955), and E. O. Kane, *Phys. Chem. Solids* **1,** 249 (1956).

The $\mathbf{k} \cdot \mathbf{p}$ perturbation does not give a term in the energy first order in $\mathbf{k}$ for either of the twofold representations. This follows because $\mathbf{p}$ transforms as the vector representation Γ_V, and $\Gamma_V \times \Gamma_6$ does not contain Γ_6; similarly $\Gamma_V \times \Gamma_7$ does not contain Γ_7. Thus first-order matrix elements $\mathbf{k} \cdot \langle \Gamma_6 | \mathbf{p} | \Gamma_6 \rangle \equiv 0$, and $\mathbf{k} \cdot \langle \Gamma_7 | \mathbf{p} | \Gamma_7 \rangle \equiv 0$. There are second-order contributions to the energy, but the degeneracy is not lifted in this order. To third order for Γ_6 or Γ_7 the energy is

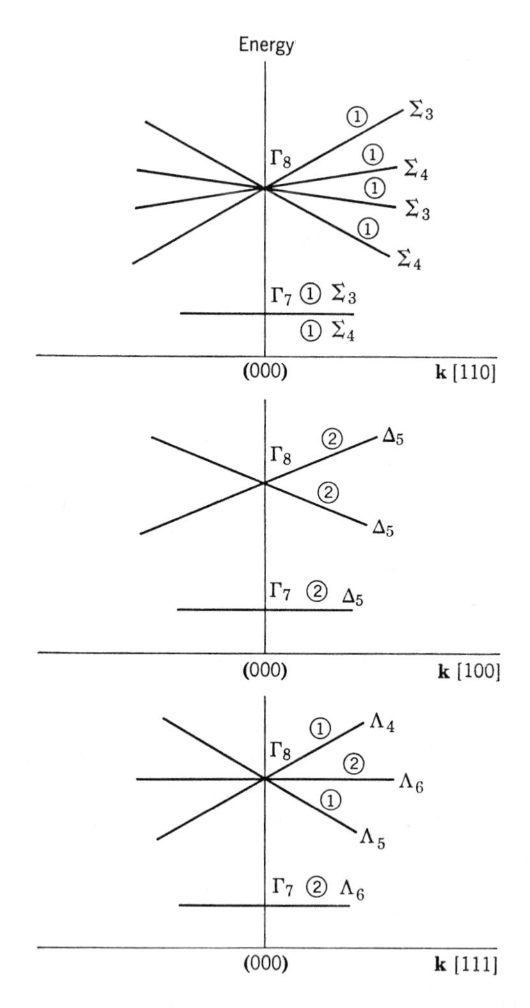

FIG. 6. Plot of energy versus wavevector in InSb showing the first-order energy for the spin-orbit split Γ'_{25} level in [100], [110], and [111] directions. The circled numbers indicate the dimension of the representation. (After Dresselhaus.)

split in all but the 100 and 111 directions:

$$(12) \quad \mathcal{E}(\mathbf{k}) = C_0k^2 \pm C_1[k^2(k_x{}^2k_y{}^2 + k_y{}^2k_z{}^2 + k_z{}^2k_x{}^2) - 9k_x{}^2k_y{}^2k_z{}^2]^{1/2}.$$

It is known from cyclotron resonance that $m^* = 1/2C_0 = 0.014m$.

The fourfold representation Γ_8 gives contributions to the energy in first order in $\mathbf{k}$, because $\Gamma_V \times \Gamma_8 = \Gamma_6 + \Gamma_7 + 2\Gamma_8$, which contains Γ_8. Thus there will be first-order matrix elements $\mathbf{k} \cdot \langle\Gamma_8|\mathbf{p}|\Gamma_8\rangle \neq 0$. *Very* close to $\mathbf{k} = 0$ the four bands at the valence-band edge have the general form, to first order in $\mathbf{k}$,

$$(13) \qquad \mathcal{E}(\mathbf{k}) = \pm C\{k^2 \pm [3(k_x{}^2k_y{}^2 + k_y{}^2k_z{}^2 + k_z{}^2k_x{}^2)]^{1/2}\}^{1/2}.$$

The four signs are independent. The splitting is shown in Fig. 6. The constant C is very small, and the linear region of (13) is soon dominated by the normal quadratic terms as in (10). The C terms shift the band edge slightly from $\mathbf{k} = 0$; the expectation is that there will be a nest of band edges in InSb along 111 directions about 0.003 of the way out to the zone boundary, with an energy at the maximum about 10^{-4} ev above the energy at $\mathbf{k} = 0$. For hole energies $\gg 10^{-4}$ ev the valence band of InSb will resemble that of Ge.

CYCLOTRON AND SPIN RESONANCE IN SEMICONDUCTORS, WITH SPIN-ORBIT COUPLING

We consider a conduction-band edge at $\mathbf{k} = 0$ in an orthorhombic crystal and suppose the band has only the twofold time reversal degeneracy. The crystal is assumed to have a center of inversion. In the absence of a magnetic field the energy to second order in $\mathbf{k}$ is

$$(14) \qquad \mathcal{E}(\mathbf{k}) = \Sigma\, D_{\alpha\beta}k_\alpha k_\beta, \qquad (\alpha, \beta = x, y, z).$$

The two-component Wannier effective wave equation (Chap. 9) in a magnetic field is written by viewing $\mathbf{k} = \mathbf{p} - (e/c)\mathbf{A}$ as an operator:

$$(15) \quad \left[\sum D_{\alpha\beta}\left(p_\alpha - \frac{e}{c}A_\alpha\right)\left(p_\beta - \frac{e}{c}A_\beta\right) - \mu_B\boldsymbol{\sigma}\cdot\mathbf{H}\right]\psi(\mathbf{x},s) = \mathcal{E}\psi(\mathbf{x},s),$$

where s is a spin coordinate; we have written the spin magnetic-moment operator as $\mu_B\boldsymbol{\sigma}$.

In using form (15) it is usually assumed implicitly that the components of $\mathbf{k}$ commute. In the absence of a magnetic field they do:

$$(16) \qquad [k_\alpha,k_\beta] = [p_\alpha,p_\beta] = 0;$$

but in a magnetic field the commutator of the k's includes $[p_\alpha, A_\beta]$

which is not necessarily zero. For the gauge $\mathbf{A} = H(0,x,0)$ we have

$$[k_x,k_z] = 0; \qquad [k_y,k_z] = 0; \tag{17}$$

$$[k_x,k_y] = \frac{eH}{c}[x,p_x] = i\frac{eH}{c}. \tag{18}$$

We need, therefore, to write (14) in a form which will permit the possibility of a contribution from the antisymmetric form $[k_x,k_y]$:

$$\varepsilon(\mathbf{k}) = \Sigma\,(D^S_{\alpha\beta}\{k_\alpha,k_\beta\} + D^A_{\alpha\beta}[k_\alpha,k_\beta]), \tag{19}$$

where in the summation each pair $\alpha\beta$ is to be taken only once; that is, $\beta\alpha$ is not counted if $\alpha\beta$ is counted. Here

$$\{k_\alpha,k_\beta\} = k_\alpha k_\beta + k_\beta k_\alpha, \tag{20}$$

D^S and D^A denote the symmetrical and antisymmetrical coefficients, respectively. In the absence of a magnetic field only the symmetrical term of (19) will contribute, as then $[k_\alpha,k_\beta] = 0$.

The coefficient $D_{\alpha\beta}$ is given by $\mathbf{k}\cdot\mathbf{p}$ perturbation theory:

$$D_{\alpha\beta} = \frac{1}{2m}\delta_{\alpha\beta} + \frac{1}{m^2}\sum_\delta{}' \frac{\langle\gamma|p_\alpha|\delta\rangle\langle\delta|p_\beta|\gamma\rangle}{\varepsilon_\gamma - \varepsilon_\delta} \equiv \frac{1}{2m}\delta_{\alpha\beta} + \frac{1}{m^2}\left\langle\gamma\left|p_\alpha\frac{1}{\varepsilon_\gamma - H_0}p_\beta\right|\gamma\right\rangle, \tag{21}$$

where γ denotes the state under consideration at $\mathbf{k} = 0$. Then

$$D^S_{\alpha\beta} = \tfrac{1}{2}(D_{\alpha\beta} + D_{\beta\alpha}); \qquad D^A_{\alpha\beta} = \tfrac{1}{2}(D_{\alpha\beta} - D_{\beta\alpha}); \tag{22}$$

$$D^A_{\alpha\beta} = \frac{1}{2m^2}\sum_\delta{}' \frac{\langle\gamma|p_\alpha|\delta\rangle\langle\delta|p_\beta|\gamma\rangle - \langle\gamma|p_\beta|\delta\rangle\langle\delta|p_\alpha|\gamma\rangle}{\varepsilon_\gamma - \varepsilon_\delta} = -D^A_{\beta\alpha}. \tag{23}$$

In our gauge and with the coordinate axes along the crystal axes, we have

$$D^S_{\alpha\beta} = D_{\alpha\alpha}\delta_{\alpha\beta}, \tag{24}$$

and the total antisymmetric contribution to $\varepsilon(\mathbf{k})$ is

$$D^A_{xy}[k_x,k_y] = D^A_{xy}\frac{ie}{c}H, \tag{25}$$

using (18). Thus

$$\varepsilon(\mathbf{k}) = D_{\alpha\alpha}k_\alpha k_\alpha + iD^A_{xy}\frac{e}{c}H - \mu_B\boldsymbol{\sigma}\cdot\mathbf{H}. \tag{26}$$

We now want to write D^A_{xy} in terms of the angular momentum component L_z.

The equation of motion of x in zero magnetic field is

$$(27) \qquad i\dot{x} = [x,H] = i\left(\frac{1}{m}\,p_x + \frac{1}{4m^2c^2}\,(\boldsymbol{\sigma} \times \text{grad } V)_x\right) \equiv i\,\frac{\pi_x}{m}.$$

The term following p_x arises from the spin-orbit term in the hamiltonian:

$$(28) \qquad H_{so} = \frac{1}{4m^2c^2}\,(\boldsymbol{\sigma} \times \text{grad } V)\cdot \mathbf{p},$$

where $V(\mathbf{x})$ is the periodic crystal potential; the commutator

$$(29) \qquad [x,H_{so}] = \frac{i}{4m^2c^2}\,(\boldsymbol{\sigma} \times \text{grad } V)_x.$$

The operator $\boldsymbol{\pi}$ is defined by (27), as in (9.29); for most purposes the term in $\boldsymbol{\sigma} \times \text{grad } V$ is a small correction to $\mathbf{p}$. The $\boldsymbol{\pi}$'s have essentially the properties with spin-orbit interaction which the $\mathbf{p}$'s have without it.

Now write (27) in a representation in which H is diagonal, for zero magnetic field:

$$(30) \qquad \frac{i}{m}\,\langle\gamma|\pi_x|\delta\rangle = \langle\gamma|x|\delta\rangle\varepsilon_\delta - \varepsilon_\gamma\langle\gamma|x|\delta\rangle.$$

If we neglect the difference between $\mathbf{p}$ and $\boldsymbol{\pi}$, then

$$(31) \qquad \sum_\delta{}' \frac{\langle\gamma|p_\alpha|\delta\rangle\langle\delta|p_\beta|\gamma\rangle - \langle\gamma|p_\beta|\delta\rangle\langle\delta|p_\alpha|\gamma\rangle}{\varepsilon_\gamma - \varepsilon_\delta}$$
$$= im \sum_\delta{}' \left(\langle\gamma|x_\alpha|\delta\rangle\langle\delta|p_\beta|\gamma\rangle - \langle\gamma|x_\beta|\delta\rangle\langle\delta|p_\alpha|\gamma\rangle\right) = im\langle\gamma|L_{\alpha\times\beta}|\gamma\rangle;$$

here we have made use of $\langle 0|x|0\rangle = 0$, which follows by parity for $\mathbf{k} = 0$ if the crystal has a center of inversion. If $\alpha \equiv x$; $\beta \equiv y$, then L_z is the component of the orbital angular momentum $\mathbf{L}$ in (31).

With (23) and (31), we have

$$(32) \qquad iD^A_{xy} = -\frac{1}{2m}\,\langle\gamma|L_z|\gamma\rangle.$$

Now if $|C\gamma\rangle$ is the state conjugate to $|\gamma\rangle$ in the sense introduced in (9.44), then

$$(33) \quad \langle C\gamma|L_z|C\gamma\rangle = \langle\gamma|C^{-1}L_zC|\gamma\rangle = -\langle\gamma|L_zC^{-1}C|\gamma\rangle = -\langle\gamma|L_z|\gamma\rangle,$$

because $CL_z = -L_z$; $C^{-1}L_z = -L_z$, according to (9.135). Therefore we may write

$$(34)\qquad \mathcal{E}(\mathbf{k}) = D_{\alpha\alpha}k_\alpha k_\alpha - \frac{e}{2mc}\langle\gamma|L_z|\gamma\rangle\sigma_z H - \mu_B\sigma_z H,$$

because γ and $C\gamma$ have opposite spins. Thus

$$(35)\qquad \boxed{\mathcal{E}(\mathbf{k}) = D_{\alpha\alpha}k_\alpha k_\alpha - \mu^*\boldsymbol{\sigma}\cdot\mathbf{H},}$$

where the anomalous magnetic moment μ^* is defined by

$$(36)\qquad \frac{\mu^*}{\mu_B} = (\langle\gamma|L_z|\gamma\rangle + 1)$$

$$= 1 + \frac{1}{im}\sum_\delta{}' \frac{\langle\gamma|p_x|\delta\rangle\langle\delta|p_y|\gamma\rangle - \langle\gamma|p_y|\delta\rangle\langle\delta|p_x|\gamma\rangle}{\mathcal{E}_\gamma - \mathcal{E}_\delta},$$

$$(37)\qquad \frac{\mu^*}{\mu_B} = 1 + \frac{1}{2im}\mathscr{I}\left(\sum_\delta{}' \frac{\langle\gamma|p_x|\delta\rangle\langle\delta|p_y|\gamma\rangle}{\mathcal{E}_\gamma - \mathcal{E}_\delta}\right),$$

where $\mathscr{I}$ signifies "the imaginary part of"; here we have used the fact that $\mathbf{p}$ is hermitian. The terms $D_{\alpha\alpha}k_\alpha k_\alpha$ in (35) gives the splitting observed in cyclotron resonance; the term $-\mu^*\boldsymbol{\sigma}\cdot\mathbf{H}$ gives the splitting observed in spin resonance.

We now consider the anomalous magnetic moment for a specific model which is quite typical of many semiconductors. In the model we are concerned with the g value and magnetic moment of the state $|0\gamma\rangle$ which we suppose is an s-like state at the conduction-band edge with spin $\uparrow$. The state lies an energy E_g above a $p_{3/2}$ valence-band edge, also at $\mathbf{k} = 0$; at an energy Δ below this there lies the level $p_{1/2}$, split off by the spin-orbit interaction. We assume that there are interactions only among these bands. In the absence of spin-orbit interaction Δ is zero and then $\langle 0|L_z|0\rangle$ vanishes because the representation can then be chosen so that either $\langle\gamma|p_x|\delta\rangle = 0$ or $\langle\gamma|p_y|\delta\rangle = 0$ for any state δ in the representation. To see this, let $\delta = x, y, z$.

We may represent schematically the states $|J;m_J\rangle$ by

(38)

$$|\tfrac{3}{2};\tfrac{3}{2}\rangle = 2^{-1/2}(x+iy)\alpha; \qquad |\tfrac{3}{2};\tfrac{1}{2}\rangle = 6^{-1/2}[2z\alpha + (x+iy)\beta];$$

$$|\tfrac{3}{2};-\tfrac{3}{2}\rangle = 2^{-1/2}(x-iy)\beta); \qquad |\tfrac{3}{2};-\tfrac{1}{2}\rangle = 6^{-1/2}[2z\beta - (x-iy)\alpha];$$

$$|\tfrac{1}{2};\tfrac{1}{2}\rangle = 3^{-1/2}[z\alpha - (x+iy)\beta]; \qquad |\tfrac{1}{2};-\tfrac{1}{2}\rangle = 3^{-1/2}[z\beta + (x-iy)\alpha].$$

The phases of these states satisfy $K|J;m_J\rangle = |J;-m_J\rangle$, where $K = -i\sigma_y K_0$ is the Kramers time reversal operator. Note that these are not the phases obtained from successive applications of the J^- lowering operator. Now

$$
\begin{aligned}
p_x|\tfrac{3}{2};\tfrac{3}{2}\rangle &= -i2^{-1/2}\alpha; & p_y|\tfrac{3}{2};\tfrac{3}{2}\rangle &= 2^{-1/2}\alpha; \\
p_x|\tfrac{3}{2};-\tfrac{1}{2}\rangle &= -i6^{-1/2}\beta; & p_y|\tfrac{3}{2};-\tfrac{1}{2}\rangle &= 6^{-1/2}\beta; \\
p_x|\tfrac{3}{2};-\tfrac{1}{2}\rangle &= i6^{-1/2}\alpha; & p_y|\tfrac{3}{2};-\tfrac{1}{2}\rangle &= 6^{-1/2}\alpha; \\
p_x|\tfrac{3}{2};-\tfrac{3}{2}\rangle &= -i2^{-1/2}\beta; & p_y|\tfrac{3}{2};-\tfrac{3}{2}\rangle &= -2^{-1/2}\beta.
\end{aligned}
\tag{39}
$$

On taking the appropriate matrix elements with $\langle\gamma|$ we see that the contribution to $\langle\gamma|L_z|\gamma\rangle$ from the states with $J = \frac{3}{2}$ is $-2/3mE_g$. The contribution from the states with $J = \frac{1}{2}$ is found using

$$
\begin{aligned}
p_x|\tfrac{1}{2};\tfrac{1}{2}\rangle &= i3^{-1/2}\beta; & p_y|\tfrac{1}{2};\tfrac{1}{2}\rangle &= -3^{-1/2}\beta; \\
p_x|\tfrac{1}{2};-\tfrac{1}{2}\rangle &= -i3^{-1/2}\alpha; & p_y|\tfrac{1}{2};-\tfrac{1}{2}\rangle &= -3^{-1/2}\alpha;
\end{aligned}
\tag{40}
$$

the contribution is $\frac{2}{3}m(E_g + \Delta)$. Thus

$$
\langle\gamma|L_z|\mu\rangle = \frac{2}{3m}\left(\frac{1}{E_g + \Delta} - \frac{1}{E_g}\right)|\langle 0|p_x|X\rangle|^2, \tag{41}
$$

where X denotes symbolically the state $x\alpha$ in the $x,y,z;\alpha,\beta$ representation of the valence-band edge.

The effective-mass tensor for the conduction-band edge is given by, for $m^* \ll m$,

$$
\frac{m}{m^*} \cong \frac{2}{m}\sum_{\delta}\frac{|\langle\gamma|p_x|\delta\rangle|^2}{\varepsilon_\gamma - \varepsilon_\delta} = \frac{2}{m}\left[\frac{2}{3E_g} + \frac{1}{3(E_g + \Delta)}\right]|\langle 0|p_x|X\rangle|^2, \tag{42}
$$

with the matrix elements used for (41). Thus we have the Roth relation

$$
\langle 0|L_z|0\rangle \cong -\frac{m}{m^*}\left(\frac{\Delta}{3E_g + 2\Delta}\right), \tag{43}
$$

for $m^* \ll m$. For InSb, $m/m^* \cong 70$; $E_g \cong 0.2$ ev; $\Delta \cong 0.9$ ev, so that $\langle 0|L_z|0\rangle \approx 25$. In Ge and Si the orbital moments are much smaller.

The g value in conduction-electron spin resonance is defined as

$$
g = 2\mu^*/\mu_B \cong -\frac{m}{m^*}\left(\frac{2\Delta}{3E_g + 2\Delta}\right) + 2 \approx -50, \tag{44}
$$

for InSb, in good agreement with experiment. Here the effective hamiltonian is

$$
H = \frac{1}{2m^*}\left(\mathbf{p} - \frac{e}{c}\mathbf{A}\right)^2 - \mu^*\boldsymbol{\sigma}\cdot\mathbf{H}, \tag{45}
$$

The crystal having the narrowest band gap known at present[4] is $Cd_xHg_{1-x}Te$; with $x = 0.136$, the energy gap between conduction and valence bands is believed to be $\leqq 0.006$ ev; further, $m^*/m \leqq 0.0004$ and $g \geqq 2500$ at the bottom of the conduction band. The low mass and high g value are direct consequences of the low value of the energy gap, according to $\mathbf{k} \cdot \mathbf{p}$ perturbation theory.

A careful detailed study of the g values of conduction electrons is given by Y. Yafet, *Solid state physics* **14**, 1 (1963). The method which led to (35) was developed by J. M. Luttinger, *Phys. Rev.* **102,** 1030 (1956), for a more difficult problem, the study of magnetic effects on the $p_{3/2}$ band edge of a diamond-type semiconductor. He found that the hamiltonian in a magnetic field to terms quadratic in $\mathbf{k}$ may be written as

$$(46)\quad H = \beta_1 k_\alpha k_\alpha + \beta_2 k_\alpha k_\alpha J_\alpha J_\alpha + 4\beta_3(\{k_x,k_y\}\{J_x,J_y\} + \{k_y,k_z\}\{J_y,J_z\} + \{k_z,k_x\}\{J_z,J_x\}) + \beta_4 H_\alpha J_\alpha + \beta_5 H_\alpha J_\alpha J_\alpha J_\alpha,$$

where J_x, J_y, J_z are 4×4 matrices which satisfy $\mathbf{J} \times \mathbf{J} = i\mathbf{J}$. In (46) the $\{\quad\}$ denote anticommutator, as usual.

Valence-Band Edge with Spin-Orbit Interaction. We want to derive the form (10) of the energy near the Γ_8 valence-band edge of a crystal with diamond structure. In the absence of a magnetic field the second-order hamiltonian has the form (46) of the 4×4 matrix

$$(47)\quad H = \beta_1 k^2 + \beta_2(k_x^2 J_x^2 + k_y^2 J_y^2 + k_z^2 J_z^2) + 4\beta_3(\{k_x,k_y\}\{J_x,J_y\} + \{k_y,k_z\}\{J_y,J_z\} + \{k_z,k_x\}\{J_z,J_x\}).$$

Here $\mathbf{J}$ is a 4×4 matrix which satisfies the angular momentum commutation relation $\mathbf{J} \times \mathbf{J} = i\mathbf{J}$. The expression (47) contains all the forms quadratic in the $\mathbf{k}$'s and $\mathbf{J}$'s which are invariant under the operations of the cubic point groups. The dimensionality of the matrix representing $\mathbf{J}$ is 4×4 because the Γ_8 state is fourfold degenerate.

We know that H is invariant under the conjugation operation, so that every root of the eigenvalue problem is double. We now give a procedure due to Hopfield for the reduction of (47) to a 2×2 matrix for the two independent roots.

In the basis $|Jm_J\rangle$ for $J = \frac{3}{2}$ as given by (38) the time reversal operator is represented by

$$(48)\qquad K = \begin{pmatrix} 0 & 0 & 0 & 1 \\ 0 & 0 & 1 & 0 \\ 0 & -1 & 0 & 0 \\ -1 & 0 & 0 & 0 \end{pmatrix} K_0,$$

[4] T. C. Harman et al., *Bull. Amer. Phys. Soc.* **7,** 203 (1962).

where K_0 denotes complex conjugation. This is easily demonstrated: if

$$\varphi = \begin{pmatrix} a \\ b \\ c \\ d \end{pmatrix} \tag{49}$$

is an eigenvector of H, then

$$K\varphi = K\begin{pmatrix} a \\ b \\ c \\ d \end{pmatrix} = \begin{pmatrix} d^* \\ c^* \\ -b^* \\ -a^* \end{pmatrix} \tag{50}$$

is an eigenvector having the same energy, because K commutes with the hamiltonian. But the states φ and $K\varphi$ are independent:

$$\begin{pmatrix} d & c & -b & -a \end{pmatrix}\begin{pmatrix} a \\ b \\ c \\ d \end{pmatrix} = 0. \tag{51}$$

We may then combine φ and $K\varphi$ to form a state having the same energy, but with one coefficient, say d, equal to zero:

$$\varphi' = (1 + \rho e^{i\alpha}K)\varphi, \tag{52}$$

where ρ and α are constants. We write

$$\varphi' = \begin{pmatrix} a' \\ b' \\ c' \\ 0 \end{pmatrix}. \tag{53}$$

Then, because

$$H_{\mu\nu}\varphi_\nu = \lambda\varphi_\nu, \tag{54}$$

we have

$$H_{41}a' + H_{42}b' + H_{43}c' = 0; \qquad c' = -\frac{H_{41}}{H_{43}}a' - \frac{H_{42}}{H_{43}}b'. \tag{55}$$

Substitute this for c' and we have the redundant 3×3 equation

$$\begin{pmatrix} H_{11} & H_{12} & H_{13} \\ H_{21} & \cdots & \cdots \\ H_{31} & \cdots & \cdots \end{pmatrix}\begin{pmatrix} a' \\ b' \\ -\dfrac{H_{41}}{H_{43}}a \quad -\dfrac{H_{42}}{H_{43}}b \end{pmatrix} = \lambda\begin{pmatrix} \cdot \\ \cdot \\ \cdot \end{pmatrix}. \tag{56}$$

The first two components of this equation are sufficient. They may be written

$$\begin{pmatrix} H_{11} - \dfrac{H_{41}}{H_{43}} H_{13} & H_{12} - \dfrac{H_{42}}{H_{43}} H_{13} \\ H_{21} - \dfrac{H_{41}}{H_{43}} H_{23} & H_{22} - \dfrac{H_{42}}{H_{43}} H_{23} \end{pmatrix} \begin{pmatrix} a \\ b \end{pmatrix} = \lambda \begin{pmatrix} a \\ b \end{pmatrix}. \tag{57}$$

With the representation (39) of the J's we have that

$$H_{41} = H_{23} = 0; \qquad H_{21} = -H_{43}; \qquad H_{42} = -H_{13}^*, \tag{58}$$

so that (57) has the solutions

$$\lambda = \tfrac{1}{2}(H_{11} + H_{22}) \pm [\tfrac{1}{4}(H_{11} - H_{22})^2 + |H_{12}|^2 + |H_{13}|^2]^{1/2}, \tag{59}$$

which is equivalent to the standard form (10).

IMPURITY STATES AND LANDAU LEVELS IN SEMICONDUCTORS

We are now concerned with the theory of the shallow donor and acceptor states associated with impurities in semiconductors, particularly trivalent and pentavalent impurities in germanium and silicon. The ionization energies of these impurities are of the order of 0.04 ev in Si and 0.01 ev in Ge. Such energies are much less than the energy gap; thus it is reasonable to expect the impurity states to be formed from one-particle states of the appropriate band, conduction or valence. The impurity states will in a sense be hydrogen-like, but loosely bound, largely because the dielectric constant ϵ of the medium is high. The rydberg constant involves $1/\epsilon^2$; for $\epsilon = 15$ the binding energy will be reduced from that of hydrogen by the factor $\frac{1}{225}$. When $m^* < m$ there is a further reduction of the binding energy.

The Wannier theorem gives us the effective hamiltonian for the problem. We treat first a simplified model of a pentavalent impurity in silicon; we consider for the conduction band the single spheroidal energy surface

$$\mathcal{E}(\mathbf{k}) = \frac{1}{2m_t}(k_x^2 + k_y^2) + \frac{1}{2m_l} k_z^2. \tag{60}$$

In the actual crystal there are six equivalent spheroids, each along a [100] axis. The Wannier equation associated with (60) is

$$\left[\frac{1}{2m_t}(p_x^2 + p_y^2) + \frac{1}{2m_l} p_y^2 - \frac{e^2}{\epsilon r}\right] F(\mathbf{x}) = EF(\mathbf{x}), \tag{61}$$

in the absence of a magnetic field.

We now examine the validity of (61). First, there is the question of the dielectric constant. It is fairly obvious that we should use the dielectric constant $\epsilon(\omega)$ or, better, $\epsilon(\omega,\mathbf{q})$ as measured at the frequency ω corresponding to the energy E of the impurity level referred to the band edge. In situations of interest to us this energy is smaller than the band gap, so that the electronic polarizability will contribute to $\epsilon(\omega)$ in full. The ionic polarizability will contribute only if the binding energy of the impurity level is small in comparison with the optical phonon frequency near $\mathbf{k} = 0$.

We now consider the validity of the effective mass approach itself. The Schrödinger equation of one electron in the perturbed periodic lattice is

$$(H_0 + V)\Psi = E\Psi, \tag{62}$$

where H_0 refers to the perfect lattice and V to the impurity. Here Ψ is just a one-electron wavefunction. Consider the solution of the unperturbed problem

$$H_0\varphi_{\mathbf{k}l} = \varepsilon_\gamma(\mathbf{k})\varphi_{\mathbf{k}l}, \tag{63}$$

where

$$\varphi_{\mathbf{k}l} \equiv |\mathbf{k}l\rangle = e^{i\mathbf{k}\cdot\mathbf{x}}u_{\mathbf{k}l}(\mathbf{x}) \tag{64}$$

is the Bloch function with wavevector $\mathbf{k}$ and band index l. We assume the band is not degenerate. The solutions $\Psi(\mathbf{x})$ of the perturbed problem may be written

$$\Psi(\mathbf{x}) = \sum_{\mathbf{k}'l'} |\mathbf{k}'l'\rangle\langle l'\mathbf{k}'|\rangle. \tag{65}$$

We substitute (65) in the Schrödinger equation (62) and take the scalar product with $\langle l\mathbf{k}|$, finding directly the secular equation

$$\varepsilon_l(\mathbf{k})\langle l\mathbf{k}|\rangle + \Sigma \langle l\mathbf{k}|V|\mathbf{k}'l'|\rangle\langle l'\mathbf{k}'|\rangle = E\langle l\mathbf{k}|\rangle. \tag{66}$$

We next expand the perturbation V in a fourier series:

$$V = \sum_{\mathbf{K}} V_{\mathbf{K}}e^{i\mathbf{K}\cdot\mathbf{x}}, \tag{67}$$

whence

$$\langle l\mathbf{k}|V|\mathbf{k}'l'\rangle = \sum_{\mathbf{K}} V_{\mathbf{K}} \int d^3x\, e^{i(\mathbf{k}'-\mathbf{k}+\mathbf{K})\cdot\mathbf{x}}u^*_{\mathbf{k}l}u_{\mathbf{k}'l'}. \tag{68}$$

Because $u_{\mathbf{k}l}(\mathbf{x})$ is periodic in the direct lattice, the integral vanishes unless

$$\mathbf{k} = \mathbf{k}' + \mathbf{K} + \mathbf{G}. \tag{69}$$

We are concerned only with small **k**, **k**′, and **K**, so that **G** = 0 for the matrix elements of interest. We note that for the coulomb potential $V_{\mathbf{K}} \propto 1/K^2$. The secular equation may be written

$$\varepsilon_l(\mathbf{k})\langle l\mathbf{k}|\rangle + \sum_{\mathbf{K}l'} V_{\mathbf{K}}\Delta^{ll'}_{\mathbf{k}+\mathbf{K},\mathbf{k}}\langle l',\mathbf{k}+\mathbf{K}|\rangle = E\langle l\mathbf{k}|\rangle, \tag{70}$$

where the function

$$\Delta^{ll'}_{\mathbf{k}+\mathbf{K},\mathbf{k}} = \int d^3x\; u^*_{\mathbf{k}+\mathbf{K},l}(\mathbf{x})u_{\mathbf{k}l'}(\mathbf{x}). \tag{71}$$

As $|\mathbf{K}| \rightarrow 0$,

$$\Delta^{ll'}_{\mathbf{k}+\mathbf{K},\mathbf{k}} \rightarrow \delta_{ll'}. \tag{72}$$

In this limit the secular equation reduces to

$$\varepsilon_l(\mathbf{k})\langle l\mathbf{k}|\rangle + \sum_{\mathbf{K}} V_{\mathbf{K}}\langle l,\mathbf{k}+\mathbf{K}|\rangle = E\langle l\mathbf{k}|\rangle. \tag{73}$$

The use of (72) is our central approximation. In this approximation the different bands are entirely independent. The secular equation (73) is precisely the Schrödinger equation in the momentum representation of the following Wannier problem in a coordinate representation:

$$[\varepsilon_l(\mathbf{p}) + V(\mathbf{x})]F_l(\mathbf{x}) = EF_l(\mathbf{x}), \tag{74}$$

where

$$F_l(\mathbf{x}) = \sum_{\mathbf{k}} e^{i\mathbf{k}\cdot\mathbf{x}}\langle l\mathbf{k}|\rangle. \tag{75}$$

We suppose we have solved (74) for $F_l(\mathbf{x})$. In a specimen of unit volume

$$\langle l\mathbf{k}|\rangle = \int d^3x\, F_l(\mathbf{x})e^{-i\mathbf{k}\cdot\mathbf{x}}, \tag{76}$$

so that the solutions $\Psi_l(\mathbf{x})$ of our original problem are

$$\Psi_l(\mathbf{x}) = \sum_{\mathbf{k}} \varphi_{\mathbf{k}l}(\mathbf{x}) \int d^3\xi\, F_l(\xi)e^{-i\mathbf{k}\cdot\xi}, \tag{77}$$

where F_l is an eigenfunction of the Wannier problem, Eq. (74).

For slowly varying perturbations only a small range of **k** will enter the solution for low-lying states in a given band. If we make the approximation that $u_{\mathbf{k}l}$ may be replaced by $u_{0l}(\mathbf{x})$ in (77), then

$$\boxed{\Psi_l(\mathbf{x}) \cong u_{0l}(\mathbf{x}) \int d^3\xi\, F_l(\xi) \sum_{\mathbf{k}} e^{i\mathbf{k}\cdot(\mathbf{x}-\xi)} = u_{0l}(\mathbf{x})F_l(\mathbf{x}),} \tag{78}$$

because

$$\sum_{\mathbf{k}} e^{i\mathbf{k}\cdot(\mathbf{x}-\boldsymbol{\xi})} \cong \frac{1}{(2\pi)^3} \int d^3k \, e^{i\mathbf{k}\cdot(\mathbf{x}-\boldsymbol{\xi})} = \delta(\mathbf{x} - \boldsymbol{\xi}). \tag{79}$$

This displays the role of $F_l(\mathbf{x})$ as a slow modulation on $u_0(\mathbf{x})$. The replacement of $u_{\mathbf{k}l}(\mathbf{x})$ by $u_{0l}(\mathbf{x})$ introduces no approximation more important than we have made already in neglecting the interband mixing terms in (72).

The Wannier equation (74) is seen to be rigorous in an approximation which can be stated precisely, namely that (73) should hold. Kittel and Mitchell, *Phys. Rev.* **96,** 1488 (1954), show for $l \neq l'$ that

$$\Delta^{l'l} \approx \left(\frac{\text{impurity ionization energy}}{\text{band gap}}\right)^{1/2}, \tag{80}$$

which may be of the order of 0.1 for Si and perhaps less for Ge.

The method is easily extended to degenerate bands, where one deals with coupled Wannier equations connecting the several degenerate components. For a discussion of the acceptor levels in Si and Ge, and for a treatment of short-range effects within the atomic core of the impurity, the reader should consult the review by W. Kohn, *Solid state physics* **5,** 258 (1957).

We go on to the solution of (61). For a spherical energy surface $m_l = m_t = m^*$, and we have exactly the hydrogen-atom problem with e^2/ϵ written for e^2 and m^* written for m. The anisotropic hamiltonian (61) is not exactly solvable in closed form. We may determine an upper bound to the ground-state energy relative to the band edge by a variational calculation. With $m_l = \alpha_l m$; $m_t = \alpha_t m$; and $r_0 = \epsilon/me^2$, we try a variational function of the form

$$F(\mathbf{x}) = (ab^2/\pi r_0{}^3)^{1/2} \exp\{-[a^2z^2 + b^2(x^2 + y^2)]^{1/2}/r_0\}. \tag{81}$$

We find on carrying out the variational calculation the following results:

For n-Ge with $\alpha_1 = 1.58$; $\alpha_2 = 0.082$; $\varepsilon = 16$:

$$E_0 = -0.00905 \text{ ev}; \qquad a^2 = 0.135; \qquad b^2 = 0.0174. \tag{82}$$

For n-Si with $\alpha_1 = 1$; $\alpha_2 = 0.2$; $\varepsilon = 12$:

$$E_0 = -0.0298 \text{ ev}; \qquad a^2 = 0.216; \qquad b^2 = 0.0729. \tag{83}$$

The theory may be generalized directly for degenerate band edges. The Wannier equation becomes an equation for a multicomponent wave function $\bar{\bar{F}}$:

$$\overline{\overline{(H(p)}} + V)\overline{\overline{F(x)}} = E\overline{\overline{F(x)}}, \tag{84}$$

where $\overline{\overline{H(p)}}$ is the square matrix from second-order $\mathbf{k} \cdot \mathbf{p}$ perturbation theory and $\overline{\overline{F(x)}}$ is a column matrix.

LANDAU LEVELS

By Landau levels we mean the quantized orbits of a free particle in a crystal in a magnetic field. In the chapter on electron dynamics in a magnetic field we gave the Landau solution for a free particle in a vacuum in a magnetic field, and we considered the semiclassical theory of magnetic orbits on general fermi surfaces. At present we consider only the quantization of a spinless electron near the non-degenerate conduction band edge of a semiconductor having the spheroidal energy surface

$$\varepsilon(\mathbf{k}) = \frac{1}{2m_t}(k_x{}^2 + k_y{}^2) + \frac{1}{2m_l}k_z{}^2, \tag{84a}$$

in the absence of the magnetic field.

The hamiltonian with the magnetic field is

$$H = \frac{1}{2m}\left(\mathbf{p} - \frac{e}{c}\mathbf{A}\right)^2 + V(\mathbf{x}); \tag{85}$$

the vector potential for a uniform magnetic field $\mathcal{H}$ in the z direction in the Landau gauge is

$$\mathbf{A} = (0,\mathcal{H}x,0). \tag{86}$$

If we write

$$s = e\mathcal{H}/c, \tag{87}$$

the hamiltonian is

$$H = H_0 - \frac{s}{m}xp_y + \frac{s^2}{2m}x^2. \tag{88}$$

The eigenfunctions of H_0 are the Bloch functions $|\mathbf{k},l\rangle$; the eigenvalues are $\varepsilon_l(\mathbf{k})$.

Because of the presence of terms in x and x^2 it is a difficult and lengthy problem to analyze the validity of the effective mass equation in a uniform magnetic field. In order to avoid singular matrix elements, we consider instead a static vector potential of the form

$$\mathbf{A}(\mathbf{x}) = i\mathbf{A}_{\mathbf{q}}(e^{i\mathbf{q}\cdot\mathbf{x}} - e^{-i\mathbf{q}\cdot\mathbf{x}}); \tag{89}$$

in the limit $\mathbf{q} \to 0$ the magnetic field will be constant over the spatial region of interest. The perturbation terms in the hamiltonian are

$$U = U_1 + U_2 = -\frac{e}{mc}\mathbf{A}(\mathbf{x}) \cdot \mathbf{p} + \frac{e^2}{2mc^2} A^2(\mathbf{x}). \tag{90}$$

We write as before in (65) the one-electron solution $\Psi(\mathbf{x})$ of the perturbed problem as

$$\Psi(\mathbf{x}) = \Sigma\, |\mathbf{k}'l'\rangle\langle l'\mathbf{k}'|\rangle; \tag{91}$$

the secular equation (66) is

$$[\varepsilon_l(\mathbf{k}) - E]\langle l\mathbf{k}|\rangle + \sum_{l'k'} \langle l\mathbf{k}|U|\mathbf{k}'l'\rangle\langle l'\mathbf{k}'|\rangle = 0. \tag{92}$$

We suppose that $\langle l'\mathbf{k}|\rangle \ll \langle l\mathbf{k}|\rangle$ for all $l' \neq l$; thus to first order in $\mathbf{A}$ the secular equation (92) has the approximate solution, for E associated with band l,

$$\langle l'\mathbf{k}'|\rangle = \frac{1}{\varepsilon_l(\mathbf{k}') - \varepsilon_{l'}(\mathbf{k}')} \sum_{\mathbf{k}''} \langle l'\mathbf{k}|U_1|\mathbf{k}''l\rangle\langle l\mathbf{k}''|\rangle, \tag{93}$$

for $l' \neq l$. Now let us substitute this in (92) for $\langle l'\mathbf{k}'|\rangle$ and solve to second order in $\mathbf{A}$ for $\langle l\mathbf{k}|\rangle$:

$$[\epsilon_l(\mathbf{k}) - E]\langle l\mathbf{k}|\rangle + \sum_{\substack{l'\neq l \\ \mathbf{k}',\mathbf{k}''}} \frac{\langle l\mathbf{k}|U_1|\mathbf{k}'l'\rangle\langle l'\mathbf{k}'|U_1|\mathbf{k}''l\rangle}{\varepsilon_l(\mathbf{k}') - \varepsilon_{l'}(\mathbf{k}')} \langle l\mathbf{k}''|\rangle + \sum_{\mathbf{k}''} \langle l\mathbf{k}|U|l\mathbf{k}''\rangle\langle l\mathbf{k}''|\rangle = 0. \tag{94}$$

We may write this equation as, on changing superscripts,

$$[\epsilon_l(\mathbf{k}) - E]\langle l\mathbf{k}|\rangle + \sum_{\mathbf{k}'} \langle l\mathbf{k}|\mathcal{P}|\mathbf{k}'l\rangle\langle \mathbf{k}'l|\rangle = 0, \tag{95}$$

with the appropriate definition of $\langle l\mathbf{k}|\mathcal{P}|\mathbf{k}'l\rangle$.

Now

$$\langle l\mathbf{k}|U_1|\mathbf{k}'l\rangle \cong -i\frac{e}{mc}\mathbf{A}_{\mathbf{q}} \cdot (\langle l\mathbf{k}|e^{-i\mathbf{q}\cdot\mathbf{x}}\mathbf{p}|\mathbf{k}'l\rangle\delta_{\mathbf{k}';\mathbf{k}+\mathbf{q}} - \langle l\mathbf{k}|e^{i\mathbf{q}\cdot\mathbf{x}}\mathbf{p}|\mathbf{k}'l\rangle\delta_{\mathbf{k}';\mathbf{k}-\mathbf{q}}. \tag{96}$$

In the limit $\mathbf{q} \to 0$ we have

$$\langle l\mathbf{k}|e^{\mp i\mathbf{q}\cdot\mathbf{x}}\mathbf{p}|\mathbf{k} \pm \mathbf{q};l\rangle \to \langle l\mathbf{k}|\mathbf{p}|\mathbf{k}l\rangle; \tag{97}$$

but we have seen in Chapter 9 that

$$\langle l\mathbf{k}|\mathbf{p}_\mu|\mathbf{k}l\rangle \cong k_\alpha\left(\frac{m}{m^*}\right)_{\alpha\mu},$$

to first order in **k**. Thus for a spherical energy surface

$$(98)\qquad \langle l\mathbf{k}|U_1|\mathbf{k}'l\rangle \cong -\frac{ie}{m^*c}\,\mathbf{k}\cdot\mathbf{A}_\mathbf{q}(\delta_{\mathbf{k}';\mathbf{k}+\mathbf{q}} - \delta_{\mathbf{k}';\mathbf{k}-\mathbf{q}}).$$

We need also

$$(99)\qquad \langle l\mathbf{k}|U_2|\mathbf{k}'l\rangle = -\frac{e^2}{2mc^2}\,|\mathbf{A}_\mathbf{q}|^2\,\langle l\mathbf{k}|e^{2i\mathbf{q}\cdot\mathbf{x}} + e^{-2i\mathbf{q}\cdot\mathbf{x}} - 2|\mathbf{k}'l\rangle;$$

in the limit $\mathbf{q}\rightarrow 0$

$$(100)\qquad \langle l\mathbf{k}|U_2|\mathbf{k}'l\rangle \cong -\frac{e^2}{2mc^2}\,\mathbf{A}_\mathbf{q}(\delta_{\mathbf{k}';\mathbf{k}+2\mathbf{q}} + \delta_{\mathbf{k}';\mathbf{k}-2\mathbf{q}} - 2\delta_{\mathbf{k}';\mathbf{k}}).$$

Similarly

$$(101)\qquad \sum{}' \frac{\langle l\mathbf{k}|U_1|\mathbf{k}'l'\rangle\langle l'\mathbf{k}'|U_1|\mathbf{k}''l\rangle}{\mathcal{E}_l(\mathbf{k}') - \mathcal{E}_{l'}(\mathbf{k}')}$$
$$= -\frac{e^2}{m^2c^2}\sum{}'\frac{|\mathbf{A}_\mathbf{q}\cdot\langle l\mathbf{k}|\mathbf{p}|\mathbf{k}l'\rangle|^2}{\mathcal{E}_l(\mathbf{k}) - \mathcal{E}_{l'}(\mathbf{k})}(\delta_{\mathbf{k}';\mathbf{k}+2\mathbf{q}} + \delta_{\mathbf{k}';\mathbf{k}-2\mathbf{q}} - 2\delta_{\mathbf{k}'\mathbf{k}})$$
$$= -\frac{e^2}{2mc^2}\left(\frac{m}{m^*} - 1\right)|\mathbf{A}_\mathbf{q}|^2(\delta_{\mathbf{k}';\mathbf{k}+2\mathbf{q}} + \delta_{\mathbf{k}';\mathbf{k}-2\mathbf{q}} - 2\delta_{\mathbf{k}'\mathbf{k}}),$$

using the f-sum rule. Thus

$$(102)\qquad \langle l\mathbf{k}|\mathcal{P}|\mathbf{k}'l\rangle = -\frac{ie}{m^*c}\,\mathbf{A}_\mathbf{q}\cdot\mathbf{k}(\delta_{\mathbf{k}';\mathbf{k}+\mathbf{q}} - \delta_{\mathbf{k}';\mathbf{k}-\mathbf{q}})$$
$$+\frac{e^2}{2m^*c^2}|\mathbf{A}_\mathbf{q}|^2(2\delta_{\mathbf{k}'\mathbf{k}} - \delta_{\mathbf{k}';\mathbf{k}-2\mathbf{q}} - \delta_{\mathbf{k}';\mathbf{k}+2\mathbf{q}}),$$

and (95) is identical with the effective mass equation in the plane wave representation

$$(103)\qquad F_l(\mathbf{x}) = \sum_\mathbf{k} e^{i\mathbf{k}\cdot\mathbf{x}}\langle l\mathbf{k}|\rangle,$$

so that $F_l(\mathbf{x})$ satisfies the Wannier equation

$$(104)\qquad \frac{1}{2m^*}\left(\mathbf{p} - \frac{e}{c}\mathbf{A}\right)^2 F_l(\mathbf{x}) = EF_l(\mathbf{x}).$$

The above derivation is due to Argyres (unpublished); the derivation of Luttinger and Kohn [*Phys. Rev.* **97,** 869 (1955)] avoids the limiting process $\mathbf{q} \rightarrow 0$ at the price of treating singular matrix elements.

PROBLEMS

1. Consider the hamiltonian

$$H = p^2 + x^2$$

of a harmonic oscillator; solve for the eigenvalues using a plane wave representation:

$$\Psi = \int dx\, e^{ikx}\langle k|\rangle,$$

with

$$\langle k'|x^2|k\rangle = -\frac{\partial^2}{\partial k^2}\,\delta(k - k'),$$

and show that one obtains the correct energy eigenvalues. This is an exercise in a method of handling matrix elements of the coordinates.

2. Show that states which transform as $J = \frac{3}{2}$ are split into two twofold levels by an axial crystal field. Evaluate the splitting for the crystal potential

$$V = a(x^2 + y^2 - z^2),$$

and the spin-orbit splitting λ between the $J = \frac{3}{2}$ and $J = \frac{1}{2}$ levels in the absence of the crystal field.

3. (*a*) Show for a cubic crystal at $\mathbf{k} = 0$ that

$$D^S_{xy} = 0; \qquad D^A_{xy} \neq 0,$$

using the definitions (21) and (22), with coordinate axes along the cube edge directions. *Hint:* Consider the effect on D^S_{xy} of rotating about the z axis by $\pi/2$. (*b*) Show that in the absence of spin-orbit coupling $D^A_{xy} = 0$.

4. In *CdS* the conduction band edge may be written in the form, to $0(k^2)$,

$$\varepsilon_{\mathbf{k}} = [A(k_x{}^2 + k_y{}^2) + Bk_z{}^2]\begin{pmatrix}1 & 0\\ 0 & 1\end{pmatrix} + C(k_x\sigma_y - k_y\sigma_x),$$

where the z direction is parallel to the symmetry axis of the crystal; the σ_x, σ_y are pauli matrices. Plot a section of a constant energy surface in the plane $k_y = 0$, for the band with spin parallel to the positive y axis. Considering *only* this spin orientation, does this band carry a net current when filled at 0°K to a level ε_F?

5. Verify the statement made in connection with 3-5 crystals that the antisymmetric crystal potential component will act to increase the splitting between Γ'_{25} and Γ_{15} states.

15 Semiconductor crystals: II. Optical absorption and excitons

DIRECT OPTICAL TRANSITIONS

In the process of direct photon absorption a photon of energy ω and wavevector $\mathbf{K}$ is absorbed by the crystal with the creation of an electron at $\mathbf{k}_{\text{el}}$ in a conduction band and a hole at $\mathbf{k}_{\text{hole}}$ in the valence band. The scale of wavevectors of optical photons is of the order of $10^4\ \text{cm}^{-1}$ and may almost always be neglected in comparison with the scale of wavevectors in the Brillouin zone, $10^8\ \text{cm}^{-1}$. The conservation of wavevector in the absorption process requires

$$\mathbf{k}_{\text{el}} + \mathbf{k}_{\text{hole}} \cong \mathbf{G}, \tag{1}$$

where $\mathbf{G}$ is a vector in the reciprocal lattice. In the reduced zone scheme $\mathbf{G} = 0$; so that $\mathbf{k}_{\text{el}} \cong -\mathbf{k}_{\text{hole}}$. This is simply interpreted: the total wavevector of the valence band when filled is zero, so that if we remove an electron from the valence band, the total wavevector of the $N - 1$ electrons left in the valence band is equal but opposite to the wavevector of the electron which was taken away. It is customary to call a transition in which

$$\mathbf{k}_{\text{el}} + \mathbf{k}_{\text{hole}} \cong 0 \tag{2}$$

a *vertical* or *direct* transition because the electron is raised vertically on an energy-band diagram, as in Fig. 1*a* and *b*. In some connections, particularly for excitons, the fact that the photon wavevector is not exactly zero is of considerable importance.

The matrix element for electric dipole allowed transitions is

$$\left\langle \delta\mathbf{k}' \left| \frac{1}{m}\mathbf{A}\cdot\mathbf{p} \right| \mathbf{k}\gamma \right\rangle; \tag{3}$$

if the fourier components of the radiation field $\mathbf{A}$ are at low wavevector,

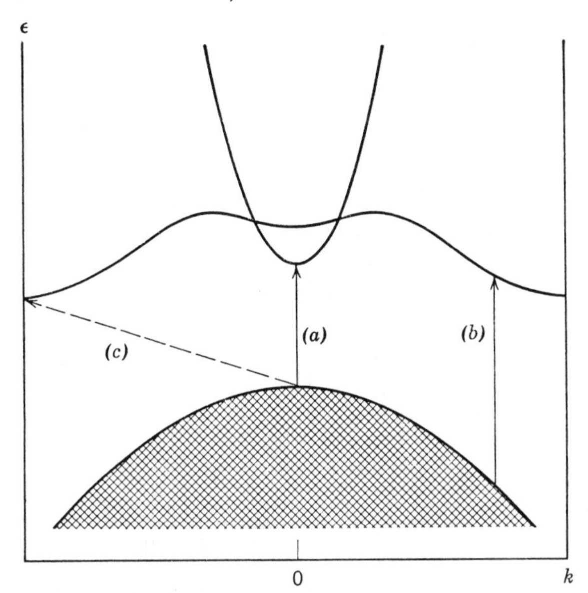

FIG. 1. Direct absorption processes in (*a*) and (*b*); the absorption process (*c*) is indirect and takes place with the emission or absorption of a phonon.

we may usually replace $\mathbf{k}'$ by $\mathbf{k}$. Thus $|\langle\delta\mathbf{k}|\mathbf{p}|\mathbf{k}\gamma\rangle|^2$ determines the intensity of the transition; this same quantity determines the mutual interaction of the two bands γ, δ in the reciprocal effective mass tensors. We see that bands which perturb each other strongly are always connected by allowed optical transitions for the direct absorption or emission of a photon.

INDIRECT OPTICAL TRANSITIONS

Sometimes, as in Si and Ge, the minimum energy difference between the valence and conduction bands does not occur for $\Delta\mathbf{k} = 0$, but the band minima fall at different $\mathbf{k}$ values and cannot be connected by an allowed optical transition. If this is true, the threshold of strong optical absorption will lie at a higher energy than the energy gap. But at energies slightly above the energy gap a weak absorption takes place with the emission or absorption of a *phonon* of wavevector $\mathbf{q}$:

$$\mathbf{k}_{\mathrm{el}} + \mathbf{k}_{\mathrm{hole}} \pm \mathbf{q} \cong 0. \tag{4}$$

If the conduction and valence band edges do not lie at the same point in $\mathbf{k}$ space, the indirect or nonvertical process will dominate the optical absorption over the appropriate energy interval. The energy balance at the indirect threshold will be

$$\omega = \varepsilon(\mathbf{k}_c) - \varepsilon(\mathbf{k}_v) \pm \omega_{\mathbf{q}}; \tag{5}$$

at absolute zero no phonon is available to be absorbed in the process and here the positive sign must be taken on the right-hand side. At higher temperatures there are thermal phonons available to be absorbed, and photon absorption may take place at an energy lower by $2\omega_{\text{phonon}}$, where the phonon has a wavevector of magnitude close to the difference $|\mathbf{k}_c - \mathbf{k}_v|$ at the band edges.

The intensity of the indirect transition[1] is determined by second-order matrix elements of the electron-phonon and electron-photon interactions. Second-order matrix elements for processes in which a phonon is absorbed involve

$$\langle \delta\mathbf{k}|\mathbf{p}\cdot\mathbf{A}|\mathbf{k}\gamma\rangle\langle\gamma;\mathbf{k};n_{\mathbf{q}} - 1|c_{\mathbf{k}}^{+}c_{\mathbf{k}-\mathbf{q}}a_{\mathbf{q}}|n_{\mathbf{q}};\mathbf{k} - \mathbf{q};\gamma\rangle,$$

and

$$\langle\delta;\mathbf{k};n_{\mathbf{q}} - 1|c_{\mathbf{k}}^{+}c_{\mathbf{k}-\mathbf{q}}a_{\mathbf{q}}|n_{\mathbf{q}};\mathbf{k} - \mathbf{q};\delta\rangle\langle\delta;\mathbf{k} - \mathbf{q}|\mathbf{p}\cdot\mathbf{A}|\mathbf{k} - \mathbf{q};\gamma\rangle.$$

Here the c's are electron operators and the a's are phonon operators. The form of the electron-phonon interaction was discussed in Chapter 7. In the process described the electron is initially at $\mathbf{k} - \mathbf{q}$ in the valence band γ and the phonon occupation number is $n_{\mathbf{q}}$ for wavevector $\mathbf{q}$. In the final state the electron is at $\mathbf{k}$ in the conduction band δ and the phonon occupation is $n_{\mathbf{q}} - 1$. The corresponding matrix element for emission of a phonon is written down by using the terms $c_{\mathbf{k}-\mathbf{q}}^{+}c_{\mathbf{k}}a_{\mathbf{q}}^{+}$ of the electron-phonon interaction.

Actually there will be a number of threshold energies because in principle every branch of the phonon spectrum will participate at the same wavevector, but at different frequencies. Optical measurements have been able in this way to determine directly the difference in the wavevectors of the conduction and valence band edges, provided that the phonon spectrum itself is known, as from inelastic neutron scattering studies.

OSCILLATORY MAGNETOABSORPTION—LANDAU TRANSITIONS

In the presence of a strong static magnetic field the optical absorption near the threshold of the direct transition in semiconductors is observed to exhibit oscillations. That is, at fixed $\mathbf{H}$ the absorption coefficient is periodic in the photon energy. In a magnetic field the interband transitions (Fig. 2) take place between the Landau magnetic levels in the valence band and the corresponding levels in the conduc-

[1] See Bardeen, Blatt, and Hall, *Proc. of Conf. on Photoconductivity, Atlantic City, 1954* (Wiley, 1956), p. 146.

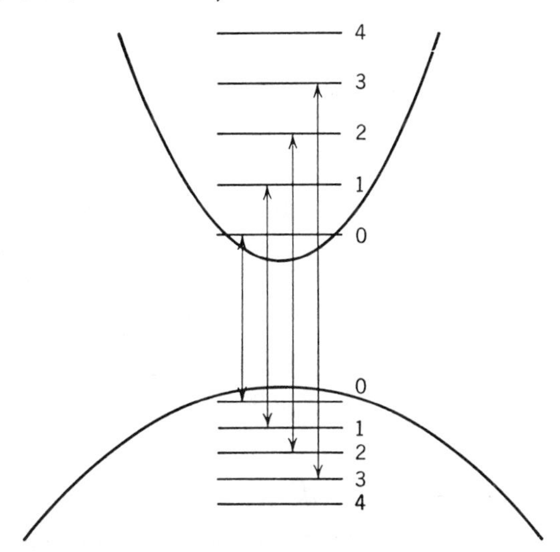

Fig. 2. Schematic diagram showing the magnetic levels for $k_z = 0$ as labeled by n ($k_z = 0$) for two simple bands. The possible transitions are shown for the case in which the direct transition is allowed by parity.

tion band. Such transitions are called *Landau transitions*. In a magnetic field parallel to the z axis the energies in the two bands, if nondegenerate, are

$$(6)\qquad \begin{aligned} \varepsilon_c(n,k_z) &= E_g + (n + \tfrac{1}{2})\omega_c + \frac{1}{2m_c} k_z^{\,2} \pm \mu_c H, \\ \varepsilon_v(n,k_z) &= -(n + \tfrac{1}{2})\omega_v - \frac{1}{2m_v} k_z^{\,2} \pm \mu_v H, \end{aligned}$$

where ω_c, ω_v are the cyclotron frequencies and μ_c, μ_v are the anomalous magnetic moments. The spatial parts of the wavefunctions in each band are of the form $\psi(x) = u_0(\mathbf{x})F(\mathbf{x})$, where $u_0(\mathbf{x})$ is the Bloch function in the appropriate band for $\mathbf{k} = 0$ and, from Eq. (11.13),

$$(7)\qquad F_n(\mathbf{x}) = e^{i(k_y y + k_z z)}\varphi_n(x - ck_y/eH)$$

in the Landau gauge. Here F_n is the solution of the appropriate Wannier equation and φ_n is the harmonic oscillator wavefunction for the nth excited state.

The matrix element for optical absorption is proportional to

$$(8)\qquad \langle n_c k_y^c k_z^c|\mathbf{p}|n_v k_y^v k_z^v\rangle \cong \int_{\text{cell}} d^3x\, u_{oc}^*(\mathbf{x})\mathbf{p}u_{ov}(\mathbf{x}) \int_{\text{crystal}} d^3x\, F_{nc}^*(\mathbf{x})F_{nv}(\mathbf{x}),$$

where we have broken up the integral by treating the F's as essentially constant over a cell. The integral involving the F's will vanish unless $k_y^c = k_y^v$; $k_z^c = k_z^v$; and $n_c = n_v$. This is analogous to the selection rules conserving $\mathbf{k}$ in the absence of a magnetic field. The equality of the n's follows by the orthogonality property of harmonic oscillator wavefunctions, noting that these do not depend on the effective mass. The allowed transitions have $\Delta n = 0$, as indicated in Fig. 2. After integrating the transition probability over the density of states for k_y and k_z, it is found that the absorption coefficient is proportional to

$$\sum_n \frac{1}{(\omega - \omega_n)^{1/2}},$$

where

$$\omega_n = E_g + (n + \tfrac{1}{2})(\omega_c + \omega_v) \pm (\mu_c - \mu_v)H. \tag{9}$$

The theory of oscillatory magnetoabsorption for degenerate bands and also for indirect transitions has been given by Roth, Lax, and Zwerdling, *Phys. Rev.* **114,** 90 (1959). We note that magnetoabsorption experiments are particularly valuable in determining the parameters of a direct conduction-band energy surface which, because it is not the band edge, cannot be kept populated sufficiently to permit a cyclotron resonance experiment to be made; further, the experiments involve the anomalous magnetic moments or g factors.

EXCITONS

An exciton is defined as a nonconducting excited electronic state in a perfect insulator, usually a nonmagnetic insulator. It is usual to speak of two types of excitons: a tightly bound or Frenkel exciton and a weakly bound or Mott exciton. Both types of excitons may be thought of as bound states of an electron and a hole; there is no sharp division between the two types. In a Frenkel exciton there is a high probability of finding the electron and hole on the same atom in the crystal; in a Mott exciton the wavefunction in the relative coordinate extends over many atoms. Frenkel excitons are realized in alkali-halide crystals and in many crystals of aromatic molecules; Mott excitons are found in semiconductor crystals having small energy gaps and high dielectric constants.

The machinery we developed for the impurity-state problem may be taken over directly to the discussion of weakly bound excitons, of radii large in comparison with a lattice constant. For this reason, and because their experimental picture is richer, we limit ourselves here to the discussion of weakly bound excitons.

If both the conduction and valence band edges are spherical, nondegenerate, and are located at $\mathbf{k} = 0$, the exciton spectrum and wavefunctions are obtained readily by an extension of the result found above for electrons bound in impurity states. We introduce the relative and center-of-mass coordinates

$$\mathbf{x} = \mathbf{x}_e - \mathbf{x}_h; \qquad \mathbf{X} = \frac{m_e\mathbf{x}_e + m_h\mathbf{x}_h}{m_e + m_h}, \tag{10}$$

where both m_e and m_h are usually positive. The effective hamiltonian in a cubic crystal is

$$H = \frac{p_e^2}{2m_e} + \frac{p_h^2}{2m_h} - \frac{e^2}{\epsilon|\mathbf{x}|} = \frac{P^2}{2(m_e + m_h)} + \frac{p^2}{2\mu} - \frac{e^2}{\epsilon|\mathbf{x}|}. \tag{11}$$

The part of the wavefunction in $\mathbf{X}$ must contain a factor $e^{i\mathbf{K}\cdot\mathbf{X}}$; the part in the relative coordinates contains a factor $F_n(\mathbf{x})$, where

$$\left(\frac{1}{2\mu}p^2 - \frac{e^2}{\epsilon|\mathbf{x}|}\right)F_n(\mathbf{x}) = E_nF_n(\mathbf{x}) \tag{12}$$

is the hydrogenic wave equation with the reduced mass

$$\frac{1}{\mu} = \frac{1}{m_e} + \frac{1}{m_h}, \tag{13}$$

and dielectric constant ϵ. In direct analogy to the treatment of impurity states in Chapter 14, the total exciton wavefunction is

$$\psi_{\mathbf{K}n}(\mathbf{X},\mathbf{x}) = e^{i\mathbf{K}\cdot\mathbf{X}}F_n(\mathbf{x})\varphi_c(\mathbf{x}_e)\varphi_v(\mathbf{x}_h), \tag{14}$$

where $\varphi_c(\mathbf{x}_e)$ is the Bloch function at $\mathbf{k} = 0$ in the conduction band and $\varphi_v(\mathbf{x}_h)$ is the valence band function at $\mathbf{k} = 0$. The excitation is propagated in the crystal as a wave of momentum $\mathbf{K}$.

The energy of the state (14) is

$$\boxed{E_{\mathbf{K}n} = E_n + \frac{1}{2(m_e + m_h)}K^2,} \tag{15}$$

referred to the conduction band edge. For bound states E_n is negative and the total exciton energy at low $\mathbf{K}$ is negative with respect to a separated hole-electron pair. For the hydrogenic hamiltonian (12) the energy is, with $\hbar$ restored,

$$E_n = -\frac{\mu e^4}{2\epsilon^2\hbar^2}\frac{1}{n^2}; \tag{16}$$

for $\epsilon = 5$ and $\mu = 0.5m$ the ionization energy ($n = 1$) of the exciton is about $\frac{1}{4}$ ev. We note that the minimum energy required to create an exciton starting from the ground state of the crystal is

$$E = E_g - \frac{\mu e^4}{2\epsilon^2\hbar^2}, \tag{17}$$

where E_g is the energy gap.

Excitons created by photon absorption from the ground state of the crystal are created near $\mathbf{K} = 0$; therefore the direct exciton absorption spectrum is a series of sharp lines below the optical absorption edge of the crystal. It is somewhat unusual to find a crystal in which there are two spherical band edges at $\mathbf{k} = 0$, but this is apparently the situation in Cu_2O, for which the exciton spectrum is closely hydrogenic.

For general energy surfaces the exciton problem is best formulated using the coordinate transformation

$$\boldsymbol{\varrho} = \tfrac{1}{2}(\mathbf{x}_e + \mathbf{x}_h); \qquad \mathbf{x} = \mathbf{x}_e - \mathbf{x}_h, \tag{18}$$

rather than with the transformation (10). It is instructive to reexamine the problem we have just solved. The hamiltonian (11) is transformed with the use of

$$\frac{\partial^2}{\partial \mathbf{x}_e^2} = \frac{1}{4}\frac{\partial^2}{\partial \boldsymbol{\varrho}^2} + \frac{\partial^2}{\partial \boldsymbol{\varrho}\partial \mathbf{x}} + \frac{\partial^2}{\partial \mathbf{x}^2}; \qquad \frac{\partial^2}{\partial \mathbf{x}_h^2} = \frac{1}{4}\frac{\partial^2}{\partial \boldsymbol{\varrho}^2} - \frac{\partial^2}{\partial \boldsymbol{\varrho}\,\partial \mathbf{x}} + \frac{\partial^2}{\partial \mathbf{x}^2}. \tag{19}$$

Thus, if $\mathbf{\Pi}$, $\mathbf{p}$ are the momenta conjugate to $\boldsymbol{\varrho}$, $\mathbf{x}$, we have for the special case of spherical surfaces

$$H = \frac{1}{8\mu}\Pi^2 + \frac{1}{2\mu}p^2 + \frac{1}{2}\left(\frac{1}{m_e} - \frac{1}{m_h}\right)\mathbf{\Pi}\cdot\mathbf{p} - \frac{e^2}{\epsilon|\mathbf{x}|}. \tag{20}$$

If we look for a wavefunction of the form

$$\psi_n(\boldsymbol{\varrho},\mathbf{x}) = e^{i\mathbf{K}\cdot\boldsymbol{\varrho}}F_n(\mathbf{x}), \tag{21}$$

the equation for $F_n(\mathbf{x})$ is

$$\left[\frac{1}{2\mu}p^2 + \frac{1}{2}\left(\frac{1}{m_e} - \frac{1}{m_h}\right)\mathbf{p}\cdot\mathbf{K} - \frac{e^2}{\epsilon|\mathbf{x}|}\right]F_n(\mathbf{x}) = \left(E_{Kn} - \frac{K^2}{8\mu}\right)F_n(\mathbf{x}). \tag{22}$$

The hamiltonian at $\mathbf{K} = 0$ has the eigenvalues

$$E_n = -\frac{\mu e^4}{2\epsilon^2}\frac{1}{n^2}. \tag{23}$$

The eigenvalue of (12) to second order in $\mathbf{K}$ is found by $\mathbf{K}\cdot\mathbf{p}$ perturbation theory:

$$E_{\mathbf{K}n} \cong E_n + \frac{1}{8\mu}K^2 + \frac{1}{4}\left(\frac{1}{m_e} - \frac{1}{m_h}\right)^2 \sum_l{}' \frac{\langle n|\mathbf{K}\cdot\mathbf{p}|l\rangle\langle l|\mathbf{K}\cdot\mathbf{p}|n\rangle}{E_n - E_l}. \tag{24}$$

But by the atomic f-sum rule on the hydrogenic states l, n,

$$\frac{2}{\mu}\sum_l{}' \frac{\langle n|p_\mu|l\rangle\langle l|p_\nu|n\rangle}{E_n - E_l} = -\delta_{\mu\nu}, \tag{25}$$

whence (24) becomes

$$E_{Kn} = E_n + \frac{1}{2(m_e + m_h)}K^2, \tag{26}$$

in agreement with (15).

The extension of the present treatment in the coordinate system (18) to ellipsoidal band edges follows directly on using the components of the reciprocal mass tensors in (19) and (20). The further extension to degenerate band edges is complicated in practice, but follows by using matrix operators for the multicomponent state functions at each band edge; see G. Dresselhaus, *Phys. Chem. Solids* **1**, 14 (1956). In practice various approximate dodges are often employed to avoid confronting the complexity of the multicomponent equations.

We now discuss the intensity of optical absorption for a process in which an allowed (electric dipole) transition creates an exciton from a filled valence band. We take the bands to be spherical about $\mathbf{k} = 0$ and nondegenerate. From (14) the exciton wavefunction at $\mathbf{K} = 0$ is

$$\psi_{0n}(\mathbf{x}) = F_n(\mathbf{x}_e - \mathbf{x}_h)\varphi_c(\mathbf{x}_e)\varphi_v(\mathbf{x}_h), \tag{27}$$

and in this scheme the wavefunction of the initial state is simply unity.

This is not the clearest way to treat a many-electron problem; it is better to use the formalism of second quantization, as in Chapter 5. We denoted the filled valence band by Φ_0; then we define

$$\Phi_\mathbf{k} \equiv \alpha_\mathbf{k}^+\beta_{-\mathbf{k}}^+\Phi_0;$$

this is a state in which an electron has been raised to the conduction band at $\mathbf{k}$, leaving a hole in the valence band at $-\mathbf{k}$. The nth exciton state for $\mathbf{K} = 0$ may be written

$$\Phi_n = \sum_\mathbf{k} \Phi_\mathbf{k}\langle\mathbf{k}|n\rangle = \sum_\mathbf{k} \alpha_\mathbf{k}^+\beta_{-\mathbf{k}}^+\Phi_0\langle\mathbf{k}|n\rangle, \tag{28}$$

The electric dipole absorption is determined by the matrix element $\langle c\mathbf{k}|\mathbf{p}|\mathbf{k}v\rangle$ of the momentum $\mathbf{p}$ between the state $\mathbf{k}$ in the valence band and the state $\mathbf{k}$ in the conduction band. In second quantization the momentum operator is

$$\mathbf{p} = \sum_{\substack{\mathbf{k} \\ ll'}} c^+_{\mathbf{k}l'} c_{\mathbf{k}l} \int d^3x\, \varphi^*_{\mathbf{k}l'}(\mathbf{x}) \mathbf{p} \varphi_{\mathbf{k}l}(\mathbf{x}), \tag{29}$$

or, with l denoting the valence band v and l' the conduction band c,

$$\mathbf{p} \cong \sum_{\mathbf{k}} \alpha^+_{\mathbf{k}} \beta^+_{-\mathbf{k}} \langle c\mathbf{k}|\mathbf{p}|\mathbf{k}v\rangle. \tag{30}$$

Then the matrix element of $\mathbf{p}$ between the vacuum and the nth exciton state is

$$\begin{aligned} \langle \Phi_n|\mathbf{p}|\Phi_0\rangle &= \sum_{\mathbf{k}'\mathbf{k}} \langle n|\mathbf{k}'\rangle \langle \Phi_0|\beta_{-\mathbf{k}'}\alpha_{\mathbf{k}'}\alpha^+_{\mathbf{k}}\beta^+_{-\mathbf{k}}|\Phi_0\rangle \langle c\mathbf{k}|\mathbf{p}|\mathbf{k}v\rangle \\ &= \sum_{\mathbf{k}} \langle n|\mathbf{k}\rangle \langle c\mathbf{k}|\mathbf{p}|\mathbf{k}v\rangle. \end{aligned} \tag{31}$$

The transition probability is proportional to

$$|\langle \Phi_n|\mathbf{p}|\Phi_0\rangle|^2 \cong |\langle c|\mathbf{p}|v\rangle|^2 \left(\sum_{\mathbf{k}} \langle n|\mathbf{k}\rangle\right) \left(\sum_{\mathbf{k}'} \langle \mathbf{k}'|n\rangle\right), \tag{32}$$

if $\langle c\mathbf{k}|\mathbf{p}|\mathbf{k}v\rangle \cong \langle c|\mathbf{p}|v\rangle$ over the range of $\mathbf{k}$ involved. But the $\langle \mathbf{k}|n\rangle$ are such that in (32)

$$F_n(\mathbf{x}) = \sum_{\mathbf{k}} e^{i\mathbf{k}\cdot\mathbf{x}} \langle \mathbf{k}|n\rangle; \tag{33}$$

whence

$$F_n(0) = \sum_{\mathbf{k}} \langle \mathbf{k}|n\rangle, \tag{34}$$

and thus the transition probability involves

$$|\langle \Phi_n|\mathbf{p}|\Phi_0\rangle|^2 \cong |\langle c|\mathbf{p}|v\rangle|^2 |F_n(0)|^2. \tag{35}$$

For spherical masses $F_n(0)$ is nonzero only for s states; for hydrogenic s states $|F_n(0)|^2 \propto n^{-3}$, if n is the principal quantum number.

"First forbidden" electric dipole transitions arise when the transition probability is proportional to $|\partial F_n(0)/\partial x|^2$, which is nonzero only for p states. Thus when electric dipole transitions are forbidden, with $F_n(0) = 0$, we may still observe excitons because $\langle c\mathbf{k}|\mathbf{p}|\mathbf{k}v\rangle \neq 0$, but the $n = 1$ exciton will be absent. There are no p-states for $n = 1$. This appears to be the picture in Cu_2O. The $n = 1$ line can actually be seen very faintly; Elliott [*Phys. Rev.* **124,** 340 (1961); **108,** 1384 (1957)] suggests the weak transition is by electric quadrupole radiation.

Longitudinal and Transverse Excitons. We have seen in Chapter 3 that the dielectric polarization field of a cubic crystal has longitudinal and transverse modes, with a frequency splitting determined by the polarizability. In a covalent crystal the polarizability is determined by the excited electronic states of the crystal; that is, the polarizability depends on the nature of the exciton states. An exciton is in fact the quantum unit of the polarization field. The polarization splitting of longitudinal and transverse excitons was derived in Chapter 3 on the assumption that the wavevector of the excitation was small so that the dispersion of the uncoupled polarization could be neglected. At the same time we supposed the wavelength was small in comparison with the dimensions of the crystal, so that shape effects could be neglected. We continue here to make the same approximation; although the wavevector of the incident photon is very small compared with the extent of the first Brillouin zone, the crystal is supposed to be large in comparison with a wavelength.

Photons are transverse and in cubic crystals couple only with transverse excitons. That is, a photon with $\mathbf{k} \parallel \hat{\mathbf{z}}$ in a crystal with an s conduction band edge and an x, y, z degenerate valence band edge will couple with the exciton bands made up from hole wavefunctions in the x or y bands and electron wavefunctions in the s band; there is no $A_z p_z$ term in the interaction. To see this we work in the gauge $\operatorname{div} \mathbf{A} = 0$ and consider the electromagnetic wave

$$\mathbf{A} = \hat{\mathbf{y}} e^{-i(\omega t - kz)}. \tag{36}$$

Then the wave is polarized in the $\hat{\mathbf{y}}$ direction:

$$\begin{aligned} \mathbf{H} &= \operatorname{curl} \mathbf{A} = -ik\hat{\mathbf{x}} e^{-i(\omega t - kz)}; \\ \mathbf{E} &= -\frac{1}{c}\frac{\partial \mathbf{A}}{\partial t} = i\frac{\omega}{c}\,\hat{\mathbf{y}} e^{-i(\omega t - kz)}, \end{aligned} \tag{37}$$

and (36) has $\mathbf{A} \cdot \mathbf{p}$ coupling only with sy excitons. The polarization associated with these excitons is purely transverse for $\mathbf{k} \parallel \hat{\mathbf{z}}$; only sz excitons have a longitudinal polarization for this direction of $\mathbf{k}$.

In uniaxial crystals the dielectric polarizability is anisotropic and a purely longitudinal exciton mode exists only in special symmetry directions of $\mathbf{k}$. We must consider depolarization effects on the exciton spectrum. Let $P_\perp$, $P_\parallel$ denote polarization components normal and parallel, respectively, to the c axis in a uniaxial crystal; $\beta_\perp$, $\beta_\parallel$ are the static polarizabilities and $\omega_\perp$, $\omega_\parallel$ are the resonance frequencies for transverse waves. We are particularly interested in the special case $\beta_\parallel \ll \beta_\perp$; that is, we consider an exciton of frequency near $\omega_\perp$ and neglect the contribution to the polarizability of the oscillators at $\omega_\parallel$. Now

$$\frac{1}{\omega_\perp{}^2}\frac{\partial^2 P_\perp}{\partial t^2} + P_\perp = \beta_\perp E_\perp, \tag{38}$$

where $E_\perp$ is the $\perp$ component of the depolarization field of a polarization wave.

We find $E_\perp$ from div $\mathbf{D} = 0$, exactly as in Chapter 4 we found the demagnetization field of a magnon. Let $\hat{\mathbf{k}}$ be the unit vector in the direction of $\mathbf{k}$. The projection of $\mathbf{P}_\perp$ on the wave normal is $\hat{\mathbf{k}} \cdot \mathbf{P}_\perp$ and the depolarization field is

$$E = -4\pi\hat{\mathbf{k}} \cdot \mathbf{P}_\perp; \qquad E_\perp = -4\pi(\hat{\mathbf{k}} \cdot \mathbf{P}_\perp) \sin \theta_{\mathbf{k}}, \tag{39}$$

where $\theta_{\mathbf{k}}$ is the angle between $\mathbf{k}$ and the c axis. Then

$$\frac{1}{\omega_\perp{}^2}\frac{\partial^2 \mathbf{P}_\perp}{\partial t^2} + \mathbf{P}_\perp = -4\pi\beta_\perp(\hat{\mathbf{k}} \cdot \mathbf{P}_\perp) \sin \theta_{\mathbf{k}}. \tag{40}$$

This has two solutions:

$$\hat{\mathbf{k}} \cdot \mathbf{P}_\perp = 0; \qquad \omega^2 = \omega_\perp{}^2; \qquad \text{transverse mode}; \tag{41}$$

$$\hat{\mathbf{k}} \cdot \mathbf{P}_\perp = \mathbf{P}_\perp \sin \theta_{\mathbf{k}}; \qquad \omega^2 = \omega_\perp{}^2(1 + 4\pi\beta_\perp \sin^2 \theta_{\mathbf{k}}); \qquad \text{mixed mode}. \tag{42}$$

We have neglected ϵ, the contribution to the dielectric properties from other modes; otherwise 4π would be replaced by $4\pi/\epsilon$. These results are due to J. J. Hopfield and D. G. Thomas, *Phys. Chem. Solids* **12,** 276 (1960).

The mixed mode is purely longitudinal for $\theta_{\mathbf{k}} = \pi/2$ and it is asymptotically transverse for $\theta_{\mathbf{k}} = 0$ on our assumption $\beta_\parallel = 0$. The photon coupling to the longitudinal or mixed mode therefore vanishes for $\theta_{\mathbf{k}} = \pi/2$, but increases sharply as $\theta_{\mathbf{k}}$ is varied from this orientation. This effect has been observed in ZnO. The observation of an energy difference between transverse and longitudinal excitons is evidence that the exciton is mobile in the sense that there is a wavevector $\mathbf{k}$ associated with the exciton.

We now discuss observations of excitons in several crystals which have been studied in detail.

Germanium.[2] Both direct and indirect excitons have been studied in germanium. The direct excitons are formed at $\mathbf{k} = 0$ by the absorption of one photon. The direct band gap is between the Γ_8 valence band edge and the Γ_2' band; the energy of the direct gap is 0.898 ev. The effective mass of the spherical Γ_2' band edge is known to be m^*/m 0.037 from experiments on Landau transitions. An approximate effective hole mass can be defined as the mass which reproduces the binding energy of the lowest acceptor state when calculated from the hydrogenic relation; this mass is $0.20m$. Thus the exciton effective mass μ is given by

$$\frac{m}{\mu} \cong \frac{1}{0.037} + \frac{1}{0.20} = \frac{1}{0.031}. \tag{43}$$

The calculated ground-state exciton energy referred to the Γ_2' edge is

$$E_1 = -\frac{\mu e^4}{2\epsilon^2\hbar^2} = -0.0017 \text{ ev}, \tag{44}$$

using $\epsilon = 16$. The observed value is -0.0025 ev.

The indirect excitons are excited across the indirect gap with the emission of a phonon of energy 0.0276 ev. The observed binding energy of the indirect exciton is 0.002(5) ev.

Cadmium Sulfide. The exciton spectrum of this crystal, including fine structure and magneto-optic effects, has been investigated rather fully; see, for example, J. J. Hopfield and D. G. Thomas, *Phys. Rev.* **122,** 35 (1961). The crystal is hexagonal and has the wurtzite structure; the energy band structure of wurtzite-type crystals is discussed by R. C. Casella, *Phys. Rev.* **114,** 1514 (1959); *Phys. Rev. Letters* **5,** 371 (1960). It is believed that the band edges in CdS, CdSe, and ZnO are similar and lie at or very near to $\mathbf{k} = 0$. The energy gap in CdS is 2.53 ev. The valence band is split at $\mathbf{k} = 0$ into three twofold degenerate states, transforming in order of increasing energy as Γ_7, Γ_7, and Γ_9, with separations of 0.057 and 0.016 ev, respectively. The conduction band edge transforms as Γ_7. For Γ_7 the energy has the form

$$\varepsilon(\mathbf{k}) = A(k_x^2 + k_y^2) + Bk_z^2 \pm C(k_x^2 + k_y^2)^{1/2}, \tag{45}$$

as in Problem 14.4. Note that the third term is linear in $\mathbf{k}$, but this term has never been detected. In CdS the conduction band edge is almost isotropic, with $m^* = 0.20m$. The hole masses for the top valence band are $m_\perp = 0.7m$ and $m_\parallel \approx 5m$; the band edge is ellipsoidal. The electronic g value is -1.8 and is very nearly isotropic; the holes

[2] Zwerdling, Lax, Roth, and Button, *Phys. Rev.* **114,** 80 (1959).

(Γ_9) have $g_\parallel = -1.15$ and $g_\perp = 0$. There are three series of exciton lines, each series associated with one of the three valence bands at $\mathbf{k} = 0$.

Perhaps the most interesting feature of the exciton spectrum in CdS is its dependence on the sense of a magnetic field perpendicular to the c axis, with the photon wavevector $\perp\mathbf{H}$ and $\perp c$. It is found that the intensities of the exciton lines vary markedly when $\mathbf{H}$ is reversed in sign, everything else remaining unchanged. That is, the effect depends on the sign of $\mathbf{q} \times \mathbf{H}$, where $\mathbf{q}$ is the photon wavevector. Such an effect is impossible for a free electron, but the absence of a center of symmetry in the crystal allows it to occur. In the reference system of the exciton wavepacket the magnetic field appears as an electric field. The observations are analyzed in the paper by Hopfield and Thomas cited previously. Only a moving exciton could experience such an effect. It would not occur for impurity absorption lines.

Cuprous Oxide. This cubic crystal exhibits beautiful hydrogenic excitons, which have been extensively studied, particularly by E. F. Gross and his school.[3] It is unfortunate that the structure of the band edges are not yet known from cyclotron resonance or other independent studies, but some strong inferences can be made from the exciton results. A striking feature of the exciton spectrum is that the optical transition from the ground state of the crystal to the $1s$ exciton state is very weak, as discussed previously.

For a discussion of excitons in ionic crystals, see D. L. Dexter, *Nuovo cimento supplemento* **7,** 245–286 (1958).

PROBLEMS

1. Discuss for a direct optical transition the dependence of the absorption coefficient on the energy difference of the photon energy from the threshold energy.

2. Show that in a uniaxial crystal with nondegenerate band edges at $\mathbf{k} = 0$ the exciton wave equation may be written as

$$\left\{-\frac{1}{2\mu_0}\nabla^2 - \frac{1}{2}\frac{2\gamma}{3}\left(\frac{1}{2}\frac{\partial^2}{\partial x^2} + \frac{1}{2}\frac{\partial^2}{\partial y^2} - \frac{\partial^2}{\partial z^2}\right) - \frac{e^2}{\epsilon_0 r}\right\}\psi = \left\{E - \frac{1}{2}\left(\frac{k_x^2}{M_\perp} + \frac{k_y^2}{M_\perp} + \frac{k_z^2}{M_\parallel}\right)\right\}\psi,$$

where

$$x = x_e - x_h; \qquad y = y_e - y_h; \qquad z = \left(\frac{\epsilon_\perp}{\epsilon_\parallel}\right)^{1/2}(z_e - z_h);$$

[3] See, for example, *Soviet Physics—Solid State Physics* **2,** 353, 1518, 2637 (1961).

$$\epsilon_0 = (\epsilon_{\parallel}\epsilon_{\perp})^{1/2}; \qquad \frac{1}{\mu_0} = \frac{2}{3}\frac{1}{\mu_{\perp}} + \frac{1}{3}\frac{1}{\mu_{\parallel}}\frac{\epsilon_{\perp}}{\epsilon}; \qquad \frac{1}{\mu_{\perp}} = \frac{1}{m_{e\perp}} + \frac{1}{m_{h\perp}};$$

$$\frac{1}{\mu_{\parallel}} = \frac{1}{m_{e\parallel}} + \frac{1}{m_{h\parallel}}; \qquad \gamma = \frac{1}{\mu_{\perp}} - \frac{1}{\mu_{\parallel}}\frac{\epsilon_{\perp}}{\epsilon_{\parallel}};$$

$$M_{\perp} = m_{e\perp} + m_{h\perp}; \qquad M_{\parallel} = m_{e\parallel} + m_{h\parallel}.$$

3. Treat the term in γ in Problem 2 as a small perturbation. Show that to first order in γ the energies of the $n = 1$ and $n = 2$ states are, with E_1 as the effective rydberg,

$$1S: \quad E_g - E_1;$$

$$2S: \quad E_g - \tfrac{1}{4}E_1;$$

$$2P_0: \quad E_g - \tfrac{1}{4}E_1(1 + \tfrac{4}{15}\gamma);$$

$$2P_{\pm 1}: \quad E_g - \tfrac{1}{4}E_1(1 + \tfrac{2}{15}\gamma).$$

4. In the magnetostark effect as discussed above for CdS, estimate the magnitude of the quasielectric field for a magnetic field of 30 kilo-oersteds.

5. Show that the transition probability for a "first forbidden" electric dipole process creating an exciton is proportional to $|(\partial F_n/\partial x)_{\substack{x=0\\r=0}}|^2$.

16 Electrodynamics of metals

Diverse and subtle phenomena are observed when a metal is coupled to an electromagnetic field. The effects often give important detailed information about the fermi surface. In this chapter we treat the anomalous skin effect; cyclotron resonance; the dielectric anomaly; magnetoplasma resonance; and spin diffusion.

ANOMALOUS SKIN EFFECT

We consider first the normal skin effect. The displacement current $\dot{\mathbf{D}}$ in a metal may usually be neglected at frequencies $\omega \ll \sigma$, where σ is the conductivity in esu. In a good conductor at room temperature $\sigma \approx 10^{18}\ \sec^{-1}$. We note that $4\pi\sigma \equiv \omega_p{}^2\tau$, where ω_p is the plasma frequency and τ the carrier relaxation time. Then the maxwell equations are

$$\text{(1)} \qquad \text{curl } \mathbf{H} = \frac{4\pi}{c}\sigma\mathbf{E}; \qquad \text{curl } \mathbf{E} = -\frac{1}{c}\mu\dot{\mathbf{H}},$$

or

$$\text{(2)} \qquad \text{curl curl } \mathbf{H} = -\nabla^2\,\mathbf{H} = -\frac{4\pi\sigma\mu}{c^2}\dot{\mathbf{H}},$$

whence we have the eddy current equation

$$\text{(3)} \qquad k^2\mathbf{H} = \frac{i4\pi\sigma\mu\omega}{c^2}\mathbf{H},$$

for $H \sim e^{i(\mathbf{k}\cdot\mathbf{x}-\omega t)}$. The wavevector has equal real and imaginary parts; if σ is real and equal to σ_0:

$$\text{(4)} \qquad k = (1+i)(2\pi\mu\sigma_0\omega/c^2)^{1/2} = (1+i)/\delta_0.$$

The imaginary part of k is the reciprocal of the classical skin depth

$$\text{(5)} \qquad \delta_0 = (c^2/2\pi\mu\sigma_0\omega)^{1/2}.$$

At room temperature in a good conductor at 3×10^{10} cps we have $\delta_0 \approx 10^{-4}$ cm; in a very pure specimen at helium temperature $\delta_0 \approx 10^{-6}$ cm. The results (4) and (5) are written on the supposition that σ is real, which means that $\omega\tau \ll 1$. The elementary result for the r-f conductivity is

$$\sigma = \frac{ne^2\tau}{m(1 - i\omega\tau)} = \frac{\sigma_0}{1 - i\omega\tau}; \tag{6}$$

for $\omega\tau \ll 1$ this reduces to the usual static value $\sigma_0 = ne^2\tau/m$. The result (6) follows immediately from a drift velocity or transport equation treatment. In pure specimens at helium temperature and microwave frequencies it is possible to have $\omega\tau \gg 1$; in this limit

$$k^2 \cong \frac{4\pi\mu\omega\sigma_0}{c^2} \frac{(i - \omega\tau)}{(\omega\tau)^2}; \qquad \mathscr{I}\{k\} \cong \frac{1}{\delta_0\omega\tau}. \tag{7}$$

But with either (5) or (7) for δ there is a serious question of the validity of the calculation at low temperatures because the carrier mean free path Λ at helium temperature may be larger than the skin depth. Typically τ might be 10^{-10} sec, so that for an electron on the fermi surface the mean free path is

$$\Lambda = v_F\tau \approx (10^8)(10^{-10}) \approx 10^{-2} \text{ cm}, \tag{8}$$

much larger than the skin depth $\delta_0 \approx 10^{-6}$ cm given by (5) or $\delta \approx 10^{-5}$ cm given by (7). When $\Lambda/\delta \gtrsim 1$ the electric current density $\mathbf{j}(\mathbf{x})$ can no longer be determined only by the local value $\mathbf{E}(\mathbf{x})$ of the electric field, and our use of $\sigma\mathbf{E}$ in the maxwell equation is invalid. The region $\Lambda > \delta$ is called the region of the *anomalous skin effect.* A long mean free path has a profound effect on the propagation characteristics of the medium.

SURFACE IMPEDANCE

The observable electrodynamic properties of a metal surface are described completely by the surface impedance $Z(\omega)$ defined as

$$\boxed{Z = R - iX = \frac{4\pi}{c} \cdot \frac{E_t}{H_t},} \tag{9}$$

where E_t, H_t are the tangential components of $\mathbf{E}$ and $\mathbf{H}$ evaluated at the surface of the metal. The real part R of Z is called the surface resistance; R determines the power absorption by the metal. The imaginary part X is called the surface reactance; it determines the frequency shift of a resonant cavity bounded by the metal. From the equation for curl **E** we have

$$i\omega\mu H_t/c = \partial E_t/\partial z, \tag{10}$$

for permeability μ. Then

$$Z = \frac{4\pi i\omega\mu}{c^2}\left(\frac{E_t}{\partial E_t/\partial z}\right)_{+0} = \frac{4\pi\omega\mu}{kc^2}, \tag{11}$$

with $\hat{\mathbf{z}}$ normal to the surface and directed inward.

For the normal skin effect $k = (1 + i)/\delta_0$, so that

$$Z = \frac{4\pi\omega\mu}{kc^2} = (1 - i)\frac{2\pi\omega\delta_0\mu}{c^2} = (1 - i)\frac{2\pi\mu}{c}\cdot\frac{\delta_0}{\lambda\!\!\!^{-}}, \tag{12}$$

where $\lambda\!\!\!^{-} = c/\omega$. Thus the ratio of the skin depth to the wavelength determines the magnitude of Z.

The rate of energy loss per unit area normal to the z direction is given by the time average of the real part of the poynting vector **S** of magnitude

$$S = \frac{c}{4\pi}|\mathbf{E}\times\mathbf{H}| = \frac{c}{4\pi}E_xH_y = \left(\frac{c}{4\pi}\right)^2 ZH_y^2, \tag{13}$$

when the fields are evaluated at $z = +0$. The time average of the real part is

$$\langle\mathcal{R}\{S\}\rangle = \tfrac{1}{2}\left(\frac{c}{4\pi}\right)^2 H^2\mathcal{R}\{Z\}, \tag{14}$$

where H is the amplitude of H_y at the surface.

We now examine the contribution of the surface to the inductance of a magnetizing circuit: consider a flat solenoid with inside it a flat slab of conductor of thickness $2d$. Let $\mathcal{J}$ be the current per unit length in the solenoid. Then the inductance per unit length is

$$L = \mathcal{R}\{\text{flux}/\mathcal{J}\} = \mathcal{R}\left\{\frac{1}{\mathcal{J}}\int_{-d}^{d} dz\, H_y(z)\mu\right\}. \tag{15}$$

Because $\partial E_x/\partial z = i\omega\mu H_y/c$, we have

$$2E_x(d) = \frac{i\omega}{c}\int_{-d}^{d} dz\, H_y(z)_\mu, \tag{16}$$

with $H(d) = 4\pi\mathcal{J}/c$. Thus, from the definition (9),

$$L = \mathcal{R}\left\{\frac{(2c/i\omega)E_x(d)}{(c/4\pi)H_y(d)}\right\} = \frac{2c}{\omega}\mathcal{R}\{-iZ\} = \frac{2c}{\omega}\mathcal{I}\{Z\}. \tag{17}$$

For a specimen forming one end surface of a rectangular cavity in the fundamental TE mode the effective change in length of the cavity can be shown to be

$$\delta l = -\frac{c}{4\pi}\lambda\mathcal{I}\{Z\}. \tag{18}$$

The extreme anomalous skin effect ($\Lambda/\delta \gg 1$) can be understood qualitatively in terms of the ineffectiveness concept of Pippard. Only those electrons traveling nearly parallel to the surface remain in the electric field long enough to absorb a significant amount of energy. We suppose that the effective electrons are those in an angle $\sim\delta/\Lambda$. Thus the concentration of effective electrons is $n_{\text{eff}} = \gamma n\delta/\Lambda$, where γ is a constant of the order of unity. Thus the effective conductivity is

$$\sigma_{\text{eff}} = \frac{n_{\text{eff}}e^2\tau}{m^*} = \gamma\frac{\delta}{\Lambda}\sigma_0. \tag{19}$$

If we use σ_{eff} for σ_0 in (5) and solve for the effective skin depth, we find

$$\delta^3 = \frac{c^2\Lambda}{2\pi\gamma\sigma_0\omega}, \tag{20}$$

and the surface resistivity $\mathcal{R}\{Z\}$ is

$$R = \frac{2\pi\omega}{c^2}\left(\frac{c^2\Lambda}{2\pi\gamma\sigma_0\omega}\right)^{1/3}. \tag{21}$$

This indeed is of the form of the correct answer derived below, with $\gamma \approx 10$.

The most important property of (21) is that the surface resistance is independent of the mean free path Λ, because $\sigma_0 \propto \Lambda$. Thus a measurement of R determines the momentum at the fermi level in the direction of the electric field:

$$\frac{\sigma_{xx}}{\Lambda} = \frac{ne^2}{m_{xx}^* v_F} = \frac{ne^2}{k_F}. \tag{22}$$

For a general fermi surface the surface resistance tensor component R_{xx} involves the form of the fermi surface through $\int |\rho(k_y)|\, dk_y$, where $\rho(k_y)$ is the radius of curvature of the slice of the fermi surface at constant k_y. This result is seen directly from Fig. 1. The effective

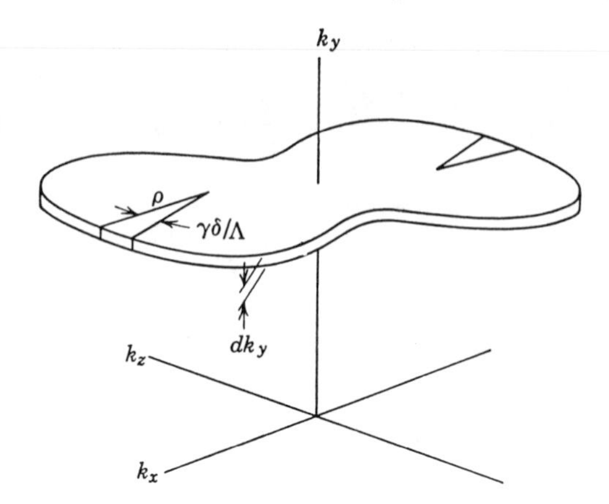

FIG. 1. Slice of a fermi surface at constant k_y, showing the effective electrons whose velocities lie within the angle $\gamma\delta/\Lambda$ of the surface. Here ρ is the radius of curvature in the slice.

current involves only the electrons in the sectors shown as lying in the appropriate angular range. In a relaxation time the electrons in the sectors advance by $\Delta k_x = e\mathcal{E}_x\tau$; the volume in **k** space which contributes to the current is

$$\Omega_{\text{eff}} = \int dk_y\, e\mathcal{E}_x\tau |\rho(k_y)| (\gamma\delta/\Lambda),$$

so that

$$J_x = ev_F \frac{1}{4\pi^3} \Omega_{\text{eff}} = \frac{e^2\mathcal{E}_x}{4\pi^3} \gamma\delta \int dk_y\, |\rho(k_y)|. \tag{23}$$

This defines the effective conductivity for a general fermi surface. Then for diffuse scattering

$$R_{xx} = \left(\frac{3^{3/2}\omega^2}{64\pi^2c^4e^2 \int dk_y\, |\rho(k_y)|}\right)^{1/3}. \tag{24}$$

The work of Pippard [*Trans. Roy. Soc. (London)* **A250,** 325 (1957)] on copper demonstrated the power of the anomalous skin effect in the determination of fermi surfaces. For further details see A. B. Pippard, "Dynamics of Conduction Electrons" in *LTP*. If the surface is polycrystalline with crystallites oriented randomly, the anomalous skin effect leads to a determination of the total area of the fermi surface.

Mathematical Theory of the Anomalous Skin Effect. The surface of a semi-infinite metal is in the xy plane, with $\hat{\mathbf{z}}$ directed towards the interior. We write

$$E(z) = E_{x0}(z)e^{-i\omega t}; \qquad H(z) = H_{y0}(z)e^{-i\omega t}. \tag{25}$$

Then the maxwell equations are

$$-\frac{dH}{dz} = -\frac{i\omega}{c}E + \frac{4\pi}{c}j; \qquad \frac{dE}{dz} = \frac{i\omega}{c}H, \tag{26}$$

where now $j(z)$ is the current density. We eliminate H:

$$\frac{d^2E}{dz^2} + \frac{\omega^2}{c^2}E = -\frac{4\pi i\omega}{c^2}j. \tag{27}$$

We treat the electrons as isotropic with mass m. The distribution function is $f = f_0 + f_1(\mathbf{v},z)$, where f_0 is the unperturbed distribution. If the relaxation time is τ, the boltzmann transport equation in lowest order for f_1 is

$$v_z\frac{\partial f_1}{dz} + \frac{1 - i\omega\tau}{\tau}f_1 = -\frac{e}{m}\frac{\partial f_0}{\partial v_x}E(z) = -e\frac{\partial f_0}{\partial \varepsilon}v_xE(z). \tag{28}$$

It is convenient to work with the fourier transforms defined by:

$$\mathcal{E}(q) = (2\pi)^{-1/2}\int_{-\infty}^{\infty} dz\, Ee^{iqz}; \qquad J(q) = (2\pi)^{-1/2}\int_{-\infty}^{\infty} dz\, je^{iqz}. \tag{29}$$

We assume specular reflection of the electrons at the surface. If we consider the remaining half of space ($z < 0$) filled with another piece of the same metal, the electrons in each half will have the same history as if the reflection were specular. We must only provide the proper electric field at $z = 0$. The gradient of this artificial field will have a cusp at $z = 0$, because the field is damped in both $\pm z$ directions:

$$\left(\frac{\partial E}{\partial z}\right)_{+0} = -\left(\frac{\partial E}{\partial z}\right)_{-0}. \tag{30}$$

This condition can be incorporated into (27) by adding a delta function:

$$\frac{d^2E}{dz^2} + \frac{\omega^2}{c^2}E = \frac{4\pi i\omega}{c^2}j + 2\left(\frac{dE}{dz}\right)_{+0}\delta(z). \tag{31}$$

We express each term as a fourier integral and find

$$-q^2\mathcal{E}(q) + (\omega/c)^2\mathcal{E}(q) = -(4\pi i\omega/c^2)J(q) + (2/\pi)^{1/2}(dE/dz)_{+0}. \tag{32}$$

The transport equation gives us another relation between $\mathcal{E}$ and J.

We define the transform

$$\Phi_1(q) = (2\pi)^{-1/2} \int_{-\infty}^{\infty} dz\, f_1 e^{iqz}. \tag{33}$$

Then (28) becomes

$$(1 - i\omega\tau + iqv_z\tau)\Phi_1(q) = -\frac{\partial f_0}{\partial \varepsilon} \tau e v_x \mathcal{E}(q) \cong \delta(\varepsilon - \varepsilon_F)\tau e v_x \mathcal{E}(q), \tag{34}$$

for a fermi gas at $k_B T \ll \varepsilon_F$. The solution is

$$\Phi_1(q) = \frac{\delta(\varepsilon - \varepsilon_F) e v_x \tau}{1 - i\omega\tau + iqv_z\tau} \mathcal{E}(q). \tag{35}$$

The electric current density

$$j(z) = \frac{e}{4\pi^3} \int d^3k\, v_x f_1 \tag{36}$$

has the fourier components

$$J(q) = \frac{e}{4\pi^3} \int d^3k\, v_x \Phi_1(q) = \mathcal{E}(q) \frac{e^2 m^3 \tau}{4\pi^3} \int d^3v\, \frac{v_x^2 \delta(\varepsilon - \varepsilon_F)}{1 - i\omega\tau + iq\tau v_z}. \tag{37}$$

We define the conductivity tensor component $\sigma_{xx}(q)$ by $J(q) = \sigma_{xx}(q)\mathcal{E}(q)$; then

$$\sigma_{xx}(q) = \frac{e^2 m^3 \tau}{4\pi^2} \int dv\, v^4\, \delta(\varepsilon - \varepsilon_F) \int_{-1}^{1} d(\cos\theta)\, \frac{\sin^2\theta}{1 - i\omega\tau + iq\tau v \cos\theta}. \tag{38}$$

The whole problem now revolves on the evaluation of the integral over $d(\cos\theta)$. The general solution is given in detail by G. E. H. Reuter and E. H. Sondheimer, *Proc. Roy. Soc.* **A195,** 336 (1948). The reader may verify that for $q \to 0$ the result (38) becomes

$$\sigma_{xx}(0) \to \frac{ne^2\tau}{m} \cdot \frac{1}{1 - i\omega\tau} \equiv \frac{\sigma_0}{1 - i\omega\tau}, \tag{39}$$

in agreement with our earlier result.

We evaluate (38) in the extreme anomalous region $\Lambda q \gg |1 - i\omega\tau|$. Let

$$\Lambda' = \frac{\Lambda}{1 - i\omega\tau} = \frac{v_F\tau}{1 - i\omega\tau}; \tag{40}$$

then the integral over $d(\cos\theta)$ is

$$\frac{1}{1 - i\omega\tau} \int_{-1}^{1} dx\, \frac{1 - x^2}{1 + i\Lambda' qx} \cong \frac{2\tan^{-1}\Lambda' q}{\Lambda q} \cong \frac{\pi}{\Lambda|q|}, \tag{41}$$

on neglecting terms of higher order in $(\Lambda' q)^{-1}$. Thus

$$\sigma_{xx}(q) \cong \frac{ne^2\tau}{m} \cdot \frac{3\pi}{4\Lambda|q|} = \sigma_0 \cdot \frac{3\pi}{4\Lambda|q|}, \tag{42}$$

on using the relation $2m\mathcal{E}_F = (3\pi^2 n)^{2/3}$. Note that this is a transverse conductivity ($\hat{\mathbf{x}} \perp \mathbf{q}$) and therefore is not identical with the longitudinal conductivity associated with the longitudinal dielectric constant of the electron gas of Chapter 6.

We combine this result with (32), neglecting the displacement current term in $(\omega/c)^2$:

$$\mathcal{E}(q)(-q^2 + 4\pi i \sigma_{xx}\omega c^{-2}) = (2/\pi)^{1/2}(dE/dz)_{+0}. \tag{43}$$

so that

$$E(z) = \frac{1}{\pi}\left(\frac{dE}{dz}\right)_{+0} \int_{-\infty}^{\infty} dq \, \frac{e^{-iqz}}{-q^2 + 2i(\sigma_{xx}/\delta^2\sigma_0)}, \tag{44}$$

where $\delta^2 = c^2/2\pi\omega\sigma_0$. Using (42) and setting $\xi = q(2\Lambda\delta^2/3\pi)^{1/3}$, the integral in (44) at $z = 0$ has the value

$$-2i(2\Lambda\delta^2/3\pi)^{1/3} \int_0^{\infty} \frac{d\xi}{1 + i\xi^3} = -\tfrac{2}{3}(2\pi^2\Lambda\delta^2/3)^{1/3}(1 + i3^{-1/2}), \tag{45}$$

using Dwight, 856.6.

The surface impedance Z_∞ in the extreme anomalous region is found from (11), (44), and (45):

$$Z_\infty = \frac{4\pi i\omega}{c^2}\left(\frac{E}{dE/dz}\right)_{+0} = \tfrac{8}{9}(3^{1/2}\pi\omega^2\Lambda/c^4\sigma_0)^{1/3}(1 - i3^{1/2}), \tag{46}$$

in agreement with result of Reuter and Sondheimer, but we have not assumed $\omega\tau \ll 1$. Because $\sigma_0 \propto \Lambda$, we see that Z_∞ is independent of the mean free path. For diffuse surface reflection it turns out that the factor $\frac{8}{9}$ is absent from Z_∞. It is believed that reflection is likely to be diffuse except under special surface conditions.

The quantum theory of the anomalous skin effect in normal and superconducting metals is treated by D. C. Mattis and J. Bardeen, *Phys. Rev.* **111,** 412 (1958).

CYCLOTRON RESONANCE IN METALS

The observation of cyclotron resonance in metals naturally requires that the surface impedance should show a resonance behavior as a function of the static magnetic field. The most suitable geometry was suggested by M. I. Azbel and E. A. Kaner [*Soviet Phys.—JETP* **3,** 772 (1956)]. The static field $\mathbf{H}_0$ lies in the plane of the sample; the rf electric field also lies in the plane of the surface and may be either parallel (longitudinal) or at right angles (transverse) to $\mathbf{H}_0$. If the relaxation time is sufficiently long, we may think of the carriers as spiraling about $\mathbf{H}_0$, dipping once each cycle in and out of the rf field localized in the skin depth. Resonant absorption of energy will occur if a carrier sees an electric field in the same phase every time the carrier is in the skin depth: thus at resonance

$$\frac{2\pi}{\omega_c} = p\frac{2\pi}{\omega}; \qquad p = \text{integer}, \tag{47}$$

or $\omega_c = \omega/p$, where p is the index of the subharmonic. Recall that in cyclotron resonance in semiconductors only the possibility $p = 1$ occurred because the rf field was assumed to penetrate the specimen uniformly.

A qualitative argument given below suggests that in the presence of the magnetic field

$$Z_\infty(H) \cong Z_\infty(0)\left[1 - \exp\left(-\frac{2\pi}{\omega_c\tau} - i\frac{2\pi\omega}{\omega_c}\right)\right]^{1/3}. \tag{48}$$

This is indeed periodic in ω/ω_c. To understand (48), consider the contribution to the current and thus to the conductivity of a single carrier having an orbital radius large in comparison with the skin depth. The change in phase of the rf field in the period $2\pi/\omega_c$ is $2\pi\omega/\omega_c$; with collisions $\omega \to \omega - i/\tau$, so that on each revolution the current is multiplied by the phase factor

$$e^{-w} \equiv \exp\left(-\frac{2\pi}{\omega_c\tau} - i\frac{2\pi\omega}{\omega_c}\right). \tag{49}$$

The phase factor of the total current arising from all cycles is

$$1 + e^{-w} + e^{-2w} + \cdots = \frac{1}{1 - e^{-w}}; \tag{50}$$

this quantity is involved in the effective conductivity, as is clear from Chamber's formulation of the transport equation. From (21) or (46) we know that $Z_\infty \propto (1/\sigma_{\text{eff}})^{1/3}$, whence (48) follows.

A result closely similar to (48) follows from the solution of the transport equation in the extreme anomalous limit in the presence of a static magnetic field parallel to the surface. We shall see that the treatment is exactly the same as without the field, but now a term $\omega_c\, \partial f_1/\partial\varphi$ is added to the left-hand side of the transport equation (28), where $\omega_c = eH/m_cc$ is the cyclotron frequency and φ is the azimuthal angle about $\mathbf{H}_0$ as polar axis.

In a static magnetic field $\mathbf{H}_0$ the linearized transport equation is

$$-i\omega f_1 + v_z \frac{\partial f_1}{\partial z} + \frac{eE}{m}\frac{\partial f_0}{\partial v_x} + \frac{e}{mc}\mathbf{v} \times \mathbf{H}_0 \cdot \frac{\partial f_1}{\partial \mathbf{v}} = -\frac{f_1}{\tau}. \tag{51}$$

We consider specifically the arrangement with $\mathbf{H}_0 \| \mathbf{E} \| \hat{\mathbf{x}}$. Introduce spherical polar coordinates in velocity space so that $\mathbf{v} \equiv (v,\theta,\varphi)$, referred to $\hat{\mathbf{x}}$ as the polar axis. Then

$$\frac{eH_0}{mc}\hat{\mathbf{x}} \cdot \mathbf{v} \times \frac{\partial f_1}{\partial \mathbf{v}} = \omega_c \hat{\mathbf{x}} \cdot \operatorname{curl} \mathbf{v} f_1 = \omega_c \frac{\partial f_1}{\partial \varphi}, \tag{52}$$

so that the transport equation is

$$(1 - i\omega\tau)f_1 + v\tau \sin\theta \sin\varphi \frac{\partial f_1}{\partial z} + \omega_c\tau \frac{\partial f_1}{\partial \varphi} = -e\tau v \cos\theta \frac{\partial f_0}{\partial \varepsilon}. \tag{53}$$

In terms of the fourier transforms,

$$\left(1 + iv\bar{\tau}q \sin\theta \sin\varphi + \omega_c\bar{\tau}\frac{\partial}{\partial\varphi}\right)\Phi_1(q) = \delta(\varepsilon - \varepsilon_F)e\bar{\tau}v\cos\theta\,\mathcal{E}(q), \tag{54}$$

where $\bar{\tau} \equiv \tau/(1 - i\omega\tau)$. This is a simple linear differential equation and has the solution

$$\begin{aligned}\Phi_1(q) &= \delta(\varepsilon - \varepsilon_F)ev\cos\theta\,\mathcal{E}(q)\omega_c^{-1} \\ &\quad \cdot \int_{-\infty}^{\varphi} d\varphi' \exp\left\{\frac{(\varphi' - \varphi) - iv\bar{\tau}q\sin\theta(\cos\varphi' - \cos\varphi)}{\omega_c\bar{\tau}}\right\},\end{aligned} \tag{55}$$

as may be verified by differentiation. The q component of the current density is

$$J(q) = \frac{em^3}{4\pi^3}\int v^3\,dv\cos\theta\sin\theta\,d\theta\,d\varphi \cdot \Phi_1(q). \tag{56}$$

The integration is not trivial; for full details see S. Rodriguez, *Phys. Rev.* **112,** 80 (1958). In the limit $v_Fq/\omega_c \gg 1$, which is equivalent to $r_c \gg \delta$, where r_c is the cyclotron radius, it is found that

$$J(q) = \frac{3\pi}{4}\frac{ne^2\mathcal{E}(q)}{mv_Fq}\coth\left(\frac{1 - i\omega\tau}{\omega_c\tau}\pi\right), \tag{57}$$

which shows periodic oscillations provided $\omega_c\tau$, $\omega\tau \gg 1$. In the same limit the surface impedance in the extreme anomalous region with specular reflection is

$$Z_\infty(H) = Z_\infty(0)\tanh^{1/3}[\pi(1 - i\omega\tau)/\omega_c\tau], \tag{58}$$

according to D. C. Mattis and G. Dresselhaus, *Phys. Rev.* **111,** 403 (1958), and S. Rodriguez, *Phys. Rev.* **112,** 1016 (1958). The periodicity of (58) is the same as the approximate form (48), because

$$\tanh (x - iy) = \frac{\sinh 2x - i \sin 2y}{\cosh 2x + \cos 2y}. \tag{59}$$

In the extreme anomalous limit the result (58) applies to both the longitudinal and transverse geometries. The boundary condition in a magnetic field is more complicated than realized by these authors: see Azbel and Kaner, *JETP* **39,** 80 (1963).

An experimental cyclotron resonance curve for copper, together with a theoretical calculation for adjusted values of m_c and τ, is shown in Fig. 2. The mass involved is given in terms of the area S of the fermi surface by the relation

$$m_c = \frac{1}{2\pi}\frac{dS}{d\varepsilon}, \tag{60}$$

from (11.54). Stationary values of $dS/d\varepsilon$ with respect to k_H define

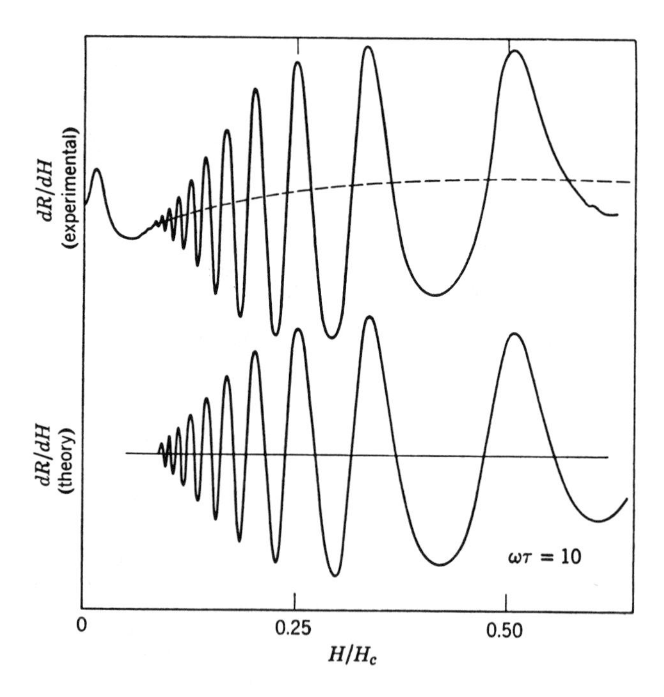

FIG. 2. Cyclotron resonance absorption in copper; comparison of calculations of the magnetic field dependence of the derivative of the surface resistivity with experimental results at 24 kMc/sec. (After Kip, Langenberg, and Moore.)

the sections of the fermi surface which contribute to the central and resolved portions of the cyclotron lines.

In the cyclotron resonance experiment in metals, we observe the resonance of a few selected electrons with orbits passing through the skin depth. These electrons do not necessarily have the same contribution to their effective mass from electron-electron interactions as in the de Haas-van Alphen effect where all the electrons participate equally. One must not expect the cyclotron experiments to reproduce the masses obtained from the de Haas-van Alphen effect, even if the fermi surface is such that $dS/d\varepsilon$ in (60) can be related unambiguously to S itself, the quantity involved in the de Haas-van Alphen effect.

The effective mass is altered also by the electron-phonon interaction, as in the polaron problem, but we expect this effect to be the same in cyclotron resonance as in other experiments involving the effective mass at the fermi surface. Optical experiments, however, may see the effective mass of the electron unclothed by the phonon; the electrons involved in an optical transition will not, however, lie only on the fermi surface.

The theory of cyclotron resonance in metals when the magnetic field is normal to the surface is discussed by P. B. Miller and R. R. Haering, *Phys. Rev.* **128,** 126 (1962); in this geometry the resonance condition contains a doppler shift.

DIELECTRIC ANOMALY

We consider now the dielectric properties of a free electron gas having n carriers per unit volume of effective mass m^* and charge e. The acceleration equation is, assuming $\omega\tau \gg 1$,

$$m^*\ddot{x} = eE; \qquad -\omega^2 m^* x = eE, \tag{61}$$

so that the free carrier contribution to the dielectric polarization is

$$P = nex = -\frac{ne^2}{m^*\omega^2} E. \tag{62}$$

Thus the dielectric constant is given by

$$\epsilon = 1 - \frac{4\pi ne^2}{m^*\omega^2} + 4\pi\chi_a, \tag{63}$$

where χ_a is the dielectric susceptibility per unit volume of whatever other material is present. The carrier contribution to ϵ may also be written as $-\omega_p{}^2/\omega^2$, where ω_p is equal to the plasma frequency.

Now at low frequencies $-\omega_p{}^2/\omega^2$ will dominate ϵ and make it negative. The dispersion relation for electromagnetic waves is, with $\mu = 1$,

$$\omega^2 = c^2k^2/\epsilon; \tag{64}$$

if $\epsilon \cong -\omega_p^2/\omega^2$, we have $-\omega_p^2 = c^2k^2$, or

$$k = i\omega_p/c. \tag{65}$$

Thus the wave is damped in a metal in a distance of the order of the Debye length, 10^{-6} cm, independent of frequency provided that $\omega\tau \gg 1$ and $\omega_p \gg \omega$. This result neglects the anomalous skin effect.

For ω higher than a root ω_0 of

$$\epsilon(\omega_0) = 0, \tag{66}$$

the dielectric constant becomes positive, the wavevector is real, and a wave may propagate in the medium. This theory accounts for the onset of transmission in the alkali metals in the ultraviolet. The change in the reflectivity of the crystal when ω passes through ω_0 is known as the *dielectric anomaly.* If the carrier concentration n is known, a determination of ω_0 gives us a value of m^*. The usefulness of the value naturally depends on the simplicity of the band edge structure.

The treatment above assumes that we are not in the region of the anomalous skin effect. It is quite possible in the visible region of the spectrum to have $\omega\tau \gg 1$ without having $\Lambda > \delta$. The situation in the visible region of the spectrum is discussed by Reuter and Sondheimer, as previously cited, and by T. Holstein, *Phys. Rev.* **88,** 1427 (1952).

With values of the electron effective masses for the alkali metals calculated by Brooks, the transparency limit should occur at the optical wavelength 1840, 2070, 2720, and 3000 A for Li, Na, K, and Rb, respectively. The values observed are 1550, 2100, 3150, and 3400 A. No correction for ion core polarizability was made; the cores should have an observable effect. We see from (49) that the transparency limit in the presence of χ_a is

$$\omega_0^2 = \frac{\omega_p^2}{1 + 4\pi\chi_a}. \tag{67}$$

In metallic silver the value of $(1 + 4\pi\chi_a)^{1/2}$ is about 2, from reflectivity experiments. This is consistent with estimates of χ_a from the refractive indices of silver halides compared with alkali halides.

MAGNETOPLASMA PROPAGATION

A striking propagation effect was observed by Bowers, Legendy, and Rose [*Phys. Rev. Letters* **7,** 339 (1961)] in high purity sodium at 4°K when placed in a static magnetic field of the order of 10^4 oersteds. It is observed that real electromagnetic waves propagate in metal at a frequency of the order of 10 cycles per second and at wavelengths of the order of 1 cm. That is, the phase velocity of light is reduced to ~ 10 cm/sec, which corresponds to a refractive index $\sim 3 \times 10^9$ and a dielectric constant $\sim 10^{19}$. The electromagnetic modes under these conditions are called *helicon* modes and were predicted by Aigrain.

We consider the equations of motion of a free electron of mass m^* in a static magnetic field H parallel to the z axis and in an rf electric field with components E_x, E_y:

$$m^*\ddot{x} = eE_x + \frac{e}{c}\dot{y}H; \qquad m^*\ddot{y} = eE_y - \frac{e}{c}\dot{x}H, \tag{68}$$

or, with $\omega_c = eH/m^*c$,

$$-\omega^2 x = \frac{e}{m^*}E_x - i\omega\omega_c y; \qquad -\omega^2 y = \frac{e}{m^*}E_y + i\omega\omega_c x. \tag{69}$$

In the conditions of the experiment $\omega \ll \omega_c$, so that the dielectric susceptibility tensor is given by, with $\omega_p^2 = 4\pi ne^2/m^*$,

$$4\pi\bar{\bar{\chi}} \cong \begin{pmatrix} 0 & -i\omega_p^2/\omega\omega_c & 0 \\ i\omega_p^2/\omega\omega_c & 0 & 0 \\ 0 & 0 & -\omega_p^2/\omega^2 \end{pmatrix} \cong \bar{\bar{\epsilon}}, \tag{70}$$

because $\omega_p^2/\omega\omega_c \gg 1$. For a wave propagating with wavevector k along the z axis the maxwell equations reduce to

$$\begin{gathered} kH_y = \frac{\omega}{c}(\epsilon_{xx}E_x + \epsilon_{xy}E_y); \qquad kH_x = -\frac{\omega}{c}(\epsilon_{yx}E_x + \epsilon_{yy}E_y); \\ kE_y = -\frac{\omega}{c}H_x; \qquad kE_x = \frac{\omega}{c}H_y. \end{gathered} \tag{71}$$

Then on forming the equation for $E_x + iE_y$ we have, with the neglect of ϵ_{xx} and ϵ_{yy},

$$k^2 = \frac{\omega\omega_p^2}{c^2\omega_c}; \qquad \text{or} \qquad \omega = \frac{k^2Hc}{4\pi ne}, \tag{72}$$

independent of the particle mass; the experiment is therefore suited to the determination of the effective charge of a carrier in a metal, although it appears fairly well established theoretically that the electron-electron interaction does not make the effective charge on a carrier differ from e. In (72) let $k = 10$ cm^{-1}; $H = 10^4$ oersteds, and $n = 10^{23}$ cm^{-3}; then $\omega \sim 60$ sec^{-1}.

For a general energy surface we can discuss the problem in the limit $\omega_c\tau \gg 1$ with $\omega\tau \ll 1$. Pippard has given the general limiting form of the static magnetoconductivity tensor. With $\mathbf{H} \parallel \hat{\mathbf{z}}$ and with no open orbits,

$$\bar{\bar{\sigma}} = \begin{pmatrix} AH^{-2} & GH^{-1} & CH^{-1} \\ -GH^{-1} & DH^{-2} & EH^{-1} \\ -CH^{-1} & -EH^{-1} & F \end{pmatrix}, \tag{73}$$

where A, C, D, E, F, G are independent of H. In general

$$\epsilon_{\mu\nu} = -i\frac{4\pi}{\omega}\sigma_{\mu\nu} + \delta_{\mu\nu}; \tag{74}$$

for $\omega \ll \omega_c \ll \omega_p$ as above, the secular equation is, approximately,

$$\begin{vmatrix} c^2k^2 & -i4\pi G\omega/H \\ i4\pi G\omega/H & c^2k^2 \end{vmatrix} = 0, \tag{75}$$

provided $G \neq 0$; the solution of (75) is

$$\omega = \pm k^2c^2H/4\pi G, \tag{76}$$

of the form of (72). We emphasize that all magnetoresistive effects are included in (75).

If there are open orbits in a general direction in the crystal, the static ($\omega\tau \ll 1$) magnetoconductivity tensor in the xy plane has the form

$$\bar{\bar{\sigma}} = \begin{pmatrix} A_1 & GH^{-1} \\ -GH^{-1} & A_2 \end{pmatrix}, \tag{77}$$

where A_1, A_2 are independent of H. If cA_1, $cA_2 \gg \omega$, then

$$\bar{\bar{\epsilon}} \cong i4\pi\bar{\bar{\sigma}}/\omega. \tag{78}$$

The general secular equation is, from (71),

$$\begin{vmatrix} \frac{c^2k^2}{\omega^2} - \epsilon_{xx} & \epsilon_{xy} \\ -\epsilon_{xy} & \frac{c^2k^2}{\omega^2} - \epsilon_{yy} \end{vmatrix} = 0. \tag{79}$$

If $A_1 = A_2 = A$, then for $\omega\tau \ll 1$

$$\omega = \pm\frac{k^2c^2H}{4\pi(G + iAH)}. \tag{80}$$

Now A is the fraction of the static conductivity (in zero magnetic field) which is carried by open orbits and G/H is some average of $\sigma_0/\omega_c\tau$. Thus in magnetic fields sufficiently strong that $\omega_c\tau \gg 1/\eta$, where η is the open-orbit fraction, the low-frequency resonance is lost. If H is not this large, the resonance will merely be damped by the conductivity associated with the open orbits.

SPIN RESONANCE IN THE NORMAL SKIN EFFECT

The surface impedance in the normal skin effect involves the magnetic permeability: from (12)

$$Z = (1 - i)\left(\frac{2\pi\omega\mu}{c^2\sigma}\right)^{\frac{1}{2}} \propto (1 - i)\mu^{\frac{1}{2}}. \tag{81}$$

Now write

$$\mu = 1 + 4\pi(\chi_1 + i\chi_2), \tag{82}$$

where $\chi_1 = \Re\{\chi\}$ and $\chi_2 = \Im\{\chi\}$. If, as in nuclear resonance, $|\chi| \ll 1$, then we may expand $\mu^{\frac{1}{2}}$ to obtain

$$Z \propto (1 - i)(1 + 2\pi\chi_1 + 2\pi i\chi_2), \tag{83}$$

whence

$$\Re\{Z\} \propto 1 + 2\pi(\chi_1 + \chi_2). \tag{84}$$

This shows that the actual power absorption near nuclear resonance in a metal specimen thick in comparison with the skin depth is determined by $\chi_1 + \chi_2$, and not by the absorptive component χ_2 alone. The skin depth itself varies as we pass through resonance and therefore the power absorbed involves χ_1 as well as χ_2.

In ferromagnetic resonance in these conditions it is not possible to take $|\chi|$ as small. We define real quantities μ_R and μ_L by the relation

$$Z \propto (1 - i)\mu^{\frac{1}{2}} = \mu_R^{\frac{1}{2}} - i\mu_L^{\frac{1}{2}}. \tag{85}$$

If

$$\mu = \mu_1 + i\mu_2, \tag{86}$$

then

$$\mu_R = (\mu_1^2 + \mu_2^2)^{\frac{1}{2}} + \mu_2; \qquad \mu_L = (\mu_1^2 + \mu_2^2)^{\frac{1}{2}} - \mu_2. \tag{87}$$

A measurement of losses gives μ_R; of inductance, μ_L.

Dysonian Line Shape. An interesting result is found in electron spin resonance in paramagnetic metals if the spins diffuse in and out of the skin depth many times in the relaxation time for spin relaxation.

This is not necessarily the same situation as the anomalous skin effect, because the translational or conductivity relaxation time for electrons in the alkali metals is very much shorter than the electron spin relaxation time. In our derivation we assume we are in the region of the normal skin effect with respect to the electrical conductivity.

The Bloch equation with diffusion and a single relaxation time T_1 is

$$\frac{\partial \mathbf{M}}{\partial t} = \gamma \mathbf{M} \times \mathbf{H} - \frac{\mathbf{M}}{T_1} + D\, \nabla^2 \mathbf{M}, \tag{88}$$

for the transverse components of magnetization. We solve this simultaneously with the eddy current equation

$$\nabla^2 \mathbf{H} = \frac{4\pi\sigma}{c^2} (\dot{\mathbf{H}} + 4\pi \dot{\mathbf{M}}). \tag{89}$$

These equations lead to a 2×2 secular equation formed from the coefficients of $M^+ = M_x + iM_y$; $H^+ = H_x + iH_y$. The roots of the secular equation give the dispersion relation $\mathbf{k}(\omega)$; there are two different roots. The two solutions at given ω must be combined in order to satisfy the surface boundary condition on the magnetization diffusion. If we suppose that the spins are not relaxed at the surface, then at the surface

$$\mathbf{k} \cdot \text{grad}\, \mathbf{M} = 0,$$

where $\mathbf{k}$ is normal to the surface. The shape of the magnetic absorption line under these conditions looks rather like a dispersion curve. The solution is discussed in detail by F. J. Dyson, *Phys. Rev.* **98,** 349 (1955).

PROBLEMS

1. Let $\mathbf{H}_0$ and $\mathbf{k}$ be $\|\hat{\mathbf{z}}$, and let there be an undamped periodic open orbit parallel to the k_x axis. Show that the dispersion relation of electromagnetic waves in the metal is

$$\omega \cong kc\omega_c\omega^*/\omega_p^2,$$

where $\omega^{*2} = 4\pi n_{\text{open}} e^2/m$ is the effective plasma frequency for open orbits.

2. Consider a film of thickness D; show that $D > c/\sigma_0$ (rather than $D > \delta_0$) is the criterion for the film to reflect most of the radiation incident normally upon it. We assume $c/\sigma_0 \ll \delta_0$; this problem is a straightforward exercise in the use of maxwell equations, although it is best solved by physical reasoning. The result means that the reflectivity in a very thin film is mainly a question of impedance mismatch, rather than power absorption in the film.

3. Show by the methods of Chapter 6 that the *transverse* dielectric constant of an electron gas at absolute zero is

$$\epsilon(\omega,\mathbf{q}) = \frac{4\pi i}{\omega}\sigma_{\mathbf{q}} \cong \frac{4\pi i}{\omega}\frac{3\pi ne^2}{4mv_Fq},$$

in the limit $\omega\tau \gg 1$ and $(kq/m) \gg \omega$. This result is in agreement with (42). *Hint:* calculate

$$(j_{\mathbf{q}})_x = \mathrm{Tr}\, e^{-i\mathbf{q}\cdot\mathbf{x}}\rho v_x = \frac{e}{m}\sum_{\mathbf{k}} k_x\langle\mathbf{k}|\delta_\rho|\mathbf{k}+\mathbf{q}\rangle - \frac{ne^2}{mc}A_{\mathbf{q}},$$

to terms of $O(A)$, where the density fluctuation δ_ρ results from the perturbation

$$H' = -\frac{e}{mc}A_xp_x; \qquad A_x = A_{\mathbf{q}}e^{iqz}e^{-i\omega t} + cc.$$

Note that $\sigma_{\mathbf{q}}$ is real even when there are no collisions ($\tau \to \infty$).

17 Acoustic attenuation in metals

Ultrasonic phonon attenuation studies in metals in static magnetic fields are a powerful tool for the study of fermi surfaces. It is equally true that theoretical formulation of the general attenuation problem in metals is unusually subtle and the area is littered with traps. The essential experimental virtue of the ultrasonic method for studying fermi surfaces is that at a given value of magnetic field intensity the frequency needed for resonance is much less than for cyclotron resonance, although this applies with less force to the subharmonics in cyclotron resonance. We shall not enter into a discussion of the experimental data; a good illustration of the application of ultrasonic measurements to the determination of fermi surface dimensions is given in the paper by H. V. Bohm and V. J. Easterling, *Phys. Rev.* **128,** 1021 (1962); see also the references cited there.

LONGITUDINAL PHONON FREE-ELECTRON GAS

We consider first the attenuation of a longitudinal phonon in a free-electron gas in the absence of a static magnetic field. The standard transition probability calculation assumes implicitly that the electron mean free path Λ from all causes is long in comparison with the phonon wavelength λ; otherwise the electron states cannot be treated as plane waves or as Bloch functions. We have seen in the chapter on the electron-phonon interaction that for a free electron gas

$$\varepsilon(k_F,\mathbf{x}) = \varepsilon_0(k_F) - \tfrac{2}{3}\varepsilon_0(k_F)\Delta(\mathbf{x}), \tag{1}$$

where Δ is the dilation and $\varepsilon_0(k_F)$ is the fermi energy ε_F. From (7.14) the perturbation is

$$H' = -\tfrac{2}{3}i\varepsilon_F \sum_{\mathbf{kq}} (1/2\,\rho\omega_{\mathbf{q}})^{1/2}|\mathbf{q}|(a_{\mathbf{q}}c^+_{\mathbf{k+q}}c_{\mathbf{k}} - a^+_{\mathbf{q}}c^+_{\mathbf{k-q}}c_{\mathbf{k}}), \tag{2}$$

where ρ is the density of the crystal. The probability per unit time that a phonon in a state $\mathbf{q}$ will be absorbed in scattering an electron from $\mathbf{k}$ to $\mathbf{k} + \mathbf{q}$ is

$$w_{(-)} = 2\pi|\langle \mathbf{k} + \mathbf{q}; n_{\mathbf{q}} - 1|H'|\mathbf{k}; n_{\mathbf{q}}\rangle|^2 \delta(\varepsilon_{\mathbf{k}} + \omega_{\mathbf{q}} - \varepsilon_{\mathbf{k}+\mathbf{q}}). \tag{3}$$

On averaging over an electron ensemble in thermal equilibrium,

$$w_{(-)} = \frac{4\pi\varepsilon_F{}^2 q}{9\rho c_s} n_{\mathbf{q}} f_0(\mathbf{k})[1 - f_0(\mathbf{k} + \mathbf{q})]\delta(\varepsilon_{\mathbf{k}} + \omega_{\mathbf{q}} - \varepsilon_{\mathbf{k}+\mathbf{q}}), \tag{4}$$

where f_0 is the fermi distribution function and c_s is the velocity of sound. Similarly, for phonon emission

$$w_{(+)} = \frac{4\pi\varepsilon_F{}^2 q}{9\rho c_s} (n_{\mathbf{q}} + 1) f_0(\mathbf{k})[1 - f_0(\mathbf{k} - \mathbf{q})]\delta(\varepsilon_{\mathbf{k}} - \omega_{\mathbf{q}} - \varepsilon_{\mathbf{k}-\mathbf{q}}). \tag{5}$$

We form, using (4) and (5),

$$\frac{d(n_{\mathbf{q}} - \bar{n}_{\mathbf{q}})}{dt} = -\frac{1}{T_{\mathbf{q}}}(n_{\mathbf{q}} - \bar{n}_{\mathbf{q}}), \tag{6}$$

where the phonon relaxation time $T_{\mathbf{q}}$ is given from (4) and (5) by

$$\frac{1}{T_{\mathbf{q}}} \cong \frac{16\pi\varepsilon_F{}^2 q}{9\rho c_s} \sum_{\mathbf{k}} f_0(\mathbf{k})[\mathbf{q} \cdot \nabla_{\mathbf{k}} f_0(\mathbf{k})]\delta\left(\frac{\mathbf{k} \cdot \mathbf{q}}{m} - \omega_{\mathbf{q}}\right), \tag{7}$$

with allowance for the two spin orientations. We have neglected the term in q^2 in the delta function, because $q/m \ll c_s$. The sum over $\mathbf{k}$ may be rewritten as

$$\begin{aligned} \sum_{\mathbf{k}} &= (2\pi)^{-3} \int dk\, 2\pi k^2 \int_0^1 d\mu\, f_0 \frac{\partial f_0}{\partial \varepsilon} \frac{qk\mu}{m} \delta\left(\frac{qk\mu}{m} - \omega_{\mathbf{q}}\right), \\ &= -(2\pi)^{-2} m c_s \int dk\, k f_0 \delta(\varepsilon - \varepsilon_F) = -m^2 c_s/8\pi^2. \end{aligned} \tag{8}$$

Thus the energy attenuation coefficient $\alpha_{\mathbf{q}}$ is given by

$$\alpha_{\mathbf{q}} \equiv \frac{1}{T_{\mathbf{q}} c_s} = \frac{2}{9\pi} \frac{\varepsilon_F{}^2 m^2}{\rho c_s{}^2} \omega_{\mathbf{q}} \qquad (\Lambda q \gg 1). \tag{9}$$

This is the correct standard result for the model. Here Λ is the mean free path of the conduction electrons as limited by all relevant processes.

The electrons which are chiefly responsible for the absorption of energy from the phonon are those with velocity component along the phonon direction equal to the phonon velocity. The statement of energy conservation is

$$\frac{k^2}{2m} + \omega_{\mathbf{q}} = \frac{(\mathbf{k} + \mathbf{q})^2}{2m},$$

whence, with the usual neglect of the term in q^2,

$$\omega_{\mathbf{q}} = c_s \mathrm{q} \cong \frac{1}{m} \mathbf{k} \cdot \mathbf{q} = v_{\mathbf{q}} \mathrm{q},$$

where $v_{\mathbf{q}}$ is the electron velocity component in the direction of $\mathbf{q}$.

In the other limit of short electron free path $\Lambda q = 2\pi\Lambda/\lambda \ll 1$, the problem is that of the damping of phonons by the viscosity of the electron gas. It is well known from acoustics[1] that the energy attenuation of gas in this limit is

$$\alpha_{\mathbf{q}} = \frac{4}{3} \frac{\eta}{\rho c_s{}^3} \omega_{\mathbf{q}}{}^2, \tag{10}$$

where the viscosity coefficient η of a fermi gas is given by[2]

$$\eta = \tfrac{2}{5} \tau_c n \varepsilon_F, \tag{11}$$

on the assumption of a constant electron relaxation time. Thus we have the result

$$\alpha_{\mathbf{q}} = \frac{8}{15} \frac{n \varepsilon_F \tau_c}{\rho c_s{}^3} \omega_{\mathbf{q}}{}^2 \qquad (\Lambda q \ll 1). \tag{12}$$

This result is of the order of Λq times the result (9). We recall that the electrical conductivity is

$$\sigma_0 = \frac{n e^2 \tau_c}{m} \tag{13}$$

on the same model, so that $\alpha_{\mathbf{q}} \propto \sigma_0$. This proportionality is confirmed by experiment.

Transport Treatment—Longitudinal Phonon. The boltzmann equation enables us to give a unified treatment of the above problem for arbitrary values of Λq. In the absence of magnetic fields the transport equation in the relaxation time approximation is

$$\frac{\partial f}{\partial t} + \mathbf{v} \cdot \frac{\partial f}{\partial \mathbf{x}} + \frac{e\mathbf{E}}{m} \cdot \frac{\partial f}{\partial \mathbf{v}} = -\frac{f - \bar{f}_0}{\tau}. \tag{14}$$

We suppose that the electron scattering is by impurities; then it

[1] See, for example, C. Kittel, *Rpts. Prog. Physics* **11,** 205 (1948).
[2] C. Kittel, *Elementary statistical physics,* Wiley, 1958, p. 206.

appears reasonable to take $\bar{f}_0$ as the equilibrium electron distribution function in the *local* coordinate system moving with the local lattice. This important point has been considered in detail by T. Holstein, *Phys. Rev.* **113,** 479 (1959).

If $\mathbf{u}(\mathbf{x},t)$ is the local lattice velocity, then

$$\bar{f}_0(\mathbf{x};\mathbf{v};t) = f_0(\mathbf{v} - \mathbf{u}(\mathbf{x},t);\varepsilon_F(\mathbf{x},t)); \tag{15}$$

the fermi energy is modified by the phonon because the local electron concentration is modified by the dilation accompanying a longitudinal phonon. The strong tendency of conduction electrons to screen fluctuations in the positive ion density means that the electrons will follow closely the lattice dilation. We write $n = n_0 + n_1(\mathbf{x},t)$ for the electron concentration; then

$$\begin{aligned} \bar{f}_0 &\cong f_0(\mathbf{v},n_0) - \mathbf{u}\cdot\frac{\partial f_0}{\partial \mathbf{v}} + n_1\frac{\partial f_0}{\partial n} \\ &= f_0 - \frac{\partial f_0}{\partial \varepsilon}\left(m\mathbf{v}\cdot\mathbf{u} + \tfrac{2}{3}\varepsilon_F{}^0\frac{n_1}{n_0}\right), \end{aligned} \tag{16}$$

using $\partial f_0/\partial\varepsilon = -\partial f_0/\partial\varepsilon_F$, and $\varepsilon_F{}^0 \propto n^{2/3}$. Thus (14) becomes, with $f = f_0 + f_1$,

$$-i\omega f_1 + i\mathbf{q}\cdot\mathbf{v}f_1 + e\mathbf{E}\cdot\mathbf{v}\frac{\partial f_0}{\partial\varepsilon} \cong -\frac{f_1}{\tau} - \frac{1}{\tau}\left(m\mathbf{v}\cdot\mathbf{u} + \tfrac{2}{3}\varepsilon_F{}^0\frac{n_1}{n_0}\right)\frac{\partial f_0}{\partial\varepsilon}, \tag{17}$$

whence

$$f_1 = -\left[\frac{\tau e\mathbf{v}\cdot[\mathbf{E} + (m\mathbf{u}/e\tau)] + \tfrac{2}{3}\varepsilon_F{}^0(n_1/n_0)}{1 - i\omega\tau + i\mathbf{q}\cdot\mathbf{v}\tau}\right]\frac{\partial f_0}{\partial\varepsilon}. \tag{18}$$

Now the electric current density $\mathbf{j}_e$ is given by

$$\mathbf{j}_e = \frac{2e}{(2\pi)^3}\int d^3k\, f_1\mathbf{v} = \boldsymbol{\sigma} : \left(\mathbf{E} + \frac{m\mathbf{u}}{e\tau}\right) + n_1 e c_s \mathbf{R}, \tag{19}$$

where the diffusion vector $\mathbf{R}$ is defined from (18) by

$$\mathbf{R} = -\frac{\varepsilon_F}{6\pi^3 n_0 c_s}\int d^3k\,\frac{\mathbf{v}}{1 - i\omega\tau + i\mathbf{q}\cdot\mathbf{v}\tau}\cdot\frac{\partial f_0}{\partial\varepsilon}, \tag{20}$$

and the conductivity tensor is

$$\sigma_{\mu\nu} = -\frac{e^2\tau}{4\pi^3}\int d^3k\,\frac{v_\mu v_\nu}{1 - i\omega\tau + i\mathbf{q}\cdot\mathbf{v}\tau}\cdot\frac{\partial f_0}{\partial\varepsilon}. \tag{21}$$

Equation (19) is the constitutive equation of the medium.

We may integrate (21) directly to find, for $\mathbf{q}$ parallel to the z axis, with $a = q\Lambda/(1 - i\omega\tau)$:

$$\sigma_{zz} = \frac{\sigma_0}{1 - i\omega\tau} \cdot \frac{3}{a^3}(a - \tan^{-1} a); \tag{22}$$

$$\sigma_{xx} = \sigma_{yy} = \frac{\sigma_0}{1 - i\omega\tau} \cdot \frac{3}{2a^3}[(1 + a^2)\tan^{-1} a - a]; \tag{23}$$

and the nondiagonal terms are zero in the absence of a static magnetic field. In the limit $a \to 0$ the diagonal components $\to \sigma_0/(1 - i\omega\tau)$. The result (23) was considered previously in connection with the anomalous skin effect. The nonvanishing component of $\mathbf{R}$ is

$$R_z = -i\frac{4\varepsilon_F^2}{3\pi^2 n_0 c_s(1 - i\omega\tau)} \cdot \frac{1}{a^2}(a - \tan^{-1} a), \tag{24}$$

so that

$$j_{ez} = \sigma_{zz}\left(E_z + \frac{mu_z}{e\tau} - \frac{iamv_F}{3e\tau} \cdot \frac{n_1}{n_0}\right). \tag{25}$$

The electric field $\mathbf{E}$ arises from the small local charge imbalance. We write

$$\mathbf{j} = \mathbf{j}_e - ne\mathbf{u}, \tag{26}$$

where $\mathbf{j}$ is the total current density and is composed of the electronic current density $\mathbf{j}_e$ and the ionic current density $(-e)n\mathbf{u}$. The electron current density satisfies the continuity equation

$$\frac{\partial \rho_e}{\partial t} + \operatorname{div} \mathbf{j}_e = 0; \qquad -\omega n_1 e + \mathbf{q} \cdot \mathbf{j}_e = 0. \tag{27}$$

The maxwell equation needed to relate j_z and E_z is

$$\operatorname{div} \mathbf{E} = 4\pi\rho; \qquad \text{or} \qquad \operatorname{div} \dot{\mathbf{E}} = -4\pi \operatorname{div} \mathbf{j}, \tag{28}$$

so that

$$\omega E_z = -4\pi i j_z = -4\pi i(j_{ez} - neu_z). \tag{29}$$

When we study the attenuation of transverse phonons, we will need an equation connecting $j_\perp$ and $E_\perp$. The maxwell equations for curl $\mathbf{E}$ and curl $\mathbf{H}$ may be combined to give, for $\varepsilon = \mu = 1$,

$$E_\perp = -\frac{4\pi i}{\omega}\frac{(c_s/c)^2}{1 - (c_s/c)^2} j_\perp, \tag{30}$$

where c_s is the acoustic velocity and c the velocity of light.

The power absorption per unit volume is given by

$$\mathcal{P} = \tfrac{1}{2}\mathcal{R}\left\{\mathbf{j}_e^* \cdot \mathbf{E} - \frac{n_0 m\mathbf{u}^*}{\tau}(\langle\mathbf{v}\rangle - \mathbf{u})\right\}. \tag{31}$$

The term in $\mathbf{j}_e^* \cdot \mathbf{E}$ is the ohmic loss of the electrons. We see that the other term is the power absorbed by the lattice from the electrons by virtue of the fact that the electrons have a mean velocity $\langle\mathbf{v}\rangle$ before scattering and $\mathbf{u}$ just after scattering. This collision drag term is important chiefly at high frequencies or high magnetic fields.

From the equation of continuity (27) we have for a longitudinal wave

$$n_1 = qj_{ez}/\omega e = j_{ez}/c_s e, \tag{32}$$

whence (25) becomes, dropping the subscript z on j, E, and u,

$$j_e = \frac{\sigma_{zz}}{1 + i(\sigma_{zz}/\sigma_0)(av_F/3c_s)}\left(E + \frac{mu}{e\tau}\right) \equiv \sigma'\left(E + \frac{mu}{e\tau}\right), \tag{33}$$

hereby defining σ'. This relation must be consistent with the maxwell equation (29); on eliminating E with the use of (29) we have

$$j_e = \frac{(4\pi\sigma' neui/\omega) + (\sigma' mu/e\tau)}{1 + (4\pi i\sigma'/\omega)} \cong neu, \tag{34}$$

for $\omega \ll \sigma'$ and $\omega \ll \omega_0^2\tau$. In this limit the total current j approximately vanishes. We obtain the electric field by substituting (34) in (33):

$$E \cong u\left(\frac{ne}{\sigma'} - \frac{m}{e\tau}\right). \tag{35}$$

The power density dissipation is given by, with the neglect of collision drag in this order,

$$\mathcal{P} \cong \tfrac{1}{2}u^*u\mathcal{R}\left\{\frac{n^2e^2}{\sigma'} - \frac{nm}{\tau}\right\} = \frac{nmu^*u}{2\tau}\,\mathcal{R}\left\{\frac{\sigma_0}{\sigma'} - 1\right\}, \tag{36}$$

whence, using the definition of σ' and taking a to be real ($\omega\tau \ll 1$),

$$\mathcal{P} \cong \frac{nmu^*u}{2\tau}\left[\frac{a^2 \tan^{-1} a}{3(a - \tan^{-1} a)} - 1\right]. \tag{37}$$

The attenuation coefficient α is equal to the power density dissipation per unit energy flux:

$$\alpha = \mathcal{P}/\tfrac{1}{2}\rho u^* u c_s, \tag{38}$$

where ρ is the density; thus for the longitudinal wave

$$\alpha = \frac{nm}{\rho c_s \tau}\left(\frac{a^2 \tan^{-1} a}{3(a - \tan^{-1} a)} - 1\right), \tag{39}$$

which is the Pippard result, with $a \cong \Lambda q$. This agrees with (9) in the limit $\Lambda q \gg 1$ and with (12) in the limit $\Lambda q \ll 1$.

Transverse Wave Attenuation. For transverse waves the local lattice velocity $\mathbf{u}$ is perpendicular to the wavevector $\mathbf{q}$ of the phonon. We take $\mathbf{u}$ in the x direction and $\mathbf{q}$ in the z direction. The density fluctuation n_1 is zero for a transverse wave. We have for the current density in the x direction

$$j_e = \sigma_{xx}\left(E + \frac{mu}{e\tau}\right), \tag{40}$$

which must be consistent with the maxwell equation (30):

$$E \cong -\frac{4\pi i}{\omega}\left(\frac{c_s}{c}\right)^2 (j_e - neu). \tag{41}$$

On eliminating E,

$$j_e = \frac{\sigma_{xx} u}{1 + (4\pi i \sigma_{xx}/\omega)(c_s/c)^2}\left(\frac{m}{e\tau} + \frac{4\pi inec_s^2}{\omega c^2}\right), \tag{42}$$

where the term in $m/e\tau$ is usually negligible. Now $4\pi i \sigma_{xx} c_s^2/\omega c^2$ is essentially $(\lambda/\delta)^2$, where λ is the acoustic wavelength and δ is the classical skin depth. Below the microwave frequency range λ is usually $\gg \delta$, so that

$$j_e \cong neu - i\frac{m\omega uc^2}{4\pi e\tau c_s^2} + i\frac{\omega neu}{4\pi\sigma_{xx}}\left(\frac{c}{c_s}\right)^2, \tag{43}$$

and

$$E \cong \frac{mu}{e\tau} + \frac{neu}{\sigma_{xx}}, \tag{44}$$

whence

$$\mathcal{P} \cong \frac{nmu^*u}{2\tau}\,\mathcal{R}\left\{\frac{\sigma_0}{\sigma_{xx}} - 1\right\}, \tag{45}$$

and, using (23) in the limit $\omega\tau \ll 1$, the attenuation of a shear wave is

$$\alpha = \frac{nm}{\rho c_s \tau}\left(\frac{1}{\zeta} - 1\right), \tag{46}$$

where

$$\zeta = \frac{3}{2a^3}\left[(1 + a^2)\tan^{-1} a - a\right]. \tag{47}$$

MAGNETIC FIELD EFFECTS ON ATTENUATION

Much valuable work in ultrasonic attenuation in metals is concerned with the study of fermi surfaces by means of periodicity effects in magnetic fields under conditions $\omega_c\tau \gg 1$ and $\Lambda q \gg 1$. The theory of the attenuation follows by an obvious but lengthy extension of the treatment we have given above: for the free-electron theory one may refer to the paper by M. H. Cohen, M. J. Harrison, and W. A. Harrison, *Phys. Rev.* **117,** 937 (1960); the problem for a general fermi surface is treated by A. B. Pippard, *Proc. Roy. Soc.* **A257,** 165 (1960), and V. L. Gurevich, *J. Exp. Th. Phys. USSR* **37,** 71 (1959). Ultrasonic studies may be done at relatively low frequencies, as compared to those needed in cyclotron resonance.

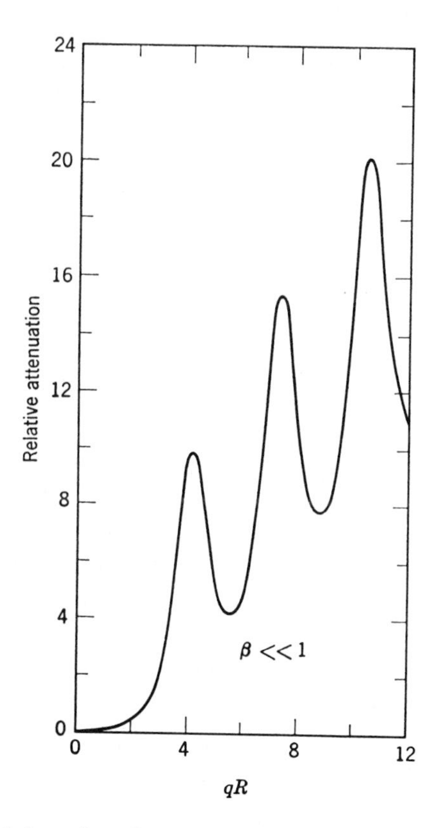

FIG. 1. The field-dependent factor in the attenuation of a transverse wave propagating perpendicular to the field, when the classical skin depth is much less than the wavelength and the cyclotron frequency is much greater than the electron scattering frequency. (After Cohen, Harrison, and Harrison.)

The resonance attenuation effects which are most studied are the geometrical resonances for which the diameter of the cyclotron orbit is an integral multiple of the half wavelength of the phonon, with **H**, **q**, and **u** mutually perpendicular. The geometrical resonances have to do with the strength of the interaction between particular orbits and the electric field in the metal. The attenuation involves the inverse of the effective conductivity, as we have seen in the results (36) and (45):

$$\alpha = \frac{nm}{\rho c_s \tau} \, \mathcal{R} \left\{ \frac{\sigma_0}{\sigma_{\text{eff}}} - 1 \right\}. \tag{48}$$

Because of the screening we have to do with a constant current system, rather than a constant voltage system.

The conductivity is reasonably expected to be a periodic function of the number of phonon wavelengths encompassed in a cyclotron orbit, for an extremal orbit. Thus $2r_c = n\lambda$ or $qr_c = n\pi$ is the periodicity condition, where $r_c = v_F/\omega_c$; and n is an integer; this may be rewritten as

$$\frac{2pc}{eH} = n\lambda, \tag{49}$$

with $p \equiv mv_F$. We see from the calculated curve in Fig. 1 that the differences in qR values at the absorption maxima are closely multiples

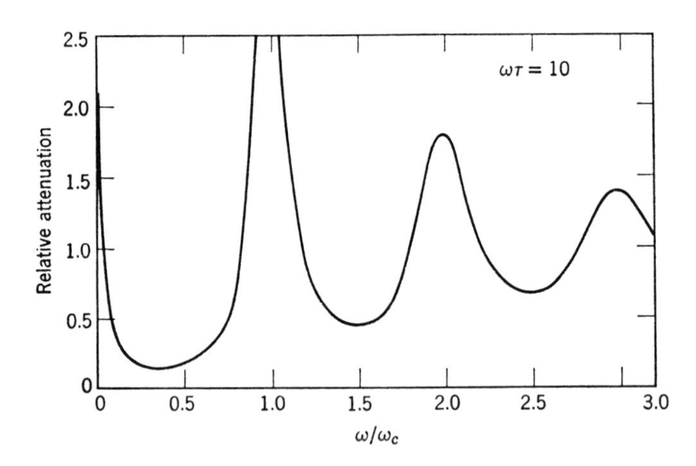

FIG. 2. The ratio of the attenuation of longitudinal waves as a function of magnetic field to that at zero field as a function of the ratio of phonon frequency to cyclotron frequency. The product of the phonon frequency and the electron scattering time $\omega\tau$ is taken equal to ten. (After Cohen, Harrison, and Harrison.)

of π. Calculations for longitudinal waves under Azbel-Kaner type resonance conditions are shown in Fig. 2.

OPEN-ORBIT MAGNETOACOUSTIC RESONANCE

A striking resonance property of periodic open orbits in a magnetic field is observed in one aspect of ultrasonic absorption in metals.[3] If the magnetic field is perpendicular to the line of an open orbit, the electron wavevector increases as

$$\dot{\mathbf{k}} = \frac{e}{c}\mathbf{v}_\perp \times \mathbf{H}, \tag{50}$$

where $\mathbf{v}_\perp = \text{grad}_\mathbf{K}\, \varepsilon(\mathbf{K},k_H)$ lies in the plane normal to $\mathbf{H}$ and through the line of the open orbit. The open-orbit absorption may be understood by an argument due to W. A. Harrison.

Consider an electron on an open orbit and moving in the x direction in the crystal. We send into the crystal a longitudinal phonon of frequency ω and wavevector $\mathbf{q}$ also in the x direction. There will be associated with the phonon an effective electric field

$$E_x = E_0 \cos(qx - \omega t). \tag{51}$$

The velocity of the electron is of the form

$$\dot{x} = v_0 + v_1 \cos \Omega t, \tag{52}$$

where v_0 is of the order of the fermi velocity, and Ω is the angular frequency associated with the motion of the electron through a reciprocal lattice vector in $\mathbf{k}$ space. That is, if $v_1 \ll v_0$,

$$\Omega = 2\pi\dot{k}/G_c \cong 2\pi e v_0 H/c\,G_c, \tag{53}$$

where G_c is the spatial period of the open orbit in $\mathbf{k}$ space.

The rate at which the phonon does work on the electron is

$$eE_x\dot{x} = eE_0 \cos(qx - \omega t)(v_0 + v_1 \cos \Omega t). \tag{54}$$

Now $x = v_0 t + (v_1/\Omega) \sin \Omega t$, so that

$$eE_x\dot{x} = eE_0[\cos\{(qv_0 - \omega)t + (qv_1/\Omega)\sin \Omega t\}][v_0 + v_1 \cos \Omega t]. \tag{55}$$

The first cosine factor may be rewritten as

$$\cos\{\cdot\cdot\cdot\} \equiv \cos(qv_0 - \omega)t \cos[(qv_1/\Omega)\sin \Omega t] - \sin(qv_0 - \omega)t \sin[(qv_1/\Omega)\sin \Omega t]. \tag{56}$$

If $qv_1/\Omega \ll 1$ we may expand (56) as a power series in this quantity.

[3] Theory: E. A. Kaner, V. G. Peschanskii, and I. A. Privorotskii, *Soviet Physics—JETP* **13,** 147 (1961); experimental observation on cadmium: J. V. Gavenda and B. C. Deaton, *Phys. Rev. Letters* **8,** 208 (1962).

From (55), to zero order,

$$(eE_x\dot{x})_0 = eE_0v_0 \cos (qv_0 - \omega)t, \tag{57}$$

which averages to zero unless $qv_0 = \omega$, a condition independent of the magnetic field. In first order

$$(eE_x\dot{x})_1 = eE_0(qv_1/\Omega)[-v_0 \sin (qv_0 - \omega)t \sin \Omega t + (\Omega/q) \cos (qv_0 - \omega)t \cos \Omega t]. \tag{58}$$

On averaging over the time this gives a contribution only if $qv_0 - \omega = \Omega$, or $q(v_0 - v_s) = \Omega$, where v_s is the phonon velocity. Because $v_s \ll v_0$, the resonance condition is

$$qv_0 \cong \Omega = 2\pi ev_0H/c\,G_c, \tag{59}$$

using (53); or, for the acoustic wavelength $\lambda_{\mathbf{q}}$,

$$\lambda_{\mathbf{q}} = \frac{c\,G_c}{eH}. \tag{60}$$

This relation is satisfied quite well by the observations on cadmium in reference 3. We notice that (60) involves e, and not e/m^*; the resonance reported is very sharp and may provide a good method for the investigation of the question of whether electron-electron interactions cause the effective carrier charge e^* to differ slightly from the electronic charge e. Theoretical arguments against the possibility that $e^* \neq e$ are given by W. Kohn, *Phys. Rev.* **115,** 1460 (1959), and J. M. Luttinger, *Phys. Rev.* **121,** 1251 (1961).

PHONON AMPLIFICATION BY ELECTRON-PHONON INTERACTIONS[4]

Let the electron density everywhere be equal to the hole density, thereby allowing the neglect of coulomb effects accompanying charge bunching. The situation in the absence of charge neutrality is considered by Weinreich, Sanders, and White in reference 4. Write $n(x,t)$ as the deviation of the particle density from the static equilibrium value. If V is the difference of the electron and hole deformation potential constants, then Ve_{xx} is the shift in relative energy of electron and hole states under a deformation e_{xx}. The equilibrium particle density is shifted by $Ve_{xx}N_F$; here N_F is the density of states at the fermi surface. Thus the transport equation for drift velocity v is

$$\frac{\partial n}{\partial t} + v\,\frac{\partial n}{\partial x} = -\,\frac{n(x,t) + N_FV(\partial u/\partial x)}{\tau_{eh}}, \tag{61}$$

[4] G. Weinreich, *Phys. Rev.* **104,** 32 (1956); G. Weinreich, T. M. Sanders, and H. G. White, *Phys. Rev.* **114,** 33 (1959); J. Hopfield, *Phys. Rev. Letters* **8,** 311 (1962); A. R. Hutson, *Phys. Rev. Letters* **7,** 237 (1961). In piezoelectric crystals the cou-

where τ_{eh} is the electron-hole recombination time. The drift velocity enters because we have in (61) already integrated the distribution function over the velocity. We assume that the drift velocity is maintained by an external source. The elastic equation of motion is

$$\rho \ddot{u} = c_{\parallel} \frac{\partial^2 u}{\partial x^2} + V \frac{\partial n}{\partial x}. \tag{62}$$

This follows from the lagrangian density

$$\mathcal{L} = \tfrac{1}{2}\rho \dot{u}^2 - \tfrac{1}{2}c_{\parallel}\left(\frac{\partial u}{\partial x}\right)^2 - Vn(x,t)\frac{\partial u}{\partial x}, \tag{63}$$

where the last term on the right-hand side is the classical transcription of the deformation potential.

We look for solutions of (61) and (62) of the form

$$n,u \sim e^{i(kx-\omega t)}. \tag{64}$$

Then

$$\begin{aligned} (-i\omega n + vikn)\tau &= -n - N_V V iku, \\ -\rho\omega^2 u &= -k^2 c_{\parallel} u + ikVn. \end{aligned} \tag{65}$$

The secular equation is

$$\begin{vmatrix} (1 - i\omega\tau + ikv\tau) & iN_F Vk \\ -ikV & k^2 c_{\parallel} - \rho\omega^2 \end{vmatrix} = 0. \tag{66}$$

If $kv > \omega$, the waves increase in amplitude—the drift velocity, if maintained, puts energy into both systems. This is essentially a traveling wave amplifier for phonons. The condition $kv > \omega$ is equivalent to $v > c_s$, where c_s is the acoustic velocity.

The approximate solution of (66) is

$$k \cong \frac{\omega}{c_s}\left\{1 + \frac{N_F V^2}{2c_{\parallel}}[1 + i(\omega - kv)\tau]\right\}, \tag{67}$$

for $(\omega - kv)\tau \ll 1$.

PROBLEM

1. The longitudinal velocity of sound in Cd perpendicular to the hexagonal axis is 3.8×10^5 cm/sec. At what acoustic frequency will the lowest open-orbit magnetoacoustic resonance occur for a magnetic field of 1 kilo-oersted?

---

pling is much stronger than that of the deformation potential at the frequencies of interest in experiments.

18 Theory of alloys

Many interesting practical and intellectual problems arise when a solid solution of one element in another is prepared. We can ask a number of questions about the alloy, including solubility limits, energy of solution, lattice dilation, electrical resistivity, magnetic moment, magnetic coupling, Knight shift, nuclear quadrupole broadening, and superconducting properties (energy gap, transition temperature, critical field). We do not have space to treat all of these, but we shall discuss several of the central aspects of dilute alloys with particular reference to the effect of the impurity atoms on the electronic structure of the host or solvent metal. We assume the impurities are substitutional for atoms of the host lattice unless otherwise specified.

In principle even a minute concentration of impurities destroys the translational periodicity of the crystal. Formally, the lattice is either periodic or it is not periodic. In the alloy the wavevector is no longer a constant of the motion and the true electronic eigenstates do not carry momentum. The actual consequences of small concentrations of impurities are often much less serious than might be imagined from this formal statement, particularly if the impurity atom has the same valence as the atoms of the host crystal. Thus in an experiment with 5 percent Si in Ge the mean free path of electrons in the conduction band at 4°K was found to be $\sim 10^{-5}$ cm, about 400 interatomic spacings.

It is also known that solid solutions or mixed crystals are often formed over very wide composition ranges without destroying the insulating or metallic nature of the materials, as the case may be. On the nearly free electron model the presence of a band gap at a zone boundary is related to an appropriate fourier component of the periodic crystalline potential; the "partial" destruction of the periodicity by the admixture of a weak spectrum of other fourier components associated with a low concentration of impurity atoms does not destroy the gap. On this particular point the reader may consult R. D. Mattuck, *Phys. Rev.* **127,** 738 (1962) and references cited therein.

We first discuss two results of wide application in the theory of alloys, the Laue theorem and the Friedel sum rule. We then treat several representative alloy problems. Some general references on the modern theory of alloys include:

[1] J. Friedel, *Phil. Mag. Supplement* **3,** 446–507 (1954).
[2] J. Friedel, *Nuovo cimento supplemento* **7,** 287–311 (1958).
[3] A. Blandin, Thesis, Paris, 1961.

LAUE THEOREM

This important theorem states that the particle density per unit energy range is approximately independent of the form of the boundary, at distances from the boundary greater than a characteristic particle wavelength at the energy considered. This result is not caused by screening, but obtains even for a noninteracting electron gas. We note that we are concerned with the product of probability density $\varphi_E^*(\mathbf{x})\varphi_E(\mathbf{x})$ and the density of states $g(E)$. In using the theorem we think of the boundary perturbation as caused by the insertion or substitution of an impurity atom in the crystal. The following proof was suggested by Dyson, unpublished.

Proof: We treat free electrons with the eigenvalue equation

$$-\frac{1}{2m}\nabla^2\varphi_{\mathbf{k}} = \varepsilon_{\mathbf{k}}\varphi_{\mathbf{k}}. \tag{1}$$

We introduce the function

$$u(\mathbf{x}t) = \sum_{\mathbf{k}} \varphi_{\mathbf{k}}^*(\mathbf{x})\varphi_{\mathbf{k}}(0)e^{-\varepsilon_{\mathbf{k}}t}, \tag{2}$$

where

$$u(\mathbf{x}0) = \sum_{\mathbf{k}} \varphi_{\mathbf{k}}^*(\mathbf{x})\varphi_{\mathbf{k}}(0) = \delta(\mathbf{x}). \tag{3}$$

Now $u(\mathbf{x}t)$ satisfies the diffusion equation

$$\frac{\partial u}{\partial t} = D\,\nabla^2 u, \tag{4}$$

with

$$D = \frac{1}{2m}. \tag{5}$$

Further, at the origin,

$$u(0t) = \sum_{\mathbf{k}} \varphi_{\mathbf{k}}^*(0)\varphi_{\mathbf{k}}(0)e^{-\varepsilon_{\mathbf{k}}t} = \int d\varepsilon_{\mathbf{k}}\, \varphi_{\mathbf{k}}^*(0)\varphi_{\mathbf{k}}(0)g(\varepsilon_{\mathbf{k}})e^{-\varepsilon_{\mathbf{k}}t} \tag{6}$$

is the laplace transform of the probability density

$$\varphi_{\mathbf{k}}^*(0)\varphi_{\mathbf{k}}(0)g(\varepsilon_{\mathbf{k}}) \tag{7}$$

per unit energy range per unit volume.

We know from the theory of diffusion that $u(0t)$ is influenced by the presence of a boundary at a distance ξ from the origin after a time t such that

$$t > t_c \cong \xi^2/D = 2m\xi^2. \tag{8}$$

Thus in the laplace transform (6) the presence of the boundary will be felt at time t_c. The components which contribute to (6) at time t_c are principally those for which $\varepsilon_{\mathbf{k}}t_c < 1$, or, using (8),

$$\varepsilon_{\mathbf{k}} < \frac{1}{2m\xi^2}. \tag{9}$$

But $\varepsilon_{\mathbf{k}} = k^2/2m$, so that if

$$k < \frac{1}{\xi} \tag{10}$$

the boundary will have an influence on the problem. In terms of the wavelength λ, if the distance from the origin

$$\xi < \lambda/2\pi, \tag{11}$$

then the boundary will have an influence.

We view an impurity atom as equivalent to a new boundary. Thus the electron density per unit energy range at the fermi surface of a metal will usually be only weakly perturbed by an impurity at distances greater than $1/k_F$, or about a lattice constant, from the point of observation.

We consider as a simple illustration of the Laue theorem the particle density on a one-dimensional line with impenetrable boundaries at $x = 0$ and $x = L$. The eigenfunctions are

$$\varphi_n(x) = \left(\frac{2}{L}\right)^{1/2} \sin\frac{n\pi}{L}x, \tag{12}$$

and the energy eigenvalues are

$$\varepsilon_n = \frac{1}{2m}\left(\frac{n\pi}{L}\right)^2. \tag{13}$$

Thus

$$\varphi_n^*(x)\varphi_n(x) = \left(\frac{2}{L}\right)\sin^2\frac{n\pi}{L}x; \tag{14}$$

if states up to n_F are filled with one particle each, the particle density is, with $k_F = n_F\pi/L$,

$$(15) \qquad \rho(x) = \left(\frac{2}{L}\right) \int_0^{n_F} dn \sin^2\left(\frac{n\pi x}{L}\right) = \frac{n_F}{L}\left\{1 - \frac{\sin 2k_F x}{2k_F x}\right\},$$

so that the density increases from zero at $x = 0$ to a value close to n_F/L in a distance of the order of $1/k_F$.

In one dimension $d\rho/dn$ or $d\rho/dE$ does not settle down to a steady value as we leave the boundary: this is because only one mode at a time is counted in the derivative. In two dimensions the situation is improved and both ρ and $d\rho/dE$ can be shown to have the expected behavior. Here

$$(16) \qquad \rho(x,y) = \frac{n_F^2}{L^2}\left\{1 - \frac{\sin 2k_F x}{2k_F x}\right\}\left\{1 - \frac{\sin 2k_F y}{2k_F y}\right\}.$$

FRIEDEL SUM RULE

We consider a free-electron gas and a spherical scattering potential $V(r)$. It is convenient, but not necessary, to assume that V has no bound states and is localized within one atomic cell. From standard collision theory we know that the solution $u(r,\theta)$ of the wave equation may be written as

$$(17) \qquad u(r,\theta) = \sum_{L=0}^{\infty} \frac{\varphi_L(r)}{r} P_L(\cos\theta),$$

where P_L is a legendre polynomial and $\varphi_L(r)$ satisfies

$$(18) \qquad \frac{d^2\varphi_L}{dr^2} + \left[k^2 - U(r) - \frac{L(L+1)}{r^2}\right]\varphi_L = 0;$$

here $U(r) = 2mV(r)$. We know that $\varphi_L(r) \to 0$ as $r \to 0$ and also that (Schiff, p. 104)

$$(19) \qquad \varphi_L(r) \to \left(\frac{1}{2\pi R}\right)^{1/2} \sin\left(kr + \eta_L(k) - \tfrac{1}{2}L\pi\right)$$

as $r \to \infty$. Here $\eta_L(k)$ is the phase shift produced by the scattering potential. The numerical factor in (19) is chosen to normalize φ_L/r in a sphere of large radius R:

$$(20) \qquad 4\pi \int_0^R \varphi_L^2(r)\, dr = \frac{2}{R}\int_0^R dr \cdot \tfrac{1}{2} = 1,$$

apart from oscillatory terms exhibited in (27). This result is written for $\eta_L(k) = 0$, and therefore is subject to a correction for changes in the wavefunction near the scattering potential.

We find the desired correction:

(*a*) Multiply (18) by $\varphi_L'(r)$, which is the solution of (18) for a wavevector k'.

(*b*) Form the similar product of $\varphi_L(r)$ multiplied by the differential equation for $\varphi_L'(r)$. Subtract this from the above result:

$$\varphi_L' \frac{d^2\varphi_L}{dr^2} - \varphi_L \frac{d^2\varphi_L'}{dr^2} + (k^2 - k'^2)\varphi_L'\varphi_L = 0. \tag{21}$$

(*c*) Integrate (21) over dr from 0 to R:

$$\int_0^R dr \left(\varphi_L' \frac{d^2\varphi_L}{dr^2} - \varphi_L \frac{d^2\varphi_L'}{dr^2}\right) = (k'^2 - k^2) \int_0^R dr\ \varphi_L'\varphi_L. \tag{22}$$

On integration by parts the left-hand side becomes

$$\left[\varphi_L' \frac{d\varphi_L}{dr} - \varphi_L \frac{d\varphi'}{dr}\right]_0^R. \tag{23}$$

If φ_L is a continuous function of k, we may write, for small $k' - k$,

$$\varphi_L' = \varphi_L + (k - k')\frac{d\varphi_L}{dk}; \qquad \frac{d\varphi_L'}{dr} = \frac{d\varphi_L}{dr} + (k - k')\frac{d^2\varphi_L}{dk\,dr}, \tag{24}$$

so that

$$\left[\frac{d\varphi_L}{dk}\frac{d\varphi_L}{dr} - \varphi_L \frac{d^2\varphi_L}{dk\,dr}\right]_0^R = 2k \int_0^R dr\ \varphi_L^2. \tag{25}$$

Using the asymptotic form (19) for large R,

$$\int_0^R dr\ \varphi_L^2 \rightarrow \frac{1}{4\pi R}\left[R + \frac{d\eta_L}{dk} - \frac{1}{2k}\sin 2(kR + \eta_L - \tfrac{1}{2}L\pi)\right]. \tag{26}$$

If φ_L^0 is the wavefunction in the absence of the potential,

$$\int_0^R dr\ (\varphi_L^0)^2 \rightarrow \frac{1}{4\pi R}\left[R - \frac{1}{2k}\sin 2(kR - \tfrac{1}{2}L\pi)\right]. \tag{27}$$

Thus the change in the number of particles in the state (k,L) inside the sphere of radius R is

$$4\pi \int_0^R dr\ [\varphi_L^2 - (\varphi_L^0)^2] = \frac{1}{R}\left[\frac{d\eta_L}{dk} - \frac{1}{k}\sin\eta_L \cos(2kR + \eta_L - L\pi)\right]. \tag{28}$$

The density of states of angular momentum L per unit wavenumber range inside a large sphere of radius R is $2(2L + 1)R/\pi$, counting both spin orientations. This follows because there are $(2L + 1)$ values of the azimuthal quantum number m_L for each value of L, and for fixed L and m_L the allowed values of k differ asymptotically by $\Delta k = \pi/R$, giving R/π different radial solutions per unit range of k. Then, using (28), the total number of particles ΔN displaced by the potential into the sphere of radius R is

$$(29)\quad \Delta N = \frac{2}{\pi}\sum_L (2L+1) \int_0^{k_F} dk \left[\frac{d\eta_L}{dk} - \frac{1}{k}\sin\eta_L \cos(2kR + \eta_L - L\pi)\right]$$

$$= \frac{2}{\pi}\sum_L (2L+1)\eta_L(k_F) + \text{oscillatory term}.$$

We note that the oscillatory term gives rise to an oscillatory variation in the local charge density:

$$(29')\quad \Delta\rho(R) = \frac{1}{4\pi R^2}\frac{\partial(\Delta N)}{\partial R} = -\frac{1}{2\pi^2 R^3} \sum_L (2L+1)(\sin\eta_L)(\cos(2k_F R + \eta_L - L\pi) - \cos(\eta_L - L\pi)),$$

for η_L independent of k.

If the impurity has valency Z relative to that of the host lattice, we know that a self-consistent potential $V(r)$ will ensure that the displaced charge $(\Delta N)e$ will cancel $Z|e|$ exactly in order that the impurity potential be screened at large distances. Thus (29) becomes, if we neglect the oscillatory term,

$$(30)\quad \boxed{Z = \frac{2}{\pi}\sum_L (2L+1)\eta_L(k_F);}$$

this is the Friedel sum rule. It is an important self-consistency condition on the potential. For many alloy problems it is not necessary to know the detailed form of the potential—we need only the first few phase shifts $\eta_0, \eta_1, \eta_2, \cdots$, chosen to satisfy (30). Often the phase shifts are negligible for $L > 3$ or 4.

Note that (30) may be checked by an elementary argument: for phase shift $\eta_L(k_F)$ the quantized values of k near k_F in a sphere of radius R are shifted by $\Delta k = -\eta_L(k_F)/R$; but there are $2(2L + 1)R/\pi$ states of given L for a unit range of k, so that the total change in the number of states below the fermi level is just $(2/\pi)\sum_L (2L+1)\eta_L(k_F)$, which must equal Z if the excess charge is to be screened.

RIGID BAND THEOREM

According to the rigid band theorem, the effect of a localized perturbation $V_P(\mathbf{x})$ in first order is to shift every energy level of the host crystal by nearly the same amount, as given by the first-order perturbation result:

$$\Delta\varepsilon_{\mathbf{k}} = \langle\mathbf{k}|V_P|\mathbf{k}\rangle = \int d^3x\, u_{\mathbf{k}}^*(\mathbf{x})V_P(\mathbf{x})u_{\mathbf{k}}(\mathbf{x}), \tag{31}$$

where the Bloch function is written $\varphi_{\mathbf{k}}(\mathbf{x}) = e^{i\mathbf{k}\cdot\mathbf{x}}u_{\mathbf{k}}(\mathbf{x})$. If $u_{\mathbf{k}}(\mathbf{x})$ does not depend strongly on $\mathbf{k}$, then

$$\Delta\varepsilon_{\mathbf{k}} \cong \int d^3x\, u_0^*(\mathbf{x})V_P(\mathbf{x})u_0(\mathbf{x}). \tag{32}$$

That is, the whole energy band is shifted without change of shape. Any extra electrons added with the impurity atom will simply fill up the band.

The validity of the result (31) is not as obvious as it may seem at first sight. We give a discussion only for free electrons, not for Bloch electrons. Outside the region of the localized potential the wave equation is unchanged by the perturbation, and thus the energy eigenvalue as a function of $\mathbf{k}$ must be unchanged. This statement appears to contradict (31). However, the allowed values of $\mathbf{k}$ are changed: it is essential to consider the effect of the over-all boundary conditions. Suppose the impurity is at the center of a large empty sphere of radius R; the electron is free within the sphere except for the impurity potential. If the boundary condition on the value of the wavefunction at R is satisfied by a particular k before the introduction of the impurity, the boundary condition will be satisfied by a wavevector k' such that

$$k'R + \eta_L = kR, \tag{33}$$

where η_L is the phase shift. The kinetic energy away from the potential has a contribution proportional to k^2, and the shift in the eigenvalue is

$$\Delta\varepsilon_k = \frac{1}{2m}(k'^2 - k^2) = \frac{1}{2m}\left[\left(k - \frac{\eta_L}{R}\right)^2 - k^2\right] \cong -\frac{k\eta_L}{mR}. \tag{34}$$

In the Born approximation for $\eta_L \ll 1$ we have (Schiff, p. 167)

$$\eta_L \cong -2km \int dr\, r^2 j_L{}^2(kr)V(r). \tag{35}$$

The unperturbed wavefunction is, from Schiff (15.8),

$$u_L \cong k\left(\frac{1}{2\pi R}\right)^{1/2} j_L(kr)Y_L^m(\theta,\varphi), \tag{36}$$

where the spherical harmonics are normalized to give 4π on integrating their square over the surface of a sphere; the radial normalization in (36) is only approximate and follows from the asymptotic form (19). Thus

$$\int dr\, r^2 j_L{}^2(kr)V_P(r) = \frac{R}{2k^2}\langle\mathbf{k}|V_P|\mathbf{k}\rangle, \tag{37}$$

so that

$$\eta_L \cong -\frac{mR}{k}\langle\mathbf{k}|V_P|\mathbf{k}\rangle, \tag{38}$$

and, from (34),

$$\Delta\varepsilon_{\mathbf{k}} = \langle\mathbf{k}|V_P|\mathbf{k}\rangle. \qquad \text{Q.E.D.}$$

A more general derivation is given in the thesis of Blandin, reference 3. The result holds only to first order in V_P. It must be noted for impurities of valency different from that of the host lattice the Friedel sum rule tells us that V_P cannot really be a weak interaction.

ELECTRICAL RESISTIVITY

Let $\sigma(\theta)$ be the cross section per unit solid angle for scattering of a conduction electron by an impurity atom. As we have seen in Chapter 7, the electrical resistivity is concerned with the change on scattering of the projection of the wavevector along the axis of current flow. Thus the effective average cross section for resistivity is

$$\langle\sigma\rangle = 2\pi\int_{-\pi}^{\pi} d\theta \sin\theta\, \sigma(\theta)(1 - \cos\theta), \tag{39}$$

where the last factor on the right-hand side weights the average according to the change of k_z. The associated relaxation frequency

$$\frac{1}{\tau_i} = \frac{v_F}{\Lambda} = n_i\langle\sigma\rangle v_F, \tag{40}$$

where Λ is the mean free path, n_i the concentration of scatterers, and v_F the fermi velocity.

The contribution $\Delta\rho$ of the scattering centers to the electrical resistivity is

$$\Delta\rho = \frac{m^*}{ne^2\tau_i} = \frac{n_i k_F}{ne^2}\langle\sigma\rangle, \tag{41}$$

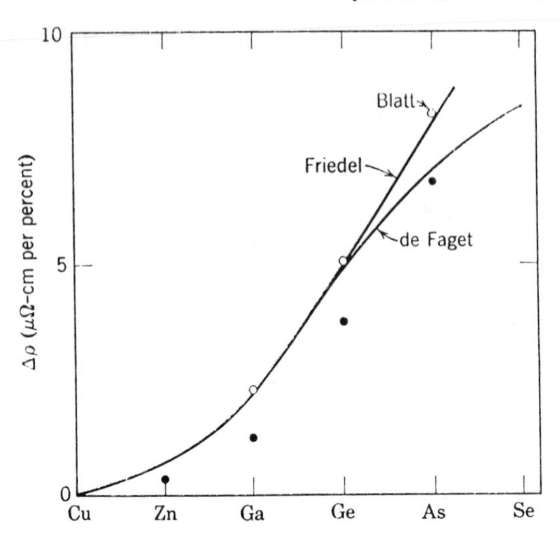

FIG. 1. Residual resistivity $\Delta\rho$ in $\mu\Omega$-cm per percent of polyvalent impurities in copper. Closed circles: experimental values; computed values by Friedel, Blatt, and de Faget de Casteljau and Friedel.

where n is the electron concentration. Now in terms of phase shifts (Schiff, p. 105)

$$\sigma(\theta) = \frac{1}{k_F^2}\left|\sum_{L=0}^{\infty} (2L + 1)e^{i\eta_L} \sin \eta_L P_L(\cos \theta)\right|^2, \tag{42}$$

whence

$$\int_{-1}^{1} d\mu \, (1 - \mu)\sigma(\theta) = \frac{2}{k_F^2} \sum_{L=0}^{\infty} (L + 1) \sin^2 (\eta_L - \eta_{L+1}), \tag{43}$$

and

$$\Delta\rho = \frac{4\pi n_i}{ne^2 k_F} \sum (L + 1) \sin^2 (\eta_L - \eta_{L+1}). \tag{44}$$

Here the phase shifts are to be taken at the fermi surface. If only one phase shift η_L is large, we have

$$\eta_L = \pi Z/(4L + 2) \tag{45}$$

from the sum rule (30), and then

$$\Delta\rho = \frac{4\pi n_i}{ne^2 k_F}(L + 1) \sin^2\left(\frac{\pi Z}{4L + 2}\right). \tag{46}$$

For $Z = 1$ and $L = 0$ we have $\langle\sigma\rangle = 4\pi/k_F{}^2 \approx 10^{-15}$ cm^2 and the resistivity is seen to be of the order of 4 $\mu\Omega$-cm per percent impurity, which is essentially an upper limit to the effect observed for nonmagnetic impurities. The results of detailed calculations of phase shifts, with allowance for the cell volume changes at the impurity, are compared with experiment in Fig. 1. The agreement is seen to be excellent and is rather insensitive to the form of the potential, provided that the value of some parameter determining the potential is chosen to make the phase shifts satisfy the sum rule.

LONG-RANGE OSCILLATIONS OF ELECTRON DENSITY

We saw in Chapter 6 that the self-consistent field solution for the screening charge density around a charged impurity contains oscillatory terms at large distances of the form $r^{-3} \cos 2k_Fr$. It is instructive to exhibit directly the long-range oscillations on the independent-particle model; the result (29′) above indicates that such terms are present also in this approximation. We reexamine the density variation here.

We write the wavefunction describing the scattering of an electron of wavevector $\mathbf{k}$ as

$$\varphi_{\mathbf{k}}(\mathbf{x}) = e^{i\mathbf{k}\cdot\mathbf{x}} + g_{\mathbf{k}}(\mathbf{x}). \tag{47}$$

Then the variation of electron density is

$$\Delta\rho_{\mathbf{k}}(\mathbf{x}) = \varphi_{\mathbf{k}}^*\varphi_{\mathbf{k}} - 1 = g_{\mathbf{k}}(\mathbf{x})e^{-i\mathbf{k}\cdot\mathbf{x}} + cc + |g_{\mathbf{k}}(\mathbf{x})|^2; \tag{48}$$

outside the central charge

$$g_{\mathbf{k}}(\mathbf{x}) \cong f_{\mathbf{k}}(\theta)\,\frac{e^{ikr}}{r}. \tag{49}$$

On averaging (48) over the surface of the sphere at $\mathbf{k}$, we have

$$\langle\Delta\rho_{\mathbf{k}}(\mathbf{x})\rangle \cong \frac{e^{ikr}}{4\pi}\left[\frac{2\pi f_{\mathbf{k}}(\pi)e^{ikr}}{ikr^2} + cc - \frac{2\pi f_{\mathbf{k}}(0)e^{-ikr}}{ikr^2} - cc + \frac{1}{r^2}\int d\Omega\,|f_{\mathbf{k}}(\theta)|^2\right]. \tag{50}$$

We achieve this result by integrating (49) by parts twice, but keeping only the dominant terms for large r. According to the optical theorem (Schiff, p. 105),

$$\int d\Omega\,|f_{\mathbf{k}}(\theta)|^2 = \frac{4\pi}{ik}\,\mathscr{I}\{f_{\mathbf{k}}(0)\}, \tag{51}$$

whence (50) simplifies to

$$\langle \Delta\rho_{\mathbf{k}}(\mathbf{x}) \rangle = \frac{f_{\mathbf{k}}(\pi)e^{2ikr}}{2kr^2 i} + cc. \tag{52}$$

Thus the charge density depends only on the backward scattering amplitude, which is determined by the phase shifts η_L. We can rewrite (52) as

$$\langle \Delta\rho_{\mathbf{k}}(\mathbf{x}) \rangle = \frac{\sin(2kr + \varphi)}{kr^2} |f_{\mathbf{k}}(\pi)|, \tag{53}$$

where φ is a constant.

The total variation $\Delta\rho(\mathbf{x})$ in electron density is obtained by weighting (53) by the density of states and integrating over the fermi sea. The asymptotic form at large r is

$$\Delta\rho(\mathbf{x}) = \frac{2}{(2\pi)^3} \int d^3k \, \langle \Delta\rho_{\mathbf{k}}(\mathbf{x}) \rangle = C_F \frac{\cos(2k_F r + \varphi_F)}{r^3}, \tag{54}$$

where C_F and φ_F are constants. We see that even on the independent-particle model there are long-range oscillatory density variations around an impurity atom; the amplitude falls off as r^{-3}. The density variation is closely similar to that shown in Fig. 6.1.

For nonmagnetic impurity atoms perhaps the strongest experimental evidence for long-range variations in the electron density among impurities is found in the quadrupolar effects in the nuclear magnetic resonance of dilute alloys. Experiments by Bloembergen and Rowland, and others, show a strong attenuation of the central intensity of the nuclear resonance lines of Cu and Al, for a small impurity atom concentration in Cu-base and Al-base alloys. The diminution of peak intensity arises because the resonance line is broadened by the strong electric field gradients associated with the screening charge, both immediately adjacent to the impurity atom and also in the long-range tail. The electric field gradients interact with the quadrupole moments of the Cu and Al nuclei and broaden the resonance.

The actual effect depends strongly on the atomic p and d character of the wavefunction at the fermi surface.[3,4] The quantitative calculations will not be reproduced here. They are in good agreement with the experimental results. The sensitivity of the experimental method is such that resonance line shape effects are seen if there is one impurity among the first 20 to 90 atoms around a given atom, according to the particular situation, particularly the difference of

[4] W. Kohn and S. H. Voski, *Phys. Rev.* **119**, 912 (1960).

valency. The effects increase with the difference in valence and with the atomic number of the matrix: the difference in valence increases the perturbation and the atomic number increases the amplitude of the Bloch functions.

VIRTUAL STATES

Let us consider the sum rule (30):

$$Z = \frac{2}{\pi} \sum_L (2L + 1)\eta_L(k_F). \tag{55}$$

There are several special situations of interest. Let Z be positive; that is, the valency of the impurity is higher than that of the host metal.

One special situation is that one of the extra electrons of the impurity remains bound in the metal to the impurity atom. The energy of this bound state must be below the minimum band edge of the conduction band of the metal, because a true bound state cannot exist anywhere in the same energy region as a continuous energy spectrum. If one electron is bound, then in effect the impurity atom acts as if possessed of an excess valency $Z - 1$, rather than Z. If we choose to do the bookkeeping to delete the one-electron bound state from (55), we replace Z by $Z - 1$ on the left-hand side and we must then delete the contribution η/π of the single bound state from the right-hand side. To be consistent, the value of η for the bound state must be π. It is known from scattering theory (Schiff, p. 113; Messiah, p. 398) that a potential well with an energy level nearly at zero will exhibit a resonance in the low-energy scattering of particles with the same L value as the energy level. An incident particle with nearly the right energy to be bound by the potential tends to concentrate there. This produces a large distortion in the wavefunction and thus a large amount of scattering. We recall (Schiff, p. 105) that the total scattering cross section is

$$\sigma = \frac{4\pi}{k^2} \sum_L (2L + 1) \sin^2 \eta_L; \tag{56}$$

at resonance for a particular L we must have $|\eta_L| = \frac{1}{2}\pi, \frac{3}{2}\pi, \cdots$.

The second special situation arises if the attractive impurity potential is reduced. The bound state is thereby increased in energy and eventually merges into the continuum of conduction band states. It is useful to think of the state as still existing as a *virtual state,* now with positive energy relative to the conduction band minimum. Somewhat below the resonance energy the phase shift is approximately $n\pi$, and somewhat above the shift is approximately $(n + 1)\pi$. The resonance energy may be defined as the energy for which $\eta_L = (n + \frac{1}{2})\pi$. In Fig. 2 we show the phase shift for p-wave scattering from a particular square well; the wavevector dependence of the partial scattering cross section is also shown. The resonance aspect is quite marked and determines the position of the virtual level.

The rapid variation of phase shift with electron energy near the virtual level makes the properties of an alloy having its fermi level in this region highly sensitive to the electron concentration. A small change in average electron concentration, as might be produced by the addition of a third component to the alloy, can sweep the fermi level through the virtual level. If the fermi level is well below the virtual level for a given L, say L', then the screening charge is made up of various L components as necessary to satisfy the sum rule (55). If the fermi level is well above the virtual level, then the phase shift η_L will be large and a major part of the screening charge may arise

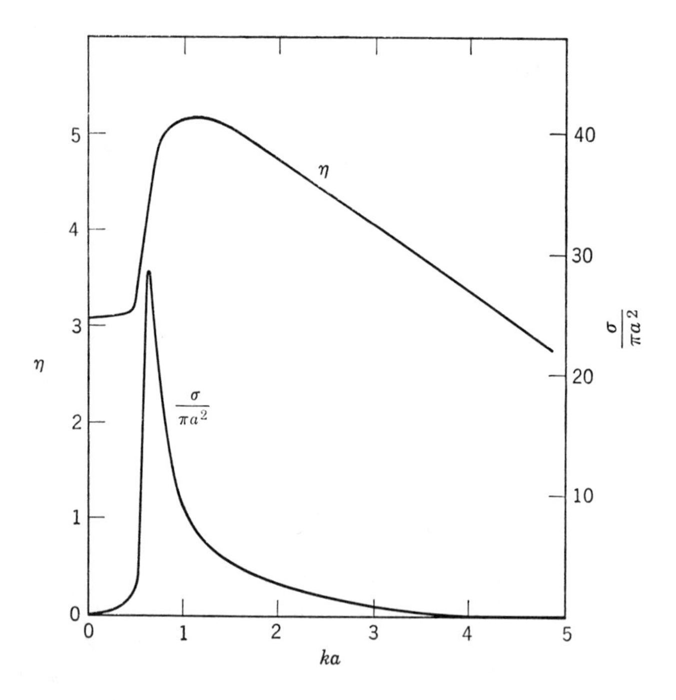

FIG. 2. Phase shift η_1 and partial cross section σ_1 for p-wave scattering from square well of radius a and depth V_0 such that $(2mV_0)^{1/2}a = 6.2$. (After Merzbacher, *Quantum mechanics*, Wiley, 1961.)

from electrons in the virtual level L'—such electrons act as quasi-bound and may be thought of as localized.

The mathematical treatment[5] of virtual states in a metal has been considered by a number of authors, and we cite only several of the early papers. We give separate, but related, treatments best suited for nonmagnetic impurities (Wolff) and for magnetic impurities (Anderson). For nonmagnetic impurities it is convenient to use the Wannier functions defined by (9.114). Let V be the impurity potential and H_0 the one-electron unperturbed hamiltonian of the host lattice. The exact solution ψ of the wave equation in the perturbed lattice may be written in the limit $s \to +0$ as

$$\psi = \varphi_{\mathbf{k}\gamma} + \frac{1}{\varepsilon - H_0 + is} V\psi. \tag{57}$$

This follows from the formal theory of scattering, as in Messiah, Chapter 19. We now make the fairly drastic assumption that V does not mix states from different bands; specifically, we assume that ψ may be expanded in terms of Wannier functions from a single band:

$$\psi = N^{-1/2} \sum_n U(\mathbf{x}_n) w_\gamma(\mathbf{x} - \mathbf{x}_n), \tag{58}$$

where the Wannier function w for band γ is defined in terms of the Bloch functions $\varphi_{\mathbf{k}\gamma}(\mathbf{x})$ by

$$w_\gamma(\mathbf{x} - \mathbf{x}_n) = N^{-1/2} \sum_n e^{-i\mathbf{k}\cdot\mathbf{x}_n} \varphi_{\mathbf{k}\gamma}(\mathbf{x}). \tag{59}$$

Our neglect of interband terms may not be too serious if both the impurity and the band states are derived chiefly from the same atomic states.

We note that the matrix element

$$\begin{aligned} \int d^3x\, \varphi^*_{\mathbf{k}'\gamma}(\mathbf{x}) V\psi \\ = N^{-1/2} \sum_{mn} \int d^3x\, w^*_\gamma(\mathbf{x} - \mathbf{x}_m) e^{-i\mathbf{k}'\cdot\mathbf{x}_m} V U(\mathbf{x}_n) w_\gamma(\mathbf{x} - \mathbf{x}_n). \end{aligned} \tag{60}$$

If V is localized at $\mathbf{x}_0$ and of such short range that it does not overlap appreciably the Wannier functions centered on nearby lattice points,

[5] G. F. Koster and J. C. Slater, *Phys. Rev.* **96,** 1208 (1954); P. A. Wolff, *Phys. Rev.* **124,** 1030 (1961); P. W. Anderson, *Phys. Rev.* **124,** 41 (1961); A. M. Clogston, *Phys. Rev.* **125,** 439 (1962); Clogston et al., *Phys. Rev.* **125,** 541 (1962). For the theory of impurity scattering effects in superconductors, see H. Suhl and B. T. Matthias, *Phys. Rev.* **114,** 977 (1959), and H. Suhl, *LTP*, pp. 233–259.

we may make the approximation

$$\int d^3x\, w_\gamma^*(\mathbf{x} - \mathbf{x}_m) V w_\gamma(\mathbf{x} - \mathbf{x}_n) = V_{\gamma\gamma}\delta_{0n}\delta_{0m}. \tag{61}$$

Thus (57) may be written as

$$\psi = \varphi_{\mathbf{k}\gamma} + \sum_{\mathbf{k}'}{}' \varphi_{\mathbf{k}'\gamma} U(\mathbf{x}_0) e^{-i\mathbf{k}'\cdot\mathbf{x}_0} N^{-\frac{1}{2}} \frac{V_{\gamma\gamma}}{\varepsilon - \varepsilon_{\mathbf{k}'\gamma} + is}; \tag{62}$$

or, using (58),

$$U(\mathbf{x}_n) = e^{i\mathbf{k}\cdot\mathbf{x}_n} + \sum_{\mathbf{k}'}{}' \frac{e^{i\mathbf{k}'\cdot(\mathbf{x}_n - \mathbf{x}_0)}}{\varepsilon - \varepsilon_{\mathbf{k}'\gamma} + is} V_{\gamma\gamma} U(\mathbf{x}_0). \tag{63}$$

For $\mathbf{x}_n = \mathbf{x}_0$,

$$U(\mathbf{x}_0) = \frac{e^{i\mathbf{k}\cdot\mathbf{x}_0}}{1 - V_{\gamma\gamma} \sum_{\mathbf{k}'} [\varepsilon - \varepsilon_{\mathbf{k}'\gamma} + is]^{-1}}. \tag{64}$$

The summation may be written as

$$\frac{1}{(2\pi)^3} \int d^3k' \frac{1}{\varepsilon - \varepsilon_{\mathbf{k}'\gamma} + is} = \int dE \frac{g_\gamma(E)}{\varepsilon - E + is} = \mathcal{P} \int dE \frac{g_\gamma(E)}{\varepsilon - E} - i\pi g(\varepsilon). \tag{65}$$

Here $g(\varepsilon)$ is the density of states in the band under consideration, and $\mathcal{P}$ denotes principal value. We denote the principal value integral by

$$F_\gamma(\varepsilon) \equiv \mathcal{P} \int dE \frac{g_\gamma(E)}{\varepsilon - E}, \tag{66}$$

so that (64) may be written as

$$\boxed{U(\mathbf{x}_0) = \frac{e^{i\mathbf{k}\cdot\mathbf{x}_0}}{1 - V_{\gamma\gamma} F_\gamma(\varepsilon) + i\pi V_{\gamma\gamma} g_\gamma(\varepsilon)}.} \tag{67}$$

The amplitude $U(\mathbf{x}_0)$ of the wavefunction on the impurity atom will be large at the roots ε_0 of

$$1 - V_{\gamma\gamma} F_\gamma(\varepsilon_0) = 0; \tag{68}$$

this is the definition on this model of the position of the virtual level. If a root ε_0 lies outside the band, then $g_\gamma(\varepsilon_0) = 0$ and $U(\mathbf{x}_0) \to \infty$; the root then represents a real bound state. If ε_0 lies within the band, $g_\gamma(\varepsilon_0)$ is finite and $U(\mathbf{x}_0)$ has its resonance maximum at ε_0.

LOCALIZED MAGNETIC STATES IN METALS

The treatment by Anderson of localized magnetic states in metals is particularly applicable when the impurity atom has an unfilled or partly filled d shell and the conduction band states of the matrix are not d-like, say s-like or an s-p mixture. We suppose there is a *single* orbital d-shell state φ_d, with two possible spin orientations. This is a nontrivial assumption, but it is shown in an appendix to the paper by Anderson not to be of major consequence. If a localized magnetic moment exists around the impurity atom, one of the two spin states (say the up state) will be filled or partly filled. Then an electron with spin down will feel the coulomb repulsion of the spin-up electron. If the unperturbed energy of the spin-up state lies a distance E' below the fermi level, the energy of the spin-down localized state will be $-E' + U$, where U is the repulsive d-d interaction. A localized moment will exist if $-E' + U$ lies above the fermi level.

A covalent mixture of free electron states with the d state will reduce the number of electrons in the spin-up state and increase the number in the spin-down state. The coupling of s and d states raises the energy of the spin-up state and lowers that of the spin-down state. With this effect and the associated broadening of the d states, the persistence of a localized moment becomes a cooperative phenomenon; also, the virtual state can contain any nonintegral number of spins. We assume a single nondegenerate d level; the degenerate situation is considered in an appendix to the original paper.

We write the hamiltonian for the Anderson model as

$$(69)\qquad H = \sum_{\mathbf{k}\sigma} \varepsilon_{\mathbf{k}} n_{\mathbf{k}\sigma} + E(n_{d\uparrow} + n_{d\downarrow}) + U n_{d\uparrow} n_{d\downarrow} + \sum_{\mathbf{k}\sigma} V_{d\mathbf{k}}(c^+_{\mathbf{k}\sigma} c_{d\sigma} + c^+_{d\sigma} c_{\mathbf{k}\sigma}).$$

Here $\varepsilon_{\mathbf{k}}$ is the energy of a free-electron state; $n_{\mathbf{k}\sigma} \equiv c^+_{\mathbf{k}\sigma} c_{\mathbf{k}\sigma}$; E is the unperturbed energy of the d state on an impurity atom; U is the coulomb repulsive energy between electrons in $d\uparrow$ and $d\downarrow$; and $V_{d\mathbf{k}}$ is the interaction energy between a d state and a Wannier function on a nearest-neighbor atom to the d-state impurity. We assume that φ_d is orthogonal to all Wannier functions of the conduction band.

Let

$$(70)\qquad \Phi_0 = \prod_{\varepsilon < \varepsilon_F} c^+_n \Phi_{\text{vac}}$$

denote the ground state in the Hartree-Fock approximation of the system with the hamiltonian written above. According to the argument of Chapter 5, this means that

$$-i\dot{c}_{n\sigma}^{+} = -\varepsilon_{n\sigma}c_{n\sigma}^{+} = [c_{n\sigma}^{+},H]_{\mathrm{av}}, \tag{71}$$

where the subscript av means that the three fermion terms in the commutator are to be reduced to single fermion terms times average values over the state Φ_0. Now write

$$c_{n\sigma}^{+} = \Big(\sum_{\mathbf{k}} \langle n|\mathbf{k}\rangle_\sigma c_{\mathbf{k}\sigma}^{+}\Big) + \langle n|d\rangle_\sigma c_{d\sigma}^{+}, \tag{72}$$

where the operators $c_{\mathbf{k}\sigma}^{+}$, $c_{d\sigma}^{+}$ refer to the unperturbed states and satisfy

$$-[c_{\mathbf{k}\sigma}^{+},H]_{\mathrm{av}} = \varepsilon_{\mathbf{k}}c_{\mathbf{k}\sigma}^{+} + V_{\mathbf{k}d}c_{\mathbf{k}\sigma}^{+}; \tag{73}$$

$$-[c_{d\sigma}^{+},H]_{\mathrm{av}} = (E - U\langle n_{d,-\sigma}\rangle)c_{d\sigma}^{+} + \sum_{\mathbf{k}} V_{d\mathbf{k}}c_{\mathbf{k}\sigma}^{+}. \tag{74}$$

Now substitute (72) in (71) and utilize the results (73) and (74); on equating the coefficients of $c_{\mathbf{k}\sigma}^{+}$ and of $c_{d\sigma}^{+}$, we find the relations

$$\varepsilon_{n\sigma}\langle n|\mathbf{k}\rangle_\sigma = \varepsilon_{\mathbf{k}}\langle n|\mathbf{k}\rangle_\sigma + V_{\mathbf{k}d}\langle n|d\rangle_\sigma; \tag{75}$$

$$\varepsilon_{n\sigma}\langle n|d\rangle_\sigma = (E + U\langle n_{d,-\sigma}\rangle)\langle n|d\rangle_\sigma + \sum_{\mathbf{k}} V_{d\mathbf{k}}\langle n|\mathbf{k}\rangle_\sigma. \tag{76}$$

What we want to calculate in this problem is

$$\rho_{d\sigma}(\varepsilon) \equiv \sum_n |\langle n|d\rangle_\sigma|^2\delta(\varepsilon - \varepsilon_n), \tag{77}$$

which is the mean density in energy of the admixture $|\langle n|d\rangle_\sigma|^2$ of the state $d\sigma$ in the continuum levels n of energy ε.

The quantity $\rho_{d\sigma}(\varepsilon)$ may be calculated very neatly. We examine the Green's function

$$G(\varepsilon + is) = \frac{1}{\varepsilon + is - H}, \tag{78}$$

which is diagonal in the representation n of the exact eigenstates:

$$G_{nn}^{\sigma}(\varepsilon + is) = \langle n\sigma|G|n\sigma\rangle = \frac{1}{\varepsilon + is - \varepsilon_{n\sigma}}. \tag{79}$$

Now

$$\mathscr{I}\{G_{nn}^{\sigma}\} = -\pi\delta(\varepsilon - \varepsilon_{n\sigma}), \tag{80}$$

so that

$$\rho_{d\sigma}(\varepsilon) = -\frac{1}{\pi}\sum_n |\langle n|d\rangle_\sigma|^2\mathscr{I}\{G_{nn}^{\sigma}\} = -\frac{1}{\pi}\mathscr{I}\{\langle d\sigma|G|d\sigma\rangle\}. \tag{81}$$

We note that the total density of states

$$\rho_\sigma(\varepsilon) = \sum_n \delta(\varepsilon - \varepsilon_{n\sigma}) = -\frac{1}{\pi} \mathcal{I}\{\mathrm{Tr}\, G^\sigma\}. \tag{82}$$

From (81) we see that our problem is reduced to the determination of the matrix elements G^σ_{dd}. The equations for the matrix elements of G are

$$\sum_\nu (\varepsilon + is - H)_{\mu\nu} G_{\nu\lambda} = \varepsilon_{\mu\lambda}, \tag{83}$$

according to the definition (78) of G. If we write

$$E_\sigma = E + U\langle n_{d,-\sigma}\rangle; \qquad \xi = \varepsilon + is, \tag{84}$$

we can find the matrix elements of $(\varepsilon + is - H)$ from (75) and (76). Thus we obtain

$$(\xi - E_\sigma)G^\sigma_{dd} - \sum_{\mathbf{k}} V_{d\mathbf{k}} G^\sigma_{\mathbf{k}d} = 1; \tag{85}$$

$$(\xi - \varepsilon_{\mathbf{k}})G^\sigma_{\mathbf{k}d} - V_{\mathbf{k}d} G^\sigma_{dd} = 0; \tag{86}$$

$$(\xi - E_\sigma)G^\sigma_{d\mathbf{k}} - \sum_{\mathbf{k}'} V_{d\mathbf{k}'} G^\sigma_{\mathbf{k}'\mathbf{k}} = 0; \tag{87}$$

$$(\xi - \varepsilon_{\mathbf{k}'})G^\sigma_{\mathbf{k}'\mathbf{k}} - V_{\mathbf{k}'d} G^\sigma_{d\mathbf{k}} = \delta_{\mathbf{k}'\mathbf{k}}. \tag{88}$$

We solve for G_{dd} from (85) and (86):

$$G^\sigma_{dd}(\xi) = \left[\xi - E_\sigma - \sum_{\mathbf{k}} \frac{|V_{d\mathbf{k}}|^2}{\xi - \varepsilon_{\mathbf{k}}}\right]^{-1}. \tag{89}$$

The sum over **k** in this result may be evaluated:

$$\lim_{s\to+0} \sum_{\mathbf{k}} \frac{|V_{d\mathbf{k}}|^2}{\varepsilon - \varepsilon_{\mathbf{k}} + is} = \mathcal{P} \sum_{\mathbf{k}} \frac{|V_{d\mathbf{k}}|^2}{\varepsilon - \varepsilon_{\mathbf{k}}} - i\pi \sum_{\mathbf{k}} |V_{d\mathbf{k}}|^2 \delta(\varepsilon - \varepsilon_{\mathbf{k}}). \tag{90}$$

The first term on the right-hand side is an energy shift which may be absorbed in E_σ. Neglecting this, the right-hand side of (90) may be written as

$$-i\pi\langle V_{d\mathbf{k}}{}^2\rangle_{\mathrm{av}}\rho(\varepsilon) \equiv i\Delta, \tag{91}$$

where $\rho(\varepsilon)$ denotes the density of states. Thus, apart from the energy shift, G^σ_{dd} behaves exactly as if there were a virtual state at

$$\xi = E_\sigma - i\Delta, \tag{92}$$

with Δ defined by (91).

If we assume Δ to be independent of E_σ, we may write (81) as

$$\rho_{d\sigma}(\varepsilon) = -\frac{1}{\pi}\mathscr{I}\left\{\frac{1}{\varepsilon - E_\sigma + i\Delta}\right\} = \frac{1}{\pi}\frac{\Delta}{(\varepsilon - E_\sigma)^2 + \Delta^2}. \tag{93}$$

The total number of d electrons of spin σ is given by

$$\langle n_{d\sigma}\rangle = \int_{-\infty}^{\varepsilon_F} d\varepsilon\, \rho_{d\sigma}(\varepsilon) = \frac{1}{\pi}\cot^{-1}\frac{E_\sigma - \varepsilon_F}{\Delta}. \tag{94}$$

But E_σ involves $\langle n_{d,-\sigma}\rangle$, according to the definition (84). We must make the values of $n_{d\uparrow}$ and $n_{d\downarrow}$ self-consistent:

$$\langle n_{d\uparrow}\rangle = \frac{1}{\pi}\cot^{-1}\frac{E - \varepsilon_F + U\langle n_{d\downarrow}\rangle}{\Delta}; \tag{95}$$

$$\langle n_{d\downarrow}\rangle = \frac{1}{\pi}\cot^{-1}\frac{E - \varepsilon_F + U\langle n_{d\uparrow}\rangle}{\Delta}. \tag{96}$$

The solutions of these equations are considered in detail by Anderson, with the results for the ground state shown in Fig. 3. If $\langle n_{d\uparrow}\rangle = \langle n_{d\downarrow}\rangle$, the solution is nonmagnetic. In the iron group U may be about 10 ev; for the s band of copper $\rho(\varepsilon)$ may be of the order of 0.1 $(\text{ev})^{-1}$. From binding energy considerations V_{av} may be estimated roughly as $\sim$2 to 3 ev, so that $\Delta = \pi\langle V^2\rangle_{\text{av}}\rho(\varepsilon) \sim 2$ to 5 ev and $\pi\Delta/U$ falls in the range 0.6 to 1.5, which covers the transition between magnetic and nonmagnetic behavior. For rare earth solutes the s-f interaction V will be much smaller, so that for the smaller Δ we expect magnetic behavior to occur frequently, as is observed.

We note that the theory does not automatically give solutions which satisfy the Friedel sum rule; this is a defect which should be remedied. The point is considered by Clogston[5] and by Blandin[3] and Friedel.

We now discuss several examples of virtual levels.

Transition Element Impurities in Aluminum. When Ti, at the beginning of the first transition group, is dissolved in Al, we expect its d shell to be relatively unstable compared with the d shell of Ni at the end of the group. When going continuously from Ti to Ni added to Al, the d-shell states of the impurity should descend and cross the fermi level of Al and give rise to virtual bound d states. The residual resistivity of these impurities in Al are plotted in Fig. 4, which shows a large broad peak around Cr. The peak is believed to be caused by resonance scattering of electrons at the fermi level of Al when this level crosses the broadened d-shell levels of the impurity.

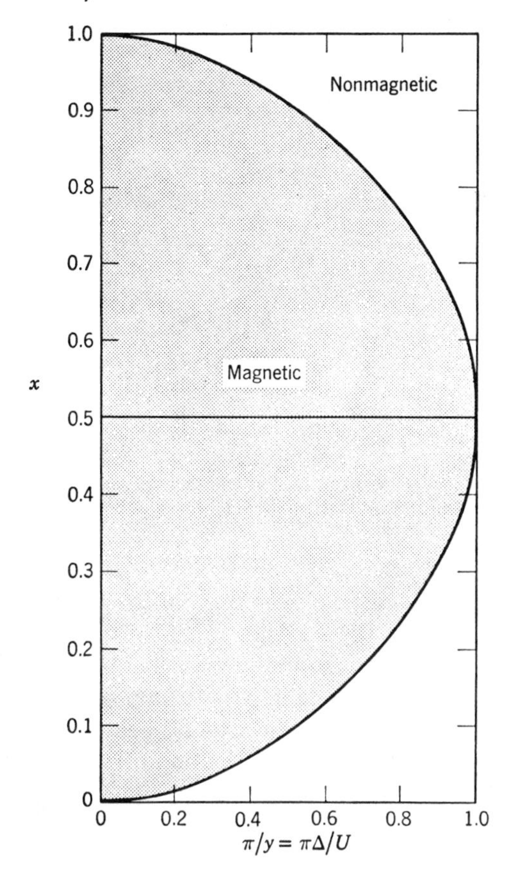

FIG. 3. Regions of magnetic and nonmagnetic behavior. (After Anderson.) Here $x \equiv (\varepsilon_F - E)/U$ and $y \equiv Y/\Delta$.

Transition Element Impurities in Copper. The exchange and particularly the coulomb interactions within a d shell tend to split it into two halves of opposite spin directions. In the free atom the first five d electrons have parallel spins, in conformity with Hund's rule; the sixth and subsequent d electrons have their spin antiparallel to the direction of the spins of the first five electrons. In the alloy this spin splitting is maintained if the associated energy is larger than the width of the virtual level and larger than the distance from the fermi level to the virtual level. The numerical criterion is discussed in detail by Anderson, reference 5. Not all transition element impurities are magnetic in copper.

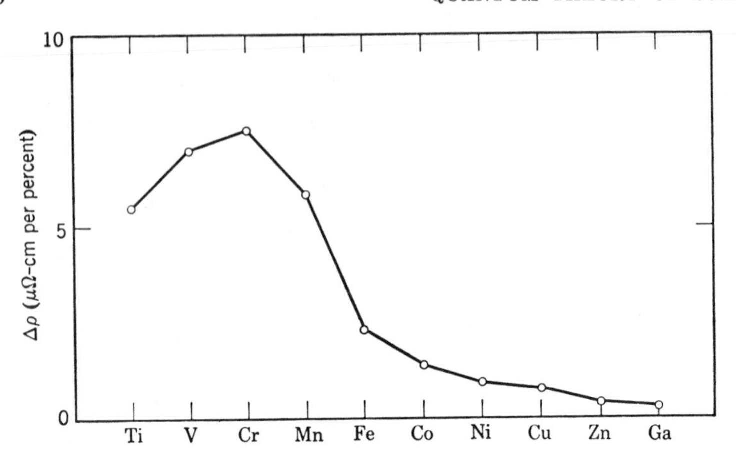

FIG. 4. Residual resistivity $\Delta\rho$ in $\mu\Omega$-cm per percent for transition element and other impurities in aluminum. (After Friedel.)

The exchange splitting explains the double peak (Fig. 5) of the residual resistivity $\Delta\rho$ in the series from Ti to Ni dissolved in Cu, Ag, or Au. Peak A corresponds to the emptying of the upper half A of the d shell, peak B to the lower half B with opposite spin direction, as shown in Fig. 6. Magnetic measurements provide a useful criterion for a localized magnetic moment: if the impurity contributes a Curie-Weiss temperature-dependent term to the magnetic susceptibility, then

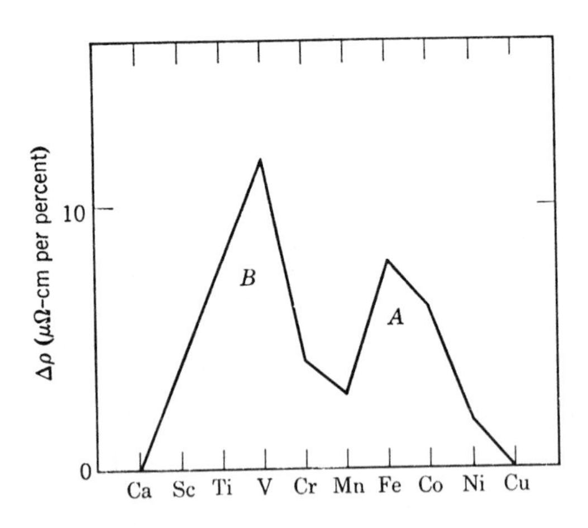

FIG. 5. Residual resistivities of transitional impurities in copper.

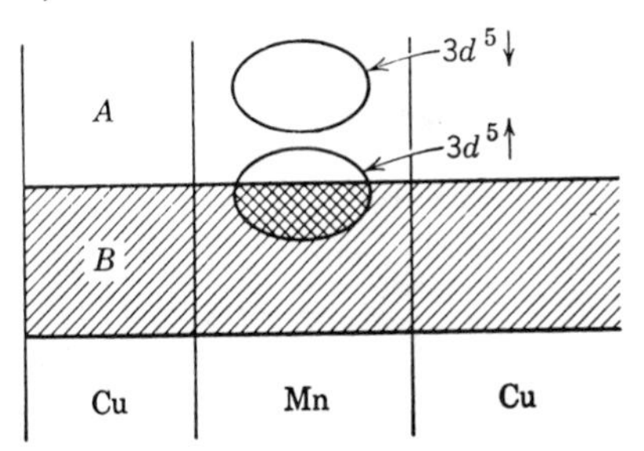

FIG. 6. *d* shell of a transitional impurity such as Mn in Cu, split into two virtual bound levels of opposite spin directions.

there is a localized magnetic moment. If there is no temperature-dependent contribution, there is no localized moment. In Fig. 7 we exhibit the local magnetic moment of Fe in various solvents as a function of the electron concentration of the solvent. Varying the electron concentration varies the width of the virtual level and the position of the fermi level. A local moment is developed when the fermi level is close to a virtual level and the virtual level is sufficiently narrow. It is a question of relative width which makes Fe magnetic in Cu, but not in Al.

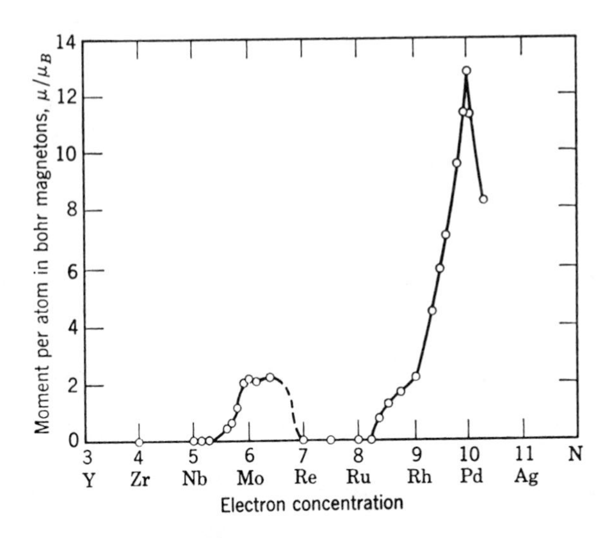

FIG. 7. Magnetic moment in bohr magnetons of an iron atom dissolved in various second row transition metals and alloys as a function of electron concentration. (After Clogston et al.)

INDIRECT EXCHANGE INTERACTION VIA CONDUCTION ELECTRONS[6, 7, 8, 9]

There are two closely related problems in metals: the indirect interaction of two nuclei via their hyperfine interaction with the conduction electron sea, and the indirect interaction of two ions via the exchange interaction of their inner shells (d or f) with the conduction electrons. The ionic interaction is of considerable importance in the study of metals with magnetic order. First we treat the nuclear problem. Our treatment neglects correlation effects; we also assume that the one-electron energy is $\propto k^2$. The essential results of our calculation below can be understood easily from (29′) or (54), but it is useful to see the detailed calculation.

The spin of one nucleus in a metal senses the spin direction of another nucleus in the following way. The contact part $\mathbf{I}_1 \cdot \mathbf{S}$ of the hyperfine coupling scatters a conduction electron having a given state of the spin $\mathbf{S}$ differently according to the state of the nuclear spin $\mathbf{I}_1$. A second nuclear spin $\mathbf{I}_2$ sees the density of the scattered electron through the interaction $\mathbf{I}_2 \cdot \mathbf{S}$, and thereby senses the state of $\mathbf{I}_1$. This process effectively couples together the two nuclear spins, $\mathbf{I}_1$ and $\mathbf{I}_2$.

The electrons are described by Bloch functions

$$\varphi_{\mathbf{k}s}(\mathbf{x}) = e^{i\mathbf{k}\cdot\mathbf{x}}u_{\mathbf{k}s}(\mathbf{x}) = \varphi_{\mathbf{k}}(\mathbf{x})|s\rangle, \tag{97}$$

normalized in unit volume; here s is the spin index and denotes $\uparrow$ or $\downarrow$ for S_z.

We calculate first the perturbation in electron density created by the hyperfine interaction of the electron with the nucleus at $\mathbf{R}_n$. For the contact interaction the hamiltonian has the form

$$H = \sum_j A(\mathbf{x}_j - \mathbf{R}_n)\mathbf{S}_j \cdot \mathbf{I}_n, \tag{98}$$

where $A(\mathbf{x}_j - \mathbf{R}_n)$ is proportional to a delta-function, according to (19.63), where $\mathbf{x}_j$ is the position of electron j. The contact interaction is in metals often the dominant part of the hyperfine coupling.

In the language of second quantization the electron field operator is

$$\Psi(\mathbf{x}) = \sum_{\mathbf{k}s} c_{\mathbf{k}s}\varphi_{\mathbf{k}s}(\mathbf{x}); \qquad \Psi^+(\mathbf{x}) = \sum_{\mathbf{k}s} c_{\mathbf{k}s}^+\varphi_{\mathbf{k}s}^*(\mathbf{x}), \tag{99}$$

where c, c^+ are fermion operators. In this representation we obtain

[6] M. A. Ruderman and C. Kittel, *Phys. Rev.* **96,** 99 (1954).
[7] K. Yosida, *Phys. Rev.* **106,** 893 (1957).
[8] J. H. Van Vleck, *Rev. Mod. Phys.* **34,** 681 (1962).
[9] A. Blandin and J. Friedel, *J. phys. rad.* **20,** 160 (1956).

the hamiltonian by the usual prescription. Thus, guided by the one-electron expectation value

$$\int d^3x\, \varphi^*(\mathbf{x}) A(\mathbf{x} - \mathbf{R}_n)\mathbf{S} \cdot \mathbf{I}_n \varphi(\mathbf{x}),$$

we obtain

$$H = \sum_{\substack{\mathbf{k}\mathbf{k}' \\ ss'}} \left[\int d^3x\, \varphi^*_{\mathbf{k}'s}(\mathbf{x}) A(\mathbf{x} - \mathbf{R}_n)\mathbf{S} \cdot \mathbf{I}_n \varphi_{\mathbf{k}s}(\mathbf{x}) \right] c^+_{\mathbf{k}'s'} c_{\mathbf{k}s}, \tag{100}$$

where $\mathbf{S}$ operates on the spin part of $\varphi_{\mathbf{k}s}$. With $\varphi_{\mathbf{k}s}(\mathbf{x}) = \varphi_{\mathbf{k}}(\mathbf{x})|s\rangle$, we get directly:

$$H = \tfrac{1}{2} \sum_{\mathbf{k}\mathbf{k}'} e^{i(\mathbf{k}-\mathbf{k}')\cdot\mathbf{R}_n} J(\mathbf{k}',\mathbf{k}) \{ I^+_n c^+_{\mathbf{k}'\downarrow} c_{\mathbf{k}\uparrow} + I^-_n c^+_{\mathbf{k}'\uparrow} c_{\mathbf{k}\downarrow} + I^z_n (c^+_{\mathbf{k}'\uparrow} c_{\mathbf{k}\uparrow} - c^+_{\mathbf{k}'\downarrow} c_{\mathbf{k}\downarrow} \}, \tag{101}$$

where

$$J(\mathbf{k}',\mathbf{k}) = \int d^3x\, \varphi^*_{\mathbf{k}'}(\mathbf{x}) A(\mathbf{x}) \varphi_{\mathbf{k}}(\mathbf{x}). \tag{102}$$

We shall assume that $J(\mathbf{k}',\mathbf{k})$ is constant, independent of $\mathbf{k}$ and $\mathbf{k}'$. We note that we may simulate a constant $J(\mathbf{k}',\mathbf{k})$ by setting

$$A(\mathbf{x}) = J\delta(\mathbf{x}); \tag{103}$$

thus

$$J(\mathbf{k}',\mathbf{k}) = J, \tag{104}$$

a constant with the dimensions energy $\times$ volume. We may view $J\delta(\mathbf{x})$ as the pseudopotential associated with the s-wave scattering length and phase shift

$$b = mJ/2\pi; \qquad \eta_0 = -kb, \tag{105}$$

as is discussed in Chapter 19.

In the Born approximation the wavefunctions to first order in J are, with $\mathbf{R}_n = 0$,

$$|\mathbf{k}\uparrow\rangle = |\mathbf{k}\uparrow\rangle_0 + \sum_{\mathbf{k}'s}{}' |\mathbf{k}'s\rangle_0 \frac{\langle \mathbf{k}'s|H|\mathbf{k}\uparrow\rangle}{\varepsilon_{\mathbf{k}} - \varepsilon_{\mathbf{k}}'} = |\mathbf{k}\uparrow\rangle_0 + \sum_{\mathbf{k}'}{}' \frac{m^*J}{k^2 - k'^2} (I^+_n|\mathbf{k}'\downarrow\rangle_0 + I^z_n|\mathbf{k}'\uparrow\rangle_0), \tag{106}$$

for the wavefunction with the electron spin mainly up, and

$$|\mathbf{k}\downarrow\rangle = |\mathbf{k}\downarrow\rangle_0 + \sum_{\mathbf{k}'}{}' \frac{m^*J}{k^2 - k'^2} (I^-_n|\mathbf{k}'\uparrow\rangle_0 - I^z_n|\mathbf{k}'\downarrow\rangle_0), \tag{107}$$

for the wavefunction with electron spin mainly down. The arrows refer to the electron spin; at this point we have not carried out any operations on the nuclear spin. The prime in Σ' means that the state $\mathbf{k}'s = \mathbf{k}\uparrow$ is to be excluded from the summation. This is a delicate point, because the adiabatic-switching form of perturbation theory tells us to exclude the energy shell $\varepsilon_{\mathbf{k}} = \varepsilon_{\mathbf{k}'}$ in the sense of taking the principal value of the integral formed from the summation above; for details of why the principal value gives the full answer in the end for this problem, see Yosida[7] and Van Vleck.[8]

If the $u_{\mathbf{k}}$ part of the Bloch function is independent of $\mathbf{k}$, we may write (106) as

$$(108)\quad |\mathbf{k}\uparrow\rangle = |\mathbf{k}\uparrow\rangle_0 + \frac{m^*J}{(2\pi)^3}\,\mathcal{P}\int d^3k'\,\frac{e^{i\mathbf{k}'\cdot\mathbf{x}}}{k^2 - k'^2}\,(I_n^+|\downarrow\rangle + I_n^z|\uparrow\rangle).$$

The integral in (108) has the value

$$(109)\quad \mathcal{P}\int = 2\pi\mathcal{P}\int_{-1}^{1} d\mu \int_0^\infty dk'\,\frac{e^{ik'r\mu}k'^2}{(k + k')(k - k')}$$

$$= 2\pi(ir)^{-1}\mathcal{P}\left\{\int_{-\infty}^\infty d\rho\,\frac{\rho e^{i\rho}}{\rho^2 - \sigma^2} - \int_{-\infty}^\infty d\rho\,\frac{\rho e^{-i\rho}}{\rho^2 - \sigma^2}\right\}$$

with $\rho = k'r$ and $\sigma = kr$. The integrands are even functions of ρ when this is real, so we may write

$$(110)\quad \mathcal{P}\int = \left(\frac{\pi}{ir}\right)\mathcal{P}\left\{\int_{-\infty}^\infty d\rho\,\frac{\rho e^{i\rho}}{\rho^2 - \sigma^2} - \int_{-\infty}^\infty d\rho\,\frac{\rho e^{-i\rho}}{\rho^2 - \sigma^2}\right\}.$$

The first integral may be evaluated by completing the contour with an infinite semicircle in the upper half-plane; the semicircle gives no contribution to the integral because the exponential is vanishingly small on this segment. The residue at the pole $\rho = \sigma$ is $\frac{1}{2}e^{i\sigma}$ and at the pole $\rho = -\sigma$ the residue is $\frac{1}{2}e^{-i\sigma}$. Thus the value of the first integral is

$$(111)\quad \tfrac{1}{2}2\pi i(\tfrac{1}{2}e^{i\sigma} + \tfrac{1}{2}e^{-i\sigma}) = \pi i \cos\sigma,$$

where the $\frac{1}{2}$ on the extreme left-hand side arises because of the principal value in (110). The contour of the second integral in (110) is completed by an infinite semicircle in the lower half-plane, and the integral has the value $-\pi i \cos\sigma$. Thus (109) assumes the value

$$(112)\quad \mathcal{P}\int = (2\pi^2/r)\cos kr,$$

whence

$$|k\uparrow\rangle = |k\uparrow\rangle_0 + \frac{m^*J\cos kr}{4\pi r}(I_n^-|\downarrow\rangle + I_n^z|\uparrow\rangle). \tag{113}$$

The electron density corresponding to (113) is, to $0(J)$,

$$\rho(\mathbf{k}\uparrow) = 1 + \frac{m^*J\cos kr}{2\pi r}\cos kx \quad I_n^z \tag{114}$$

also

$$\rho(\mathbf{k}\downarrow) = 1 - \frac{m^*J\cos kr}{2\pi r}\cos kx \quad I_n^z \tag{115}$$

It is interesting to sum $\rho(\mathbf{k}\uparrow)$ over the ground state fermi sea:

$$\rho(\uparrow) = \frac{1}{(2\pi)^3}\int d^3k\rho(\mathbf{k}\uparrow) = \frac{k_F^3}{6\pi^2} + \frac{m^*JI_n^z}{4(2\pi)^3r^4}\int_0^{2k_Fr} dx\ x\sin x, \tag{116}$$

or

$$\rho(\uparrow) = \frac{k_F^3}{6\pi^2}\left\{1 - \frac{3m^*JI_n^zk_F}{\pi}F(2k_Fr)\right\}, \tag{117}$$

where

$$F(x) = \frac{x\cos x - \sin x}{x^4}. \tag{118}$$

If the electron concentration is $2n$, of which n are of each spin, we may summarize (117) and the corresponding result for $\rho(\downarrow)$ by*

$$\rho_\pm(\mathbf{x}) = n\left[1 \mp \frac{9n}{\varepsilon_F}\pi JF(2k_Fr)I_n^z\right]. \tag{119}$$

We see that the nuclear moment perturbs the electron spin polarization in an oscillatory fashion. The net polarization has at $k_Fr \gg 1$ the asymptotic form

$$\rho_\downarrow - \rho_\uparrow \cong \frac{9\pi n^2}{4\varepsilon_F}JI_n^z\frac{\cos 2k_Fr}{(k_Fr)^3}. \tag{120}$$

Now suppose a second nuclear moment $\mathbf{I}_m$ is added at a lattice point m a distance r from n. The nucleus at m will be perturbed by the spin

* The second term here differs by a factor of 2 from Eq. (2.23) in the paper by Yosida;[7] the difference is merely a matter of definition, as is seen on examining his Eq. (3.1).

polarization due to the nucleus at n, and *vice versa*, thereby leading to an effective indirect interaction between the two moments, via the conduction electrons.

The second-order interaction between the two nuclear spins is given by

$$H''(\mathbf{x}) = \sum_{\substack{\mathbf{kk}' \\ ss'}}' \frac{\langle \mathbf{k}s|H|\mathbf{k}'s'\rangle\langle \mathbf{k}'s'|H|\mathbf{k}s\rangle}{\varepsilon_{\mathbf{k}} - \varepsilon_{\mathbf{k}'}}; \tag{121}$$

using (101) for H, we have, with the assumption (104),

$$H''(\mathbf{x}) = \sum_{s} (\mathbf{S} \cdot \mathbf{I}_n)(\mathbf{S} \cdot \mathbf{I}_m) m^* J^2 (2\pi)^{-6} \rho \int_0^{k_F} d^3k \int_{k_F}^{\infty} d^3k' \frac{e^{-i(\mathbf{k}-\mathbf{k}')\cdot\mathbf{x}}}{k^2 - k'^2} + cc. \tag{122}$$

The sum over electron spin states is carried out with the help of a standard relation (Schiff, p. 333) between pauli operators:

$$(\boldsymbol{\sigma} \cdot \mathbf{I}_n)(\boldsymbol{\sigma} \cdot \mathbf{I}_m) = \mathbf{I}_n \cdot \mathbf{I}_m + i\boldsymbol{\sigma} \cdot \mathbf{I}_n \times \mathbf{I}_m. \tag{123}$$

Now the trace of any component of $\boldsymbol{\sigma}$ vanishes, so that

$$\sum_{s} (\mathbf{S} \cdot \mathbf{I}_n)(\mathbf{S} \cdot \mathbf{I}_m) = \tfrac{1}{2}\mathbf{I}_n \cdot \mathbf{I}_m. \tag{124}$$

Thus the indirect interaction resulting from the isotropic hyperfine coupling has the form of an isotropic exchange interaction between the two nuclear spins.

The integrations are similar to those we have already carried out. The exclusion principle actually plays no part in the k' integration—the value of the total integral is not changed by carrying the integration at the lower limit down to $k' = 0$. The final form of (122) is

$$\boxed{H''(\mathbf{x}) = \mathbf{I}_n \cdot \mathbf{I}_m \frac{4J^2 m^* k_F{}^4}{(2\pi)^3} F(2k_F r),} \tag{125}$$

where the range function $F(x)$ is given by (118). For small x, $F(x) \to -1/6x$, so that the oscillatory interaction (125) for x less than the first zero of $F(x)$ is ferromagnetic. The first zero occurs at $x = 4.49$. The magnitude of the interaction (125) is supported by experimental results on nuclear resonance line widths in pure metals. The evaluation of J is discussed in reference 6.

It is tempting to extend the use of the result (125) to the indirect exchange coupling of paramagnetic ions in metals. Now the operator $\mathbf{I}$

is the electronic spin of the paramagnetic ion; the coupling J between $\mathbf{I}$ and a conduction electron spin $\mathbf{S}$ is the exchange interaction, instead of the contact hyperfine interaction. The prototype system much studied experimentally is the CuMn system at low concentration of Mn in Cu. The results have been analyzed in detail by Blandin and Friedel.[9] They find that all the alloy properties may be accounted for by an interaction of the form of (125) but they require a considerably stronger interaction than is provided by values of the coupling deduced from atomic spectroscopy. This is not surprising because the phase shift (105) from the delta-function potential is seen to be of the order of 0.1 (atomic exchange energy/fermi energy), whereas the Friedel rule for d states and $Z = 1$ demands $\eta_2 = \pi/10$, which is much larger. They go on to show that a self-consistent spin polarization around the Mn ion will lead to an interaction considerably stronger than from a bound d shell.

The oscillatory character of the indirect exchange interaction leads to a large variety of possible ordered spin structures in magnetic crystals, including spirals. The nature of the spin structure is determined by the value of k_F and thus by the electron concentration; see D. Mattis and W. E. Donath, *Phys. Rev.* **128**, 1618 (1962), and papers cited there.

The metals of the rare earth (lanthanide) group have very small $4f^n$ magnetic cores immersed in a sea of conduction electrons from the $6s$-$6p$ bands; the core diameters are about 0.1 of the interatomic spacings. The magnetic properties of these metals can be understood in detail in terms of an indirect exchange interaction between the magnetic cores via the conduction electrons. The ion cores are too far apart in relation to their radii for direct exchange to be significant. The fact that the curie temperatures are much higher in the metals than in the oxides is consistent with the role we ascribe to the conduction electrons.

The indirect exchange interaction will have the form (125), but with the ionic spins $\mathbf{S}$ written for $\mathbf{I}$:

$$H''(\mathbf{x}) = \Gamma_{\mathbf{S}}\, \mathbf{S}_n \cdot \mathbf{S}_m F(2k_F r). \tag{126}$$

It is an experimental fact that the coupling $\Gamma_{\mathbf{S}}$ is roughly constant for most of the rare earth metals; this is compatible with the indirect exchange model and is quite encouraging. In the limit of strong spin orbit coupling the results are more naturally analyzed in terms of a $\mathbf{J}_n \cdot \mathbf{J}_m$ interaction:

$$H''(\mathbf{x}) = \Gamma_{\mathbf{J}}\, \mathbf{J}_n \cdot \mathbf{J}_m F(2k_F r), \tag{127}$$

but now it is found that a constant Γ_J does *not* fit the various experimental data.

We recall that g for free atoms or ions is defined so that

$$g\mu_B\mathbf{J} = \mu_B(\mathbf{L} + 2\mathbf{S}), \tag{128}$$

or

$$g\mathbf{J} = \mathbf{L} + 2\mathbf{S}. \tag{129}$$

But

$$\mathbf{J} = \mathbf{L} + \mathbf{S}, \tag{130}$$

so that

$$(g - 1)\mathbf{J} = \mathbf{S}, \tag{131}$$

whence

$$\Gamma_S(g - 1)^2 = \Gamma_J. \tag{132}$$

This relation, due to de Gennes, is very well satisfied, in the sense that a constant Γ_S will reproduce well the Γ_J deduced from experiment. For example, experiments on the reduction in the superconducting transition temperature of lanthanum caused by the solution of various rare earth elements are compatible with the value $\Gamma_S \cong 5.1$ ev-A^3 [H. Suhl and B. T. Matthias, *Phys. Rev.* **114,** 977 (1959)]. A review of the theory of the structural and magnetic properties of the rare earth metals is given by Y. A. Rocher, *Adv. in Physics* **11,** 233 (1963).

On a molecular field model the ferromagnetic curie temperature is proportional to $J(J + 1)\Gamma_S^2(g - 1)^2$. The theoretical values of the curie temperature in the following table were calculated by Rocher using the value $\Gamma_S = 5.7$ ev-A^3 which fits the observed curie temperature of Gd.

	Gd	Tb	Dy	Ho	Er	Tm	Yb	Lu	
T_c(exp)	300	237	154	85	41	20	0	0	deg K
T_c(theo)	300	200	135	85	48	25	0	0	deg K

The agreement is quite satisfying.

PROBLEMS

1. Show, using (30), (34), and the result following (28) for the density of states, that in the Born approximation the screening charge

$$Z = \rho_F\langle\mathbf{k}|V_P|\mathbf{k}\rangle, \tag{133}$$

where ρ_F is the unperturbed density of states per unit energy range. From this result it follows that the fermi level is not changed by the presence of isolated impurities, although they cause the bottom of the band to be shifted.

2. Suppose that the density of states $g(\varepsilon)$ in a band is equal to a constant g_0 for $0 < \varepsilon < \varepsilon_1$ and is zero elsewhere. Find from (67) the positions of the virtual levels for V positive and for V negative.

3. (*a*) Show for the localized magnetic state problem that

$$G^\sigma_{\mathbf{kk}} = \frac{1}{\xi - \varepsilon_\mathbf{k}} + \frac{|V_{d\mathbf{k}}|^2}{(\xi - \varepsilon_\mathbf{k})^2(\xi - E_\sigma + i\Delta)}. \tag{134}$$

(*b*) Show that the free-electron density in energy

$$\begin{aligned} \rho^\sigma_{\text{free}}(\varepsilon) &\equiv -\frac{1}{\pi}\left(\frac{1}{2\pi}\right)^3 \int d^3k \, \mathscr{I}\{G^\sigma_{\mathbf{kk}}(\varepsilon)\} \\ &\cong \rho^\sigma_0(\varepsilon) + \frac{d\rho^\sigma_0}{d\varepsilon_\mathbf{k}} \frac{|V_{d\mathbf{k}}|^2(\varepsilon - E_\sigma)}{(\varepsilon - E_\sigma)^2 + \Delta^2}, \end{aligned} \tag{135}$$

where $\rho^\sigma_0(\varepsilon_\mathbf{k})$ is the density of states in the unperturbed problem. Note that if ρ^σ_0 is independent of $\varepsilon_\mathbf{k}$, the virtual d state does not change the free-electron density; in this situation there will be no set polarization of the free electrons. This is known as the compensation theorem.

4. Evaluate $\Delta\rho(R)$ for large R from (29) for a delta-function potential with $\eta_L = 0$ except for η_0, which is described by the scattering length b given by (105). According to the discussion following (19.19) we have $\eta_0 = -kb$.

5. Express the coefficients C_F and φ_F of (54) in terms of $f_\mathbf{k}$ by carrying out the integration over $\mathbf{k}$. Note that $\tan \varphi_\mathbf{k} = \mathscr{I}\{f_\mathbf{k}(\pi)\}/\mathscr{R}\{f_\mathbf{k}(\pi)\}$ and integrate by parts, keeping only lowest order terms in $1/r$. Now assume that only the phase shifts η_0 and η_1 are important: write down equations for C_F and φ_F in terms of the residual resistance and the valence difference between solute and solvent. Solve these equations for small η_0 and η_1.

19 correlation functions and neutron diffraction by crystals

We consider a crystalline system bombarded by an incident particle which interacts weakly with the crystal. The incident particles of most interest to us are x-ray photons and slow neutrons. Suppose that in a single scattering event the incident particle is scattered from a state $|\mathbf{k}\rangle$ to a state $|\mathbf{k}'\rangle$, and the state of the crystal is changed from $|i\rangle$ of energy ε_i to $|f\rangle$ of energy ε_f. We are particularly interested in the excitation of phonons or magnons in the crystal as a means of studying the dispersion relations over the entire Brillouin zone.

BORN APPROXIMATION

According to the Born approximation the inelastic differential scattering cross section per unit solid angle per unit energy range is, per unit volume of specimen,

$$\frac{d^2\sigma}{d\omega\, d\Omega} = \frac{k'}{k}\left(\frac{M}{2\pi}\right)^2 |\langle \mathbf{k}'f|H'|i\mathbf{k}\rangle|^2 \delta(\omega + \varepsilon_i - \varepsilon_f), \tag{1}$$

where H' describes the interaction of the particle with the target; ω is the energy transfer to the target; M is the reduced mass of the particle; Ω here denotes solid angle and not volume. This relation is derived in the standard texts on quantum mechanics.

In the first Born approximation without spin the state $|\mathbf{k}\rangle = e^{i\mathbf{k}\cdot\mathbf{x}}$ and $|\mathbf{k}'\rangle = e^{i\mathbf{k}'\cdot\mathbf{x}}$. Here $\mathbf{x}$ is the position of the incident particle. Then

$$\langle \mathbf{k}'f|H'|i\mathbf{k}\rangle = \left\langle f\right| \int d^3x\, e^{i\mathbf{K}\cdot\mathbf{x}} H' \left|i\right\rangle, \tag{2}$$

where

$$\mathbf{K} = \mathbf{k} - \mathbf{k}' \tag{3}$$

is the change in wavevector of the incident particle. If the interaction H' is the sum of two-particle interactions between the incident particle and the particles of the target,

$$H' \equiv \sum_j V(\mathbf{x} - \mathbf{x}_j), \qquad j = 1, \cdots, N; \tag{4}$$

and, using (2),

$$\langle \mathbf{k}'f|H'|i\mathbf{k}\rangle = V_{\mathbf{K}} \sum_j \langle f|e^{i\mathbf{K}\cdot\mathbf{x}_j}|i\rangle, \tag{5}$$

with

$$V_{\mathbf{K}} = \int d^3x\, e^{i\mathbf{K}\cdot\mathbf{x}} V(\mathbf{x}). \tag{6}$$

Using (5) and assuming a statistical distribution of initial target states with the probability p_i of finding the target initially in $|i\rangle$, we have

$$\frac{d^2\sigma}{d\varepsilon\, d\Omega} = \frac{k'}{k}\left(\frac{M}{2\pi}\right)^2 |V_{\mathbf{K}}|^2 \sum_{ifjl} p_i \langle i|e^{-i\mathbf{K}\cdot\mathbf{x}_j}|f\rangle\langle f|e^{i\mathbf{K}\cdot\mathbf{x}_l}|i\rangle \cdot \delta(\omega + \varepsilon_i - \varepsilon_f). \tag{7}$$

With the integral representation of the delta function, we arrive at an important form due to Van Hove:

$$\frac{d^2\sigma}{d\varepsilon\, d\Omega} = \frac{k'}{2\pi k}\left(\frac{M}{2\pi}\right)^2 |V_{\mathbf{K}}|^2 \sum_{ifjl} p_i \int_{-\infty}^{\infty} dt\, e^{-i(\omega+\varepsilon_i-\varepsilon_f)t} \cdot \langle i|e^{-i\mathbf{K}\cdot\mathbf{x}_j}|f\rangle\langle f|e^{i\mathbf{K}\cdot\mathbf{x}_l}|i\rangle, \tag{8}$$

or

$$\boxed{\frac{d^2\sigma}{d\varepsilon\, d\Omega} = \frac{k'}{2\pi k}\left(\frac{M}{2\pi}\right)^2 |V_{\mathbf{K}}|^2 \int_{-\infty}^{\infty} dt\, e^{-i\omega t} \sum_{jl} \langle e^{-i\mathbf{K}\cdot\mathbf{x}_j(0)} e^{i\mathbf{K}\cdot\mathbf{x}_l(t)}\rangle_T,} \tag{9}$$

where $\langle \cdots \rangle_T$ denotes quantum and ensemble average over a canonical ensemble at temperature T.

In the last step we have transformed $\mathbf{x}_l(t)$ to the Heisenberg representation:

$$e^{-i(\varepsilon_i-\varepsilon_f)t}\langle f|e^{i\mathbf{K}\cdot\mathbf{x}_l}|i\rangle \equiv \langle f|e^{iH_0t}e^{i\mathbf{K}\cdot\mathbf{x}_l(0)}e^{-iH_0t}|i\rangle \equiv \langle f|e^{i\mathbf{K}\cdot\mathbf{x}_l(t)}|i\rangle. \tag{10}$$

We have also used

$$\sum_f \langle i|e^{-i\mathbf{K}\cdot\mathbf{x}_j(0)}|f\rangle\langle f|e^{i\mathbf{K}\cdot\mathbf{x}_l(t)}|i\rangle = \langle i|e^{-i\mathbf{K}\cdot\mathbf{x}_j(0)}e^{i\mathbf{K}\cdot\mathbf{x}_l(t)}|i\rangle. \tag{11}$$

Both exponentials are quantum operators and commute only at identical times; we cannot in general write the product of the two exponentials as a single exponential. The statistical average in (9) is defined as

$$\sum_i p_i \langle i | e^{-i\mathbf{K}\cdot\mathbf{x}_j(0)} e^{i\mathbf{K}\cdot\mathbf{x}_l(t)} | i \rangle \equiv \langle e^{-i\mathbf{K}\cdot\mathbf{x}_j(0)} e^{i\mathbf{K}\cdot\mathbf{x}_l(t)} \rangle_T, \tag{12}$$

in thermal equilibrium.

The result (9) is conveniently written as

$$\frac{d^2\sigma}{d\Omega\, d\varepsilon} = A_{\mathbf{K}} \mathcal{S}(\omega,\mathbf{K}), \tag{13}$$

where

$$A_{\mathbf{K}} = \frac{k'}{k} \left(\frac{M}{2\pi}\right)^2 |V_{\mathbf{K}}|^2 \tag{14}$$

depends essentially only on the two-body potential, and

$$\mathcal{S}(\omega,\mathbf{K}) = \frac{1}{2\pi} \int_{-\infty}^{\infty} dt\, e^{-i\omega t} \sum_{jl} \langle e^{-i\mathbf{K}\cdot\mathbf{x}_j(0)} e^{i\mathbf{K}\cdot\mathbf{x}_l(t)} \rangle_T, \tag{15}$$

in conformity with (6.64), is the time fourier transform of a correlation function describing the system. It is revealing to introduce the particle density operator

$$\rho(\mathbf{x},t) = \sum_j \delta[\mathbf{x} - \mathbf{x}_j(t)], \tag{16}$$

so that we have $\mathcal{S}(\omega,\mathbf{K})$ in terms of the space-time fourier transform of a density correlation function:

$$\mathcal{S}(\omega,\mathbf{K}) = \frac{1}{2\pi} \int d^3x\, d^3x'\, e^{i\mathbf{K}\cdot(\mathbf{x}-\mathbf{x}')} \int dt\, e^{-i\omega t} \langle \rho(\mathbf{x}'0)\rho(\mathbf{x}t) \rangle_T. \tag{17}$$

This is not a particularly handy form for actual computation, but it exhibits clearly the dependence of the differential scattering cross section on the density correlation function $\langle \rho(\mathbf{x}'0)\rho(\mathbf{x}t) \rangle_T$.

NEUTRON DIFFRACTION

The theory of x-ray scattering by crystals is closely similar to the theory of neutron scattering. Both subjects are important in solid state physics, but we shall develop the neutron theory, with emphasis on the determination of phonon and magnon dispersion relation by inelastic neutron scattering. In solid state physics there is as much interest in inelastic scattering processes, with the excitation of phonons and magnons, as there is in the classical applications of elastic scattering to the determination of crystal and magnetic structures. Several general references are:

[1] L. S. Kothari and K. S. Singwi, *Solid state physics* **8,** 109 (1959).
[2] W. Marshall and R. D. Lowde, *Repts. Prog. Phys.* **31,** pt. 2, 1968.
[3] C. G. Shull and E. O. Wollan, *Solid state physics* **2,** 137 (1956).
[4] L. Van Hove, *Phys. Rev.* **95,** 249, 1374 (1954.)

The concept of the *scattering length* is useful for the description of *s*-wave scattering of low-energy incident neutrons which interact with a deep and narrow potential well. In the conditions assumed the regular solution of the wave equation in the interior of the well is insensitive to small changes in the energy of the incident particle;

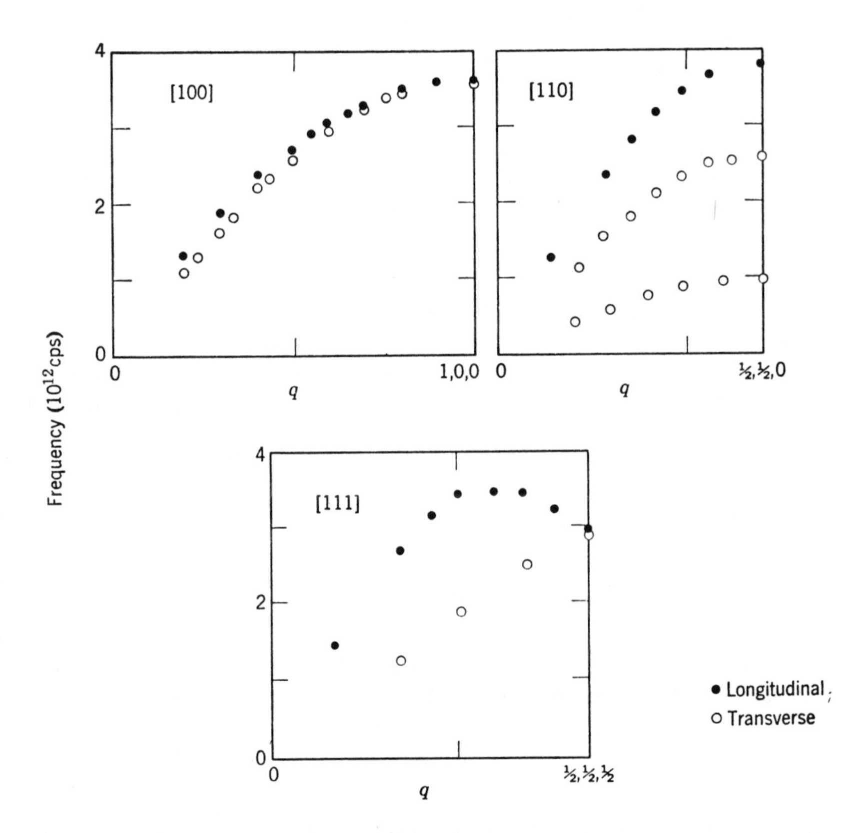

FIG. 1. The dispersion curves of sodium in the [001], [110], and [111] directions at 90°K as determined by inelastic scattering of neutrons by Woods, Brockhouse, March, and Bowers, *Proc. Phys. Soc.* **79,** pt. 2. 440 (1962).

the energy of the incident particle; thus the logarithmic derivative at the edge of the well is insensitive to the incident energy. This is a very useful property. In terms of the s-wave phase shift η_0, the wavefunction $\varphi(\mathbf{x})$ just outside the well has the form

$$r\varphi = B\sin(kr + \eta_0) \cong B(kr + \eta_0), \tag{18}$$

which may be written in terms of quantities C and b which are approximately constant with energy at low energies

$$r\varphi = C(r - b), \tag{19}$$

according to our argument above. Thus $B = C/k$ and $\eta_0 = -kb$. Here b is called the *scattering length* or *scattering amplitude;* it is the intercept shown in the figure. But the standard result for the s-wave elastic scattering cross section in terms of η_0 is

$$\frac{d\sigma}{d\Omega} = \frac{\sin^2\eta_0}{k^2}, \tag{20}$$

so that in our approximation

$$\frac{d\sigma}{d\Omega} = b^2, \tag{21}$$

approximately independent of scattering angle and energy, as long as $k|b| \ll 1$. Note that b here is a length, not a field operator.

The cross section in (21) in the scattering length approximation may be simulated in the Born approximation by a suitable choice of an effective potential or pseudopotential. We consider the fictitious potential

$$\tilde{V}(\mathbf{x}) = (2\pi/M)b\delta(\mathbf{x}); \tag{22}$$

then

$$V_{\mathbf{K}} = \frac{2\pi b}{M}\int d^3x\, e^{i\mathbf{K}\cdot\mathbf{x}}\delta(\mathbf{x}) = \frac{2\pi b}{M}, \tag{23}$$

independent of $\mathbf{K}$. Thus, in (14), for elastic scattering $A = b^2$. Now for elastic scattering from a single nucleus the form factor (15) is

$$\mathcal{S}(\omega,\mathbf{K}) = \frac{1}{2\pi}\int_{-\infty}^{\infty} dt\, e^{-i\omega t}\langle e^{-i\mathbf{K}\cdot\mathbf{x}(0)}e^{i\mathbf{K}\cdot\mathbf{x}(0)}\rangle_T = \delta(\omega), \tag{24}$$

so that, from (13) with $k = k'$ for elastic scattering, (14), (23), and (24),

$$\frac{d^2\sigma}{d\omega\, d\Omega} = b^2\delta(\omega); \qquad \frac{d\sigma}{d\Omega} = b^2. \tag{25}$$

COHERENT AND INCOHERENT ELASTIC NUCLEAR SCATTERING

We suppose that the target contains N particles. The jth particle is located at $\mathbf{x}_j$ and is characterized by the scattering amplitude b_j. The pseudopotential of the target system is

$$\tilde{V}(\mathbf{x}) = (2\pi/M) \sum_j b_j \delta(\mathbf{x} - \mathbf{x}_j), \tag{26}$$

where for elastic scattering from a macroscopic target we may take M as the mass of the neutron. From (23) and (26),

$$V_{\mathbf{K}} = (2\pi/M) \sum_j b_j e^{i\mathbf{K}\cdot\mathbf{x}_j}; \qquad |V_{\mathbf{K}}|^2 = (2\pi/M)^2 \sum_{lm} b_l^* b_m e^{i\mathbf{K}\cdot(\mathbf{x}_m - \mathbf{x}_l)}. \tag{27}$$

Here we have included for convenience the exponential factors in $V_{\mathbf{K}}$ rather than in the form factor $\mathcal{S}(\omega,\mathbf{K})$. Thus

$$\frac{d\sigma}{d\Omega} = \Big| \sum_j b_j e^{i\mathbf{K}\cdot\mathbf{x}_j} \Big|^2 = \sum_{lm} b_l^* b_m e^{i\mathbf{K}\cdot(\mathbf{x}_m - \mathbf{x}_l)}, \tag{28}$$

for scattering with the scattering vector $\mathbf{K}$.

If the values b_l, b_m are uncorrelated, the ensemble average for $l \neq m$ is

$$\langle b_l^* b_m \rangle = |\langle b \rangle|^2, \tag{29}$$

or, more generally,

$$\langle b_l^* b_m \rangle = |\langle b \rangle|^2 + \delta_{lm}(\langle |b|^2 \rangle - |\langle b \rangle|^2). \tag{30}$$

Thus from an ensemble of scattering centers

$$\frac{d\sigma}{d\Omega} = \underbrace{|\langle b \rangle|^2 \Big| \sum_l e^{i\mathbf{K}\cdot\mathbf{x}_l} \Big|^2}_{\text{coherent}} + \underbrace{N(\langle |b|^2 \rangle - |\langle b \rangle|^2)}_{\text{incoherent}}. \tag{31}$$

The incoherent scattering term is seen to be isotropic; it arises from the presence of different nuclear isotopes of the scattering atoms, and also from different directions of the nuclear spin relative to the spin direction of the incident particle.

The coherent scattering cross section per atom is defined from (31) as

$$\sigma_{\text{coh}} = 4\pi \Big(\sum_j p_j b_j\Big)^2 \equiv 4\pi |\langle b \rangle|^2, \tag{32}$$

where p_j is the *probability* that an atom will have the scattering amplitude b_j. The total scattering per atom must be the sum of the contributions to the scattered intensity from all sources, so that per atom

$$\sigma_{\text{total}} = 4\pi \sum_j p_j b_j^2 \equiv 4\pi \langle |b|^2 \rangle. \tag{33}$$

The incoherent scattering cross section per atom is

$$\sigma_{\text{incoh}} = \sigma_{\text{total}} - \sigma_{\text{coh}} = 4\pi(\langle|b|^2\rangle - |\langle b\rangle|^2). \tag{34}$$

The coherent scattering from a target involves $\left|\sum_l e^{i\mathbf{K}\cdot\mathbf{x}_l}\right|^2$, as we see in (31). The sum vanishes unless $\mathbf{K}\cdot\mathbf{x}_l$ is an integral multiple of 2π for all sites l. If the $\mathbf{x}_l$ are at lattice points

$$\mathbf{x}_l = u_l\mathbf{a} + v_l\mathbf{b} + w_l\mathbf{c}, \tag{35}$$

where u, v, w are integers and $\mathbf{a}$, $\mathbf{b}$, $\mathbf{c}$ are the crystal axes, then coherent scattering occurs when

$$\mathbf{K} = l\mathbf{a}^* + m\mathbf{b}^* + n\mathbf{c}^* \equiv \mathbf{G}, \tag{36}$$

where $\mathbf{a}^*$, $\mathbf{b}^*$, $\mathbf{c}^*$ are the basis vectors of the reciprocal lattice and l, m, n are integers. For then

$$\mathbf{K}\cdot\mathbf{x}_l = 2\pi(ul + vm + wn) = 2\pi \times \text{integer}.$$

We thus have the Bragg condition that coherent scattering occurs when the scattering vector $\mathbf{K}$ is equal to a vector $\mathbf{G}$ in the reciprocal lattice.

We evaluate the lattice sum in (31) for a one-dimensional crystal of lattice constant a, and then write down the result for three dimensions. For a crystal of N atoms, with $x_l = la$, where l is an integer between 0 and $N - 1$,

$$\sum_{l=0}^{N-1} e^{ilKa} = \frac{1 - e^{iNKa}}{1 - e^{iKa}},$$

and

$$\left|\sum_{l=0}^{N-1} e^{ilKa}\right|^2 = \frac{1 - \cos NKa}{1 - \cos Ka} = \frac{\sin^2 \frac{1}{2}NKa}{\sin^2 \frac{1}{2}Ka} \cong \frac{2\pi N}{a}\sum_G \delta(K - G). \tag{37}$$

We notice that the left-hand side is large when the denominator $\sin^2 \frac{1}{2}Ka$ is small, that is, when $\frac{1}{2}Ka = n\pi$ or $K = 2\pi n/a \equiv G$. Thus the left-hand side represents a periodic delta function. To obtain the normalization we let $K = G + \eta$, where η is a small quantity, and calculate

$$\int_{-\infty}^{\infty} d\eta\, \frac{\sin^2 \frac{1}{2}NKa}{\sin^2 \frac{1}{2}Ka} = \frac{\cos^2 \frac{1}{2}NGa}{\cos^2 \frac{1}{2}Ga}\int_{-\infty}^{\infty} d\eta\, \frac{\sin^2 \frac{1}{2}N\eta a}{\sin^2 \frac{1}{2}\eta a} \cong \int_{-\infty}^{\infty} d\eta\, \frac{\sin^2 \frac{1}{2}N\eta a}{\frac{1}{4}\eta^2 a^2}$$
$$= \frac{2\pi N}{a} = \frac{2\pi N}{a}\int d\eta\, \delta(\eta),$$

where we have implicitly restricted K to lie near G; we have also used the fact that $\frac{1}{2}Ga = \pi \times$ integer. By extension to three dimensions

$$\left|\sum_l e^{i\mathbf{K}\cdot\mathbf{x}_l}\right|^2 = (2\pi)^3(N/V_c)\sum_{\mathbf{G}}\delta(\mathbf{K}-\mathbf{G}) = NV_c^*\sum_{\mathbf{G}}\delta(\mathbf{K}-\mathbf{G}), \tag{38}$$

where V_c is the volume $\mathbf{a}\cdot\mathbf{b}\times\mathbf{c}$ of a primitive cell and V_c^* is the cell volume in the reciprocal lattice. Thus the coherent scattering cross section of the crystal is

$$\left(\frac{d\sigma}{d\Omega}\right)_{\text{coh}} = NV_c^*|\langle b\rangle|^2\sum_{\mathbf{G}}\delta(\mathbf{K}-\mathbf{G}). \tag{39}$$

INELASTIC LATTICE SCATTERING

We suppose that all atoms in the crystal have the identical scattering length b, which is assumed to be real. The form factor (15) involves

$$\begin{aligned} F(\mathbf{K},t) &= \sum_{jl}\langle e^{-i\mathbf{K}\cdot\{\mathbf{x}_j(0)+\mathbf{u}_j(0)\}}e^{i\mathbf{K}\cdot\{\mathbf{x}_l(0)+\mathbf{u}_l(t)\}}\rangle_T \\ &= \sum_{jl} e^{i\mathbf{K}\cdot(\mathbf{x}_l-\mathbf{x}_j)}\langle e^{-i\mathbf{K}\cdot\mathbf{u}_j(0)}e^{i\mathbf{K}\cdot\mathbf{u}_l(t)}\rangle_T, \end{aligned} \tag{40}$$

where $\mathbf{x}_l$, $\mathbf{x}_j$ are now the coordinates of the undisplaced atoms and $\mathbf{u}_l$, $\mathbf{u}_j$ are the displacements referred to $\mathbf{x}_l$, $\mathbf{x}_j$. The phonon operators enter through the expansion of the $\mathbf{u}$'s in phonon coordinates.

The second factor on the right-hand side of (40) is quite famous; we shall discuss it in detail for $l = j$ in the following chapter on recoilless emission; for the present we take over the results (20.51) and (20.53), with the appropriate modifications for $l \neq j$. Thus

$$\langle e^{-i\mathbf{K}\cdot\mathbf{u}_l(0)}e^{i\mathbf{K}\cdot\mathbf{u}_l(t)}\rangle_T = e^{-Q_{lj}(t)}, \tag{41}$$

where

$$Q_{lj}(t) = \tfrac{1}{2}K^2(\langle\{u_j(t) - u_l(0)\}^2\rangle_T - [u_l(0),u_j(t)]); \tag{42}$$

for simplicity it is assumed that all phonon modes of given wavevector $\mathbf{q}$ are degenerate and we arrange always to take the polarization of one of the modes $\mathbf{q}$ along $\mathbf{K}$.

In (42) we expand the displacement in phonon modes, following (2.33),

$$u_j(t) = \sum_{\mathbf{q}}(2NM\omega_{\mathbf{q}})^{-1/2}(a_{\mathbf{q}}e^{i(\mathbf{q}\cdot\mathbf{x}_j-\omega_{\mathbf{q}}t)} + a_{\mathbf{q}}^+e^{-i(\mathbf{q}\cdot\mathbf{x}_j-\omega_{\mathbf{q}}t)}), \tag{43}$$

where $a_{\mathbf{q}}$, $a_{\mathbf{q}}^+$ are phonon operators. Then

$$\begin{aligned} u_j(t) - u_l(0) = \sum_{\mathbf{q}}(2NM\omega_{\mathbf{q}})^{-1/2}(a_{\mathbf{q}}(e^{-i(\mathbf{q}\cdot\mathbf{x}_j-\omega_{\mathbf{q}}t)} - e^{i\mathbf{q}\cdot\mathbf{x}_l}) \\ + a_{\mathbf{q}}^+(e^{-i(\mathbf{q}\cdot\mathbf{x}_j-\omega_{\mathbf{q}}t)} - e^{-i\mathbf{q}\cdot\mathbf{x}_l})), \end{aligned} \tag{44}$$

whence, exhibiting only terms diagonal in the phonon occupancy,

$$(45)\quad \{u_j(t) - u_l(0)\}^2 = (1/2NM) \sum_{\mathbf{q}} \omega_{\mathbf{q}}^{-1}(2 - e^{i\theta_{lj}} - e^{-i\theta_{lj}})(a_{\mathbf{q}}^+ a_{\mathbf{q}} + a_{\mathbf{q}} a_{\mathbf{q}}^+) + \text{nondiagonal terms};$$

where

$$(46)\quad \theta_{lj} = \omega_{\mathbf{q}} t + \mathbf{q} \cdot (\mathbf{x}_l - \mathbf{x}_j).$$

Also

$$(47)\quad [u_l(0), u_j(t)] = (1/2NM) \sum_{\mathbf{q}} \omega_{\mathbf{q}}^{-1}\{[a_{\mathbf{q}}, a_{\mathbf{q}}^+]e^{i\theta_{lj}} + [a_{\mathbf{q}}^+, a_{\mathbf{q}}]e^{-i\theta_{lj}}\} = (i/NM) \sum_{\mathbf{q}} \omega_{\mathbf{q}}^{-1} \sin \theta_{lj}.$$

Thus, from (42),

$$(48)\quad Q_{lj}(t) = \frac{K^2}{2NM} \sum_{\mathbf{q}} \omega_{\mathbf{q}}^{-1}((2\langle n_{\mathbf{q}}\rangle + 1)(1 - \cos \theta_{lj}) - i \sin \theta_{lj}).$$

If we split $Q_{lj}(t)$ up into time-dependent and time-independent terms, we have

$$(49)\quad e^{-Q_{lj}(t)} = \exp\left\{-(K^2/2NM) \sum_{\mathbf{q}} \omega_{\mathbf{q}}^{-1}(2\langle n_{\mathbf{q}}\rangle + 1)\right\} \cdot \exp\left\{(K^2/2NM) \sum_{\mathbf{q}} \omega_{\mathbf{q}}^{-1}[(2\langle n_{\mathbf{q}}\rangle + 1) \cos \theta_{lj} + i \sin \theta_{lj}]\right\}.$$

Each term in the exponential is small by order N^{-1}, so the second factor on the right-hand side may be expanded in a power series as

$$(50)\quad \exp\{\quad\} = 1 + \sum_{\mathbf{q}} (K^2/2NM\omega_{\mathbf{q}})[(2\langle n_{\mathbf{q}}\rangle + 1) \cos \theta_{lj} + i \sin \theta_{lj}] + \cdots.$$

The neglected terms represent multiple phonon effects. We write the first factor on the right-hand side of (49) as e^{-2W}; then

$$F(\mathbf{K},t)/e^{-2W} = \sum_{jl} \Big\{e^{i\mathbf{K}\cdot(\mathbf{x}_j - \mathbf{x}_l)} + \sum_{\mathbf{q}} (K^2/2NM\omega_{\mathbf{q}})(\langle n_{\mathbf{q}} + 1\rangle e^{i\omega_{\mathbf{q}} t} e^{i(\mathbf{q}-\mathbf{K})\cdot(\mathbf{x}_l - \mathbf{x}_j)} + \langle n_{\mathbf{q}}\rangle e^{-i\omega_{\mathbf{q}} t} e^{-i(\mathbf{q}+\mathbf{K})\cdot(\mathbf{x}_l - \mathbf{x}_j)})\Big\}.$$

We may rewrite this as, using (38),

$$(51)\quad F(\mathbf{K},t) = NV_c^* e^{-2W} \Big\{\sum_{\mathbf{G}} \delta(\mathbf{K} - \mathbf{G}) + \sum_{\mathbf{q}} (K^2/2NM\omega_{\mathbf{q}}) \Big(\langle n_{\mathbf{q}} + 1\rangle e^{i\omega_{\mathbf{q}} t} \sum_{\mathbf{G}} \delta(\mathbf{K} - \mathbf{q} - \mathbf{G}) + \langle n_{\mathbf{q}}\rangle e^{-i\omega_{\mathbf{q}} t} \sum_{\mathbf{G}} (\mathbf{K} + \mathbf{q} - \mathbf{G})\Big)\Big\}.$$

The first term on the right-hand side represents elastic scattering at any **G**; the second term represents inelastic scattering $\mathbf{K} = \mathbf{G} + \mathbf{q}$ in which a phonon **q** is emitted by the neutron; in the last term a phonon **q** is absorbed by the neutron, with $\mathbf{K} = \mathbf{G} - \mathbf{q}$.

The energy changes in the inelastic scattering processes are found from the time factors. The time integrand in (15) for the form factor will involve for phonon emission the factor $e^{-i\omega t}e^{i\omega_{\mathbf{q}}t}$, so that the neutron must lose energy $\omega_{\mathbf{q}}$. For phonon absorption the integrand is $e^{-i\omega t}e^{-i\omega_{\mathbf{q}}t}$, and the neutron energy is increased by $\omega_{\mathbf{q}}$.

This argument tells us that inelastic neutron scattering occurs with the emission or absorption of one phonon, in first order of the expansion (50). Experimental results for ω versus **k** for phonons in metallic sodium are given in Fig. 1. Of the values $\mathbf{G} + \mathbf{q}$ which we can enumerate for given **G**, we will observe only those phonons for which energy is conserved for the whole system. The requirements of energy and

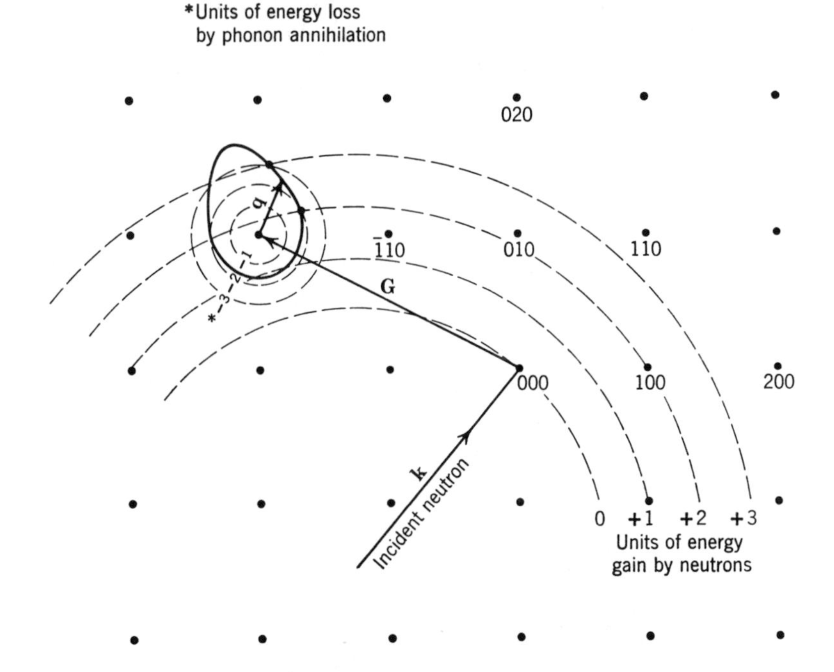

FIG. 2. The inelastic scattering surface near the point $\bar{2}10$ in reciprocal space for neutrons **k** incident on a target; phonons **q** on the scattering surface (shown as a solid curve) are absorbed by the neutrons. On the scattering surface the scattered neutrons have wavevector $\mathbf{k} + \mathbf{G}(\bar{2}10) + \mathbf{q}$, and the energy of the scattered neutrons has been increased by $\omega_{\mathbf{q}}$ over the energy of the incident neutrons.

wavevector conservation restrict severely the energies and directions of the inelastically scattered neutrons, for given **k**. In Fig. 2 we show schematically the energy balance for processes in which a phonon is absorbed. Scattering surfaces generally similar to that shown will be found centered about every reciprocal lattice point.

It is often convenient to work with very slow neutrons. If the incident neutron energy can be neglected, the neutron cannot emit phonons, but can absorb them. Further, the slow neutrons cannot be scattered elastically. Thus the entire scattered spectrum in the one-phonon approximation may arise from neutrons for which $\mathbf{K} = -\mathbf{q}$ and

$$\frac{1}{2M} q^2 = \omega_{\mathbf{q}}. \tag{52}$$

It is possible with high stiffness constants for **K** to be given by $-\mathbf{q} + \mathbf{G}$. Solutions of (52) are shown in one dimension in Fig. 3.

By inelastic neutron scattering experiments it is possible to determine the dispersion relations of all branches of the acoustic and optical phonon spectrum. This is at present the only known general method for determining the dispersion relations. It is equally possible to

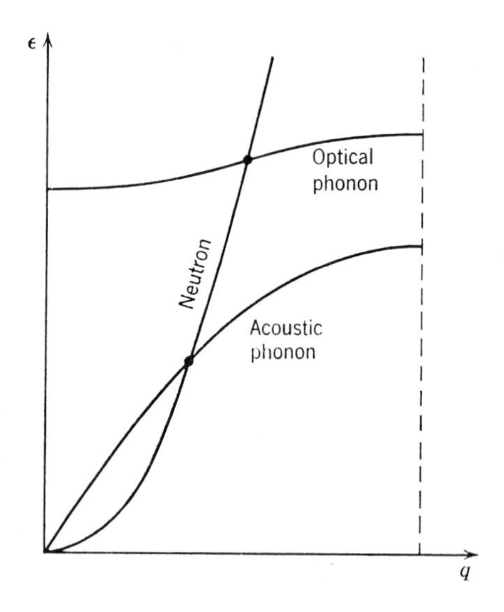

FIG. 3. The solid circles indicate the possible scattered energies in one dimension for neutrons with $k \rightarrow 0$ scattered with absorption of a phonon from a crystal with one acoustic phonon branch and one optical phonon branch.

determine the dispersion relations of magnons. Further, from the thickness of the observed scattering surfaces one may estimate the relaxation times of the phonons and magnons.

Debye-Waller Factor. We can see from (51) that the width of the peaks of elastic scattering do not broaden as the temperature increases, but the peak height decreases as the factor e^{-2W}, where

$$2W = (K^2/2NM) \sum_{\mathbf{q}} \omega_{\mathbf{q}}^{-1}(2\langle n_{\mathbf{q}}\rangle + 1). \tag{53}$$

The factor e^{-2W} is known as the Debye-Waller factor. Taking $\mathbf{x}_j = 0$ for convenience, we note from (43) that

$$\langle u_j^2\rangle = (1/2NM) \sum_{\mathbf{q}} \omega_{\mathbf{q}}^{-1}(2\langle n_{\mathbf{q}}\rangle + 1), \tag{54}$$

where here the sum is over $3N$ normal modes, whereas the sum in (53) refers to the N modes which may be chosen with the polarization vector parallel to $\mathbf{K}$. If we redefine the sum in (53) to refer to all $3N$ modes, then

$$2W = \tfrac{1}{3}(K^2/2NM) \sum \omega_{\mathbf{q}}^{-1}(2\langle n_{\mathbf{q}}\rangle + 1) = \tfrac{1}{3}K^2\langle u_j^2\rangle. \tag{55}$$

Thus W is proportional to the mean square oscillation amplitude $\langle u^2\rangle$ of an atom in thermal equilibrium at temperature T. It is shown in the following chapter that for a Debye phonon spectrum

$$\tfrac{1}{3}K^2\langle u_j^2\rangle = \frac{3K^2}{4Mk_B\Theta}\left\{1 + \frac{2\pi^2}{3}\left(\frac{T}{\Theta}\right)^2 + \cdots\right\} = 2W. \tag{56}$$

The term 1 in the curly brackets arises from zero-point motion, that is, from the integrated effects at absolute zero of phonon emission by the incident neutron. For $K \to 0$, $e^{-2W} \to 1$. We note that at absolute zero W is approximately the ratio of the neutron recoil energy $K^2/2M$ to the Debye energy $k_B\Theta$.

The Debye-Waller factor arose originally in connection with x-ray diffraction. The factor is a measure of the effect of thermal motion in reducing the apparent periodicity of the lattice. A full discussion of the experimental situation concerning the Debye-Waller factor in x-ray diffraction is given by R. W. James, *Optical principles of the diffraction of x-rays*, Bell, London, 1950, Chapter 5. For a detailed discussion of inelastic neutron scattering, see R. Weinstock, *Phys. Rev.* **65**, 1 (1944).

ELECTRONIC MAGNETIC SCATTERING OF NEUTRONS

The magnetic interaction of the electronic magnetic moment of a paramagnetic atom with the magnetic moment of a slow neutron leads to scattering lengths of the same order of magnitude as the purely nuclear scattering we considered above. The scattering cross sections for bound nuclei are given in the following table:

Nucleus	Coherent Nuclear Cross Section (barns)	Paramagnetic Ion	Magnetic Cross Section (barns)
Mn^{55}	2	Mn^{++}	21
Fe^{56}	13	Fe^{+++}	21
Ni^{58}	26	Ni^{++}	5
Co^{59}	6	Co^{++}	9

The magnetic values refer to forward scattering.

We now treat the magnetic scattering, first calculating the matrix elements of the neutron-electron magnetic interaction in the Born approximation. The hamiltonian is

$$H = -\boldsymbol{\mu}_e \cdot \mathbf{H}_n, \tag{57}$$

where $\mathbf{H}_n$ is the magnetic field caused by the magnetic moment $\boldsymbol{\mu}_n$ of the neutron. We neglect the magnetic field of the orbital motion of the electron; this is a good approximation for the elements of the iron group. Thus if $\mathbf{r} = \mathbf{x}_n - \mathbf{x}_e$,

$$\mathbf{H}_n = \operatorname{curl} \frac{\boldsymbol{\mu}_n \times \mathbf{r}}{r^3} = -\operatorname{curl} \boldsymbol{\mu}_n \times \nabla \frac{1}{r}, \tag{58}$$

but by standard vector algebra

$$\operatorname{curl} \boldsymbol{\mu}_n \times \nabla \frac{1}{r} = -(\boldsymbol{\mu}_n \cdot \nabla)\nabla \frac{1}{r} + \boldsymbol{\mu}_n \nabla^2 \frac{1}{r}. \tag{59}$$

Now

$$\nabla^2 \frac{1}{r} = -4\pi\delta(\mathbf{r}), \tag{60}$$

so that

$$H = -\boldsymbol{\mu}_e \cdot \nabla\left(\boldsymbol{\mu}_n \cdot \nabla \frac{1}{r}\right) - 4\pi \boldsymbol{\mu}_e \cdot \boldsymbol{\mu}_n \delta(\mathbf{r}). \tag{61}$$

This is usually written as the sum of

$$H_{\text{dipole}} = \frac{\boldsymbol{\mu}_e \cdot \boldsymbol{\mu}_n}{r^3} - \frac{3(\boldsymbol{\mu}_e \cdot \mathbf{r})(\boldsymbol{\mu}_n \cdot \mathbf{r})}{r^5}, \tag{62}$$

and

$$H_{\text{contact}} = -\frac{8\pi}{3} \boldsymbol{\mu}_e \cdot \boldsymbol{\mu}_n \delta(\mathbf{r}). \tag{63}$$

In the Born approximation we require the matrix elements of H between the spin states a, b and between the spatial states $\mathbf{k}$ and $\mathbf{k}' = \mathbf{k} - \mathbf{K}$. Here $\mathbf{r} = \mathbf{x}_n - \mathbf{x}_e$ is the difference between the nuclear and electron positions. We want to separate the spatial and spin parts of the matrix element in order to exhibit the spin function explicitly. It is convenient to choose the origin of the coordinate system at the center-of-mass of the atom. The initial state of the system may be written as

$$|i\rangle = e^{i\mathbf{k}\cdot\mathbf{x}_n}\varphi_i(\mathbf{x}_e)|a_i\rangle, \tag{64}$$

where a_i contains the electron and neutron spin coordinates and $\varphi_i(\mathbf{x}_e)$ is the initial spatial electronic wavefunction. The final state is

$$|f\rangle = e^{i\mathbf{k}'\cdot\mathbf{x}_n}\varphi_f(\mathbf{x}_e)|a_f\rangle, \tag{65}$$

where the f's denote final. Then

$$\begin{aligned}\langle f|H|i\rangle &= \langle a_f|\int d^3x_n\, d^3x_e\, \varphi_f^*(\mathbf{x}_e)\varphi_i(\mathbf{x}_e)e^{i\mathbf{K}\cdot\mathbf{x}_n}H|a_i\rangle \\ &= \langle a_f|\int d^3r\, e^{i\mathbf{K}\cdot\mathbf{r}}H|a_i\rangle\int d^3x_e\,\varphi_f^*(\mathbf{x}_e)\varphi_i(\mathbf{x}_e)e^{i\mathbf{K}\cdot\mathbf{x}_e},\end{aligned} \tag{66}$$

where we have made a coordinate transformation having a functional determinant of value unity. The part of $\langle a_f|\cdots|a_i\rangle$ in $\delta(\mathbf{r})$ gives $-4\pi\boldsymbol{\mu}_e\cdot\boldsymbol{\mu}_n$ after integration; the other part of (61) gives, on integration by parts,

$$\begin{aligned}-\int d^3r\, e^{i\mathbf{K}\cdot\mathbf{r}}\,\boldsymbol{\mu}_e\cdot\nabla\left(\boldsymbol{\mu}_n\cdot\nabla\frac{1}{r}\right) &= i\mathbf{K}\cdot\boldsymbol{\mu}_e\int d^3r\, e^{i\mathbf{K}\cdot\mathbf{r}}\boldsymbol{\mu}_n\cdot\nabla\frac{1}{r} \\ &= (\mathbf{K}\cdot\boldsymbol{\mu}_e)(\mathbf{K}\cdot\boldsymbol{\mu}_n)\int d^3r\,\frac{1}{r}\,e^{i\mathbf{K}\cdot\mathbf{r}} = (\mathbf{K}\cdot\boldsymbol{\mu}_e)(\mathbf{K}\cdot\boldsymbol{\mu}_n)(4\pi/K^2),\end{aligned} \tag{67}$$

Thus

$$\langle f|H|i\rangle = -4\pi\langle a_f|\boldsymbol{\mu}_e\cdot\boldsymbol{\mu}_n - (\boldsymbol{\mu}_e\cdot\mathbf{K})(\boldsymbol{\mu}_n\cdot\mathbf{K})K^{-2}|a_i\rangle F(\mathbf{K}), \tag{68}$$

where the *magnetic form factor* $F(\mathbf{K})$ is defined by

$$F(\mathbf{K}) = \int d^3x_e\,\varphi_f^*(\mathbf{x}_e)\varphi_i(\mathbf{x}_e)e^{i\mathbf{K}\cdot\mathbf{x}_e}. \tag{69}$$

For transitions in which the spatial state is unchanged $\varphi_f = \varphi_i$; we note that then $F(0) = 1$. With

$$\boldsymbol{\mu}_e = -\frac{|e|}{m_ec}\mathbf{S}_e; \qquad \boldsymbol{\mu}_n = g\frac{|e|}{m_nc}\mathbf{S}_n, \qquad g = -1.91,$$

we may write

$$\langle f|H|i\rangle = 4\pi g\frac{e^2}{m_em_nc^2}\langle a_f|\mathbf{S}_e\cdot\mathbf{S}_n - (\mathbf{S}_e\cdot\mathbf{K})(\mathbf{S}_n\cdot\mathbf{K})K^{-2}|a_i\rangle F(\mathbf{K}). \tag{70}$$

Note that if $\hat{\mathbf{K}}$ is a unit vector in the direction of $\mathbf{K}$, the function of the spins in (70) may be written as

$$\mathbf{S}_n \cdot [\hat{\mathbf{K}} \times [\mathbf{S}_e \times \hat{\mathbf{K}}],$$

so that the scattering matrix element is proportional to the component of $\mathbf{S}_e$ normal to $\mathbf{K}$. If we write $\mathbf{P}_\perp$ as the vector of length $S_{e\perp}$ and in the direction $\hat{\mathbf{K}} \times [\mathbf{S}_e \times \hat{\mathbf{K}}]$, then the matrix element involves $\mathbf{S}_n \cdot \mathbf{P}_\perp$.

The scattering cross section is proportional to, with s, s' as the neutron spin quantum numbers and q, q' the electron spin quantum numbers,

$$(71)\quad \sum_{\substack{q's'\\ qs}} \langle qs|S_n^\alpha P_\perp^\alpha|q's'\rangle\langle q's'|S_n^\beta P_\perp^\beta|qs\rangle\langle qs|\rho|qs\rangle$$
$$= \sum_{\substack{q's'\\ qs}} \langle s|S_n^\alpha|s'\rangle\langle s'|S_n^\beta|s\rangle\langle q|P_\perp^\alpha|q'\rangle\langle q'|P_\perp^\beta|q\rangle\langle qs|\rho|qs\rangle,$$

where ρ is the spin density matrix for the initial state. Now

$$(72)\quad \sum_{s'} \langle s|S_n^\alpha|s'\rangle\langle s'|S_n^\beta|s\rangle\langle s|\rho|s\rangle = \langle s|S_n^\alpha S_n^\beta|s\rangle\langle s|\rho|s\rangle = \tfrac{1}{4}\delta_{\alpha\beta}\langle s|\rho|s\rangle.$$

If the neutron beam is unpolarized, all $\langle s|\rho|s\rangle = \frac{1}{2}$, and

$$(73)\quad \sum_s \tfrac{1}{4}\delta_{\alpha\beta}\langle s|\rho|s\rangle = \tfrac{1}{4}\delta_{\alpha\beta}.$$

We suppose below that the neutrons are unpolarized. Thus the quantity in (71) becomes, for elastic processes with all $\langle q|\rho|q\rangle$ equal,

$$(74)\quad \tfrac{1}{4}\delta_{\alpha\beta}\langle q|P_\perp^\alpha P_\perp^\beta|q\rangle = \tfrac{1}{4}\langle q|P_\perp^\alpha P_\perp^\alpha|q\rangle.$$

To include inelastic processes we must follow the time-dependent generalization introduced earlier for neutron-phonon scattering. We combine (13), (70), and (71) to give

$$(75)\quad \frac{d^2\sigma}{d\Omega\, d\varepsilon} = \frac{k'}{k}\left(\frac{m_n}{2\pi}\right)^2\left(\frac{e^2}{m_e m_n c^2}\right)^2\frac{(4\pi g)^2}{4(2\pi)}|F(\mathbf{K})|^2 \int dt\, e^{-i\omega t}\langle \mathbf{P}_\perp(0)\cdot\mathbf{P}_\perp(t)\rangle_T$$
$$= \frac{1}{2\pi}(gr_0)^2\frac{k'}{k}|F(\mathbf{K})|^2\int dt\, e^{-i\omega t}\langle \mathbf{P}_\perp(0)\cdot\mathbf{P}_\perp(t)\rangle_T,$$

where $r_0 = e^2/mc^2 = 2.82 \times 10^{-13}$ cm is the classical radius of the electron.

Paramagnetic Scattering. We now consider several applications of (75). For elastic scattering by an isolated paramagnetic ion of spin S, in the absence of a magnetic field,

$$\frac{d\sigma}{d\Omega} = (gr_0)^2|F(\mathbf{K})|^2\langle P_\perp^2\rangle_T, \tag{76}$$

where for random electronic spin orientation

$$\langle P_\perp^2\rangle_T = \tfrac{2}{3}S(S+1), \tag{77}$$

because all component pairs $S^\alpha S^\alpha$ contribute equally except the component parallel to $\mathbf{K}$. Thus

$$\frac{d\sigma}{d\Omega} = \tfrac{2}{3}(gr_0)^2|F(\mathbf{K})|^2S(S+1). \tag{78}$$

The only dependence on scattering angle enters from the form factor. The magnetic cross section is somewhat larger than the square of the classical radius of the electron. Paramagnetic scattering is often used to determine the magnetic form factor, $F(\mathbf{K})$.

Elastic Ferromagnetic Scattering. In a ferromagnetic crystal with spins $\mathbf{S}_l$ at lattice points $\mathbf{x}_l$, we may define

$$\mathbf{P}_\perp(\mathbf{K}) = \sum_l e^{i\mathbf{K}\cdot\mathbf{x}_l}\mathbf{S}_{\perp l}. \tag{79}$$

Then

$$\langle \mathbf{P}_\perp(0)\cdot\mathbf{P}_\perp(t)\rangle_T = \sum_{jl} e^{i\mathbf{K}\cdot(\mathbf{x}_l-\mathbf{x}_j)}\langle \mathbf{S}_{\perp j}(0)\cdot\mathbf{S}_{\perp l}(t)\rangle_T, \tag{80}$$

if the lattice is rigid.

For elastic scattering we may replace the trace in (80) by its time average:

$$\text{time average of } [S_j^\alpha(0)S_l^\beta(t)\rangle_T = \langle S\rangle_T{}^2\delta_{\alpha\sigma}\delta_{\beta\sigma}, \tag{81}$$

where $\langle S\rangle$ is a function of temperature; $\boldsymbol{\sigma}$ is defined as a unit vector along the axis of magnetic orientation. Then (80) becomes, using (38) and the result of Problem 4,

$$NV_c^*\langle S\rangle^2\{1-(\hat{\mathbf{K}}\cdot\boldsymbol{\sigma})^2\}\sum_{\mathbf{G}}\delta(\mathbf{K}-\mathbf{G}),$$

and, from (75),

$$\frac{d\sigma}{d\Omega} = (gr_0)^2NV_c^*\langle S\rangle_T{}^2\sum_{\mathbf{G}}|F(\mathbf{K})|^2\{1-(\hat{\mathbf{K}}\cdot\boldsymbol{\sigma})^2\}\delta(\mathbf{K}-\mathbf{G}). \tag{82}$$

This gives magnetic Bragg scattering for a ferromagnet at exactly the same reciprocal lattice points as nuclear scattering. For an antiferromagnet the definitions of **G** are different for the magnetic and the atomic primitive nuclear cells; thus the Bragg reflections are different for magnetic and for nuclear scattering.

We note from (82) that because of the factor $\langle S\rangle_T{}^2$ the coherent magnetic elastic scattering will fall off rapidly with temperature near the curie temperature; further, the magnetic scattering vanishes for **K** parallel to the magnetization. Magnetic scattering involves a form factor $F(\mathbf{K})$, whereas the corresponding nuclear form factor is unity apart from the effect of lattice vibrations.

The theory of inelastic ferromagnetic scattering is quite similar to that for inelastic lattice scattering. The theory follows on developing (80) in magnon operators. The results of the experiments can be used to determine the dispersion relations and relaxation times of magnons in a wavevector region at present entirely inaccessible by other methods.

PROBLEMS

1. Suppose that the state is represented by an incident wave e^{ikz} and an isotropic scattered wave $(\alpha/r)e^{ikr}$:

$$\varphi = e^{ikz} + \frac{\alpha}{r}e^{ikr}; \tag{83}$$

by averaging over a sphere show that the s-wave part of φ is

$$\varphi_s = \frac{1}{kr}\sin kr + \frac{\alpha}{r}e^{ikr}. \tag{84}$$

We can express φ_s in the form

$$\varphi_s = e^{i\eta_0}\frac{\sin(kr+\eta_0)}{kr}, \tag{85}$$

as usual in scattering problems; here η_0 is the s-wave phase shift. Show that

$$\alpha = \frac{1}{k}e^{i\eta_0}\sin\eta_0 = -e^{i\eta_0}b, \tag{86}$$

where b is the scattering length.

2. Suppose that the scattering nuclei consist of a single isotope with spin I; let b_+ be the scattering length in the state $I + \frac{1}{2}$ of nucleus plus neutron, and let b_- be the scattering length in the state $I - \frac{1}{2}$. Show that

$$\langle b\rangle = \left\{\frac{I+1}{2I+1}b_+ + \frac{I}{2I+1}b_-\right\}; \tag{87}$$

$$\langle |b|^2\rangle = \left\{\frac{I+1}{2I+1}|b_+|^2 + \frac{I}{2I+1}|b_-|^2\right\}. \tag{88}$$

The scattering from natural iron is almost entirely coherent because of the high abundance of an isotope of zero spin. The scattering from vanadium is almost entirely incoherent. The scattering from hydrogen also is almost entirely incoherent, but not from deuterons.

3. The result (79) was derived for a single paramagnetic ion; show that the identical result applies, per ion, to a paramagnetic crystal if the spins are uncorrelated; that is, $\langle S_i^\alpha S_j^\beta \rangle = \delta_{ij}\langle S_i^\alpha S_i^\beta \rangle$, where i and j are the labels of any two ion sites in the lattice.

4. Show that

$$P_\perp^\alpha P_\perp^\alpha = (\delta_{\alpha\beta} - \hat{K}^\alpha \hat{K}^\beta) S^\alpha S^\beta; \tag{89}$$

the right-hand side gives the form employed by Van Hove and others.

5. Show that if $n(\mathbf{x})$ is the concentration of nuclei of scattering length b, then the differential elastic scattering cross section is given by

$$\frac{d^2\sigma}{d\Omega\, d\varepsilon} = b^2 \delta(\varepsilon) \left| \int d^3x\, e^{i\mathbf{K}\cdot\mathbf{x}}\, n(\mathbf{x}) \right|^2. \tag{90}$$

For a liquid $n(\mathbf{x})$ = constant, so that the cross section involves $\delta(\mathbf{K})$, but this corresponds to no scattering. Thus in a strict sense there is no elastic scattering from a liquid.

20 Recoilless emission

When a low-energy γ ray is emitted by the nucleus of an isolated atom, the atom recoils and the energy of the emitted γ ray is reduced by the amount of the recoil energy. If the atom is initially at rest, the final velocity of the atom of mass M is obtained directly from the principle of conservation of momentum:

$$0 = M\mathbf{v} + \mathbf{K}, \tag{1}$$

where $\mathbf{K}$ is the wavevector of the γ ray; $|\mathbf{K}| = \omega/c$. One may alternatively simply say that the wavevector of the recoiling atom is $-\mathbf{K}$. From (1),

$$-v = \omega/Mc; \tag{2}$$

for a γ ray of energy $\sim$100 kev or 10^{-7} erg emitted by an atom of mass 10^{-23} g, we have $v \sim 10^{-7}/(10^{-23})(3 \times 10^{10}) \sim 3 \times 10^5$ cm/sec, of the order of thermal velocities.

The recoil energy R is

$$R = K^2/2M = E_0{}^2/2Mc^2, \tag{3}$$

where E_0 is the energy of the γ ray. In the present example $R \sim 10^{-14}/2(10^{-23})(10^{21}) \sim 5 \times 10^{-12}$ ergs, which corresponds to a frequency shift of the γ ray

$$\Delta\omega = R/\hbar \sim 10^{15}\ \text{sec}^{-1}. \tag{4}$$

A shift of this magnitude may sometimes be greater than the natural radiative width of the γ-ray line; some γ-ray lines of interest have widths as narrow as 10^7 sec^{-1}. Thus because of recoil the emitted γ ray may not have the frequency needed to be reabsorbed by a nucleus of the same kind. The recoil may therefore quench the γ-ray resonance fluorescence.

From a free atom in thermal equilibrium the γ-ray line will be broadened by the doppler effect as a result of the thermal distribution of velocities. The mean square doppler width $\langle(\Delta\omega)^2\rangle$ is given approximately by

$$\langle(\Delta\omega)^2\rangle/\omega^2 \approx \langle v^2\rangle/c^2, \tag{5}$$

where $\langle v^2\rangle$ is the mean square thermal velocity of the atom. If we define $\Delta^2 = \langle(\Delta\omega)^2\rangle$, then from (5),

$$\Delta^2 \approx \frac{\omega^2}{c^2}\langle v^2\rangle = \frac{K^2}{2M}\,2M\langle v^2\rangle, \tag{6}$$

or

$$\Delta^2 \approx R \cdot k_BT, \tag{7}$$

where R is the recoil energy and k_B is the boltzmann constant. For the numerical values above, the width Δ is of the order of the shift R. The doppler width may therefore be much greater than the natural line width of a γ emitter having a long lifetime.

When the radiating atoms are embedded in a crystal a fraction of the emission of γ rays take place with no perceptible recoil energy, and with a width close to the natural width. This is known as the Mössbauer effect. In a solid one may find both a sharp γ-ray line, essentially unshifted in frequency (as distinguished by the cross section for reabsorption) and a broad shifted background. The relative proportions are temperature-dependent, the proportion of the unshifted radiation increasing as the temperature is lowered, but never becoming unity. The width of the sharp line is independent of temperature. The natural width is usually dominant.

The shifted portions of the γ-ray spectrum, the portions with recoil, arise when the emission of the γ ray is accompanied by the emission or absorption of phonons in the crystal. The energy of the phonons now plays the part of the translation recoil energy in the free atom.

From our experience with x-ray diffraction by crystals we should not be entirely surprised at the existence of γ-ray transitions in a crystal not accompanied by the excitation of phonons; after all, the Bragg reflections are recoilless or elastic in the same sense. The inelastic diffuse reflections of x rays are entirely analogous to that portion of the γ-ray spectrum which is accompanied by the emission or absorption of phonons. Indeed, the usual energies of x rays utilized in diffraction work are of the same order as the energies of interest in the study of the recoilless emission of γ rays. Thus the fraction of elastic events may be the same in a diffraction experiment with 20 kev x rays as in the emission of a 20 kev γ ray. We have seen in the last chapter that the intensity of a Bragg-reflected x-ray line is proportional to

$$e^{-(1/3)K^2\langle u^2\rangle_T};$$

here $\mathbf{K}$ is the wavevector of the x ray; and $\langle u^2\rangle_T$ is the mean square displacement of an atom under both thermal motion and zero-point motion. The Debye-Waller factor describes the temperature-dependence of the elastic reflections. The same factor determines the proportion of recoilless events in γ-ray emission of absorption in a crystal.

A review of the applications of recoilless emission of γ rays to solid state problems is given by A. Abragam in *LTP*; see also the reprint collection edited by H. Frauenfelder: *The Mössbauer effect*, Benjamin, New York, 1962.

TRANSITION MATRIX ELEMENT

We first consider the emission or absorption of a gamma ray by a free nucleus not bound in a lattice. This transition is described by a matrix element M of an appropriate operator A between the initial state $|i\rangle$ and the final state $|f\rangle$ of the nucleus:

$$M = \langle f|A(\mathbf{x}_i,\mathbf{p}_i,\sigma_i)|i\rangle. \tag{8}$$

The operator A depends upon the coordinates, momenta, and spins of the particles in the nucleus. Let us now express A in terms of the center-of-mass coordinate of the nucleus $\mathbf{x}$ and the relative coordinates $\mathbf{q}$ which include spins. The dependence of A upon the center-of-mass coordinate $\mathbf{x}$ is determined completely by the requirements of translational and Galilean invariance; that is, by the requirements that momentum should be conserved and that the transition probability for a moving nonrelativistic observer should not depend upon the velocity of the observer. For the emission of a γ ray of momentum $-\mathbf{K}$, the above requirements are satisfied only if the operator A has the form

$$A = e^{i\mathbf{K}\cdot\mathbf{x}}a(q), \tag{9}$$

where the operator $a(q)$ depends only upon the relative variables and spins of the particles and has an explicit form depending upon the nature of the transition (electric, magnetic, dipole, quadrupole, etc.). The explicit form of $a(q)$ is of no interest for our present purposes. The factor $e^{i\mathbf{K}\cdot\mathbf{x}}$ allows nonvanishing matrix elements with a γ-ray wave function having a spatial variation $e^{-i\mathbf{K}\cdot\mathbf{x}}$.

We now consider the emission or absorption of a γ ray by a nucleus bound in a crystal. The operator describing the transition is the same operator A, but we must take the matrix element between initial and final states of the whole lattice, rather than of the free nucleus. We can now write down an expression for the matrix element describing the transition in which a γ ray of momentum $\mathbf{K}$ is emitted by a nucleus whose center-of-mass coordinate is $\mathbf{x}$, while the lattice goes from a state specified by quantum numbers n_i to a state specified by quantum numbers n_f, and the internal state of the emitting nucleus changes from $|i\rangle$ to $|f\rangle$. This matrix element is

$$M_L = \langle n_f|\exp(i\mathbf{K}\cdot\mathbf{x})|n_i\rangle \cdot \langle f|a(q)|i\rangle. \tag{10}$$

The matrix element thus separates into the product of a factor depending only upon the lattice and a factor depending only upon the internal structure of the nucleus.

The transition probability depends upon the square of the matrix element. We are interested mainly in the fraction of the total number of transitions which are recoilless, that is, in which the state of the lattice is unchanged. The probability $P(n_f,n_i)$ of a transition in which the phonon quantum numbers change from n_i to n_f is just

$$P(n_f,n_i) = |\langle n_f|e^{i\mathbf{K}\cdot\mathbf{x}}|n_i\rangle|^2, \tag{11}$$

because, from any initial state n_i of the lattice,

$$\begin{aligned}\sum_{n_f} P(n_f,n_i) &= \sum_{n_f} \langle n_i|e^{-i\mathbf{K}\cdot\mathbf{x}}|n_f\rangle\langle n_f|e^{i\mathbf{K}\cdot\mathbf{x}}|n_i\rangle \\ &= \langle n_i|e^{-i\mathbf{K}\cdot\mathbf{x}}e^{i\mathbf{K}\cdot\mathbf{x}}|n_i\rangle = \langle n_i|n_i\rangle = 1.\end{aligned} \tag{12}$$

This confirms the normalization of (11); that is, the total probability of anything happening or not happening is unity.

There is a powerful sum rule due to Lipkin which states that the average energy transferred to the lattice is just the energy the individual nucleus would have if it recoiled freely. The sum rule also gives a quick way of seeing at least that recoilless emission can occur.

If the binding forces of the crystal depend only on the positions of the atoms and not on their velocities, then the only term in the hamiltonian H of the crystal which does not commute with $\mathbf{x}$ is the kinetic energy of the same nucleus, namely $p^2/2M$. Then from the result of Problem 1.4,

$$[H,e^{i\mathbf{K}\cdot\mathbf{x}}] = \frac{1}{2M}\,e^{i\mathbf{K}\cdot\mathbf{x}}(K^2 + 2\mathbf{K}\cdot\mathbf{p}). \tag{13}$$

Further

$$[[H,e^{i\mathbf{K}\cdot\mathbf{x}}],e^{-i\mathbf{K}\cdot\mathbf{x}}] = -K^2/M, \tag{14}$$

Note that

$$(15)\qquad [[H,e^{i\mathbf{K}\cdot\mathbf{x}}],e^{-i\mathbf{K}\cdot\mathbf{x}}] = 2H - e^{i\mathbf{K}\cdot\mathbf{x}}He^{-i\mathbf{K}\cdot\mathbf{x}} - e^{-i\mathbf{K}\cdot\mathbf{x}}He^{i\mathbf{K}\cdot\mathbf{x}}.$$

Now write the ii diagonal matrix element of (15) in the representation in which the phonon populations are diagonal:

$$\begin{aligned}(16)\quad \langle n_i|[[H,e^{i\mathbf{K}\cdot\mathbf{x}}], e^{-i\mathbf{K}\cdot\mathbf{x}}]|n_i\rangle &= 2\langle n_i|H|n_i\rangle \\ &- \sum_f \{\langle n_i|e^{i\mathbf{K}\cdot\mathbf{x}}|n_f\rangle\langle n_f|e^{-i\mathbf{K}\cdot\mathbf{x}}|n_i\rangle\langle n_f|H|n_f\rangle \\ &- \langle n_i|e^{-i\mathbf{K}\cdot\mathbf{x}}|n_f\rangle\langle n_f|e^{i\mathbf{K}\cdot\mathbf{x}}|n_i\rangle\langle n_f|H|n_f\rangle\} \\ &= 2\sum_f (\varepsilon_i - \varepsilon_f)P(n_f,n_i),\end{aligned}$$

using the result (11). Thus, with (14),

$$(17)\qquad \sum_f \{\varepsilon(n_f) - \varepsilon(n_i)\}P(n_f,n_i) = K^2/2M = R,$$

where R is exactly the free atom recoil energy, according to (3).

The probability of recoilless emission is given by $P(n_i,n_i)$. For a single harmonic oscillator of frequency ω, the left-hand side of (17) is larger than

$$\omega \sum_{f\neq i} P(n_f,n_i) = \omega\{1 - P(n_i,n_i)\},$$

so that

$$(18)\qquad P(n_i,n_i) > 1 - (R/\omega);$$

this inequality demonstrates the necessity for recoilless emission, at least when $\omega > R$. The ensemble average of $P(n_i,n_i)$ is essentially equivalent to the Debye-Waller factor.

RECOILLESS EMISSION IN A CRYSTAL LATTICE, ABSOLUTE ZERO

The simplest problem is that of emission in a crystal lattice at absolute zero. We assume that the crystal binding is purely harmonic. We introduce normal coordinates for a bravais lattice of N identical atoms of mass M. If $\mathbf{x}_0$ is the equilibrium position of a radioactive nucleus, then we may write

$$(19)\qquad \mathbf{x} = \mathbf{x}_0 + \mathbf{u},$$

with $\mathbf{u}$ as the displacement from equilibrium. Then, from the phonon expansion of Chapter 2,

$$(20)\quad \mathbf{u} = N^{-\frac{1}{2}} \sum_{\mathbf{q}j} (1/2M\omega_{\mathbf{q}j})^{\frac{1}{2}}\mathbf{e}_{\mathbf{q}j}(a_{\mathbf{q}j}e^{i\mathbf{q}\cdot\mathbf{x}_0}e^{-i\omega_{\mathbf{q}j}t} + a_{\mathbf{q}j}^{+}e^{-i\mathbf{q}\cdot\mathbf{x}_0}e^{i\omega_{\mathbf{q}j}t}),$$

where $\mathbf{e}_{\mathbf{q}j}$ is a unit vector for the jth polarization component of the mode $\mathbf{q}$. It is convenient to simplify the notation by setting $\mathbf{x}_0 = 0$

and by using s to denote the pair of indices $\mathbf{q}j$. Thus

$$\mathbf{u} = N^{-1/2} \sum_s (1/2M\omega_s)^{1/2} \mathbf{e}_s (a_s e^{-i\omega_s t} + a_s^+ e^{i\omega_s t}), \tag{21}$$

which may be expressed as

$$\mathbf{u} = N^{-1/2} \sum_s Q_s \mathbf{e}_s, \tag{22}$$

with Q_s as the amplitude of the normal mode s. The associated momentum is

$$P_s = M\dot{Q}_s = -iM\omega_s(1/2M\omega_s)^{1/2}(a_s e^{-i\omega_s t} - a_s^+ e^{i\omega_s t}). \tag{23}$$

The hamiltonian for the normal modes is

$$H = \tfrac{1}{2} \sum \left\{ M\omega_s^2 Q_s^2 + \frac{1}{M} P_s^2 \right\}. \tag{24}$$

The normalized ground-state wavefunction in the coordinate representation for the mode s is just the harmonic oscillator result:

$$\langle x|0_s\rangle = \alpha_s^{1/2} \pi^{-1/4} e^{-\alpha_s^2 Q_s^2/2}, \qquad \alpha_s^2 = M\omega_s. \tag{25}$$

The probability of recoilless emission from the ground state is

$$P(0,0) = |\langle 0|e^{i\mathbf{K}\cdot\mathbf{u}}|0\rangle|^2 = \Big|\prod_s \langle 0_s|e^{i(\mathbf{K}\cdot\mathbf{e}_s)Q_s N^{-1/2}}|0_s\rangle\Big|^2, \tag{26}$$

where 0 signifies that all lattice oscillators are in their ground states. Using (25),

$$P(0,0) = \Big|\prod_s \left\{ (\alpha_s/\pi^{1/2}) \int_{-\infty}^{\infty} e^{-\alpha_s^2 Q_s^2} e^{i\beta_s Q_s}\, dQ_s \right\}\Big|^2, \tag{27}$$

where $\beta_s = (\mathbf{K}\cdot\mathbf{e}_s)N^{-1/2}$. We note that the quantity within the curly brackets has the value $\exp(-\beta_s^2/4\alpha_s^2)$, so that

$$\begin{aligned} P(0,0) &= \Big|\prod_s e^{-\beta_s^2/4\alpha_s^2}\Big|^2 = \prod_s e^{-\beta_s^2\langle Q_s^2\rangle} \\ &= e^{-\sum_s (K\cdot e_s)^2 \langle Q_s^2\rangle/N}, \end{aligned} \tag{28}$$

where we have used the result for the ground state that

$$\langle Q_s^2\rangle = \langle 0_s|Q_s^2|0_s\rangle = 1/2\alpha_s^2. \tag{29}$$

If we assume that $\langle Q_{\mathbf{q}j}^2\rangle = \langle Q_{\mathbf{q}}^2\rangle$, independent of the polarization j, then

$$\sum_j (K\cdot e_s)^2 \langle Q_s^2\rangle/N = K^2\langle Q_{\mathbf{q}}^2\rangle/N. \tag{30}$$

Now

$$\langle u^2 \rangle = N^{-1} \sum_s \langle Q_s{}^2 \rangle = 3N^{-1} \sum_{\mathbf{q}} \langle Q_{\mathbf{q}}{}^2 \rangle, \tag{31}$$

so that, combining (28), (30), (31),

$$P(0,0) = e^{-K^2\langle u^2 \rangle/3}. \tag{32}$$

By the properties of hermite polynomials or otherwise (see Messiah, pp. 449–451), the identical result follows for a harmonic system not in the ground state of all phonons, but with the phonons in a canonical distribution. Thus the probability f of recoilless emission with phonons distributed according to the canonical distribution at temperature T is

$$f = P(n_T, n_T) = e^{-K^2\langle u^2 \rangle_T/3}. \tag{33}$$

The same result obtains even for anharmonic lattice oscillators, provided that the total number of atoms $\gg 1$, according to a theorem due to R. J. Glauber, *Phys. Rev.* **84,** 395 (1951).

For a harmonic oscillator

$$M\omega_s{}^2\langle Q_s{}^2 \rangle = (n_s + \tfrac{1}{2})\hbar\omega_s, \tag{34}$$

so that, from (31),

$$\tfrac{1}{3}K^2\langle u^2 \rangle = \frac{K^2}{2M} \cdot \frac{2}{3N} \sum_s \frac{(n_s + \frac{1}{2})}{\omega_s} = \frac{2R}{3N} \sum_s \frac{(n_s + \frac{1}{2})}{\omega_s}, \tag{35}$$

where $R = K^2/2M$ is the free-atom recoil energy.

At absolute zero $n_s = 0$ and we are concerned with $(R/3N) \sum_s \omega_s{}^{-1}$. For a Debye solid $\omega = vq$, where v is the velocity of sound and q is the wavevector. We note that $(1/3N) \sum_s (1/q_s)$ is just the mean value of $1/q$ over the $3N$ phonon modes. Thus

$$\frac{1}{3N} \sum \frac{1}{q_s} = \frac{\int_0^{q_m} q\,dq}{\int_0^{q_m} q^2\,dq} = \frac{3}{2} \cdot \frac{1}{q_m} = \frac{3v}{2\omega_m} = \frac{3v}{2k_B\Theta}, \tag{36}$$

using the standard definition of the Debye temperature $\Theta = \omega_m/k_B$. Therefore

$$\frac{R}{3N} \sum \frac{1}{\omega_s} = \frac{3R}{2k_B\Theta},$$

and

$$P(0,0) = e^{-3R/2k_B\Theta}. \tag{37}$$

Thus the fraction of emission events at absolute zero which are recoilless is substantial if the free-atom recoil energy is less than the maximum possible phonon energy.

At a nonzero temperature we must include the equilibrium value of n_s; it is easily seen that we need the quantity

$$\frac{3k_B{}^2T^2}{2\pi^2 v^3}\int_0^{\Theta/T}\frac{x}{e^x-1}\,dx,$$

which, for $T \ll \Theta$, has the value

$$\frac{k_B{}^2T^2}{4v^3} = \tfrac{3}{2}\pi^2 N\left(\frac{T}{\Theta}\right)^2. \tag{38}$$

The total exponent is, for $T \ll \Theta$,

$$\tfrac{1}{3}K^2\langle u^2\rangle_T = \frac{3R}{2k_B\Theta}\left\{1+\frac{2\pi^2}{3}\left(\frac{T}{\Theta}\right)^2\right\}, \tag{39}$$

including the absolute zero contribution.

TIME CORRELATIONS IN RECOILLESS EFFECTS AND THE LINE SHAPE

Our object now is to calculate the shape of the emitted γ-ray line. We consider the emission of a γ ray by a nucleus bound harmonically in a lattice of N atoms each of mass M. We assume that the shape of the line when emitted from a nucleus held at rest is described by

$$I(E) = \frac{\Gamma^2}{4}\,\frac{1}{(E-E_0)^2+\frac{1}{4}\Gamma^2}, \tag{40}$$

where Γ is the lifetime; note that $I(E_0) = 1$. If the lattice is initially in the state i of energy ε_i and after emission is in the state f of energy ε_f, the line shape function is

$$I(E) = \frac{\Gamma^2}{4}\,\frac{1}{(E-E_0+\varepsilon_f-\varepsilon_i)^2+\frac{1}{4}\Gamma^2}. \tag{41}$$

If $\rho(i)$ is the probability that the system is initially in the state i, and $P(fi) = |\langle f|e^{i\mathbf{K}\cdot\mathbf{u}}|i\rangle|^2$ is the probability that the phonon state will change from i to f, then the shape function is given by

$$I(E) = \frac{\Gamma^2}{4}\sum_{i,f}\rho(i)\,\frac{|\langle f|e^{i\mathbf{K}\cdot\mathbf{u}}|i\rangle|^2}{(E-E_0+\varepsilon_f-\varepsilon_i)^2+\frac{1}{4}\Gamma^2}. \tag{42}$$

Now

$$\int_{-\infty}^{\infty} dt\,\exp\left\{-i(E-E_0)t-\tfrac{1}{2}\Gamma|t|-i(\varepsilon_f-\varepsilon_i)t\right\} = \frac{\Gamma}{(E-E_0+\varepsilon_f-\varepsilon_i)^2+\frac{1}{4}\Gamma^2}. \tag{43}$$

We may rearrange some of the factors in (42):

$$(44)\quad \sum_f e^{i\varepsilon_i t}\langle i|e^{-i\mathbf{K}\cdot\mathbf{u}}|f\rangle e^{-i\varepsilon_f t}\langle f|e^{i\mathbf{K}\cdot\mathbf{u}}|i\rangle = \langle i|e^{iHt}e^{-i\mathbf{K}\cdot\mathbf{u}}e^{-iHt}e^{i\mathbf{K}\cdot\mathbf{u}}|i\rangle,$$

where H is the hamiltonian of the phonons. In the Heisenberg representation $\mathbf{u}(t) = e^{iHt}\mathbf{u}(0)e^{-iHt}$, and the right side of (44) becomes

$$\langle i|e^{i\mathbf{K}\cdot\mathbf{u}(t)}e^{i\mathbf{K}\cdot\mathbf{u}(0)}|i\rangle.$$

We assume ρ is the canonical distribution, and we introduce the notation for thermal average:

$$(45)\qquad \langle e^{-i\mathbf{K}\cdot\mathbf{u}(t)}e^{i\mathbf{K}\cdot\mathbf{u}(0)}\rangle_T = \sum_i \rho(i)\langle i|e^{-i\mathbf{K}\cdot\mathbf{u}(t)}e^{i\mathbf{K}\cdot\mathbf{u}(0)}|i\rangle.$$

Then, from (42) and (43),

$$(46)\qquad I(E) = \frac{\Gamma}{4}\int_{-\infty}^{\infty} dt\; e^{-i(E-E_0)t-(\frac{1}{2})\Gamma|t|}\langle e^{-i\mathbf{K}\cdot\mathbf{u}(t)}e^{i\mathbf{K}\cdot\mathbf{u}(0)}\rangle_T.$$

We observe that $[\mathbf{u}(t),\mathbf{u}(0)] \neq 0$ for $t \neq 0$. This follows because, in terms of creation and annihilation operators a^+, a

$$(47)\qquad \mathbf{u}(t) = N^{-\frac{1}{2}}\sum_s (1/2M\omega_s)^{\frac{1}{2}}\mathbf{e}_s(a_s e^{-i\omega_s t} + a_s^+ e^{i\omega_s t}).$$

Then

$$(48)\quad [\mathbf{u}(0),\mathbf{u}(t)] = (1/2NM)\sum_s \omega_s^{-1}\{[a_s,a_s^+]e^{i\omega_s t} + [a_s^+,a_s]e^{-i\omega_s t}\};$$

the term in the curly brackets has the value $2i \sin \omega_s t$, so that

$$(49)\qquad [\mathbf{u}(0),\mathbf{u}(t)] = \frac{i}{NM}\sum_s \frac{\sin \omega_s t}{\omega_s}.$$

The commutator is therefore a c-number.

We now recall the theorem (Messiah, p. 442) that if $[A,B]$ commutes with A and with B, then

$$(50)\qquad e^A e^B = e^{A+B}e^{[A,B]/2}.$$

It follows that

$$(51)\qquad \langle e^{-i\mathbf{K}\cdot\mathbf{u}(t)}e^{i\mathbf{K}\cdot\mathbf{u}(0)}\rangle_T = \langle e^{-i\mathbf{K}\cdot\{\mathbf{u}(t)-\mathbf{u}(0)\}}\rangle_T e^{\frac{1}{2}[\mathbf{K}\cdot\mathbf{u}(t),\mathbf{K}\cdot\mathbf{u}(0)]}.$$

If we make the assumption that all phonon polarization modes of given wavevector $\mathbf{q}$ are degenerate, we may without further assumption arrange to take one of the three polarization directions j always along

K. Then we may rewrite the right side of (51) as

$$\langle e^{-i\mathbf{K}\cdot\mathbf{u}(t)}e^{i\mathbf{K}\cdot\mathbf{u}(0)}\rangle_T = \langle e^{-iK\{u(t)-u(0)\}}\rangle_T e^{(\frac{1}{2})K^2[u(t),u(0)]}, \tag{52}$$

where u denotes the component of $\mathbf{u}$ along $\mathbf{K}$.

Using the exact harmonic oscillator result given in Messiah, pp. 382–383, we have

$$\langle e^{iK\{u(t)-u(0)\}}\rangle_T = e^{-(\frac{1}{2})K^2\langle\{u(t)-u(0)\}^2\rangle_T}, \tag{53}$$

so that

$$I(E) = \frac{\Gamma}{4}\int_{-\infty}^{\infty} dt\, e^{-i(E-E_0)t-(\frac{1}{2})\Gamma|t|-Q(t)}, \tag{54}$$

where

$$Q(t) = \tfrac{1}{2}K^2\{\langle\{u(t) - u(0)\}^2\rangle_T + [u(0),u(t)]\}. \tag{55}$$

With one of the polarization directions along **K**, we have, from (47),

$$u(t) = N^{-\frac{1}{2}}\sum_{\mathbf{q}}(1/2M\omega_{\mathbf{q}})^{\frac{1}{2}}(a_{\mathbf{q}}e^{-i\omega_{\mathbf{q}}t} + a_{\mathbf{q}}^+e^{i\omega_{\mathbf{q}}t}). \tag{56}$$

Then

$$\begin{aligned}\{u(t) - u(0)\}^2 = (1/2NM)\sum_{\mathbf{q}}\omega_{\mathbf{q}}^{-1}\{&a_{\mathbf{q}}a_{\mathbf{q}}(1 - e^{-i\omega_{\mathbf{q}}t})^2 \\ &+ a_{\mathbf{q}}^+a_{\mathbf{q}}^+(1 - e^{i\omega_{\mathbf{q}}t})^2 + 2(a_{\mathbf{q}}a_{\mathbf{q}}^+ + a_{\mathbf{q}}^+a_{\mathbf{q}})(1 - \cos\omega_{\mathbf{q}}t)\},\end{aligned} \tag{57}$$

whence

$$\langle\{u(t) - u(0)\}^2\rangle_T = (1/NM)\sum_{\mathbf{q}}\omega_{\mathbf{q}}^{-1}(2\langle n_{\mathbf{q}}\rangle + 1)(1 - \cos\omega_{\mathbf{q}}t). \tag{58}$$

As in the last chapter, we separate $e^{-Q(t)}$ into two factors, one time-dependent and one time-independent. With $R = K^2/2M$,

$$\begin{aligned}e^{-Q(t)} = \exp[-(R/N)\sum_{\mathbf{q}}\omega_{\mathbf{q}}^{-1}(2\langle n_{\mathbf{q}}\rangle + 1]\exp\{(R/N)\sum_{\mathbf{q}}\omega_{\mathbf{q}}^{-1} \\ \cdot[(2\langle n_{\mathbf{q}}\rangle + 1)\cos\omega_{\mathbf{q}}t + i\sin\omega_{\mathbf{q}}t]\}.\end{aligned} \tag{59}$$

The time-dependent factor contains N terms each of order $1/N$; we expand this exponential as

$$1 + \{\quad\} + \tfrac{1}{2}\{\quad\}^2 + \cdots;$$

all the terms in $\{\quad\}$ are time-dependent; in $\{\quad\}^2$ there are N^2 terms of order $1/N^2$, of which only N terms are independent of the time. Therefore for the time-independent parts of a large system we may replace $\exp\{\quad\}$ by unity. The time-dependent parts do not contribute significantly to $I(E_0)$. The usual argument leading to the same result is that, with a finite collision time for the phonons, the system must forget at *long times* the value of $u(0)$, which may therefore be set equal to zero.

The intensity at the peak $E = E_0$ is, from (54), given essentially by the time-independent terms of $Q(t)$:

$$I(E_0) \cong \exp\left\{-\frac{2R}{3N}\sum_s \frac{(\langle n_s\rangle + \frac{1}{2})}{\omega_s}\right\}, \tag{60}$$

in exact agreement with our earlier result (33) and (35); the factor $\frac{1}{3}$ in (60) arises because the sum is written over the $3N$ modes s, whereas in (59) only the N modes $\mathbf{q}$ appear. The line shape $I(E)$ off the peak will have a contribution whenever $E - E_0$ is equal to some $\omega_{\mathbf{q}}$; thus there will be a broad continuous background in addition to the sharp central peak of width Γ.

PROBLEMS

1. Find an approximate relation for the proportion of recoilless emissions in the limit $T \gg \Theta$ for a Debye solid.

2. Study the line width and recoil shift of γ rays from a free thermalized atom in a gas, using the method of Eqs. (54) and (55). Neglecting collisions, we have $\langle\{u(t) - u(0)\}^2\rangle_T = \langle v^2\rangle_T t^2$; further, $[u(t),u(0)] = M^{-1}[tP,u] = -itM^{-1}$. Thus

$$Q(t) = \tfrac{1}{2}K^2\{\langle v^2\rangle t^2 + itM^{-1}\},$$

so that the line is shifted by an energy $\frac{1}{2}K^2M^{-1}$, which is just the recoil energy R of Eq. (3). Go on to estimate the line width.

3. Study the line width of γ rays from atoms in a gas when the collision frequency ρ of the atoms is much greater than the line width Γ.

For Brownian motion of a free particle we have the well-known result

$$\langle\{u(t) - u(0)\}^2\rangle_T = 2Dt,$$

where the diffusivity $D = k_BT/\rho M$; for the derivation, see Kittel, *Elementary statistical physics*, pp. 153–156. The result tells us that the line width is Γ if $K^2D \ll \Gamma/2$, or

$$\Delta^2 = Rk_BT \ll 4\Gamma\rho,$$

where Δ is the doppler width. Thus for rapid relaxation the doppler broadening is suppressed.

21 Green's functions—application to solid state physics

There has been a rapid growth in the use of Green's functions in published work in theoretical solid state physics, particularly in connection with many-body problems. The Green's function method is in principle merely a particular unified formulation of the quantum mechanical problem. It has often the advantage of great directness and flexibility; occasionally, as in some spin problems, it has the disadvantage of concealing the physical nature of approximations which may be made in using the method. It does allow us to introduce in a natural way matrix elements connecting states differing in the number of particles, as we were led to handle in connection with superfluidity in Chapter 2 and superconductivity in Chapter 8. Our object here is to give the reader some familiarity with the properties of Green's functions, an idea of how they are used, and an impression of their direct relevance to many-body problems. General references include the following.

A. A. Abrikosov, L. P. Gorkov, E. Dzyaloshinsky, *Methods of the quantum theory of fields in statistical physics*, Prentice-Hall, Englewood Cliffs, New Jersey, 1963.

L. P. Kadanoff and G. Baym, *Quantum statistical mechanics*, Benjamin, New York, 1962.

P. Nozières, *Le problème à N corps*, Dunod, Paris, 1963.

Various reprints in the collection cited as Pines, particularly those by Beliaev, Galitskii, and Migdal.

A one-particle Green's function describes the motion of one particle added to a many-particle system. A two-particle Green's function describes the motion of two added particles. We say that the Green's function is a thermodynamic or thermal Green's function if the system is in a grand canonical ensemble at a nonzero temperature.

We have seen (Problem 5.4) that the operator $\Psi^+(\mathbf{x})$ adds a particle at $\mathbf{x}$ to the vacuum state on which it operates. We give the proof here. The particle density operator is

$$\rho(\mathbf{x}') = \int d^3x''\, \Psi^+(\mathbf{x}'')\delta(\mathbf{x}' - \mathbf{x}'')\Psi(\mathbf{x}''); \tag{1}$$

thus

$$\begin{aligned}\rho(\mathbf{x}')\Psi^+(\mathbf{x})|\text{vac}\rangle &= \int d^3x''\, \Psi^+(\mathbf{x}'')\delta(\mathbf{x}' - \mathbf{x}'')\Psi(\mathbf{x}'')\Psi^+(\mathbf{x})|\text{vac}\rangle \\ &= \int d^3x''\, \Psi^+(\mathbf{x}'')\delta(\mathbf{x}' - \mathbf{x}'')\delta(\mathbf{x} - \mathbf{x}'')|\text{vac}\rangle,\end{aligned} \tag{2}$$

because $\Psi(\mathbf{x}'')|\text{vac}\rangle = 0$. Then

$$\rho(\mathbf{x}')\Psi^+(\mathbf{x})|\text{vac}\rangle = \delta(\mathbf{x} - \mathbf{x}')\Psi^+(\mathbf{x})|\text{vac}\rangle, \tag{3}$$

which is the desired result.

The notation $\Psi^+(\mathbf{x}t)$ for the operator in the Heisenberg representation tell us that the added particle is at $\mathbf{x}$ at time t. The operator $\Psi(\mathbf{x}t)$ applied to the state $\Psi^+(\mathbf{x}t)|\text{vac}\rangle$ will remove the added particle. The operator $\Psi(\mathbf{x}'t')$ at another point $\mathbf{x}'$ and a later time $t' > t$ measures the probability amplitude that the added particle has moved from $\mathbf{x}$ at t to $\mathbf{x}'$ at t'. Important aspects of the dynamical behavior of an added particle in a state of the system thus are described by, for $t' > t$,

$$\Psi(\mathbf{x}'t')\Psi^+(\mathbf{x}t)|\rangle.$$

It is convenient to work with the expectation value

$$\langle|\Psi(\mathbf{x}'t')\Psi^+(\mathbf{x}t)|\rangle, \tag{4}$$

or with the average of this quantity over a grand canonical ensemble:

$$\langle\Psi(\mathbf{x}'t')\Psi^+(\mathbf{x}t)\rangle = \text{Tr}\,\rho\Psi(\mathbf{x}'t')\Psi^+(\mathbf{x}t), \tag{5}$$

where ρ is the appropriate statistical operator. The quantity in (4) or (5) is essentially a correlation function, and it is well suited to the description of a quasiparticle.

There is some motivation for working with a generalization of (4) or (5). To exhibit one aspect, we consider as the state of the system the unperturbed ground state of a fermi gas. We decompose Ψ^+ into two parts: $\Psi^+ = \Psi_e^+ + \Psi_h$, where Ψ_e^+ is defined for electron states, $k > k_F$, and Ψ_h is the hole annihilation operator defined for $k < k_F$. Then

$$\begin{aligned}\Psi(1')\Psi^+(1) = \Psi_e(1')\Psi_e^+(1) + \Psi_h^+(1')\Psi_h(1) + \Psi_e(1')\Psi_h(1) \\ + \Psi_h^+(1')\Psi_e^+(1),\end{aligned} \tag{6}$$

where $1' \equiv \mathbf{x}'t'$; $1 \equiv \mathbf{x}t$. The only term which contributes to the expectation value of (6) in the unperturbed ground state is $\Psi_e(1')\Psi_e^+(1)$, so that the product $\Psi(1')\Psi^+(1)$ is adapted to the study of the motion of an added electron, but not to the motion of an added hole. However,

$$\Psi^+(1)\Psi(1') = \Psi_e^+(1)\Psi_e(1') + \Psi_h(1)\Psi_h^+(1') + \Psi_h(1)\Psi_e(1') + \Psi_e^+(1)\Psi_h^+(1') \tag{7}$$

is adapted to the study of the motion of an added hole, provided $t > t'$, because $\Psi_h(1)\Psi_h^+(1')$ has a nonvanishing expectation value. Now note that

$$P(\Psi(\mathbf{x}'t')\Psi^+(\mathbf{x}t)) = \begin{cases} \Psi(\mathbf{x}'t')\Psi^+(\mathbf{x}t) & \text{if } t' > t \\ \Psi^+(\mathbf{x}t)\Psi(\mathbf{x}'t') & \text{if } t > t', \end{cases} \tag{8}$$

where P is the Dyson chronological operator which orders earlier times to the right. For fermion fields it is more satisfactory to take account of the equal time anticommutation relation by using the Wick chronological operator T defined to give

$$T(\Psi(\mathbf{x}'t')\Psi^+(\mathbf{x}t)) = \begin{cases} \Psi(\mathbf{x}'t')\Psi^+(\mathbf{x}t) & \text{if } t' > t; \\ -\Psi^+(\mathbf{x}t)\Psi(\mathbf{x}'t') & \text{if } t > t'. \end{cases} \tag{9}$$

If Ψ and Ψ^+ anticommute we have always that $T(\Psi(\mathbf{x}'t')\Psi^+(\mathbf{x}t)) = \Psi(\mathbf{x}'t')\Psi^+(\mathbf{x}t)$. For boson fields T is identical with P.

The one-particle Green's function is defined* over the whole time domain by

$$\boxed{G(\mathbf{x}'t';\mathbf{x}t) = -i\langle T(\Psi(\mathbf{x}'t')\Psi^+(\mathbf{x}t))\rangle;} \tag{10}$$

in the ground state

$$G(\mathbf{x}'t';\mathbf{x}t) = -i\langle 0|T(\Psi(\mathbf{x}'t')\Psi^+(\mathbf{x}t))|0\rangle. \tag{11}$$

The field operators are in the Heisenberg representation. We have seen that we get more information by using the entire time domain than by restricting t' to be $>t$. The two-particle Green's function is similarly defined by

$$K(1234) = \langle T(\Psi(1)\Psi(2)\Psi^+(3)\Psi^+(4))\rangle, \tag{12}$$

where 1 denotes $\mathbf{x}_1t_1$, etc.

* Some workers use i in place of $-i$ in the definition of the one-particle Green's function. With our factor $-i$ the Green's function for a free-electron gas satisfies, using (14), (17), and (18),

$$\left(i\frac{\partial}{\partial t} + \frac{1}{2m}\frac{\partial^2}{\partial \mathbf{x}^2}\right) G_0(\mathbf{x}t) = \delta(\mathbf{x})\delta(t),$$

which shows that G_0 is like a Green's function for the Schrödinger equation.

If the system has galilean invariance and also rotational invariance the Green's functions may be written in terms of relative coordinates; thus (10) becomes

$$G(\mathbf{x}t) = -i\langle T(\Psi(\mathbf{x}t)\Psi^+(00))\rangle. \tag{13}$$

As a trivial example, consider a one-dimensional system of noninteracting fermions in the true vacuum state $|\text{vac}\rangle$. The natural expansion for $\Psi(xt)$ is in terms of the free particle eigenfunctions:

$$\Psi(xt) = \sum_k c_k(t)e^{ikx} = \sum_k c_k e^{-i\omega_k t}e^{ikx}, \tag{14}$$

where $\omega_k = k^2/2m$. Then for the vacuum state the one-particle Green's function is

$$\begin{aligned} G_v(xt) &= -i\langle\text{vac}|\Psi(xt)\Psi^+(00)|\text{vac}\rangle \\ &= i\langle\text{vac}|T\left(\sum_{kk'} c_k c_{k'}^+ e^{-i\omega_k t}e^{ikx}\right)|\text{vac}\rangle \\ &= -i\sum_k e^{ikx}e^{-i\omega_k t}, \end{aligned} \tag{15}$$

for $t > 0$ and zero otherwise. We observe that $G_v(x,+0) = -i\sum_k e^{ikx} = -i\delta(x)$. On replacing the summation by an integral,

$$\begin{aligned} G_v(xt) &= -i\int_{-\infty}^{\infty} dk\, \exp\left[i\left(kx - \frac{1}{2m}k^2t\right)\right] \\ &= e^{-i3\pi/4}(2\pi m/t)^{1/2}e^{imx^2/2t}, \end{aligned} \tag{16}$$

for $t > 0$. Note that this is a solution of the time-dependent Schrödinger equation.

The evaluation of $G(xt)$ for the ground state $|0\rangle$ of a noninteracting fermi gas in one dimension is the subject of Problem 1. We note here merely that for this problem

$$\begin{aligned} G_0(x,+0) &= -\frac{i}{2\pi}\left[\int_{k_F}^{\infty} + \int_{-\infty}^{-k_F}\right] dk\, e^{ikx} \\ &= -\frac{i}{2\pi}\left[2\pi\delta(x) - \int_{-k_F}^{k_F} dk\, e^{ikx}\right] \\ &= -\frac{i}{2\pi}\left[2\pi\delta(x) - 2\frac{\sin k_F x}{x}\right]; \end{aligned} \tag{17}$$

$$G_0(x,-0) = \frac{i}{2\pi}\int_{-k_F}^{k_F} dk\, e^{ikx} = i\,\frac{\sin k_F x}{\pi x}. \tag{18}$$

Observe the difference between (15) and these results; (15) was for the vacuum.

FOURIER TRANSFORMS

The theory makes abundant use of the transforms $G(\mathbf{k}t)$ and $G(\mathbf{k}\omega)$. We define

$$G(\mathbf{k}t) = \int d^3x\, e^{-i\mathbf{k}\cdot\mathbf{x}} G(\mathbf{x}t); \tag{19}$$

the inverse transformation is

$$G(\mathbf{x}t) = \frac{1}{(2\pi)^3}\int d^3k\, e^{i\mathbf{k}\cdot\mathbf{x}} G(\mathbf{k}t). \tag{20}$$

Further,

$$G(\mathbf{k}\omega) = \int dt\, e^{i\omega t} G(\mathbf{k}t), \tag{21}$$

with the inverse

$$G(\mathbf{k}t) = \frac{1}{2\pi}\int d\omega\, e^{-i\omega t} G(\mathbf{k}\omega). \tag{22}$$

NONINTERACTING FERMION GAS

The one-particle Green's function $G_0(\mathbf{k}t)$ in the ground state of a noninteracting fermion gas is

$$\begin{aligned} G_0(\mathbf{k}t) &= -i\langle 0|T\Big(\sum_{\mathbf{k}'} c_{\mathbf{k}'}c^+_{\mathbf{k}'}e^{-i\omega_{\mathbf{k}'}t}\int d^3x\, e^{i(\mathbf{k}'-\mathbf{k})\cdot\mathbf{x}}\Big)|0\rangle \\ &= \begin{cases} -i\langle 0|c_{\mathbf{k}}c^+_{\mathbf{k}}|0\rangle e^{-i\omega_{\mathbf{k}}t}, & t>0; \\ i\langle 0|c^+_{\mathbf{k}}c_{\mathbf{k}}|0\rangle e^{-i\omega_{\mathbf{k}}t}, & t<0. \end{cases} \end{aligned} \tag{23}$$

If $n_{\mathbf{k}} = 1, 0$ is the population of the state $\mathbf{k}$ in the ground state,

$$G_0(\mathbf{k}t) = \begin{cases} -ie^{-i\omega_{\mathbf{k}}t}(1-n_{\mathbf{k}}), & t>0; \\ ie^{-i\omega_{\mathbf{k}}t}n_{\mathbf{k}}, & t<0; \end{cases} \tag{24}$$

thus

$$G_0(\mathbf{k},+0) - G_0(\mathbf{k},-0) = -i.$$

For a quasiparticle excitation in a real (interacting) fermion gas we expect the time dependence of $G(\mathbf{k}t)$ for $t > 0$ to be of the form

$$e^{-i\omega_{\mathbf{k}}t}e^{-\Gamma_{\mathbf{k}}t},$$

over a limited interval of time; here $1/\Gamma_{\mathbf{k}}$ is the lifetime for the decay of a quasiparticle excitation into other excitations.

We now establish that

$$G_0(\mathbf{k}\omega) = \lim_{s_\mathbf{k}\to 0} \frac{1}{\omega - \omega_\mathbf{k} + is_\mathbf{k}}, \tag{25}$$

where

$s_\mathbf{k}$ positive for $k > k_F$;

$s_\mathbf{k}$ negative for $k < k_F$.

We show that this expression for $G_0(\mathbf{k}\omega)$ is consistent with the result (24) for $G_0(\mathbf{k}t)$. Consider the situation for $k > k_F$; for this region (24) tells us that

$$G_0(\mathbf{k}t) = \begin{cases} -ie^{-i\omega_\mathbf{k}t} & t > 0; \\ 0 & t < 0. \end{cases} \tag{26}$$

Now from (25)

$$\begin{aligned} G_0(\mathbf{k}t) &= \frac{1}{2\pi}\int_{-\infty}^{\infty} d\omega\, e^{-i\omega t} G_0(\mathbf{k}\omega) \\ &= \frac{1}{2\pi}\int_{-\infty}^{\infty} d\omega\, e^{-i\omega t}\left[\frac{\mathcal{P}}{\omega - \omega_\mathbf{k}} - i\pi\delta(\omega - \omega_\mathbf{k})\right], \end{aligned} \tag{27}$$

written for $k > k_F$. For $t > 0$ a contour integral may be completed by an infinite semicircle in the lower half-plane. We see that

$$\mathcal{P}\int_{-\infty}^{\infty} d\omega\, \frac{e^{-i\omega t}}{\omega - \omega_\mathbf{k}} = -i\pi e^{-i\omega_\mathbf{k}t}, \qquad t > 0; \tag{28}$$

further,

$$-i\pi\int_{-\infty}^{\infty} d\omega\, e^{-i\omega t}\delta(\omega - \omega_\mathbf{k}) = -i\pi e^{-i\omega_\mathbf{k}t}, \tag{29}$$

and we have the result

$$G_0(\mathbf{k}t) = -ie^{-i\omega_\mathbf{k}t}, \qquad t > 0; \qquad k > k_F. \tag{30}$$

For $t < 0$ the contour in (28) must be completed by a semicircle in the upper half-plane:

$$\mathcal{P}\int_{-\infty}^{\infty} d\omega\, \frac{e^{-i\omega t}}{\omega - \omega_\mathbf{k}} = i\pi e^{-i\omega_\mathbf{k}t}, \qquad t < 0; \tag{31}$$

thus

$$G_0(\mathbf{k}t) = 0, \qquad t < 0; \qquad k > k_F. \tag{32}$$

For $k < k_F$ we change the sign of $s_{\mathbf{k}}$ in (25); this changes the sign in front of the delta function in (27) and we have

$$G_0(\mathbf{k}t) = \begin{cases} 0, & t > 0; \quad k < k_F; \\ ie^{-i\omega_{\mathbf{k}}t}, & t < 0; \quad k < k_F. \end{cases} \tag{33}$$

INTERACTING FERMION GAS

We exhibit the form of the Green's function in the exact ground state $|0\rangle$ of a fermion gas with interactions:

$$\begin{aligned} G(\mathbf{k}t) &= -i\langle 0|T(c_{\mathbf{k}}(t)c_{\mathbf{k}}^{+}(0))|0\rangle \\ &= \begin{cases} -i\langle 0|e^{iHt}c_{\mathbf{k}}e^{-iHt}c_{\mathbf{k}}^{+}|0\rangle, & t > 0; \\ i\langle 0|c_{\mathbf{k}}^{+}e^{iHt}c_{\mathbf{k}}e^{-iHt}|0\rangle, & t < 0. \end{cases} \end{aligned} \tag{34}$$

Here H is the exact hamiltonian. Let $E_0{}^N$ denote the exact ground state energy of the N particle system. Then

$$G(\mathbf{k}t) = \begin{cases} -i\langle 0|c_{\mathbf{k}}e^{-iHt}c_{\mathbf{k}}^{+}|0\rangle e^{iE_0{}^Nt}, & t > 0; \\ i\langle 0|c_{\mathbf{k}}^{+}e^{iHt}c_{\mathbf{k}}|0\rangle e^{-iE_0{}^Nt}, & t < 0. \end{cases} \tag{35}$$

Let the excited states of the systems of $(N + 1)$ or $(N - 1)$ particles be denoted by the index n. Then (35) may be rewritten as

$$G(\mathbf{k}t) = \begin{cases} -i\sum_n \langle 0|c_{\mathbf{k}}e^{-iHt}|n\rangle\langle n|c_{\mathbf{k}}^{+}|0\rangle e^{iE_0{}^Nt}, & t > 0; \\ i\sum_n \langle 0|c_{\mathbf{k}}^{+}e^{iHt}|n\rangle\langle n|c_{\mathbf{k}}|0\rangle e^{-iE_0{}^Nt}, & t < 0. \end{cases} \tag{36}$$

For $t > 0$ the states n are excited states of a system of $(N + 1)$ particles; for $t < 0$ the states n refer to excited states of a system of $(N - 1)$ particles. We may arrange (36) as

$$G(\mathbf{k}t) = \begin{cases} -i\sum_n |\langle n|c_{\mathbf{k}}^{+}|0\rangle|^2 e^{-i(E_n{}^{N+1}-E_0{}^N)t}, & t > 0; \\ i\sum_n |\langle n|c_{\mathbf{k}}|0\rangle|^2 e^{i(E_n{}^{N-1}-E_0{}^N)t}, & t < 0. \end{cases} \tag{37}$$

The exponents in (37) may be rearranged:

$$\begin{aligned} E_n^{N+1} - E_0^N &= (E_n^{N+1} - E_0^{N+1}) + (E_0^{N+1} - E_0^N) \cong \omega_n + \mu; \\ E_n^{N-1} - E_0^N &= (E_n^{N-1} - E_0^{N-1}) + (E_0^{N-1} - E_0^N) \cong \omega_n - \mu, \end{aligned} \tag{38}$$

where

$$\omega_n \cong E_n^{N\pm 1} - E_0^{N\pm 1} \tag{39}$$

is the excitation energy referred to the ground state of the system of $(N \pm 1)$ particles, and

$$\mu = \frac{\partial E}{\partial N} \cong E_0^{N+1} - E_0^N \cong E_0^N - E_0^{N-1} \tag{40}$$

is the chemical potential. If the system is large ($N \gg 1$), we do not need to add superscripts to ω_n or μ to distinguish $N \pm 1$ from N. Then (37) becomes

$$G(\mathbf{k}t) = \begin{cases} -i \sum_n |\langle n|c_{\mathbf{k}}^+|0\rangle|^2 e^{-i(\omega_n+\mu)t}, & t > 0; \\ i \sum_n |\langle n|c_{\mathbf{k}}|0\rangle|^2 e^{i(\omega_n-\mu)t}, & t < 0. \end{cases} \tag{41}$$

SPECTRAL DENSITY AND THE LEHMANN REPRESENTATION

We introduced the spectral density functions defined by

$$\rho^+(\mathbf{k}\omega) \equiv \sum_n |\langle n|c_{\mathbf{k}}^+|0\rangle|^2 \delta(\omega - \omega_n); \tag{42}$$

$$\rho^-(\mathbf{k}\omega) \equiv \sum_n |\langle n|c_{\mathbf{k}}|0\rangle|^2 \delta(\omega - \omega_n). \tag{43}$$

We also encounter the notation $A(\mathbf{k}\omega) \equiv \rho^+(\mathbf{k}\omega)$; $B(\mathbf{k}\omega) \equiv \rho^-(\mathbf{k}\omega)$. With our notation (41) becomes

$$G(kt) = \begin{cases} -i \int_0^\infty d\omega\, \rho^+(\mathbf{k}\omega) e^{-i(\omega+\mu)t}, & t > 0; \\ i \int_0^\infty d\omega\, \rho^-(\mathbf{k}\omega) e^{i(\omega-\mu)t}, & t < 0. \end{cases} \tag{44}$$

The integral over $d\omega$ need include only positive ω because ω_n is always positive.

For noninteracting fermions the spectral densities reduce to single delta functions:

$$\rho^+(\mathbf{k}\omega) = (1 - n_{\mathbf{k}})\delta(\omega - \omega_{\mathbf{k}} + \mu); \tag{45}$$

$$\rho^-(\mathbf{k}\omega) = n_{\mathbf{k}}\, \delta(\omega - \mu + \omega_{\mathbf{k}}), \tag{46}$$

where $n_{\mathbf{k}} = 1,\ 0$ is the ground-state occupancy and $\omega_{\mathbf{k}} = k^2/2m$. Then (44) reduces to

$$G_0(\mathbf{k}t) = \begin{cases} -ie^{-i\omega_{\mathbf{k}}t}(1 - n_{\mathbf{k}}), & t > 0; \\ ie^{-i\omega_{\mathbf{k}}t} n_{\mathbf{k}}, & t < 0, \end{cases} \tag{47}$$

in agreement with (24).

We now exhibit an important result known as the Lehmann representation:

$$(48)\quad G(\mathbf{k}\omega) = \lim_{s\to+0} \int_0^\infty d\omega' \left[\frac{\rho^+(\mathbf{k}\omega')}{(\omega-\mu)-\omega'+is} + \frac{\rho^-(\mathbf{k}\omega')}{(\omega-\mu)+\omega'-is} \right].$$

We verify this result by taking the transform of (44):

$$(49)\quad G(\mathbf{k}\omega) = \int_{-\infty}^{\infty} dt\, e^{i\omega t} G(\mathbf{k}t)$$
$$= \lim_{s\to+0} \left[-i \int_0^\infty dt\, e^{i(\omega+is)t} \int_0^\infty d\omega'\, \rho^+(\mathbf{k}\omega') e^{-i(\omega'+\mu)t} + i \int_{-\infty}^0 dt\, e^{i(\omega-is)t} \int_0^\infty d\omega'\, \rho^-(\mathbf{k}\omega') e^{i(\omega'-\mu)t} \right].$$

The first integral involves

$$(50)\qquad -i \int_0^\infty dt\, e^{i(\omega-\omega'-\mu+is)t} = \frac{1}{(\omega-\mu)-\omega'+is};$$

the second integral involves

$$(51)\qquad i \int_{-\infty}^0 dt\, e^{i(\omega+\omega'-\mu-is)t} = \frac{1}{(\omega-\mu)+\omega'-is}.$$

Combining (50) and (51) with (49), we obtain the result (48).

DISPERSION RELATIONS

We see from their definitions (42) and (43) that ρ^+ and ρ^- are real. Using

$$(52)\qquad \lim_{s\to+0} \frac{1}{\varepsilon \pm is} = \frac{\mathcal{P}}{\varepsilon} \mp i\pi\delta(\varepsilon)$$

in the Lehmann representation, we can separate the real and imaginary parts of $G(\mathbf{k}\omega)$. Thus

$$(53)\quad \mathcal{I}\{G(\mathbf{k}\omega)\} = -\pi \int_0^\infty d\omega' \left[\rho^+(\mathbf{k}\omega')\delta(\omega-\mu-\omega') - \rho^-(\mathbf{k}\omega')\delta(\omega-\mu+\omega')\right]$$
$$= \begin{cases} -\pi\rho^+(\mathbf{k},\omega-\mu), & \omega > \mu; \\ \pi\rho^-(\mathbf{k},\mu-\omega), & \omega < \mu. \end{cases}$$

The separation occurs because the integral is only over positive values of ω'. We note from their definitions that ρ^+, ρ^- are non-negative; thus $\mathscr{I}\{G\}$ changes sign at $\omega = \mu$.

We now calculate the real part of $G(\mathbf{k}\omega)$ and express the result in terms of the imaginary part. The result is the analog of the Kramers-Kronig relation. Using (52),

$$(54) \quad \mathscr{R}\{G(\mathbf{k}\omega)\} = \mathscr{P}\int_0^\infty d\omega' \left[\frac{\rho^+(\mathbf{k}\omega')}{(\omega-\mu)-\omega'} + \frac{\rho^-(\mathbf{k}\omega')}{(\omega-\mu)+\omega'}\right];$$

with (53),

$$(55) \quad \mathscr{R}\{G(\mathbf{k}\omega)\} = \frac{\mathscr{P}}{\pi}\int_0^\infty d\omega' \left[-\frac{\mathscr{I}\{G(\mathbf{k},\omega'+\mu)\}}{(\omega-\mu)-\omega'} + \frac{\mathscr{I}\{G(\mathbf{k},\mu-\omega')\}}{(\omega-\mu)+\omega'}\right].$$

We make the substitutions $\omega'' = \omega' + \mu$ and $\omega'' = \mu - \omega'$ in the appropriate parts of the integrand; thus

$$(56) \quad \mathscr{R}\{G(\mathbf{k}\omega)\} = -\frac{\mathscr{P}}{\pi}\int_\mu^\infty d\omega'' \frac{\mathscr{I}\{G(\mathbf{k}\omega'')\}}{\omega-\omega''} + \frac{\mathscr{P}}{\pi}\int_\mu^{-\infty} (-d\omega'') \frac{\mathscr{I}\{G(\mathbf{k}\omega'')\}}{\omega-\omega''},$$

$$(57) \quad \boxed{\mathscr{R}\{G(\mathbf{k}\omega)\} = \frac{\mathscr{P}}{\pi}\left[\int_\mu^\infty - \int_{-\infty}^\mu\right] d\omega'' \frac{\mathscr{I}\{G(\mathbf{k}\omega'')\}}{\omega''-\omega}.}$$

GROUND-STATE ENERGY

If the system has only two-particle interactions, the ground-state energy is determined by the one-particle Green's function.

Proof: We note from (34) that the expectation value $n_\mathbf{k}$ of the occupancy of the state $\mathbf{k}$ is given by

$$(58) \quad n_\mathbf{k} = \langle 0|c_\mathbf{k}^+ c_\mathbf{k}|0\rangle = -iG(\mathbf{k},-0).$$

Further, using (22),

$$(59) \quad n_\mathbf{k} = \frac{1}{2\pi i}\int_c d\omega\, G(\mathbf{k}\omega),$$

where c is a contour consisting of the real axis and a semicircle at infinity in the upper half-plane, because $t = -0$.

We write $H = H_0 + H_1$, with

$$(60) \quad H_0 = \sum \omega_\mathbf{k} c_\mathbf{k}^+ c_\mathbf{k};$$

$$(61) \quad H_1 = \sum V(\mathbf{k}_1, \cdots, \mathbf{k}_4) c_{\mathbf{k}_1}^+ c_{\mathbf{k}_2}^+ c_{\mathbf{k}_3} c_{\mathbf{k}_4}.$$

Now we see readily that

$$\sum_{\mathbf{k}} c_{\mathbf{k}}^{+}[H_0,c_{\mathbf{k}}] = -H_0 \tag{62}$$

and

$$\sum_{\mathbf{k}} c_{\mathbf{k}}^{+}[H_1,c_{\mathbf{k}}] = -2H_1, \tag{63}$$

so that

$$\sum_{\mathbf{k}} \langle 0|c_{\mathbf{k}}^{+}[H,c_{\mathbf{k}}]|0\rangle = -\langle 0|H_0 + 2H_1|0\rangle. \tag{64}$$

The ground-state energy E_0 may be written

$$E_0 = \langle 0|H_0 + H_1|0\rangle = \tfrac{1}{2}\sum_{\mathbf{k}} (\omega_{\mathbf{k}} - \langle 0|c_{\mathbf{k}}^{+}[H,c_{\mathbf{k}}]|0\rangle). \tag{65}$$

We note now that, from (42),

$$\langle 0|c_{\mathbf{k}}^{+}[H,c_{\mathbf{k}}]|0\rangle = \int_0^{\infty} d\omega\, \rho^{-}(\mathbf{k}\omega)(\omega - \mu), \tag{66}$$

and, from (44) and (22),

$$\int_0^{\infty} d\omega\, \rho^{-}(\mathbf{k}\omega)(\omega - \mu) = -\left[\frac{dG\,(\mathbf{k}t)}{dt}\right]_{t=-0} = \frac{i}{2\pi}\int_c d\omega\, \omega G(\mathbf{k}\omega). \tag{67}$$

Thus (65) may be written as

$$E_0 = \frac{1}{4\pi i}\sum_{\mathbf{k}} \int_c d\omega\, (\omega_{\mathbf{k}} + \omega)G(\mathbf{k}\omega). \qquad \text{Q.E.D.} \tag{68}$$

This result is exact.

THERMAL AVERAGES

The Green's function method has directed attention to several neat tricks in taking ensemble averages. As a simple example, consider the following exact and concise derivation of the boson and fermion distribution laws. We calculate the thermal average occupancy

$$n = \langle a^{+}a\rangle = \frac{\operatorname{Tr} e^{-\beta\hat{H}}a^{+}a}{\operatorname{Tr} e^{-\beta\hat{H}}}, \tag{69}$$

where $\hat{H} = H - \mu\hat{N}$; $\beta = 1/k_BT$, and a^{+}, a satisfy the commutation relation

$$aa^{+} - \eta a^{+}a = 1, \tag{70}$$

with $\eta = 1$ for bosons and -1 for fermions. Now by the invariance of the trace under cyclic permutations,

$$\text{Tr } e^{-\beta\hat{H}}a^+a = \text{Tr } ae^{-\beta\hat{H}}a^+ = \text{Tr } e^{-\beta\hat{H}}ae^{-\beta\hat{H}}a^+e^{\beta\hat{H}}. \tag{71}$$

But on any eigenstate Φ of the system of noninteracting particles

$$e^{-\beta\hat{H}}a^+e^{\beta\hat{H}}\Phi = e^{-\beta(\omega-\mu)}a^+\Phi, \tag{72}$$

so that

$$\begin{aligned} \text{Tr } e^{-\beta\hat{H}}a^+a &= e^{-\beta(\omega-\mu)} \text{ Tr } e^{-\beta\hat{H}}aa^+ \\ &= e^{-\beta(\omega-\mu)} \text{ Tr } e^{-\beta\hat{H}}(1 + \eta a^+a). \end{aligned} \tag{73}$$

Thus

$$n = e^{-\beta(\omega-\mu)}(1 + \eta n), \tag{74}$$

or

$$n = \frac{1}{e^{\beta(\omega-\mu)} - \eta}, \tag{75}$$

which is the standard result.

In the same way it is a trivial matter to show that the average over a canonical ensemble of two arbitrary operators A and B satisfies the relation

$$\langle A(t)B(0)\rangle = \langle B(0)A(t + i\beta)\rangle. \tag{76}$$

It is instructive to rederive the dispersion relation (57) now with averages taken over a grand canonical ensemble, with Ω the grand potential. We first note that

$$|\langle m|c_{\mathbf{k}}^+|n\rangle|^2 = |\langle n|c_{\mathbf{k}}|m\rangle|^2; \tag{77}$$

then our previous result (37) may be written, with the introduction of the statistical operator

$$\rho = e^{\beta(\Omega+\mu\hat{N}-H)}, \tag{78}$$

as, with $\omega_{nm} = E_n - E_m$,

$$G(\mathbf{k}t) = \begin{cases} -i \sum\limits_{nm} e^{\beta(\Omega+\mu N_n-E_n)}|\langle n|c_{\mathbf{k}}|m\rangle|^2 e^{i\omega_{nm}t}, & t > 0; \\ i \sum\limits_{nm} e^{\beta(\Omega+\mu N_n-E_n)}|\langle m|c_{\mathbf{k}}|n\rangle|^2 e^{i\omega_{mn}t}, & t < 0. \end{cases} \tag{79}$$

Now in the result for $t < 0$ we interchange the indices n and m; we obtain

$$G(\mathbf{k}t) = i \sum_{nm} e^{\beta(\Omega+\mu N_m-E_m)}|\langle n|c_{\mathbf{k}}|m\rangle|^2 e^{i\omega_{nm}t}, \qquad t < 0. \tag{80}$$

Because $N_n = N_m - 1$, this may be written as

$$(81)\quad G(\mathbf{k}t) = i \sum_{nm} e^{\beta(\Omega+\mu N_n - E_n)} e^{\beta(\omega_{nm}+\mu)} |\langle n|c_\mathbf{k}|m\rangle|^2 e^{i\omega_{nm}t}, \qquad t < 0.$$

We now take the time fourier transform, with separate regions of integration from $-\infty$ to 0 and from 0 to ∞. We employ the identities

$$(82)\qquad \lim_{s\to+0} \int_0^\infty dx\, e^{i(\alpha+is)x} = \pi\delta(\alpha) + i\frac{\mathcal{P}}{\alpha};$$

$$(83)\qquad \lim_{s\to+0} \int_{-\infty}^0 dx\, e^{i(\alpha-is)x} = \pi\delta(\alpha) - i\frac{\mathcal{P}}{\alpha}.$$

Thus

$$(84)\qquad G(\mathbf{k}\omega) = -\sum_{nm} e^{\beta(\Omega+\mu N_n - E_n)} |\langle n|c_\mathbf{k}|m\rangle|^2 \cdot$$

$$\left[i\pi\delta(\omega - \omega_{mn})(1 - e^{\beta(\mu-\omega_{mn})}) + \frac{\mathcal{P}}{\omega_{mn} - \omega}(1 + e^{\beta(\mu-\omega_{mn})}) \right].$$

We may separate the real and imaginary parts:

$$(85)\quad \mathcal{R}\{G(\mathbf{k}\omega)\} = -\sum_{nm} e^{\beta(\Omega+\mu N_n - E_n)} |\langle n|c_\mathbf{k}|m\rangle|^2 \frac{\mathcal{P}}{\omega_{mn} - \omega}(1 + e^{\beta(\mu-\omega_{mn})})$$

$$(86)\quad \mathcal{I}\{G(\mathbf{k}\omega)\} = -\pi \sum_{nm} e^{\beta(\Omega+\mu N_n - E_n)} |\langle n|c_\mathbf{k}|m\rangle|^2 \delta(\omega - \omega_{mn})(1 - e^{\beta(\mu-\omega_{mn})}).$$

Note now that

$$(87)\qquad \frac{1 + e^{\beta(\mu-\omega_{mn})}}{1 - e^{\beta(\mu-\omega_{mn})}} = \coth \tfrac{1}{2}\beta(\omega_{mn} - \mu),$$

and form

$$(88)\qquad \frac{\mathcal{P}}{\pi} \int_{-\infty}^\infty d\omega' \coth \tfrac{1}{2}\beta(\omega' - \mu) \frac{\mathcal{I}\{G(\mathbf{k}\omega')\}}{\omega' - \omega};$$

this is equal, by (86), to

$$(89)\qquad -\mathcal{P} \sum_{nm} e^{\beta(\Omega+\mu N_n - E_n)} (1 + e^{\beta(\mu-\omega_{mn})}) |\langle n|c_\mathbf{k}|m\rangle|^2 \frac{1}{\omega_{mn} - \omega}.$$

Thus we have the dispersion relation

$$(90)\qquad \mathcal{R}\{G(\mathbf{k}\omega)\} = \frac{\mathcal{P}}{\pi} \int_{-\infty}^\infty d\omega' \coth \tfrac{1}{2}\beta(\omega' - \mu) \frac{\mathcal{I}\{G(\mathbf{k}\omega')\}}{\omega' - \omega}.$$

This agrees in the limit $T \to 0$ with (57).

EQUATION OF MOTION

From (5.38), (5.42), and (5.43) we have the exact equations of motion

$$(91)\quad \left(i\frac{\partial}{\partial t}-\frac{p^2}{2m}\right)\Psi(\mathbf{x}t)=\int d^3y\,V(\mathbf{y}-\mathbf{x})\Psi^+(\mathbf{y}t)\Psi(\mathbf{y}t)\Psi(\mathbf{x}t);$$

$$(92)\quad \left(-i\frac{\partial}{\partial t}-\frac{p^2}{2m}\right)\Psi^+(\mathbf{x}t)$$
$$=\Psi^+(\mathbf{x}t)\int d^3y\,V(\mathbf{y}-\mathbf{x})\Psi^+(\mathbf{x}t)\Psi^+(\mathbf{y}t)\Psi(\mathbf{y}t).$$

It is revealing to study the meaning of the Hartree-Fock approximation in terms of Green's functions. From (91) we form

$$(93)\quad -i\left\langle T\left(i\frac{\partial}{\partial t'}-\frac{(p')^2}{2m}\right)\Psi(\mathbf{x}'t')\Psi^+(\mathbf{x}t)\right\rangle$$
$$=-i\int d^3y\,V(\mathbf{y}-\mathbf{x}')\langle T(\Psi^+(\mathbf{y}t')\Psi(\mathbf{y}t')\Psi(\mathbf{x}'t')\Psi^+(\mathbf{x}t))\rangle$$
$$=\int d^3y\,V(\mathbf{y}-\mathbf{x}')K(\mathbf{y}t';\mathbf{x}'t';\mathbf{y}t'_+;\mathbf{x}t),$$

with the two-particle Green's function K defined by (12). Here t'_+ is infinitesimally larger than t', and is used to maintain the order of the factors.

Now

$$(94)\quad \frac{\partial}{\partial t'}\langle T(\Psi(\mathbf{x}'t')\Psi^+(\mathbf{x}t))\rangle-\left\langle T\left(\frac{\partial}{\partial t'}\Psi(\mathbf{x}'t')\Psi^+(\mathbf{x}t)\right)\right\rangle$$
$$=\delta(t'-t)\delta(\mathbf{x}'-\mathbf{x}),$$

where we have used the equal time commutation relations. Thus we may rewrite (93) as

$$(95)\quad \left(i\frac{\partial}{\partial t'}-\frac{(p')^2}{2m}\right)G(\mathbf{x}'t';\mathbf{x}t)=\delta(t'-t)\delta(\mathbf{x}'-\mathbf{x})$$
$$-i\int d^3y\,V(\mathbf{y}-\mathbf{x}')K(\mathbf{y}t';\mathbf{x}'t';\mathbf{y}t'_+;\mathbf{x}t).$$

This is an exact equation which relates the single-particle Green's function to the two-particle Green's function.

In the Hartree approximation we solve (95) under the assumption that the two added particles in the two-particle Green's function propagate through the system entirely independently of each other, so that the basic assumption of the Hartree approximation is

$$K(1234) \cong G(13)G(24), \tag{96}$$

provided the spin indices $s_1 = s_3$ and $s_2 = s_4$. But this does not take into account the identity of the particles; we cannot in principle distinguish processes in which the particle added at 4 appears at 2 from processes in which it appears at 1. Thus in the Hartree-Fock approximation the assumption is made that, for fermions,

$$K(1234) \cong G(13)G(24) - G(14)G(23). \tag{97}$$

The relative signs on the right-hand side are fixed by the property $K(1234) = -K(2134)$ for fermions.

SUPERCONDUCTIVITY

It is instructive to develop the BCS theory of superconductivity using the Green's function approach as developed by L. P. Gorkov [*Soviet Physics—JETP* **34,** 505 (1958); reprinted in Pines]. The treatment is closely similar to the equation-of-motion method given in Chapter 8.

The effective hamiltonian (8.31) may be written as

$$H = \int d^3x\, \Psi_\alpha^+(\mathbf{x}t)\, \frac{p^2}{2m}\, \Psi_\alpha(\mathbf{x}t) - \tfrac{1}{2}V \int d^3x\, d^3y\, \Psi_\alpha^+(\mathbf{x}t)\Psi_\beta^+(\mathbf{y}t)\,\delta(\mathbf{x}-\mathbf{y})\Psi_\beta(\mathbf{y}t)\Psi_\alpha(\mathbf{x}t). \tag{98}$$

Here α, β are spin indices—it is important to display them explicitly; repeated indices are to be summed. We observe that an interaction independent of $\mathbf{k}$ may be represented by a delta function potential. We adopt, however, the convention that V is zero except in an energy shell of thickness $2\omega_D$, centered about the fermi surface, where V is positive for an attractive interaction. Note that the potential term in (98) automatically vanishes for parallel spins ($\alpha = \beta$).

The one-particle Green's function is defined as

$$G_{\alpha\beta}(\mathbf{x}t) = -i\langle T(\Psi_\alpha(\mathbf{x}t)\Psi_\beta^+(00))\rangle. \tag{99}$$

The equations of motion are obtained in the familiar way:

$$\left(i\frac{\partial}{\partial t} - \frac{p^2}{2m}\right)\Psi_\alpha(x) + V\Psi_\beta^+(x)\Psi_\beta(x)\Psi_\alpha(x) = 0; \tag{100}$$

$$\left(-i\frac{\partial}{\partial t} - \frac{p^2}{2m}\right)\Psi_\alpha^+(x) + V\Psi_\alpha^+(x)\Psi_\beta^+(x)\Psi_\beta(x) = 0, \tag{101}$$

We have used the notation $x \equiv \mathbf{x}t$. On operating from the left with $\Psi_\beta^+(x')$ and forming the thermal average, we have

$$(102) \quad \left(i\frac{\partial}{\partial t} - \frac{p^2}{2m}\right)G_{\alpha\beta}(x,x') - iV\langle T(\Psi_\gamma^+(x)\Psi_\gamma(x)\Psi_\alpha(x)\Psi_\beta^+(x'))\rangle = \delta(x - x')\delta_{\alpha\beta}.$$

Now in the Hartree-Fock approximation we made the following assumption for the two-particle Green's function:

$$(103) \quad \langle T(\Psi_\alpha(x_1)\Psi_\beta(x_2)\Psi_\gamma^+(x_3)\Psi_\delta^+(x_4))\rangle = \langle T(\Psi_\alpha(x_1)\Psi_\delta^+(x_4))\rangle\langle T(\Psi_\beta(x_2)\Psi_\gamma^+(x_3))\rangle - \langle T(\Psi_\alpha(x_1)\Psi_\gamma^+(x_3))\rangle\langle T(\Psi_\beta(x_2)\Psi_\delta^+(x_4))\rangle.$$

But the ground state of the superconducting system is characterized by bound pairs of electrons. The number of such pairs is a variable of the problem, so that we should add to the right-hand side of (103) the term

$$(104) \quad \langle N|T(\Psi_\alpha(x_1)\Psi_\beta(x_2))|N + 2\rangle\langle N + 2|T(\Psi_\gamma^+(x_3)\Psi_\delta^+(x_4))|N\rangle.$$

This is a very natural addition to (103). The states $|N\rangle$, $|N + 2\rangle$ are corresponding states for N and $N + 2$ particles. If $|N\rangle$ is in the ground state, so is $|N + 2\rangle$.

We may write the factors in (104) in the form

$$(105) \quad \langle N|T(\Psi_\alpha(x)\Psi_\beta(x'))|N + 2\rangle = e^{-2i\mu t}F_{\alpha\beta}(x - x'),$$

$$(106) \quad \langle N + 2|T(\Psi_\alpha^+(x)\Psi_\beta^+(x'))|N\rangle = e^{2i\mu t}F_{\alpha\beta}^+(x - x').$$

We have assumed galilean invariance. Here μ is the chemical potential and enters the problem because, for an arbitrary operator $O(t)$,

$$(107) \quad i\frac{\partial}{\partial t}\langle N|O(t)|N + 2\rangle = \langle N|[O,H]|N + 2\rangle = (E_{N+2} - E_N)\langle N|O(t)|N + 2\rangle \cong 2\mu\langle N|O(t)|N + 2\rangle,$$

using the definition $\mu = \partial E/\partial N$.

The following equation follows directly from the equation of motion (102) and the definitions of F and G:

$$(108) \quad \left(i\frac{\partial}{\partial t} - \frac{p^2}{2m}\right)G_{\alpha\beta}(x - x') - iVF_{\alpha\gamma}(+0)F_{\gamma\beta}^+(x - x') = \delta(x - x')\delta_{\alpha\beta}.$$

This follows because the potential term in (102) is, by (104) with the neglect of the ordinary Hartree-Fock terms which may be included in μ,

$$\text{(109)}\quad \langle T(\Psi_\gamma^+(x)\Psi_\gamma(x)\Psi_\alpha(x)\Psi_\beta^+(x'))\rangle \cong -\langle\Psi_\alpha(x)\Psi_\gamma\rangle\langle T(\Psi_\gamma^+(x)\Psi_\beta^+(x'))\rangle = -F_{\alpha\gamma}(+0)F_{\gamma\alpha}^+(x-x');$$

here

$$\text{(110)}\quad F_{\alpha\gamma}(+0) \equiv \lim_{\substack{\mathbf{x}\to\mathbf{x}'\\ t\to t'+0}} F_{\alpha\gamma}(x-x') = e^{2i\mu t}\langle\Psi_\alpha(x)\Psi_\gamma(x)\rangle.$$

On inserting (109) into (102), we obtain (108).

Note that $F_{\alpha\gamma}(+0) = -F_{\gamma\alpha}(+0)$; because of spin pairing $\alpha \neq \gamma$. We may write the matrix

$$\text{(111)}\quad \hat{F}(+0) = J\begin{pmatrix} 0 & 1 \\ -1 & 0 \end{pmatrix} = J\hat{A},$$

where J is a c number; further, from (106),

$$\text{(112)}\quad (F_{\alpha\beta}^+(\mathbf{x}-\mathbf{x}',0))^* = -F_{\alpha\beta}(\mathbf{x}-\mathbf{x}',0),$$

so that we may write

$$\text{(113)}\quad \hat{F}^+(x-x') = \hat{A}F^+(x-x'); \qquad \hat{F} = -\hat{A}F(x-x');$$
$$(F^+(\mathbf{x}-\mathbf{x}',0))^* = F(\mathbf{x}-\mathbf{x}',0).$$

We record that $\hat{A}^2 = -\hat{I}$, where $\hat{I}$ is the unit matrix.

We may write (108) in matrix form:

$$\text{(114)}\quad \left(i\frac{\partial}{\partial t} - \frac{p^2}{2m}\right)\hat{G}(x-x') - iV\hat{F}(+0)\hat{F}^+(x-x') = \delta(x-x')\hat{I},$$

but it is readily verified that the off-diagonal components of $\hat{G}$ are zero; thus with

$$\text{(115)}\quad \hat{G}_{\alpha\beta}(x-x') \equiv \delta_{\alpha\beta}G(x-x'),$$

we have

$$\text{(116)}\quad \left(i\frac{\partial}{\partial t} - \frac{p^2}{2m}\right)G(x-x') + iVF(+0)F^+(x-x') = \delta(x-x').$$

By operating on (101) from the right with Ψ_β^+, we obtain the equation

$$\text{(117)}\quad \left(-i\frac{\partial}{\partial t} - \frac{p^2}{2m} + 2\mu\right)\hat{F}^+(x-x') + iV\hat{F}^+(+0)\hat{G}(x-x') = 0.$$

which reduces to

$$(118)\quad \left(-i\frac{\partial}{\partial t} - \frac{p^2}{2m} + 2\mu\right)F^+(x - x') + iVF^+(+0)G(x - x') = 0.$$

We analyze (116) and (118) in fourier components $G(\mathbf{k}\omega)$ and $F^+(\mathbf{k}\omega)$:

$$(119)\quad \left(\omega - \frac{k^2}{2m}\right)G(\mathbf{k}\omega) + iVF(+0)F^+(\mathbf{k}\omega) = 1;$$

$$(120)\quad \left(\omega + \frac{k^2}{2m} - 2\mu\right)F^+(\mathbf{k}\omega) - iVF^+(+0)G(\mathbf{k}\omega) = 0.$$

Write $\omega' = \omega - \mu$; $\varepsilon_\mathbf{k} = (k^2/2m) - \mu$; $\Delta^2 = V^2F(+0)F^+(+0)$; then (119) and (120) have the solutions

$$(121)\quad G(\mathbf{k}\omega) = \frac{\omega' + \varepsilon_\mathbf{k}}{\omega'^2 - \lambda_\mathbf{k}{}^2};\qquad F^+(\mathbf{k}\omega) = i\,\frac{VF^+(+0)}{\omega'^2 - \lambda_\mathbf{k}{}^2},$$

with

$$(122)\quad \lambda_\mathbf{k}{}^2 = \varepsilon_\mathbf{k}{}^2 + \Delta^2,$$

which is the quasiparticle energy of (8.78).

The solution (121) for $F^+(\mathbf{k}\omega)$ must be consistent with the value of $F^+(+0)$. To investigate this we form

$$(123)\quad F^+(x) = \frac{1}{(2\pi)^4}\int d^3k\, d\omega e^{i(\mathbf{k}\cdot\mathbf{x}-\omega t)}F^+(\mathbf{k}\omega),$$

or, for $\mathbf{x} = 0$,

$$(124)\quad F^+(0t) = \frac{iVF^+(+0)}{(2\pi)^4}\int d^3k\, d\omega\, \frac{e^{-i\omega t}}{\omega'^2 - \lambda_\mathbf{k}{}^2}.$$

The integral over $d\omega$ is, for $t \to +0$,

$$(125)\quad \lim_{s\to+0}\frac{1}{2\lambda_\mathbf{k}}\int_{-\infty}^{\infty} d\omega\, e^{-i\omega t}\left(\frac{1}{\omega' - \lambda_\mathbf{k} + is} - \frac{1}{\omega' + \lambda_\mathbf{k} - is}\right) = -\frac{i\pi}{\lambda_\mathbf{k}},$$

where we have replaced $\lambda_\mathbf{k}$ by $\lim_{s\to+0}(\lambda_\mathbf{k} - is)$ to represent the effect of collisions. Thus

$$(126)\quad F^+(+0) = \frac{VF^+(+0)}{2(2\pi)^3}\int d^3k\,\frac{1}{\lambda_\mathbf{k}} = \tfrac{1}{2}VF^+(+0)\sum_\mathbf{k}\frac{1}{(\varepsilon_\mathbf{k} + \Delta^2)^{1/2}},$$

or

$$(127)\quad 1 = \tfrac{1}{2}V\sum_\mathbf{k}\frac{1}{(\varepsilon_\mathbf{k} + \Delta^2)^{1/2}}.$$

This is just the fundamental BCS equation, as in (8.53).

We may rewrite $G(\mathbf{k}\omega)$ from (121) as

$$G(\mathbf{k}\omega) = \frac{u_{\mathbf{k}}^2}{\omega' - \lambda_{\mathbf{k}} + is} + \frac{v_{\mathbf{k}}^2}{\omega' + \lambda_{\mathbf{k}} - is}, \tag{128}$$

where, as in (8.93) and (8.94),

$$u_{\mathbf{k}}^2 = \tfrac{1}{2}\left(1 + \frac{\varepsilon_{\mathbf{k}}}{\lambda_{\mathbf{k}}}\right); \qquad v_{\mathbf{k}}^2 = \tfrac{1}{2}\left(1 - \frac{\varepsilon_{\mathbf{k}}}{\lambda_{\mathbf{k}}}\right). \tag{129}$$

We observe in the equation

$$(\omega'^2 - \lambda_{\mathbf{k}}^2)F^+(\mathbf{k}\omega) = iVF^+(+0), \tag{130}$$

which leads to the second part of (121), we may add to $F^+(\mathbf{k}\omega)$ a term of the form

$$B(\mathbf{k})\delta(\omega'^2 - \lambda_{\mathbf{k}}^2),$$

where $B(\mathbf{k})$ is arbitrary. This is equivalent to adding an arbitrary imaginary part to $G(\mathbf{k}\omega)$; this part was determined by Gorkov by using the dispersion relation.

PERTURBATION EXPANSION FOR GREEN'S FUNCTIONS

At this point the theory becomes intricate except for students with a technical knowledge of quantum field theory. We shall only summarize the central results for fermion systems. We write $H = H_0 + V$, and choose H_0 so that the one-particle Green's function appropriate to H_0 can be found explicitly. The unitary operator $U(t,t')$ was defined in Chapter 1; if V is bilinear in the fermion operators Ψ, Ψ^+, then (1.56) may be written as

$$U(t,t') = \sum_n \frac{(-1)^n}{n!} \int_{t'}^{t} \cdots \int_{t'}^{t} dt_1 \cdots dt_n \, T\{V(t_1)V(t_2) \cdots V(t_n)\}, \tag{131}$$

where T is the Wick time-ordering operator. The S matrix is defined as

$$S \equiv U(\infty, -\infty). \tag{132}$$

Let Φ_0 denote the ground state of H_0, denoted by $|0\rangle$ in Chapter 1. We make the adiabatic hypothesis, so that $S\Phi_0$ can differ from Φ_0 only by a phase factor. The $V(t)$ in (131) are in the interaction representation.

The first result (which we do not derive here) is that the exact one-particle Green's function is given by

$$G(\mathbf{k}t) = -i\,\frac{\langle\Phi_0|T(c_{\mathbf{k}}(t)c_{\mathbf{k}}^+(0)S)|\Phi_0\rangle}{\langle\Phi_0|S|\Phi_0\rangle}, \tag{133}$$

where all quantities are in the interaction representation. It is useful to combine (131) and (133):

$$(134)\quad G(\mathbf{k}t) = -\frac{i}{\langle\Phi_0|S|\Phi_0\rangle}\sum_n^\infty \frac{(-1)^n}{n}\int_{-\infty}^{\infty}\cdots\int_{-\infty}^{\infty} dt_1\cdots dt_n \langle\Phi_0|T(c_\mathbf{k}(t)c_\mathbf{k}^+(0)V(t_1)\cdots V(t_n))|\Phi_0\rangle.$$

The evaluation of the nth order term in the perturbation expansion of G is usually accomplished with the help of Wick's theorem which gives a systematic reduction procedure to products of expectation values involving one c and one c^+. A full discussion is given in the book by Abrikosov et al. cited at the beginning of this chapter.

PROBLEMS

1. Find the expressions for $G(\mathbf{x}t)$ for a fermi gas in one dimension for the ground state with all one-electron states filled up to k_F and zero for $k > k_F$. Consider both $t > 0$ and $t < 0$. The result may be expressed in terms of the fresnel integrals.

2. Show that the particle current density in the absence of a magnetic field may be written as, on taking account of the two spin orientations,

$$(135)\quad \mathbf{j}(\mathbf{x}t) = -\frac{1}{m}\lim_{\substack{\mathbf{x}\to\mathbf{x}'\\ t\to t'+0}}(\text{grad}_{\mathbf{x}'} - \text{grad}_\mathbf{x})G(\mathbf{x}'t';\mathbf{x}t).$$

Note that the particle density is

$$(136)\quad n(\mathbf{x}t) = -i\lim_{\substack{\mathbf{x}\to\mathbf{x}'\\ t\to t'+0}} 2G(\mathbf{x}'t';\mathbf{x}t).$$

Appendix

PERTURBATION THEORY AND THE ELECTRON GAS

In Chapter 6 we saw that the second-order coulomb energy (6.4) between two electrons diverged at low values of the momentum transfer $\mathbf{q}$. In this appendix we examine the situation in some detail and discuss the Brueckner technique for summing the perturbation theory series in the high density limit. We first show that the divergence is weakened, but not removed, on summing the second-order energy $\varepsilon_{ij}^{(2)}$ over all pairs of electrons ij in the fermi sea. We sum only over unoccupied virtual states. We calculate here only the direct coulomb terms for electron pairs of antiparallel spin; for simplicity we shall omit further reference to the spin. The sum is, with a factor $\frac{1}{2}$ because of double counting,

$$(1)\quad E_2 = m \sum_{k_1,k_2<k_F} \sum_{k_3,k_4>k_F} \frac{\langle 12|V|34\rangle\langle 34|V|12\rangle}{k_1{}^2 + k_2{}^2 - k_3{}^2 - k_4{}^2}$$

$$= -m\left(\frac{4\pi e^2}{\Omega}\right)^2 \left(\frac{\Omega}{(2\pi)^3}\right)^3 \int d^3q \int d^3k_1 \int d^3k_2 \times \frac{1}{2q^4[q^2 + \mathbf{q}\cdot(\mathbf{k}_2 - \mathbf{k}_1)]}.$$

The limits on the integrals in (1) are $k_1,\ k_2 < k_F$; $|\mathbf{k}_1 - \mathbf{q}| > k_F$; $|\mathbf{k}_2 + \mathbf{q}| > k_F$. In applying perturbation theory to a many-body problem it is clear from the second-quantized formulation that occupied states are not to be counted among the intermediate states. Now let

$$(2)\qquad \mathbf{q}\cdot\mathbf{k}_1 = -qk_1\xi_1; \qquad \mathbf{q}\cdot\mathbf{k}_2 = qk_2\xi_2,$$

where the limits imply that ξ_1, ξ_2 are positive. We are concerned with the behavior of the integrand for small q, because we want to examine the convergence here. Now

$$(3)\qquad k_F{}^2 < k_1{}^2 + q^2 - 2\mathbf{k}_1\cdot\mathbf{q} = k_1{}^2 + q^2 + 2qk_1\xi_1,$$

so that for small q

$$2qk_1\xi_1 > (k_F - k_1)(k_F + k_1) \approx 2k_F(k_F - k_1), \tag{4}$$

or

$$k_F > k_1 > k_F - q\xi_1, \tag{5}$$

and similarly for ξ_2. We use these relations, valid for small q, to express the limits on the integrations over k_1 and k_2.

For a part of (1) evaluated at small q, recalling that ξ_1, ξ_2 are positive,

$$\int d^3k_1 \int d^3k_2 \frac{1}{q^4[q^2 + \mathbf{q} \cdot (\mathbf{k}_2 - \mathbf{k}_1)]} \tag{6}$$

$$= (2\pi)^2 \int_0^1 d\xi_1 \int_0^1 d\xi_2 \int_{k_F - q\xi_1}^{k_F} k_1^2\, dk_1 \int_{k_F - q\xi_2}^{k_F} k_2^2\, dk_2 \frac{1}{q^4(q^2 + qk_1\xi_1 + qk_2\xi_2)}$$

$$\cong \frac{(2\pi)^2}{q^4} k_F^4 \int_0^1 d\xi_1 \int_0^1 d\xi_2 \int_{k_F - q\xi_1}^{k_F} dk_1 \int_{k_F - q\xi_2}^{k_F} dk_2 \frac{1}{qk_F\xi_1 + qk_F\xi_2}$$

$$= \frac{(2\pi)^2}{q^3} k_F^3 \int_0^1 d\xi_1 \int_0^1 d\xi_2 \frac{\xi_1\xi_2}{\xi_1 + \xi_2}.$$

The integrals over ξ_1, ξ_2 are definite, so that the integral in (1) over d^3q for small q involves $\int q^{-1}\, dq$, which diverges logarithmically. Thus the perturbation calculation breaks down even in second order. We now show that if the perturbation calculation is carried out to infinite order, it becomes possible to sum an important class of contributions to obtain a nondivergent result.

BRUECKNER METHOD[1]

We need an abridged notation for the terms in Rayleigh-Schrödinger perturbation expansion. We write $H = H_0 + V$, where V is the coulomb interaction:

$$V = \tfrac{1}{2} \sum_{lm} V_{lm} \tag{7}$$

over all pairs of electrons treated as spinless. Let

$$\frac{1}{b} \equiv \frac{1 - P_0}{E_0 - H_0}, \tag{8}$$

[1] For full details see the review by K. A. Brueckner in *The many-body problem*, Wiley, New York, 1959.

where P_0 is the projection operator arranged as $1 - P_0$ to eliminate (from the summation over intermediate states) terms in the original state, as denoted here by the subscript zero.

We consider, to illustrate the notation, an unperturbed system containing only two electrons, in states $\mathbf{k}_1$, $\mathbf{k}_2$. The first-order perturbation energy is

$$\varepsilon^{(1)} = \langle V \rangle = \langle 12|V_{12}|12\rangle = \langle 12|V|12\rangle, \tag{9}$$

where, for convenience, we drop the electron indices from V. The second-order perturbation energy is

$$\begin{aligned}
\varepsilon^{(2)} &= \left\langle V \frac{1}{b} V \right\rangle \\
&= \sum_{\substack{1'2' \\ 1''2''}} \langle 12|V|1''2''\rangle \left\langle 1''2'' \left| \frac{1 - P_{12}}{E_{12} - H_0} \right| 1'2' \right\rangle \langle 1'2'|V|12\rangle \\
&= \sum_{1'2'}{}' \langle 12|V|1'2'\rangle \frac{1}{E_{12} - E_{1'2'}} \langle 1'2'|V|12\rangle,
\end{aligned} \tag{10}$$

as in (6.1). Here we have used primes to denote intermediate states in order to save numbers for use later with other electrons. The prime on the summation indicates, as before, that the state $\mathbf{k}_{1'} = \mathbf{k}_1$; $\mathbf{k}_{2'} = \mathbf{k}_2$ is to be omitted from the summation.

The third-order energy correction is given in standard texts:

$$\varepsilon^{(3)} = \left\langle V \frac{1}{b} V \frac{1}{b} V \right\rangle - \left\langle V \frac{1}{b^2} V \right\rangle \langle V \rangle, \tag{11}$$

where the second term on the right-hand side represents the effect of the normalization correction to the first-order wavefunction. The first term on the right-hand side may be written out as

$$\begin{aligned}
&\left\langle V \frac{1}{b} V \frac{1}{b} V \right\rangle \\
&= \sum_{\substack{34 \\ 56}}{}' \langle 12|V|56\rangle \frac{1}{E_{12} - E_{56}} \langle 56|V|34\rangle \frac{1}{E_{12} - E_{34}} \langle 34|V|12\rangle.
\end{aligned} \tag{12}$$

This term can couple more than two electrons, as we shall see.

First we consider the form of (12) when only *two* electrons $\mathbf{k}_1$, $\mathbf{k}_2$ are present. There are three matrix elements in each contribution to (12). For the coulomb interaction the value of each matrix element is described completely by the momentum transfer $\mathbf{q}$. In $\langle 1'2'|V|12\rangle$ the value of $\mathbf{q}$ is given by $\mathbf{k}_1 - \mathbf{k}_{1'}$ or by $\mathbf{k}_{2'} - \mathbf{k}_2$. In the most general

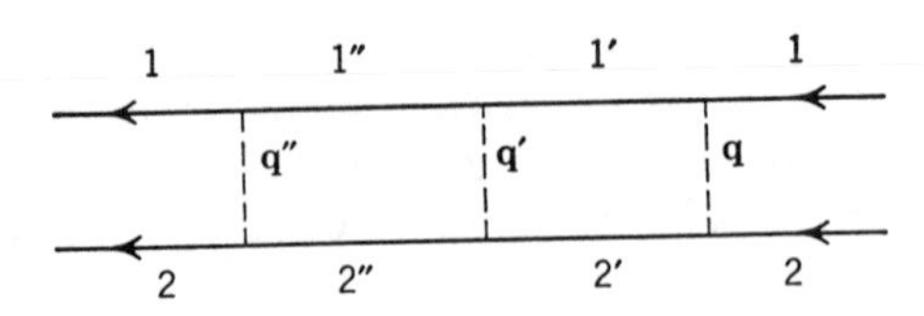

FIG. 1. Graph for a third-order energy correction involving two electrons.

term of (12) for two electrons the three matrix elements will each involve a different momentum transfer, say $\mathbf{q}$, $\mathbf{q}'$, $\mathbf{q}''$, but we must have zero total momentum transfer between the initial state $\mathbf{k}_1$ and the final state $\mathbf{k}_1$. In other words, we must have

$$(13)\quad (\mathbf{k}_1 - \mathbf{k}_{1'}) + (\mathbf{k}_{1'} - \mathbf{k}_{1''}) + (\mathbf{k}_{1''} - \mathbf{k}_1) = \mathbf{q} + \mathbf{q}' + \mathbf{q}'' = 0$$

in order to carry $|12\rangle$ back to $\langle 12|$ after the three scattering events contained in $\langle 12|V|1''2''\rangle\langle 1''2''|V|1'2'\rangle\langle 1'2'|V|12\rangle$.

We represent the structure of this contribution to the third-order energy correction by a diagram, as in Fig. 1. There are a number of ways of drawing graphs to illustrate perturbation theory: the diagrams used in this appendix are different from the Goldstone diagrams used in Chapter 6. For the present we show all electrons as solid lines directed on the graph from right to left with interactions following the order of the terms in (12) or similar perturbation expansions. The interactions are represented by dashes, labeled with the momentum transfer in the collision. Note that the solid lines on the graph are not shown as bent in the collision region—no attempt is made to simulate the scattering angles.

When more than two electrons are present, other types of graphs are possible. Consider the terms with the three electrons 1, 2, 3; then

$$(14)\qquad V_{12} + V_{13} + V_{23}$$

gives the three two-particle interactions. The development of $\left\langle V\frac{1}{b}V\frac{1}{b}V\right\rangle$ is now more complicated. The graph of Fig. 1 involves two electrons in the initial states 1 and 2. The graph in Fig. 2 involves three electrons in initial states 1, 2, 3. We label the intermediate states as 1′, 2′, 3. The term in the energy corresponding to the graph of Fig. 2 has the structure

$$(15)\quad \langle 13|V_{13}|1'3'\rangle\frac{1}{E_{13} - E_{1'3'}}\langle 3'2|V_{23}|32'\rangle\frac{1}{E_{12} - E_{1'2'}}\langle 1'2'|V_{12}|12\rangle.$$

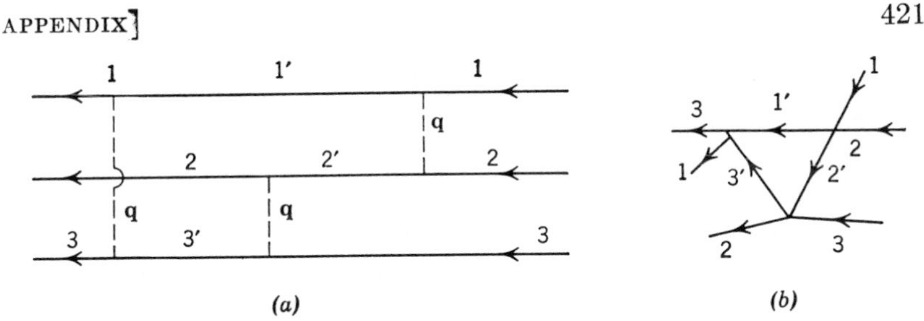

FIG. 2. (*a*) Graph for a third-order energy correction involving three electrons. This particular graph is called a ring graph; it is equivalent to the graph (*b*), drawn in a slightly different way to emphasize the ring structure. The momentum transfer at all vertices must be equal in a ring graph; otherwise all the electrons would not be carried back to their initial states.

This term is *linked* in the sense that the diagram may not be split into two parts with no interaction lines between the parts. This particular diagram is a special case of a linked graph known as a ring graph. *A ring graph is a graph in which one new particle enters and one old particle leaves at each vertex*, except at the initial and final vertices. In a ring graph the momentum changes are equal at all vertices.

The structure of perturbation theory requires that all terms in the energy be constructed to get particles back to their initial states. In third order we have, besides the terms corresponding to Figs. 1 and 2, terms such as

$$(16)\quad \langle 12|V_{12}|1'2'\rangle \frac{1}{E_{12} - E_{1'2'}} \langle 32'|V_{23}|32'\rangle \frac{1}{E_{12} - E_{1'2'}} \langle 1'2'|V_{12}|12\rangle,$$

as represented in Fig. 3. In the coulomb interaction there is no term for $\mathbf{q} = 0$, therefore $\langle 32'|V_{23}|32'\rangle = 0$, and this process does not contribute to the energy. For a general interaction this term does contribute; the scattering of $2'$ by 3 is described as *forward scattering by an unexcited particle,* if 3 lies within the fermi sea.

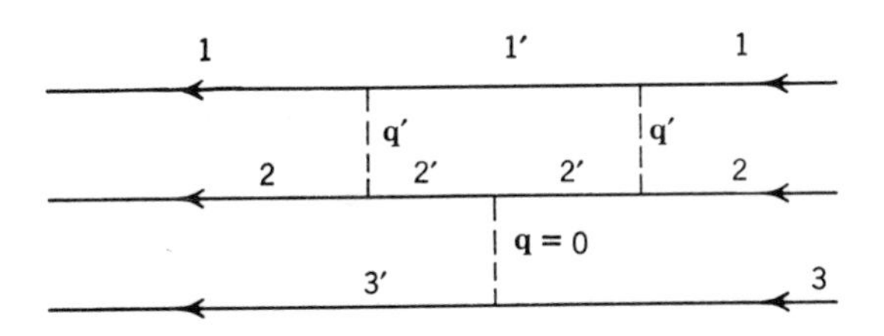

FIG. 3. Forward scattering by an unexcited particle.

In third order there are terms of the structure

$$\text{(17)}\quad \langle 12|V_{12}|1'2'\rangle \frac{1}{E_{12} - E_{1'2'}} \langle 34|V_{34}|34\rangle \frac{1}{E_{12} - E_{1'2'}} \langle 1'2'|V_{12}|12\rangle,$$

involving four electrons, as illustrated by Fig. 4. The graph is called an *unlinked* graph because it can be separated into two noninteracting parts. For the coulomb interaction $\langle 34|V_{34}|34\rangle$ vanishes, but the contribution (17) vanishes also for a more general reason: the term $-\left\langle V\frac{1}{b^2}V\right\rangle\langle V\rangle$ on the right-hand side of the expression (11) for the third-order energy exactly cancels the unlinked term (17), for we may write

$$\text{(18)}\quad -\left\langle V\frac{1}{b^2}V\right\rangle\langle V\rangle = -\langle 12|V_{12}|1'2'\rangle \frac{1}{(E_{12} - E_{1'2'})^2} \langle 1'2'|V_{12}|12\rangle\langle 34|V_{34}|23\rangle.$$

Goldstone has shown in general that *the cancellation of unlinked graphs always occurs exactly in every order;* the proof is given in Chapter 6.

We may thus restrict our calculation to linked graphs. There is a further, but approximate, simplification at high densities ($r_s < 1$) in the electron gas problem. At high densities ring graphs are dominant; they are also divergent in each order at low **q**, as we saw explicitly for the second-order energy correction, which corresponds to a ring graph. But, and this is the most important feature, the ring graphs may be summed to all orders to give a *convergent* sum even at low **q**. By carrying out the summation we obtain the correlation energy in the high-density limit. We note that the kinetic (fermi) energy is dom-

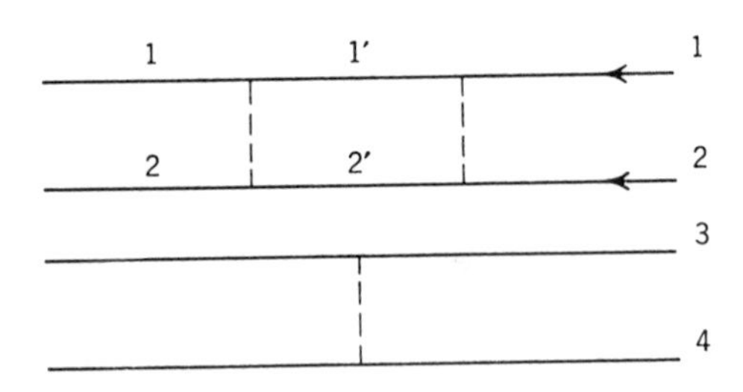

FIG. 4. An unlinked graph: the part 3,4 is not connected to the part 1,2.

inant as $r_s \to 0$, because the kinetic energy increases faster than the coulomb energy as the density is increased.

The reason for the dominance of the ring graphs at high densities may be understood from an example. Let us compare the density dependence of all third-order terms of the structure of Fig. 1 with those of the ring structure, Fig. 2, both for a system of N electrons in volume Ω. In Fig. 2 the incident particles can be chosen in N^3 ways and $\mathbf{q}$ in a number of ways proportional to Ω. In Fig. 1 there are N^2 ways of picking the incident particles, but $\mathbf{q}$ and $\mathbf{q}'$ are independent and may be picked in Ω^2 ways. Thus the ratio of the number of terms for Fig. 1 to those for Fig. 2 is $\Omega/N \propto r_s^3$, so that the ring graphs dominate as $r_s \to 0$.

The third-order ring graph contribution to the correlation energy of the fermi gas is, using (15),

$$(19) \quad E_3 = \sum_{1,2,3<k_F} \sum_{1',2',3'>k_F} 2m^2 \frac{\langle 13|V|1'3'\rangle\langle 3'2|V|32'\rangle\langle 1'2'|V|12\rangle}{(k_1^2 + k_3^2 - k_{1'}^2 - k_{3'}^2)(k_1^2 + k_2^2 - k_{1'}^2 - k_{2'}^2)}$$

times the number of possible sequences or equivalent graphs, which is two in third order. Thus

$$(20) \quad E_3 = 4m^2 \left(\frac{4\pi e^2}{\Omega}\right)^3 \left(\frac{\Omega}{(2\pi)^3}\right)^4 \int d^3q \int d^3k_1 \int d^3k_2 \int d^3k_3 \cdot \frac{1}{q^6} \cdot \frac{1}{q^2 + \mathbf{q}\cdot(\mathbf{k}_2 - \mathbf{k}_1)} \cdot \frac{1}{q^2 + \mathbf{q}\cdot(\mathbf{k}_3 - \mathbf{k}_1)},$$

where $k_1 < k_F$ and $|\mathbf{k}_1 + \mathbf{q}| > k_F$, and similarly for k_2 and k_3. Following the argument of Eq. (6), in the small q limit the integral over $\mathbf{q}$ ends up as

$$(21) \qquad E_3 \sim \int \frac{d^3q}{q^5} \sim \int \frac{dq}{q^3},$$

which diverges quadratically for small q.

Gell-Mann and Brueckner [*Phys. Rev.* **106,** 364 (1957)] have determined the form of the nth-order ring diagram contribution E_n, and they have shown that the ring diagram contributions can be summed to infinite order. We shall not repeat their calculation here, but we shall indicate schematically how an infinite series of divergent terms can be summed to give a convergent result.

Suppose that E_n were given by the divergent integral

$$E_n = \frac{(-1)^n r_s^{n-2}}{n} \int_0^\infty \frac{dq}{q^{2n-3}}, \tag{22}$$

which is correct with respect to the important exponents. Then the direct contribution to the correlation energy E_c would be

$$\begin{aligned} E_c = \sum_{n=2}^{\infty} E_n &= \int_0^\infty dq \sum_{n=2}^{\infty} \frac{(-1)^n r_s^{n-2}}{nq^{2n-3}} \\ &= \frac{1}{r_s^2} \int_0^\infty dq\, q^3 \sum_{n=2}^{\infty} \frac{(-1)^n}{n} \left(\frac{r_s}{q^2}\right)^n, \end{aligned} \tag{23}$$

where

$$\sum_{n=2}^{\infty} = \frac{r_s}{q^2} - \log\left(1 + \frac{r_s}{q^2}\right). \tag{24}$$

The integral $r_s \int dq\, q$ is convergent at the lower limit; further,

$$\begin{aligned} \int dq \cdot q^3 \cdot \log\left(1 + \frac{r_s}{q^2}\right) = \int dq \cdot q^3 \log(q^2 + r_s) \\ - 2 \int dq \cdot q^3 \cdot \log q \end{aligned} \tag{25}$$

is composed of two parts, each convergent as $q \to 0$; recall that $\lim_{x \to 0} x^m \log x = 0$, for $m > 0$.

We have shown that the ring graphs may be summed, but we must show why the ring graphs are in every order the most important graphs for a coulomb interaction at $r_s \to 0$. The reason is simple: a ring graph has a contribution $1/q^2$ from every vertex V, so that in the nth order the vertices give $1/q^{2n}$. At small q this divergence is stronger than that from any other graph of the same order, because no other graph gives the same factor at every vertex. Thus the graph of Fig. 1 has contributions to the integrand of $\varepsilon^{(3)}$ of the form

$$\frac{1}{q^2} \cdot \frac{1}{q'^2} \cdot \frac{1}{(\mathbf{q} + \mathbf{q}')^2},$$

which $\to \infty$ only as q^{-2} when $q \to 0$ independent of q'; the ring graph has a corresponding contribution $1/q^6$. A detailed analysis of all other integrals, including the exchange integrals, shows that r_s appears as a factor for all graphs except the ring graphs, so that the ring graphs are dominant as $r_s \to 0$.

There are also exchange contributions to the correlation energy.

When an exchange interaction appears, instead of a direct interaction, then $1/q^2$ is replaced by

$$\frac{1}{(\mathbf{q} + \mathbf{k}_1 - \mathbf{k}_2)^2},$$

which is regular as $q \to 0$. We see the substitution on comparing the direct matrix element

$$\langle 1'2'|V|12\rangle = \frac{1}{\Omega^2}\int d^3x\, d^3y\, V(\mathbf{x} - \mathbf{y})e^{-i\mathbf{q}\cdot(\mathbf{x}-\mathbf{y})} \propto 1/q^2, \tag{26}$$

with the exchange matrix element

$$\begin{aligned}\langle 2'1'|V|12\rangle &= \frac{1}{\Omega^2}\int d^3x\, d^3y\, e^{-i(\mathbf{k}_2-\mathbf{q})\cdot\mathbf{x}}e^{-i(\mathbf{k}_1+\mathbf{q})\cdot\mathbf{y}}V(\mathbf{x} - \mathbf{y}) \\ &\qquad \times e^{i\mathbf{k}_1\cdot\mathbf{x}}e^{i\mathbf{k}_2\cdot\mathbf{y}} \\ &= \frac{1}{\Omega^2}\int d^3x\, d^3y\, V(\mathbf{x} - \mathbf{y})e^{i(\mathbf{k}_1-\mathbf{k}_2+\mathbf{q})\cdot(\mathbf{x}-\mathbf{y})} \propto \frac{1}{(\mathbf{q} + \mathbf{k}_1 - \mathbf{k}_2)^2}.\end{aligned} \tag{27}$$

Appendix Solutions

CHAPTER 1

1. $$\int d^3k\, e^{i\mathbf{k}\cdot\mathbf{r}} = \int_0^{k_F} k^2\, dk\, d\Omega\, e^{ikr\cos\theta}$$

$$= 2\pi \int_0^{k_F} k^2\, dk \frac{e^{ikr} - e^{-ikr}}{ikr}$$

$$= \frac{2\pi}{ir} \int_0^{k_F} k\, dk (e^{ikr} - e^{-ikr})$$

$$= \frac{4\pi}{r} \left\{ -\frac{k}{r} \cos kr \Big|_0^{k_F} + \frac{1}{r} \int_0^{k_F} \cos kr\, dk \right\}$$

$$= \frac{4\pi}{r} \left\{ -\frac{k_F}{r} \cos k_F r + \frac{1}{r^2} \sin k_F r \right\}$$

$$= \frac{4\pi \{\sin k_F r - k_F r \cos k_F r\}}{r^3}$$

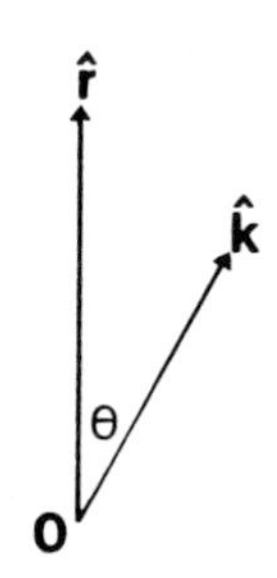

FIG. 1.1. The orientation of integration variable $\mathbf{k}$ with respect to $\hat{r}$.

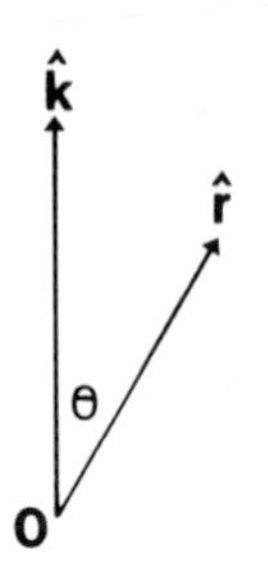

FIG. 1.2. The orientation of integration variable $\mathbf{r}$ with respect to $\hat{\mathbf{k}}$.

2.

$$I = \int d^3x\, e^{i\mathbf{k}\cdot\mathbf{r}} \frac{x_k}{r^3} = 2\pi \int r^2\, dr \sin\theta\, d\theta\, e^{irk\cos\theta} \frac{\cos\theta}{r^2}$$

$$= 2\pi \int_0^\infty dr \int_{-1}^1 d\mu\, \mu e^{ikr\mu}, \qquad \mu = \cos\theta$$

$$= 2\pi \int_0^\infty dr \left[\frac{\mu}{ikr} e^{ikr\mu} \Big|_{-1}^1 - \frac{1}{ikr} \int_{-1}^1 e^{ikr\mu}\, d\mu \right]$$

$$= 2\pi \int_0^\infty dr \left\{ \frac{2\cos kr}{ikr} - \frac{2}{ik^2r^2} \sin kr \right\}$$

$$= 4\pi \left\{ \int_0^\infty dr \frac{\cos kr}{ikr} - \frac{1}{ik^2} \int_0^\infty \frac{dr}{r^2} \sin kr \right\}$$

$$\int_0^\infty \frac{dr}{r^2} \sin kr = -\frac{1}{r} \sin kr \Big|_0^\infty + k \int_0^\infty \frac{\cos kr}{r}\, dr.$$

$$I = 4\pi \left\{ \int_0^\infty dr \frac{\cos kr}{ikr} \right.$$

$$\left. - \frac{1}{ik} \int_0^\infty \frac{\cos kr}{r}\, dr + \frac{1}{irk^2} \sin kr \Big|_0^\infty \right\}$$

$$= 4\pi \left(-\frac{kr}{ik^2 r} \right),$$

expansion of sin kr with small kr is used.

$$= \frac{4\pi i}{k}.$$

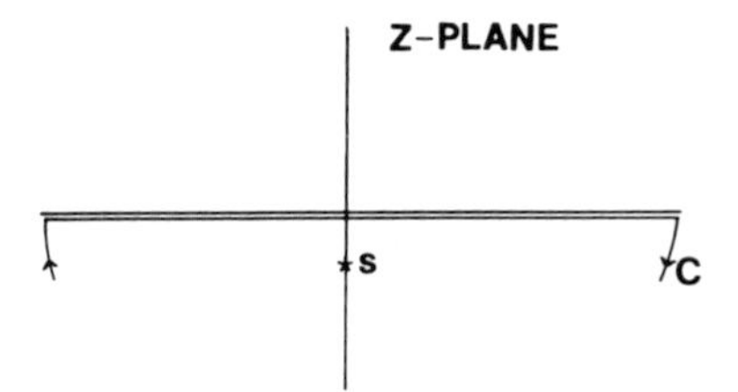

FIG. 1.3. The contour of c.

3. $\theta(t) = \lim_{s \to 0^+} \frac{i}{2\pi} \int_{-\infty}^{\infty} dx \, \frac{e^{-ixt}}{x + is}.$

Consider:

$$\oint_c \frac{e^{-izt}}{z + is} \, dz$$

$t > 0$, c is the closed contour with the great circle in the lower half plane (Fig. 1.3).

$$\oint_c \frac{e^{-izt}}{z + is} \, dz = \int_{-\infty}^{\infty} dx \, \frac{e^{-ixt}}{x + is} + \int_{\cup} \frac{e^{-izt}}{z + is} \, dz$$

$$= -2\pi i e^{st} \underset{s \to 0^+}{=} -2\pi i,$$

where

$$\int_{\cup} \frac{e^{-izt}}{z + is} \, dz \le \left| \int_0^{\pi} e^{-rt\sin\theta} \, d\theta \right| \le \left| 2 \int_0^{\pi/2} e^{-rt2\theta/\pi} \, d\theta \right|$$

$$\underset{r \to \infty}{=} 2\left(\frac{\pi}{2rt}\right)(1 - e^{-rt}) \to 0.$$

$t < 0$, we have to close the contour in the upper half plane.

$$\oint_c \frac{e^{-izt}}{z + is} \, dz = \int_{-\infty}^{\infty} dx \, \frac{1}{x + is} \, e^{-ixt} + \int_{\cap} \frac{1}{z + is} \, e^{-izt} \, dz = 0.$$

It is easy to show $\int_{\cap} \frac{e^{-izt}}{z+is}\,dz \to 0$, so,

$$\theta(t) = 1 \qquad t > 0,$$
$$= 0 \qquad t < 0.$$

4. (a) $$\left[e^{-i\mathbf{k}\cdot\mathbf{x}}, \mathbf{p}\right]\psi = e^{-i\mathbf{k}\cdot\mathbf{x}}\mathbf{p}\psi - \mathbf{p}\left(e^{-i\mathbf{k}\cdot\mathbf{x}}\psi\right)$$
$$= e^{-i\mathbf{k}\cdot\mathbf{x}}\mathbf{p}\psi + \hbar\mathbf{k}e^{-i\mathbf{k}\cdot\mathbf{x}}\psi$$
$$- e^{-i\mathbf{k}\cdot\mathbf{x}}\mathbf{p}\psi = \hbar\mathbf{k}e^{-i\mathbf{k}\cdot\mathbf{x}}\psi.$$

ψ is arbitrary, $[e^{-i\mathbf{k}\cdot\mathbf{x}}, \mathbf{p}] = \hbar\mathbf{k}e^{-i\mathbf{k}\cdot\mathbf{x}}$.

(b) $$\left[e^{-i\mathbf{k}\cdot\mathbf{x}}, p^2\right]\psi = e^{-i\mathbf{k}\cdot\mathbf{x}}p^2\psi - p^2\left(e^{-i\mathbf{k}\cdot\mathbf{x}}\psi\right)$$
$$= e^{-i\mathbf{k}\cdot\mathbf{x}}p^2\psi - \mathbf{p}\cdot\left(-\hbar\mathbf{k}e^{-i\mathbf{k}\cdot\mathbf{x}}\psi + e^{-i\mathbf{k}\cdot\mathbf{x}}\mathbf{p}\psi\right)$$
$$= e^{-i\mathbf{k}\cdot\mathbf{x}}p^2\psi - \{(-\hbar\mathbf{k})\cdot(-\hbar\mathbf{k})e^{-i\mathbf{k}\cdot\mathbf{x}}\psi$$
$$-2\hbar\mathbf{k}e^{-i\mathbf{k}\cdot\mathbf{x}}\cdot\mathbf{p}\psi + e^{-i\mathbf{k}\cdot\mathbf{x}}p^2\psi\}$$
$$= e^{-i\mathbf{k}\cdot\mathbf{x}}\{2\hbar\mathbf{k}\cdot\mathbf{p} - (\hbar^2k^2)\}\psi.$$
$$\left[e^{-i\mathbf{k}\cdot\mathbf{x}}, p^2\right] = e^{-i\mathbf{k}\cdot\mathbf{x}}\left(2\hbar\mathbf{k}\cdot\mathbf{p} - \hbar^2k^2\right).$$

CHAPTER 2

1. $$i\hbar\dot{\psi} = [\psi, H].$$
$$H = \int\left[\frac{1}{2\rho}\pi^2 + \frac{1}{2}T\left(\frac{d\psi}{dx}\right)^2\right]dx.$$
$$[\psi, H] = \left[\psi, \int\left[\frac{1}{2\rho}\pi^2 + \frac{1}{2}T\left(\frac{d\psi}{\partial x'}\right)^2\right]dx'\right] = \left[\psi, \int\frac{1}{2\rho}\pi^2\,dx'\right]$$
$$= \int\left[\psi, \frac{1}{2\rho}\pi\right]\pi\,dx' + \frac{1}{2\rho}\int\pi[\psi, \pi]\,dx'$$
$$= \frac{i\hbar}{2\rho}\pi(x) + \frac{i\hbar}{2\rho}\pi(x) = \frac{i\hbar}{\rho}\pi(x).$$
$$\dot{\psi} = \frac{1}{\rho}\pi.$$

2. $$i\hbar\dot{\pi} = [\pi, H] = \left[\pi, \int\left\{\frac{1}{2\rho}\pi^2 + \frac{1}{2}T\left(\frac{\partial\psi}{\partial x'}\right)^2\right\} dx'\right]$$

$$= \frac{T}{2}\int\left[\pi, \frac{\partial\psi}{\partial x'}\right]\frac{\partial\psi}{\partial x'}\, dx' + \frac{T}{2}\int\left(\frac{\partial\psi}{\partial x'}\right)\left[\pi, \frac{\partial\psi}{\partial x'}\right] dx'$$

$$= \frac{T}{2}\int\left\{\frac{\partial}{\partial x'}[\pi, \psi]\right\}\frac{\partial\psi}{\partial x'}\, dx' + \frac{T}{2}\int\left(\frac{\partial\psi}{\partial x'}\right)\frac{\partial}{\partial x'}[\pi, \psi]\, dx'$$

$$= \frac{T}{2}i\hbar\frac{\partial^2\psi}{\partial x^2} + \frac{T}{2}i\hbar\frac{\partial^2\psi}{\partial x^2} = i\hbar T\frac{\partial^2\psi}{\partial x^2}.$$

$$\dot{\pi} = T\frac{\partial^2\psi}{\partial x^2}.$$

From

$$\ddot{\psi} = \frac{\dot{\pi}}{\rho},$$

we get $\ddot{\psi} = \dfrac{T}{\rho}\dfrac{\partial^2\psi}{\partial x^2}$.

4. $$E = \sum_{\mathbf{q}}\hbar\omega_{\mathbf{q}}\frac{1}{e^{\beta\hbar\omega_{\mathbf{q}}} - 1} = \frac{V}{(2\pi)^3}\int q^2\, dq\, d\Omega\frac{\hbar\omega_{\mathbf{q}}}{e^{\beta\hbar\omega_{\mathbf{q}}} - 1}.$$

There are three acoustical modes, so

$$E = \frac{V}{(2\pi)^3}\cdot 4\pi\frac{1}{v_l^3}\int\frac{\hbar\omega^3\, d\omega}{e^{\beta\hbar\omega} - 1} + \frac{V}{(2\pi)^3}\cdot 4\pi\cdot\frac{2}{v_t^3}\int\frac{\hbar\omega^3\, d\omega}{e^{\beta\hbar\omega} - 1}$$

$$= \frac{V}{2\pi^2}\frac{\hbar}{v_l^3}\left(\frac{1}{\hbar\beta}\right)^4\int_0^{X_D}\frac{x^3\, dx}{e^x - 1} + \frac{V}{2\pi^2}\frac{2\hbar}{v_t^3}\left(\frac{1}{\hbar\beta}\right)^4\int_0^{X_D}\frac{x^3\, dx}{e^x - 1}$$

$x = \hbar\omega\beta$, and $X_D = \hbar\omega_D\beta$

$$= \frac{V}{2\pi^2}(k_BT)^4\left\{\frac{1}{\hbar^3 v_l^3} + \frac{2}{\hbar^3 v_t^3}\right\}\int_0^{X_D}\frac{x^3\, dx}{e^x - 1}.$$

Use: $k_B\theta_l = \hbar v_l(6\pi^2 n)^{1/3}$ and $k_B\theta_t = \hbar v_t(6\pi^2 n)^{1/3}$,

$$\varepsilon = \frac{E}{V} = \frac{1}{2\pi^2}\left(k_B T^4\right)\left\{\frac{6\pi^2 n}{k_B^3\theta_l^3} + \frac{6\pi^2 n \times 2}{k_B^3\theta_t^3}\right\}\int_0^{X_D}\frac{x^3\,dx}{e^x - 1}$$

$$= \frac{k_B}{2\pi^2}T^4 6\pi^2 n\frac{3}{\theta^3}\int_0^{X_D}\frac{x^3\,dx}{e^x-1}, \qquad \frac{3}{\theta^3} = \frac{1}{\theta_l^3} + \frac{2}{\theta_t^3}.$$

$$\int_0^{X_D}\frac{x^3\,dx}{e^x-1} \underset{T\ll\theta_D}{=} \int_0^{\infty}\frac{x^3\,dx}{e^x-1} = \frac{\pi^4}{15}, \qquad k_B\theta_D = \hbar\omega_D.$$

$$\varepsilon = \frac{3nk_B\pi^4}{5\theta^3}T^4.$$

$$C_v = \frac{\partial\varepsilon}{\partial T} = \frac{12\pi^4 nk_B}{5\theta^3}T^3.$$

5. $H_{\mathbf{k}} = \omega_0\left(a_{\mathbf{k}}^\dagger a_{\mathbf{k}} + a_{-\mathbf{k}}^\dagger a_{-\mathbf{k}}\right) + \omega_1\left(a_{\mathbf{k}}a_{-\mathbf{k}} + a_{-\mathbf{k}}^\dagger a_{\mathbf{k}}^\dagger\right).$

$$a_{\mathbf{k}} = u_{\mathbf{k}}\alpha_{\mathbf{k}} + v_{\mathbf{k}}\alpha_{-\mathbf{k}}^\dagger, \qquad a_{\mathbf{k}}^\dagger = u_{\mathbf{k}}\alpha_{\mathbf{k}}^\dagger + v_{\mathbf{k}}\alpha_{-\mathbf{k}}.$$

(a) We need to show $[\alpha_k^\dagger, H] = -\lambda\alpha_{\mathbf{k}}^\dagger$ and $[\alpha_k, H] = \lambda\alpha_{\mathbf{k}}$. From Eq. (95),

$$\begin{aligned}
\left[\alpha_k^\dagger, H\right] &= u_{\mathbf{k}}\left(-\omega_0 a_{\mathbf{k}}^\dagger - \omega_1 a_{-\mathbf{k}}\right) - v_{\mathbf{k}}\left(\omega_0 a_{-\mathbf{k}} + \omega_1 a_{\mathbf{k}}^\dagger\right)\\
&= u_{\mathbf{k}}\left\{-\omega_0\left(u_{\mathbf{k}}\alpha_{\mathbf{k}}^\dagger + v_{\mathbf{k}}\alpha_{-\mathbf{k}}\right) - \omega_1\left(u_{\mathbf{k}}\alpha_{-\mathbf{k}} + v_{\mathbf{k}}\alpha_{\mathbf{k}}^\dagger\right)\right\}\\
&\quad - v_{\mathbf{k}}\left\{\omega_0\left(u_{\mathbf{k}}\alpha_{-\mathbf{k}} + v_{\mathbf{k}}\alpha_{\mathbf{k}}^\dagger\right) + \omega_1\left(u_{\mathbf{k}}\alpha_{\mathbf{k}}^\dagger + v_{\mathbf{k}}\alpha_{-\mathbf{k}}\right)\right\}\\
&= -\omega_0\left(u_{\mathbf{k}}^2 + v_{\mathbf{k}}^2\right)\alpha_{\mathbf{k}}^\dagger - 2u_{\mathbf{k}}v_{\mathbf{k}}\omega_0\alpha_{-\mathbf{k}}\\
&\quad - \omega_1\left(u_{\mathbf{k}}^2 + v_{\mathbf{k}}^2\right)\alpha_{-\mathbf{k}} - 2\omega_1 u_{\mathbf{k}}v_{\mathbf{k}}\alpha_{\mathbf{k}}^\dagger\\
&= \cosh 2\chi_{\mathbf{k}}\left\{-\omega_0 - \omega_1\tanh 2\chi_{\mathbf{k}}\right\}\alpha_{\mathbf{k}}^\dagger\\
&\quad - \cosh 2\chi_{\mathbf{k}}\left\{\omega_1 + \omega_0\tanh 2\chi_{\mathbf{k}}\right\}\alpha_{-\mathbf{k}}.
\end{aligned}$$

But,

$$-\omega_0 - \omega_1 \tanh 2\chi_{\mathbf{k}} = -(\varepsilon_{\mathbf{k}} + NV_{\mathbf{k}}) + \frac{(NV_{\mathbf{k}})^2}{\varepsilon_{\mathbf{k}} + NV_{\mathbf{k}}}$$

$$= -\frac{\varepsilon_{\mathbf{k}}^2 + 2\varepsilon_{\mathbf{k}} NV_{\mathbf{k}}}{\varepsilon_{\mathbf{k}} + NV_{\mathbf{k}}}.$$

$$\omega_1 + \omega_0 \tanh 2\chi_{\mathbf{k}} = NV_{\mathbf{k}} - \frac{(\varepsilon_{\mathbf{k}} + NV_{\mathbf{k}})NV_{\mathbf{k}}}{\varepsilon_{\mathbf{k}} + NV_{\mathbf{k}}} = 0.$$

$$[\alpha_{\mathbf{k}}^\dagger, H] = -\cosh 2\chi_{\mathbf{k}} \left\{ \frac{\varepsilon_{\mathbf{k}}^2 + 2\varepsilon_{\mathbf{k}} NV_{\mathbf{k}}}{\varepsilon_{\mathbf{k}} + NV_{\mathbf{k}}} \right\} \alpha_{\mathbf{k}}^\dagger = -\lambda \alpha_{\mathbf{k}}^\dagger.$$

Following the same procedure, one can get

$$[\alpha, H] = \cosh 2\chi_{\mathbf{k}} \left\{ \frac{\varepsilon_{\mathbf{k}}^2 + 2\varepsilon_{\mathbf{k}} NV_{\mathbf{k}}}{\varepsilon_{\mathbf{k}} + NV_{\mathbf{k}}} \right\} \alpha_{\mathbf{k}} = \lambda \alpha_{\mathbf{k}}.$$

(b)
$$\begin{aligned}
a_{\mathbf{k}}^\dagger a_{\mathbf{k}} &= (u_{\mathbf{k}}\alpha_{\mathbf{k}}^\dagger + v_{\mathbf{k}}\alpha_{-\mathbf{k}})(u_{\mathbf{k}}\alpha_{\mathbf{k}} + v_{\mathbf{k}}\alpha_{-\mathbf{k}}^\dagger) \\
&= u_{\mathbf{k}}^2\alpha_{\mathbf{k}}^\dagger\alpha_{\mathbf{k}} + u_{\mathbf{k}}v_{\mathbf{k}}\alpha_{\mathbf{k}}^\dagger\alpha_{-\mathbf{k}}^\dagger + u_{\mathbf{k}}v_{\mathbf{k}}\alpha_{-\mathbf{k}}\alpha_{\mathbf{k}} + v_{\mathbf{k}}^2\alpha_{-\mathbf{k}}\alpha_{-\mathbf{k}}^\dagger \\
&= u_{\mathbf{k}}^2\alpha_{\mathbf{k}}^\dagger\alpha_{\mathbf{k}} + u_{\mathbf{k}}v_{\mathbf{k}}(\alpha_{\mathbf{k}}^\dagger\,\alpha_{-\mathbf{k}}^\dagger + \alpha_{-\mathbf{k}}\alpha_{\mathbf{k}}) \\
&\quad + v_{\mathbf{k}}^2(1 + \alpha_{-\mathbf{k}}^\dagger\alpha_{-\mathbf{k}}) \\
&= u_{\mathbf{k}}^2\alpha_{\mathbf{k}}^\dagger\alpha_{\mathbf{k}} + v_{\mathbf{k}}^2 + v_{\mathbf{k}}^2\alpha_{-\mathbf{k}}^\dagger\alpha_{-\mathbf{k}} \\
&\quad + u_{\mathbf{k}}v_{\mathbf{k}}(\alpha_{\mathbf{k}}^\dagger\,\alpha_{-\mathbf{k}}^\dagger + \alpha_{-\mathbf{k}}\alpha_{\mathbf{k}}).
\end{aligned}$$

(c) $\alpha_{\mathbf{k}}\Phi_0 = 0.$

$$\langle a_{\mathbf{k}}^\dagger a_{\mathbf{k}}\rangle_0 = \langle\Phi_0|a_{\mathbf{k}}^\dagger a_{\mathbf{k}}|\Phi_0\rangle = \langle\Phi_0|v_{\mathbf{k}}^2 + u_{\mathbf{k}}v_{\mathbf{k}}\alpha_{\mathbf{k}}^\dagger\alpha_{-\mathbf{k}}^\dagger|\Phi_0\rangle.$$

Because $\langle\Phi_0|\alpha_{\mathbf{k}}^\dagger = (\alpha_{\mathbf{k}}|\Phi_0\rangle)^\dagger$, so

$$\langle a_{\mathbf{k}}^\dagger a_{\mathbf{k}}\rangle_0 = v_{\mathbf{k}}^2 = \sinh^2\chi_{\mathbf{k}} = \tfrac{1}{2}(\cosh 2\chi_{\mathbf{k}} - 1).$$

$$\langle a_{\mathbf{k}}^\dagger a_{\mathbf{k}}\rangle_0 = \frac{1}{2}\left(\frac{\dfrac{\hbar^2k^2}{2m} + NV_{\mathbf{k}}}{\sqrt{\left(\dfrac{\hbar^2k^2}{2m}\right)^2 + 2\left(\dfrac{\hbar^2k^2}{2m}\right)NV_{\mathbf{k}}}} - 1 \right).$$

As $k \to 0, \quad \langle a_{\mathbf{k}}^\dagger a_{\mathbf{k}}\rangle_0 \to \infty.$

$k \to \infty, \quad \langle a_{\mathbf{k}}^\dagger a_{\mathbf{k}}\rangle_0 \to 0.$

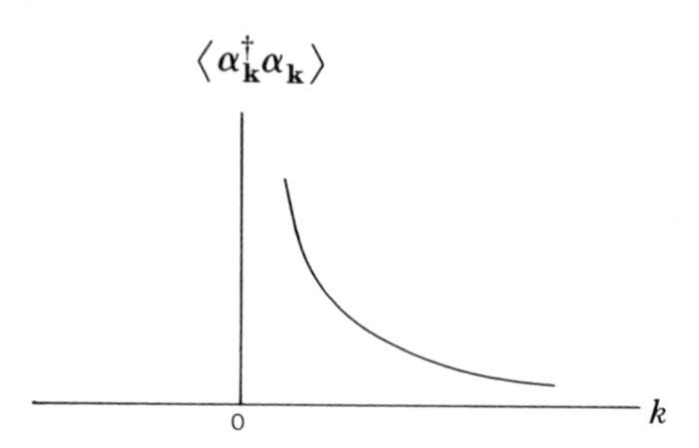

FIG. 2.1. The sketch of $\langle a_{\mathbf{k}}^{\dagger}\alpha_{\mathbf{k}}\rangle_0$ as a function of k.

7. $H = \hbar\omega a^{\dagger}a + \varepsilon(ab^{\dagger} + ba^{\dagger}) = \omega a^{\dagger}a + \varepsilon(ab^{\dagger} + ba^{\dagger})$. Let $\alpha = ua + vb$, $[\alpha, H] = (u\omega + \varepsilon v)a + u\varepsilon b = \lambda\alpha = \lambda(ua + vb)$, where λ is an eigenvalue. Then, $u\omega - \lambda u + \varepsilon v = 0$ and $u\varepsilon - \lambda v = 0$. In matrix form,

$$\begin{pmatrix} \omega - \lambda & \varepsilon \\ \varepsilon & -\lambda \end{pmatrix}\begin{pmatrix} u \\ v \end{pmatrix} = 0.$$

Solve for λ, $\lambda = \frac{1}{2}\{\omega \pm \sqrt{\omega^2 + 4\varepsilon^2}\}$. Take $\lambda_- = \frac{1}{2}\{\omega - \sqrt{\omega^2 + 4\varepsilon^2}\}$,

$$\frac{v}{u} = \frac{1}{\varepsilon}\{\lambda_- - \omega\} = -\frac{1}{\varepsilon}\frac{\omega + \sqrt{\omega^2 + 4\varepsilon^2}}{2}.$$

$$\begin{aligned}
[\alpha, H] &= u\left[\omega + \varepsilon\frac{v}{u}\right]a + u\varepsilon b \\
&= u\left[\omega - \frac{\omega + \sqrt{\omega^2 + 4\varepsilon^2}}{2}\right]a + u\varepsilon b \\
&= u \cdot \frac{\omega - \sqrt{\omega^2 + 4\varepsilon^2}}{2}\left\{a + \frac{2\varepsilon}{\omega - \sqrt{\omega^2 + 4\varepsilon^2}}b\right\} \\
&= u \cdot \frac{\omega - \sqrt{\omega^2 + 4\varepsilon^2}}{2}\left\{a - \frac{\omega + \sqrt{\omega^2 + 4\varepsilon^2}}{2\varepsilon}b\right\} \\
&= u\lambda\left(a + \frac{v}{u}b\right) = \lambda\alpha.
\end{aligned}$$

Let

$$u = 1 \qquad v = -\frac{1}{2\varepsilon}\left(\omega + \sqrt{\omega^2 + 4\varepsilon^2}\right).$$

$$\alpha = a - \frac{1}{2\varepsilon}\left(\omega + \sqrt{\omega^2 + 4\varepsilon^2}\right)b \qquad \text{is the transformation.}$$

For

$$\lambda_+ = \tfrac{1}{2}\left\{\omega + \sqrt{\omega^2 + 4\varepsilon^2}\right\},$$

$$\beta = a - \frac{1}{2\varepsilon}\left(\omega - \sqrt{\omega^2 - 4\varepsilon^2}\right)b.$$

$$H = \lambda_+ \beta^\dagger \beta + \lambda_- \alpha^\dagger \alpha.$$

CHAPTER 3

1. From Eq. (15), $P_{\text{elec}} = -\dfrac{ne^2E}{m\omega^2}$. From the atomic system $P_{\text{atomic}} = \chi_a n_a E$.

$$P = P_{\text{elec}} + P_{\text{atomic}} = \left(-\frac{ne^2}{m\omega^2} + n_a\chi_a\right)E.$$

$$\varepsilon = 1 + \left(-\frac{4\pi ne^2}{m\omega^2} + 4\pi n_a\chi_a\right) = 1 - \frac{4\pi ne^2}{m\omega^2} + 4\pi n_a\chi_a.$$

$$\varepsilon = 0, \qquad \frac{4\pi ne^2}{m\omega^2} = (1 + 4\pi n_a\chi_a).$$

$$\omega^2 = \frac{4\pi ne^2}{m} \cdot \frac{1}{1 + 4\pi n_a\chi_a} = \frac{\omega_p^2}{1 + 4\pi n_a\chi_a}.$$

From Phys. Rev. *92*, p. 890, $\chi_a \simeq 2.4 \times 10^{-24}$ cm³.

$$n_a = \frac{10.5(\text{g/cm}^3)6 \times 10^{23}}{108\ \text{g}} \simeq 6 \times 10^{22}/\text{cm}^3.$$

$$4\pi n_a\chi_a = 1.8, \qquad \omega = 0.6\omega_p.$$

2. Outside the sphere $\nabla^2\phi_e = 0$, $\phi_e = \sum_{L,m} A_L \frac{1}{r^{L+1}} P_L^{|m|} e^{im\phi}$, $\mathbf{E}_e = \mathbf{D}_e$. Inside the sphere, we treat as a dielectric sphere with $\varepsilon = 1 - \frac{\omega_P^2}{\omega^2}$.

$$\mathbf{D}_i = \varepsilon \mathbf{E}_i, \qquad \mathbf{E}_i = -\nabla\phi_i, \qquad \phi_i = \sum_{L,m} B_L r^L P_L^{|m|} e^{im\phi}.$$

At $r = a$, the normal component of $\mathbf{D}$ is continuous.

$$D_{eL} = A_L \frac{L+1}{a^{L+2}} = D_{iL} = -LB_L a^{L-1}\varepsilon. \qquad \text{(i)}$$

Also, $r = a$, the tangential component of $\mathbf{E}$ is continuous

$$A_L / a^{L+1} = B_L a^L,$$

$$\text{so} \qquad A_L = B_L a^{2L+1}. \qquad \text{(ii)}$$

From (i) and (ii), $\frac{L+1}{L} A_L = -B_L a^{2L+1}\varepsilon$, $\qquad \frac{L+1}{L} = -1 + \frac{\omega_p^2}{\omega_L^2}$.

$$\omega_L^2 = \frac{L}{2L+1}\omega_p^2.$$

CHAPTER 4

1. $H = -J \sum_{j,\delta} \mathbf{S}_j \cdot \mathbf{S}_{j+\delta} - 2\mu_B H_0 \sum_j S_{jz}$.

$$S_z = \sum_j S_{jz}.$$

$$[S_z, H] = \left[S_z, -J\sum_{j,\delta} \mathbf{S}_j \cdot \mathbf{S}_{j+\delta}\right]$$

$$= \left[\sum_i S_{iz}, -J\sum_{j,\delta} (S_{jx}S_{j+\delta x} + S_{jy}S_{j+\delta y})\right],$$

$$\text{using } \left[S_z, \sum_j S_{jz}\right] = 0.$$

$$= -J \sum_{i,j,\delta} \left\{ \left[S_{iz}, S_{jx} S_{j+\delta x} \right] + \left[S_{iz}, S_{jy} S_{j+\delta y} \right] \right\}$$

$$= -J \sum_{i,j,\delta} \{ i\hbar\, \delta_{i,j} S_{jy} S_{j+\delta x} + i\hbar\, \delta_{i,j+\delta} S_{j+\delta y} S_{jx}$$

$$- i\hbar\, \delta_{i,j} S_{jx} S_{j+\delta y} - i\hbar\, \delta_{i,j+\delta} S_{jy} S_{j+\delta x} \}$$

$$= -i\hbar J \left\{ \sum_{j,\delta} S_{jy} S_{j+\delta x} + \sum_{j,\delta} S_{jx} S_{j+\delta y} \right.$$

$$\left. - \sum_{j,\delta} S_{jx} S_{j+\delta y} - \sum_{j,\delta} S_{jy} S_{j+\delta x} \right\} = 0.$$

Other commutators can be evaluated by following the foregoing approach.

2.
$$S_j^+ = \sqrt{2S}\left(1 - a_j^\dagger a_j/2S\right)^{1/2} a_j, \quad S_j = \sqrt{2S}\, a_j^\dagger \left(1 - a_j^\dagger a_j/2S\right).$$

$$S_x = \tfrac{1}{2}\left(S_j^+ + S_j^-\right), \qquad S_y = \frac{1}{2i}\left(S_j^+ - S_j^-\right).$$

$$\left[S_x, S_y\right] = \frac{1}{4i}\left[\left(S_j^+ + S_j^-\right), \left(S_{j'}^+ - S_{j'}^-\right)\right] = \frac{-1}{2i}\left[S_j^+, S_{j'}^-\right].$$

$$\left[S_j^+, S_{j'}^-\right] = \left(2S - a_j^\dagger a_j\right)^{1/2} a_j a_{j'}^\dagger \left(2S - a_{j'}^\dagger a_{j'}\right)^{1/2}$$

$$- a_{j'}^\dagger \left(2S - a_{j'}^\dagger a_{j'}\right)^{1/2} \left(2S - a_j^\dagger a_j\right)^{1/2} a_j$$

$$= \sqrt{2S}\left(1 - \frac{1}{2} \cdot \frac{1}{2S} a_j^\dagger a_j\right) a_j a_{j'}^\dagger \sqrt{2S}\left(1 - \frac{1}{2} \cdot \frac{1}{2S} a_{j'}^\dagger a_{j'}\right)$$

$$- a_{j'}^\dagger \sqrt{2S}\left(1 - \frac{1}{2} \cdot \frac{1}{2S} a_{j'}^\dagger a_{j'}\right) \sqrt{2S}\left(1 - \frac{1}{2} \cdot \frac{1}{2S} a_j^\dagger a_j\right) a_j$$

$$= 2S \left\{ a_j a_{j'}^\dagger - \frac{1}{4S} a_j^\dagger a_j a_j a_{j'}^\dagger - \frac{1}{4S} a_j a_{j'}^\dagger a_{j'}^\dagger a_{j'} \right.$$

$$\left. - a_{j'}^\dagger a_j + \frac{1}{4S} a_{j'}^\dagger a_{j'}^\dagger a_{j'} a_j + \frac{1}{4S} a_{j'}^\dagger a_j^\dagger a_j a_j \right\}$$

$$= 2S\hbar\delta_{j'j} - \hbar a_j^\dagger a_j \delta_{jj'} - \tfrac{1}{2} a_j^\dagger \left(\hbar\, \delta_{jj'} + a_{j'}^\dagger a_j\right) a_j$$

$$- \tfrac{1}{2} a_{j'}^\dagger \left(\hbar\, \delta_{jj'} + a_{j'}^\dagger a_j\right) a_{j'}$$

$$+ \tfrac{1}{2} a_{j'}^\dagger a_{j'}^\dagger a_{j'} a_j + \tfrac{1}{2} a_{j'}^\dagger a_j^\dagger a_j a_j$$

$$= 2S\hbar\, \delta_{jj'} - 2a_j^\dagger a_j \delta_{jj'} = 2S_{jz}\hbar\, \delta_{jj'}.$$

$$\left[S_x, S_y\right] = i\hbar S_z.$$

3. $$S^2 = \sum_i \mathbf{S}_i \cdot \sum_j \mathbf{S}_j = \sum_{i,j} \{S_{ix}S_{jx} + S_{iy}S_{jy} + S_{iz}S_{jz}\}$$

$$= \sum_{i,j} \{ \tfrac{1}{4}(S_i^+ + S_i^-)(S_j^+ + S_j^-)$$

$$- \tfrac{1}{4}(S_i^+ - S_i^-)(S_j^+ - S_j^-) + S_{iz}S_{jz}\}$$

$$= \sum_{i,j} \{ \tfrac{1}{2}(S_i^+ S_j^- + S_i^- S_j^+) + S_{iz}S_{jz}\}$$

$$= \tfrac{1}{2}(2NS)(b_0 b_0^\dagger + b_0^\dagger b_0)$$

$$+ \left(NS - \sum_k b_\mathbf{k}^\dagger b_\mathbf{k}\right)\left(NS - \sum_{k'} b_{\mathbf{k}'}^\dagger b_{\mathbf{k}'}\right)$$

$$= (NS)^2 - 2NS\sum_\mathbf{k} b_\mathbf{k}^\dagger b_\mathbf{k} + NS(1 + b_0^\dagger b_0 + b_0^\dagger b_0)$$

$$= (NS)^2 + NS - 2NS \sum_{\mathbf{k}\neq 0} b_\mathbf{k}^\dagger b_\mathbf{k}.$$

A magnon at $\mathbf{k} = 0$ corresponds to all spins moving in phase so that the total spin is unchanged, although the $\hat{z}$-component of the spin does change.

5. The ground state is Φ_0 with all the spins aligned along one direction. The one magnon state is

$$\psi_\mathbf{k} = b_\mathbf{k}^\dagger \Phi_0 = \frac{1}{\sqrt{N}} \sum_j e^{-i\mathbf{k}\cdot\mathbf{x}_j} a_j^\dagger \Phi_0.$$

$S_j^- = \sqrt{2S}\, a_j^\dagger(1 - a_j^\dagger a_j \sqrt{2S})^{1/2}$. Since $a_j \Phi_0 = 0$, $S_j^- \Phi_0 = \sqrt{2S}\, a_j^\dagger \Phi_0$.

$$\psi_\mathbf{k} = \frac{1}{\sqrt{N}} \sum_j e^{-i\mathbf{k}\cdot\mathbf{x}_j} \frac{1}{\sqrt{2S}} S_j^- \Phi_0.$$

$$\Phi_0 = |S, S, \ldots, S_j, S..\rangle.$$

$$\psi_\mathbf{k} = \frac{1}{\sqrt{2SN}} \sum_j e^{-i\mathbf{k}\cdot\mathbf{x}_j} \sqrt{2S}\, |S \ldots S, S_j - 1, S\ldots\rangle$$

$$= \frac{1}{\sqrt{N}} \sum_j e^{-i\mathbf{k}\cdot\mathbf{x}_j} |S\ldots, S_j - 1,.\rangle.$$

6. $$H = -J\sum_{j,\delta} \mathbf{S}_j \cdot \mathbf{S}_{j+\delta} - 2\mu_0 H_0 \sum_j S_{jz}.$$

$$i\hbar\dot{\mathbf{S}}_j = [\mathbf{S}_j, H].$$

$$i\hbar\dot{S}_{jx} = [S_{jx}, H] = \left[S_{jx}, -J\sum_{l,\delta} \mathbf{S}_l \cdot \mathbf{S}_{l+\delta} - 2\mu_0 H_0 \sum_l S_{lz}\right].$$

$$\left[S_{jx}, \sum_{l,\delta} \mathbf{S}_l \cdot \mathbf{S}_{l+\delta}\right] = \sum_{l,\delta} \{i\hbar\,\delta_{l,j} S_{lz} S_{l+\delta y} + i\hbar\,\delta_{j,l+\delta} S_{ly} S_{l+\delta z} - i\hbar\,\delta_{j,l} S_{ly} S_{l+\delta z} - i\hbar\,\delta_{j,l+\delta} S_{lz} S_{l+\delta y}\}$$

$$= i\hbar\sum_\delta S_{jz} S_{j+\delta y} + i\hbar\sum_\delta S_{j-\delta y} S_{jz} - i\hbar\sum_\delta S_{jy} S_{j+\delta z} - i\hbar\sum_\delta S_{j-\delta z} S_{jy}$$

$$= -i\hbar\sum_\delta (\mathbf{S}_j \times \mathbf{S}_{j+\delta})_x + i\hbar\sum_\delta (\mathbf{S}_{j-\delta} \times \mathbf{S}_j)_x.$$

$$\left[S_{jy}, \sum_{l,\delta} \mathbf{S}_l \cdot \mathbf{S}_{l+\delta}\right] = i\hbar\sum_{l,\delta} \{-\delta_{j,l} S_{jz} S_{l+\delta x} - \delta_{j,l+\delta} S_{lx} S_{l+\delta z} + \delta_{j,l} S_{jx} S_{l+\delta z} + \delta_{j,l+\delta} S_{lz} S_{l+\delta x}\}$$

$$= -i\hbar\sum_\delta S_{jz} S_{j+\delta x} - i\hbar\sum_\delta S_{j-\delta x} S_{jz} + i\hbar\sum_\delta S_{jx} S_{j+\delta z} + i\hbar\sum_\delta S_{j-\delta z} S_{jx}.$$

$$\left[S_{jz}, \sum_{l,\delta} \mathbf{S}_l \cdot \mathbf{S}_{l+\delta}\right] = i\hbar\sum_\delta S_{jy} S_{j+\delta x} + i\hbar\sum_\delta S_{j-\delta x} S_{jy} - i\hbar\sum_\delta S_{jx} S_{j+\delta y} - i\hbar\sum_\delta S_{j-\delta y} S_{jx}.$$

So, $$\left[\mathbf{S}_j, \sum_{l,\delta} \mathbf{S}_l \cdot \mathbf{S}_{l,\delta}\right] = i\hbar(\mathbf{S}_{j-\delta} \times \mathbf{S}_j - \mathbf{S}_j \times \mathbf{S}_{j+\delta}).$$

$$\left[S_{jx}, \sum_l S_{lz}\right] = -i\hbar S_{jy},$$

$$\left[S_{jy}, \sum_l S_{lz}\right] = i\hbar S_{jx}.$$

$$\mathbf{H} = H_0\hat{z}, \qquad \dot{\mathbf{S}}_j = -J\sum_\delta (\mathbf{S}_{j-\delta} \times \mathbf{S}_j - \mathbf{S}_j \times \mathbf{S}_{j+\delta}) + 2\mu_0 \mathbf{S}_j \times \mathbf{H}.$$

Taking the x and y components,

$$\dot{S}_j^{\pm} = \mp iJ\sum_{\delta}\{S_j^{\pm}S_{j-\delta z} - S_{j-\delta}^{\pm}S_{jz} - S_{j+\delta}^{\pm}S_{jz} + S_j^{\pm}S_{j+\delta z}\} \mp i2\mu_0 H_0 S_j^{\pm}.$$

Let $\Delta_{\delta}\mathbf{S}_j = \mathbf{S}_{j+\delta} + \mathbf{S}_{j-\delta}$,

$$\dot{S}_j^{\pm} = (\mp i)\left\{J\sum_{\delta}\left(S_j^{\pm}\,\Delta_{\delta}S_{jz} - S_{jz}\,\Delta_{\delta}S_j^{\pm}\right) + 2\mu_0 H_0 S_j^{\pm}\right\}.$$

For spin wave mode, $S_z \simeq \langle S_z\rangle \simeq S$. For $|S^{\pm}/S| \ll 1$,

$$\dot{S}_j^{\pm} \simeq (\mp i)\left[JS\sum_{\delta}\left(2S_j^{\pm} - \Delta_{\delta}S_j^{\pm}\right) + 2\mu_0 H_0 S_j^{\pm}\right].$$

Let $S_j^{\pm} = \Delta S e^{\pm i(\mathbf{k}\cdot\mathbf{r}_j - \omega t)}$, then

$$\omega = 2JS\sum_{\delta}(1 - \cos(\mathbf{k}\cdot\boldsymbol{\delta})) + 2\mu_0 H_0.$$

For long wavelength case,

$$\mathbf{S}_{j+\delta} = \mathbf{S}_j + (\boldsymbol{\delta}\cdot\nabla)\mathbf{S}_j + \tfrac{1}{2}(\boldsymbol{\delta}\cdot\nabla)^2\mathbf{S}_j,$$

$$\mathbf{S}_{j-\delta} = \mathbf{S}_j - (\boldsymbol{\delta}\cdot\nabla)\mathbf{S}_j + \tfrac{1}{2}(\boldsymbol{\delta}\cdot\nabla)^2\mathbf{S}_j,$$

$$\mathbf{S}_{j+\delta} + \mathbf{S}_{j-\delta} = z\mathbf{S}_j + (\boldsymbol{\delta}\cdot\nabla)^2\mathbf{S}_j,$$ where z is the coordination number.

For simple cubic case,

$$\sum_{\delta}(\boldsymbol{\delta}\cdot\nabla)^2 = 2a^2\left(\frac{\partial^2}{\partial x^2} + \frac{\partial^2}{\partial y^2} + \frac{\partial^2}{\partial z^2}\right) = 2a^2\nabla^2.$$

$$\dot{\mathbf{S}} = 2Ja^2\mathbf{S}\times\nabla^2\mathbf{S} + 2\mu_0\mathbf{S}\times\mathbf{H},$$ classically $\mathbf{S}\times\mathbf{S} = 0$.

8. $$\frac{1}{Z}\beta = \frac{1}{N}\sum_k\left[1 - \left(1 - \gamma_k^2\right)^{1/2}\right]$$

$$= \frac{1}{N}\int_{-\pi/a}^{\pi/a}[1 - |\sin ka|]\frac{L}{2\pi}\,dk = \frac{a}{\pi}\int_0^{\pi/a}[1 - \sin ka]\,dk,$$

$$\frac{L}{N} = \frac{Na}{N} = a.$$

$$= \frac{a}{\pi}\int_0^{\pi/2a}(1-\sin ka)\,dk + \frac{a}{\pi}\int_{\pi/2a}^{\pi/a}(1-\sin ka)\,dk$$

$$= \frac{2a}{\pi}\int_0^{\pi/2a}(1-\sin ka)\,dk = \frac{2a}{\pi}\left[\frac{\pi}{2a}-\frac{1}{a}\right] = 0.363.$$

$$Z = 2, \qquad \beta = 0.726.$$

10. $H = \sum_{\mathbf{k}}\{\hbar\omega_{\mathbf{k}}^m a_{\mathbf{k}}^\dagger a_{\mathbf{k}} + \hbar\omega_{\mathbf{k}}^P b_{\mathbf{k}}^\dagger b_{\mathbf{k}} + c_{\mathbf{k}}(a_{\mathbf{k}} b_{\mathbf{k}}^\dagger + a_{\mathbf{k}}^\dagger b_{\mathbf{k}})\}.$

Define

$$a_{\mathbf{k}}^\dagger = A_{\mathbf{k}}^\dagger\cos\theta_{\mathbf{k}} + B_{\mathbf{k}}^\dagger\sin\theta_{\mathbf{k}}, \qquad a_{\mathbf{k}} = A_{\mathbf{k}}\cos\theta_{\mathbf{k}} + B_{\mathbf{k}}\sin\theta_{\mathbf{k}}.$$

$$b_{\mathbf{k}}^\dagger = B_{\mathbf{k}}^\dagger\cos\theta_{\mathbf{k}} - A_{\mathbf{k}}^\dagger\sin\theta_{\mathbf{k}}, \qquad b_{\mathbf{k}} = B_{\mathbf{k}}\cos\theta_{\mathbf{k}} - A_{\mathbf{k}}\sin\theta_{\mathbf{k}}.$$

$$A_{\mathbf{k}}^\dagger = a_{\mathbf{k}}^\dagger\cos\theta_{\mathbf{k}} - b_{\mathbf{k}}^\dagger\sin\theta_{\mathbf{k}}, \qquad B_{\mathbf{k}}^\dagger = a_{\mathbf{k}}^\dagger\sin\theta_{\mathbf{k}} + b_{\mathbf{k}}^\dagger\cos\theta_{\mathbf{k}}.$$

$$A_{\mathbf{k}} = a_{\mathbf{k}}\cos\theta_{\mathbf{k}} - b_{\mathbf{k}}\sin\theta_{\mathbf{k}}, \qquad B_{\mathbf{k}} = a_{\mathbf{k}}\sin\theta_{\mathbf{k}} + b_{\mathbf{k}}\cos\theta_{\mathbf{k}}.$$

$$\left[A_{\mathbf{k}}^\dagger, B_{\mathbf{k}'}\right] = \sin\theta_{\mathbf{k}}\cos\theta_{\mathbf{k}}\left\{\left[a_{\mathbf{k}}^\dagger, a_{\mathbf{k}'}\right] - \left[b_{\mathbf{k}}^\dagger, b_{\mathbf{k}'}\right]\right\} = 0.$$

$$\left[A_{\mathbf{k}}, A_{\mathbf{k}'}^\dagger\right] = \cos^2\theta_{\mathbf{k}}\left[a_{\mathbf{k}}, a_{\mathbf{k}'}^\dagger\right] + \sin^2\theta_{\mathbf{k}}\left[b_{\mathbf{k}}, b_{\mathbf{k}'}^\dagger\right] = \delta_{\mathbf{k},\mathbf{k}'}.$$

Similarly, the other commutators can be worked out. Substituting $a_{\mathbf{k}}^\dagger$, $a_{\mathbf{k}}$, $b_{\mathbf{k}}^\dagger$, $b_{\mathbf{k}}$ into H, the cross terms must vanish:

$$\begin{aligned}
&\hbar\omega_{\mathbf{k}}^m\left(A_{\mathbf{k}}^\dagger B_{\mathbf{k}}\cos\theta_{\mathbf{k}}\sin\theta_{\mathbf{k}} + B_{\mathbf{k}}^\dagger A_{\mathbf{k}}\sin\theta_{\mathbf{k}}\cos\theta_{\mathbf{k}}\right)\\
&\quad + \hbar\omega_{\mathbf{k}}^P\left(-A_{\mathbf{k}}^\dagger B_{\mathbf{k}}\cos\theta_{\mathbf{k}}\sin\theta_{\mathbf{k}} - B_{\mathbf{k}}^\dagger A_{\mathbf{k}}\sin\theta_{\mathbf{k}}\cos\theta_{\mathbf{k}}\right)\\
&\quad + c_{\mathbf{k}}\left(-A_{\mathbf{k}}^\dagger B_{\mathbf{k}}\sin^2\theta_{\mathbf{k}} + B_{\mathbf{k}}^\dagger A_{\mathbf{k}}\cos^2\theta_{\mathbf{k}}\right)\\
&\quad + c_{\mathbf{k}}\left(A_{\mathbf{k}}^\dagger B_{\mathbf{k}}\cos^2\theta_{\mathbf{k}} - B_{\mathbf{k}}^\dagger A_{\mathbf{k}}\sin^2\theta_{\mathbf{k}}\right) = 0.
\end{aligned}$$

So, $\frac{1}{2}\sin 2\theta_{\mathbf{k}}\hbar(\omega_{\mathbf{k}}^m - \omega_{\mathbf{k}}^P) + c_{\mathbf{k}}\cos 2\theta_{\mathbf{k}} = 0$ and

$$\tan 2\theta_{\mathbf{k}} = \frac{2c_{\mathbf{k}}}{\hbar\left(\omega_{\mathbf{k}}^P - \omega_{\mathbf{k}}^m\right)}.$$

$$\begin{aligned}
H = \sum_{\mathbf{k}}\Big\{&\left(\hbar\omega_{\mathbf{k}}^m\cos^2\theta_{\mathbf{k}} + \hbar\omega_{\mathbf{k}}^P\sin^2\theta_{\mathbf{k}} - 2c_{\mathbf{k}}\sin\theta_{\mathbf{k}}\cos\theta_{\mathbf{k}}\right)A_{\mathbf{k}}^\dagger A_{\mathbf{k}}\\
&+ \left(\hbar\omega_{\mathbf{k}}^m\sin^2\theta_{\mathbf{k}} + \hbar\omega_{\mathbf{k}}^P\cos^2\theta_{\mathbf{k}} + 2c_{\mathbf{k}}\sin\theta_{\mathbf{k}}\cos\theta_{\mathbf{k}}\right)B_{\mathbf{k}}^\dagger B_{\mathbf{k}}\Big\}.
\end{aligned}$$

At cross-over, $\omega_{\mathbf{k}}^m = \omega_{\mathbf{k}}^P$, $\tan 2\theta_{\mathbf{k}} \to \infty$, $2\theta_{\mathbf{k}} = \pi/2$, $\theta_{\mathbf{k}} = \pi/4$. So, $\sin\theta_{\mathbf{k}} = \cos\theta_{\mathbf{k}} = \dfrac{1}{\sqrt{2}}$.

Let

$$\omega_{\mathbf{k}}^{m}=\omega_{\mathbf{k}}^{P}=\omega,\ H=\sum_{\mathbf{k}}\left\{\hbar(\omega-c_{\mathbf{k}})A_{\mathbf{k}}^{\dagger}A_{\mathbf{k}}+\hbar(\omega+c_{\mathbf{k}})B_{\mathbf{k}}^{\dagger}B_{\mathbf{k}}\right\}.$$

$$\omega_A=\omega-c_{\mathbf{k}},\ \omega_B=\omega+c_{\mathbf{k}},\ a_{\mathbf{k}}^{\dagger}=\frac{1}{\sqrt{2}}(A_{\mathbf{k}}^{\dagger}+B_{\mathbf{k}}^{\dagger}),\ a_{\mathbf{k}}=\frac{1}{\sqrt{2}}(A_{\mathbf{k}}+B_{\mathbf{k}}).$$

$$b_{\mathbf{k}}^{\dagger}=\frac{1}{\sqrt{2}}(B_{\mathbf{k}}^{\dagger}-A_{\mathbf{k}}^{\dagger}),\ b_{\mathbf{k}}=\frac{1}{\sqrt{2}}(B_{\mathbf{k}}-A_{\mathbf{k}}).$$

CHAPTER 5

1. a. $$\left\{c_j,c_k^{\dagger}\right\}=T_1\cdots T_{j-1}\begin{pmatrix}0&0\\1&0\end{pmatrix}_j T_1\cdots T_{k-1}\begin{pmatrix}0&1\\0&0\end{pmatrix}_k+T_1\cdots T_{k-1}$$
$$\begin{pmatrix}0&1\\0&0\end{pmatrix}_k T_1\cdots T_{j-1}\begin{pmatrix}0&0\\1&0\end{pmatrix}_j=\begin{pmatrix}0&0\\1&0\end{pmatrix}_j T_j\cdots T_{k-1}$$
$$\begin{pmatrix}0&1\\0&0\end{pmatrix}_k+T_j\cdots T_{k-1}\begin{pmatrix}0&1\\0&0\end{pmatrix}_k\begin{pmatrix}0&0\\1&0\end{pmatrix}_j,$$

for $j<k$. Let $a_j=\begin{pmatrix}0&0\\1&0\end{pmatrix}_j$, then $[a_j,a_i]=0$, $\{a_j,T_j\}=0$ and $[a_j,T_i]=0$, $j\neq i$; $T_j^2=1$, then $\{c_j,c_k^{\dagger}\}=0$.
Similarly, $\{c_j,c_k^{\dagger}\}=0$, for $j>k$.

$$j=k,\qquad \left\{c_j,c_j^{\dagger}\right\}=\begin{pmatrix}0&0\\1&0\end{pmatrix}_j\begin{pmatrix}0&1\\0&0\end{pmatrix}_j+\begin{pmatrix}0&1\\0&0\end{pmatrix}_j\begin{pmatrix}0&0\\1&0\end{pmatrix}_j$$
$$=\begin{pmatrix}1&0\\0&1\end{pmatrix}=1.$$

$$\left\{c_j,c_k^{\dagger}\right\}=\delta_{jk}.$$

$$\{c_j,c_k\}=T_1\cdots T_{j-1}\begin{pmatrix}0&0\\1&0\end{pmatrix}_j T_1\cdots T_{k-1}\begin{pmatrix}0&0\\1&0\end{pmatrix}_k$$
$$+T_1\cdots T_{k-1}\begin{pmatrix}0&0\\1&0\end{pmatrix}_k T_1\cdots T_{j-1}\begin{pmatrix}0&0\\1&0\end{pmatrix}_j.$$
$$j<k,\quad=\begin{pmatrix}0&0\\1&0\end{pmatrix}_j T_j\cdots T_{k-1}\begin{pmatrix}0&0\\1&0\end{pmatrix}_k$$
$$+T_j\cdots T_{k-1}\begin{pmatrix}0&0\\1&0\end{pmatrix}_k\begin{pmatrix}0&0\\1&0\end{pmatrix}_j$$
$$=-T_j\begin{pmatrix}0&0\\1&0\end{pmatrix}_j\cdots T_{k-1}\begin{pmatrix}0&0\\1&0\end{pmatrix}_k$$
$$+T_j\cdots T_{k-1}\begin{pmatrix}0&0\\1&0\end{pmatrix}_k\begin{pmatrix}0&0\\1&0\end{pmatrix}_j$$

$$= -T_j \cdot \cdot T_{k-1}\begin{pmatrix}0 & 0\\ 1 & 0\end{pmatrix}_j \begin{pmatrix}0 & 0\\ 1 & 0\end{pmatrix}_k$$

$$+ T_j \cdot \cdot T_{k-1}\begin{pmatrix}0 & 0\\ 1 & 0\end{pmatrix}_k \begin{pmatrix}0 & 0\\ 1 & 0\end{pmatrix}_j = 0.$$

Similarly,

$$\{c_j, c_k\} = 0, \qquad j > k.$$

$$j = k, \qquad \{c_j, c_k\} = \begin{pmatrix}0 & 0\\ 1 & 0\end{pmatrix}\begin{pmatrix}0 & 0\\ 1 & 0\end{pmatrix} + \begin{pmatrix}0 & 0\\ 1 & 0\end{pmatrix}\begin{pmatrix}0 & 0\\ 1 & 0\end{pmatrix} = 0.$$

Similarly, it is easy to show $\{c_j^\dagger, c_k^\dagger\} = 0$.

b. $$c_j^\dagger | \cdot \cdot n_j \cdot \cdot \rangle = (1 - n_j)\theta^j | \cdots 1_j \cdots \rangle$$

$$= T_1 \cdots T_{j-1}\begin{pmatrix}0 & 1\\ 0 & 0\end{pmatrix}_j (\ )_1 (\ )_2 \cdots (\ )_j$$

If jth state is occupied, $\begin{pmatrix}0 & 1\\ 0 & 0\end{pmatrix}_j \begin{pmatrix}1\\ 0\end{pmatrix}_j = 0.$

jth state is unoccupied, $\begin{pmatrix}0 & 1\\ 0 & 0\end{pmatrix}_j \begin{pmatrix}0\\ 1\end{pmatrix}_j = \begin{pmatrix}1\\ 0\end{pmatrix}_j = |1_j\rangle.$

$$i < j, \qquad T_i \begin{pmatrix}1\\ 0\end{pmatrix}_i = -\begin{pmatrix}1\\ 0\end{pmatrix}_i, \ T_i \begin{pmatrix}0\\ 1\end{pmatrix}_i = \begin{pmatrix}0\\ 1\end{pmatrix}_i.$$

$$T_1 \cdot \cdot T_{j-1}\begin{pmatrix}0 & 1\\ 0 & 0\end{pmatrix}_j (\ )_1 \cdot \cdot (\ )_j = \theta_j | \cdots 1_j \cdot \cdot \rangle (1 - n_j).$$

Similarly, it is easy to show the equivalent between Eqs. (23) and (27).

2. $$\int d^3 r F^2(k_F r) = \int d^3 r \frac{1}{N^2 (2\pi)^6} \int d^3 k \int d^3 k' e^{i(\mathbf{k} - \mathbf{k}') \cdot \mathbf{r}}$$

$$= \frac{1}{N^2 (2\pi)^3} \int d^3 k \int d^3 k' \,\delta(\mathbf{k} - \mathbf{k}')$$

$$= \frac{1}{N^2 (2\pi)^3} \int d^3 k = \frac{4\pi}{3} k_F^3 \cdot \frac{1}{N^2 (2\pi)^3}$$

$$= \frac{N}{N^2} = \frac{1}{N.}$$

4. Let $\rho(\mathbf{x}) = \int d^3 x'' \psi^\dagger(\mathbf{x}'') \,\delta(\mathbf{x} - \mathbf{x}'') \psi(\mathbf{x}'')$ and consider

$$\rho(\mathbf{x}) \psi^\dagger(\mathbf{x}') | \text{vac} \rangle = \int d^3 x'' \psi^\dagger(\mathbf{x}'') \,\delta(\mathbf{x} - \mathbf{x}'') \psi(\mathbf{x}'') \psi^\dagger(\mathbf{x}') | \text{vac} \rangle$$

$$= \int d^3x'' \psi^\dagger(\mathbf{x}'')$$

$$\times \delta(\mathbf{x} - \mathbf{x}'') \left[\delta(\mathbf{x}'' - \mathbf{x}') - \psi^\dagger(\mathbf{x}') \psi(\mathbf{x}'') \right] |\text{vac}\rangle .$$

$\psi |\text{vac}\rangle = 0$, so

$$\rho(\mathbf{x}) \psi^\dagger(\mathbf{x}') |\text{vac}\rangle = \delta(\mathbf{x} - \mathbf{x}') \psi^\dagger(\mathbf{x}') |\text{vac}\rangle .$$

So, $\psi^\dagger(\mathbf{x}')|\text{vac}\rangle$ is an eigenvector of $\rho(\mathbf{x})$ with eigenvalue $\delta(\mathbf{x} - \mathbf{x}')$. $\psi^\dagger(\mathbf{x}')|\text{vac}\rangle$ is a state for an electron at $\mathbf{x}'$.

6. $N = \int d^3x' \psi^\dagger(\mathbf{x}') \psi(\mathbf{x}')$.

Fermion: $$\psi(\mathbf{x}) N = \int d^3x' \psi(\mathbf{x}) \psi^\dagger(\mathbf{x}') \psi(\mathbf{x}')$$

$$= \int d^3x' \left[\psi(\mathbf{x}') \, \delta(\mathbf{x} - \mathbf{x}') - \psi^\dagger(\mathbf{x}') \psi(\mathbf{x}) \psi(\mathbf{x}') \right]$$

$$= \psi(\mathbf{x}) + \int d^3x' \psi^\dagger(\mathbf{x}') \psi(\mathbf{x}') \psi(\mathbf{x})$$

$$= \psi(\mathbf{x}) + N\psi(\mathbf{x}) = (N+1) \psi(\mathbf{x}).$$

Bosons: $$\psi(\mathbf{x}) N = \int d^3x' \psi(\mathbf{x}) \psi^\dagger(\mathbf{x}') \psi(\mathbf{x}')$$

$$= \int d^3x' \left[\psi(\mathbf{x}') \, \delta(\mathbf{x} - \mathbf{x}') + \psi^\dagger(\mathbf{x}') \psi(\mathbf{x}) \psi(\mathbf{x}') \right]$$

$$= \psi(\mathbf{x}) + N\psi(\mathbf{x}) = (N+1) \psi(x).$$

7. $$H = \sum_s \int d^3x' \psi_s^\dagger(\mathbf{x}') \frac{p^2}{2m} \psi_s(\mathbf{x}') + \tfrac{1}{2} \sum_{s,s'} \int d^3x' \int d^3y \psi_s^\dagger(\mathbf{x}') \psi_{s'}^\dagger(\mathbf{y}) g$$

$$\times \delta(\mathbf{x}' - \mathbf{y}) \psi_{s'}(\mathbf{y}) \psi_s(\mathbf{x}').$$

$$i\hbar \frac{\partial \psi_\alpha(\mathbf{x})}{\partial t} = [\psi_\alpha, H].$$

$$\left[\psi_\alpha(\mathbf{x}), \sum_s \int d^3x' \psi_s^\dagger(\mathbf{x}') \frac{p^2}{2m} \psi_s(\mathbf{x}') \right]$$

$$= \sum_s \int d^3x' \left[\delta(\mathbf{x} - \mathbf{x}') \, \delta_{\alpha s} - \psi_s^\dagger(\mathbf{x}') \psi_\alpha(\mathbf{x}) \right] \frac{p^2}{2m} \psi_s(\mathbf{x}')$$

$$- \sum_s \int d^3x' \psi_s^\dagger(\mathbf{x}') \frac{p^2}{2m} \psi_s(\mathbf{x}') \psi_\alpha(\mathbf{x})$$

$$= \frac{p^2}{2m}\psi_\alpha(\mathbf{x}) + \sum_s \int d^3x' \psi_s^\dagger(\mathbf{x}')\frac{p^2}{2m}\psi_s(\mathbf{x}')\psi_\alpha(\mathbf{x})$$

$$- \sum_s \int d^3x' \psi_s^\dagger(\mathbf{x}')\frac{p^2}{2m}\psi_s(\mathbf{x}')\psi_\alpha(\mathbf{x})$$

$$= \frac{p^2}{2m}\psi_\alpha(\mathbf{x}).$$

$$I = \left[\psi_\alpha(\mathbf{x}), \tfrac{1}{2}\sum_{s,s'}\int d^3x' \int d^3y \psi_s^\dagger(\mathbf{x}')\psi_{s'}^\dagger(\mathbf{y}) g\,\delta(\mathbf{x}'-\mathbf{y})\psi_{s'}(\mathbf{y})\psi_s(\mathbf{x}')\right]$$

$$= \frac{g}{2}\sum_{s'}\psi_{s'}^\dagger(\mathbf{x})\psi_{s'}(\mathbf{x})\psi_\alpha(\mathbf{x}) - \frac{g}{2}\sum_s \psi_s^\dagger(\mathbf{x})\psi_\alpha(\mathbf{x})\psi_s(\mathbf{x})$$

$$+ \tfrac{1}{2}\sum_{s,s'}\int d^3x' \int d^3y \{\psi_s^\dagger(\mathbf{x}')\psi_{s'}^\dagger(y)$$

$$\times \psi_\alpha(\mathbf{x}) g\,\delta(\mathbf{x}'-\mathbf{y})\psi_{s'}(\mathbf{y})\psi_s(\mathbf{x}')$$

$$-\psi_s^\dagger(\mathbf{x}')\psi_{s'}^\dagger(\mathbf{y})\,\delta(\mathbf{x}'-\mathbf{y})\psi_{s'}(\mathbf{y})\psi_s(\mathbf{x}')\psi_\alpha(\mathbf{x})\}.$$

Use $\{\psi_s(\mathbf{x}), \psi_{s'}(\mathbf{x}')\} = 0$, $I = g\psi_\beta^\dagger(\mathbf{x})\psi_\beta(\mathbf{x})\psi_\alpha(\mathbf{x})$, $\alpha \neq \beta$. $i\hbar\dot{\psi}_\alpha(\mathbf{x}) = \frac{p^2}{2m}\psi_\alpha(\mathbf{x}) + g\psi_\beta^\dagger(\mathbf{x})\psi_\beta(\mathbf{x})\psi_\alpha(\mathbf{x})$. For $i\hbar\dot{\psi}_\alpha^\dagger(\mathbf{x}) = [\psi_\alpha^\dagger(\mathbf{x}), H]$, because $\psi_\alpha^\dagger(\mathbf{x})\psi_s^\dagger(\mathbf{x}')\psi_s(\mathbf{x}') = -\psi_s^\dagger(\mathbf{x}')\psi_\alpha^\dagger(\mathbf{x})\psi_s(\mathbf{x}')$, there is an extra "$-$" sign. Otherwise, the steps are the same to prove the result.

CHAPTER 6

4. $\ddot{x} + \eta\dot{x} = eE/m$ Let $x = x_0 e^{-i\omega t}$,

$$-\omega^2 x_0 - i\omega\eta x_0 = eE/m, \qquad x_0 = \frac{-eE/m}{\omega^2 + i\eta\omega}.$$

$$P = nex_0 = \frac{-ne^2E/m}{\omega^2 + i\eta\omega} = \alpha E, \qquad \alpha = \frac{-ne^2/m}{\omega^2 + i\eta\omega}.$$

$$D = E + 4\pi P = \left(1 - \frac{4\pi ne^2/m}{\omega^2 + i\eta\omega}\right)E.$$

$$\varepsilon(\omega) = 1 - \frac{\omega_p^2}{\omega^2 + i\eta\omega}.$$

$$\frac{1}{\varepsilon(\omega)} = \frac{\omega^2 + i\eta\omega}{\omega^2 + i\eta\omega - \omega_p^2} = \frac{(\omega + i\eta)\omega}{(\omega+\omega_p)(\omega-\omega_p) + i\eta\omega}.$$

$$\frac{1}{\varepsilon(\omega)} \simeq \frac{\omega(\omega+i\eta)}{2\omega(\omega-\omega_p)+i\eta\omega}$$

$$= \frac{1}{2}(\omega+i\eta)\left\{\mathscr{P}\frac{1}{\omega-\omega_p} - i2\pi\,\delta(\omega-\omega_p)\right\},$$

where $\mathscr{P}$ is the principal value.

$$\lim_{\eta\to 0}\mathscr{I}\left(\frac{1}{\varepsilon}\right) = -\pi\omega\,\delta(\omega-\omega_p).$$

$$E_{\text{int}} = -\sum_{\mathbf{q}}\left\{\frac{1}{2\pi}\int_0^\infty d\omega\,\mathscr{I}\left(\frac{1}{\varepsilon}\right) + \frac{2\pi ne^2}{q^2}\right\}$$

$$= -\sum_{\mathbf{q}}\left\{\frac{-1}{2\pi}\cdot\pi\omega_p + \frac{2\pi ne^2}{q^2}\right\} = \sum_{\mathbf{q}}\left\{\frac{\omega_p}{2} - \frac{2\pi ne^2}{q^2}\right\}.$$

6. $$\frac{1}{\varepsilon(\omega,\mathbf{q})} = 1 - \frac{4\pi e^2}{q^2}\sum_n |\langle n|\rho_{\mathbf{q}}|0\rangle|^2$$

$$\cdot\left\{\frac{1}{\omega+\omega_{no}+is} + \frac{1}{-\omega+\omega_{no}-is}\right\}$$

$$\simeq 1 - \frac{4\pi e^2}{q^2}\sum_n |\langle n|\rho_{\mathbf{q}}|0\rangle|^2$$

$$\cdot\left\{\frac{1}{\omega}\left(1-\frac{\omega_{no}}{\omega}\right) - \frac{1}{\omega}\left(1+\frac{\omega_{no}}{\omega}\right)\right\}$$

$$= 1 + \frac{4\pi e^2}{\omega^2 q^2}\sum_n\left\{\omega_{no}|\langle n|\rho_{\mathbf{q}}|0\rangle|^2 + \omega_{no}|\langle n|\rho_{-\mathbf{q}}|0\rangle|^2\right\}$$

$$= 1 + \frac{4\pi e^2}{\omega^2 q^2}\cdot\frac{n}{m}q^2 = 1 + \frac{4\pi ne^2}{m\omega^2},\qquad \text{from Eq. (121).}$$

$$\varepsilon(\omega,\mathbf{q}) = 1 - \frac{4\pi ne^2}{m\omega^2},\qquad \text{for large } \omega.$$

7. $\displaystyle\int_0^\infty d\omega\,\omega\,\mathscr{I}\left(\frac{1}{\varepsilon(\omega,\mathbf{q})}\right)$

$$= \int_0^\infty d\omega\cdot\omega\left\{\frac{4\pi e^2}{q^2}\sum_n\Big[|\langle n|\rho_{\mathbf{q}}|0\rangle|^2\pi\,\delta(\omega+\omega_{no}) - |\langle n|\rho_{\mathbf{q}}|0\rangle|^2\pi\,\delta(\omega-\omega_{no})\Big]\right\}$$

$$= -\frac{\pi}{2}\frac{4\pi e^2}{q^2}\sum_n\omega_{no}\Big(|\langle n|\rho_{\mathbf{q}}|0\rangle|^2 + |\langle n|\rho_{-\mathbf{q}}|0\rangle|^2\Big)$$

$$= -\frac{\pi}{2}\frac{4\pi e^2}{q^2}\cdot\frac{n}{m}q^2 = -\frac{\pi}{2}\frac{4\pi n e^2}{m} = -\frac{\pi}{2}\omega_p^2.$$

8. $\displaystyle\int_0^\infty d\sigma_1(\omega,\mathbf{q}) = \int_0^\infty d\omega\cdot\frac{\omega}{4\pi}\varepsilon_2(\omega,\mathbf{q})$

$$= -\frac{i}{8\pi}\int_{-\infty}^\infty d\omega\,\omega\,\varepsilon(\omega,\mathbf{q})$$

Using result in Prob. 6 and taking a contour along the real axis with a great circle closed in the upper half plane, we have

$$\int_{-\infty}^\infty d\omega\cdot\omega\cdot\varepsilon(\omega,\mathbf{q}) + iR^2\int_0^\pi d\theta e^{i2\theta}\left(1-\frac{\omega_p^2}{R^2}e^{-i2\theta}\right) = 0$$

where R is the radius of the circle.

$$\int_{-\infty}^\infty d\omega\cdot\omega\cdot\varepsilon(\omega,\mathbf{q}) = -iR^2\left(-\frac{\omega_p^2}{R^2}\right)\pi \underset{R\to\infty}{=} i\pi\omega_p^2$$

$$\int_0^\infty d\omega\,\sigma_1(\omega,\mathbf{q}) = \frac{1}{8}\omega_p^2$$

9. From Eq. (64),

$$\mathscr{S}(\omega,\mathbf{q}) = \sum_n |\langle n|\rho_{\mathbf{q}}^\dagger|0\rangle|^2\,\delta(\omega-\omega_{no}).$$

$$\int_0^\infty \mathscr{S}(\omega,\mathbf{q})\,d\omega = \int_0^\infty\sum_n|\langle n|\rho_{\mathbf{q}}^\dagger|0\rangle|^2\,\delta(\omega-\omega_{no})\,d\omega$$

$$= \sum_n \langle 0|\rho_{\mathbf{q}}|n\rangle\langle n|\rho_{\mathbf{q}}^\dagger|0\rangle = \langle 0|\rho_{\mathbf{q}}\rho_{\mathbf{q}}^\dagger|0\rangle$$

$$= N\mathscr{S}(\mathbf{q}).$$

10. $\int_0^\infty d\omega\, \omega\, \mathscr{S}(\omega,\mathbf{q})$

$$= \int_0^\infty d\omega\, \omega \sum_n \left|\langle n|\rho_{\mathbf{q}}^\dagger|0\rangle\right|^2 \delta(\omega - \omega_{no}),$$

from Eq. (64)

$$= \frac{1}{2}\left\{\int_0^\infty d\omega\, \omega \sum_n \left|\langle n|\rho_{\mathbf{q}}^\dagger|0\rangle\right|^2 \delta(\omega - \omega_{no}) + \int_0^\infty d\omega\, \omega \sum_n \left|\langle n|\rho_{-q}^\dagger|0\rangle\right|^2 \delta(\omega - \omega_{no})\right\},$$

$$\mathscr{S}(\omega,\mathbf{q}) = \mathscr{S}(\omega,-\mathbf{q}).$$

$$= \frac{1}{2}\sum_n \left|\langle n|\rho_{\mathbf{q}}^\dagger|0\rangle\right|^2 \omega_{no} + \frac{1}{2}\sum_n \left|\langle n|\rho_{-\mathbf{q}}^\dagger|0\rangle\right|^2 \omega_{no}$$

$$= \frac{1}{2}\frac{n}{m}q^2, \qquad \text{from Eq. (121).}$$

11. $\left[A_{\mathbf{k}}^\dagger(\mathbf{q}), A_{\mathbf{k}'}^\dagger(\mathbf{q}')\right] = \alpha_{\mathbf{k}+\mathbf{q}}^\dagger \beta_{-\mathbf{k}}^\dagger \alpha_{\mathbf{k}'+\mathbf{q}'}^\dagger \beta_{-\mathbf{k}'}^\dagger - \alpha_{\mathbf{k}'+\mathbf{q}'}^\dagger \beta_{-\mathbf{k}'}^\dagger \alpha_{\mathbf{k}+\mathbf{q}}^\dagger \beta_{-\mathbf{k}}^\dagger$

$$= -\alpha_{\mathbf{k}'+\mathbf{q}'}^\dagger \alpha_{\mathbf{k}+\mathbf{q}}^\dagger \beta_{-\mathbf{k}'}^\dagger \beta_{-\mathbf{k}}^\dagger + \alpha_{\mathbf{k}'+\mathbf{q}'}^\dagger \alpha_{\mathbf{k}+\mathbf{q}}^\dagger \beta_{-\mathbf{k}'}^\dagger \beta_{-\mathbf{k}}^\dagger = 0.$$

$\{\alpha^\dagger, \alpha^\dagger\} = 0$, $\{\alpha^\dagger, \beta^\dagger\} = 0$, and $\{\beta^\dagger, \beta^\dagger\} = 0$ are used.

$\left[A_{\mathbf{k}}(\mathbf{q}), A_{\mathbf{k}'}^\dagger(\mathbf{q}')\right]$

$$= \beta_{-\mathbf{k}}\alpha_{\mathbf{k}+\mathbf{q}}\alpha_{\mathbf{k}'+\mathbf{q}'}^\dagger \beta_{-\mathbf{k}'}^\dagger - \alpha_{\mathbf{k}'+\mathbf{q}'}^\dagger \beta_{-\mathbf{k}'}^\dagger \beta_{-\mathbf{k}}\alpha_{\mathbf{k}+\mathbf{q}}$$

$$= \beta_{-\mathbf{k}}\left(\delta_{\mathbf{k}+\mathbf{q},\mathbf{k}'+\mathbf{q}'} - \alpha_{\mathbf{k}'+\mathbf{q}'}^\dagger \alpha_{\mathbf{k}+\mathbf{q}}\right)\beta_{-\mathbf{k}'}^\dagger - \alpha_{\mathbf{k}'+\mathbf{q}'}^\dagger \beta_{-\mathbf{k}'}^\dagger \beta_{-\mathbf{k}}\alpha_{\mathbf{k}+\mathbf{q}}$$

$$= \delta_{\mathbf{k}+\mathbf{q},\mathbf{k}'+\mathbf{q}'}\left(\delta_{\mathbf{k},\mathbf{k}'} - \beta_{-\mathbf{k}'}^\dagger \beta_{-\mathbf{k}}\right) - \alpha_{\mathbf{k}'+\mathbf{q}'}^\dagger \beta_{-\mathbf{k}}\beta_{-\mathbf{k}'}^\dagger \alpha_{\mathbf{k}+\mathbf{q}} - \alpha_{\mathbf{k}'+\mathbf{q}'}^\dagger \beta_{-\mathbf{k}'}^\dagger \beta_{-\mathbf{k}}\alpha_{\mathbf{k}+\mathbf{q}}$$

$$= \delta_{\mathbf{k}+\mathbf{q},\mathbf{k}'+\mathbf{q}'}\,\delta_{\mathbf{k},\mathbf{k}'} - \delta_{\mathbf{k}+\mathbf{q},\mathbf{k}'+\mathbf{q}'}\beta_{-\mathbf{k}'}^\dagger \beta_{-\mathbf{k}} - \alpha_{\mathbf{k}'+\mathbf{q}'}^\dagger\left(\delta_{\mathbf{k},\mathbf{k}'} - \beta_{-\mathbf{k}'}^\dagger \beta_{-\mathbf{k}}\right)\alpha_{\mathbf{k}+\mathbf{q}} - \alpha_{\mathbf{k}'+\mathbf{q}'}^\dagger \beta_{-\mathbf{k}'}^\dagger \beta_{-\mathbf{k}}\alpha_{\mathbf{k}+\mathbf{q}}$$

$$= \delta_{\mathbf{k}+\mathbf{q},\mathbf{k}'+\mathbf{q}'}\,\delta_{\mathbf{k},\mathbf{k}'} - \delta_{\mathbf{k}+\mathbf{q},\mathbf{k}'+\mathbf{q}'}\beta_{-\mathbf{k}'}^\dagger \beta_{-\mathbf{k}} - \delta_{\mathbf{k},\mathbf{k}'}\alpha_{\mathbf{k}'+\mathbf{q}'}^\dagger \alpha_{\mathbf{k}+\mathbf{q}}.$$

For unperturbed vacuum state $\beta^{\dagger}_{-\mathbf{k}}\beta_{-\mathbf{k}}|0\rangle = 0$,
$\alpha^{\dagger}_{\mathbf{k}'+\mathbf{q}'}\alpha_{\mathbf{k}'+\mathbf{q}'}|0\rangle = 0$.
So, $[A_{\mathbf{k}}(\mathbf{q}), A^{\dagger}_{\mathbf{k}'}(\mathbf{q}')] \simeq \delta_{\mathbf{k},\mathbf{k}'}\delta_{\mathbf{q},\mathbf{q}'}$.

CHAPTER 7

1. From Eq. (21), $\langle N\rangle = \dfrac{1}{\pi^2}\dfrac{M^{*2}C_1^2}{\rho c_s \hbar^3}\displaystyle\int_0^{q_m}\dfrac{q}{(q+q_c)^2}\,dq$, $\quad M^*$ is the mass of the proton. Now, $q_c \sim 2M^* c_s/\hbar$, $\quad q_m/q_c \sim \dfrac{10^8}{10^9} \sim \dfrac{1}{10}$.

$$\langle N\rangle \simeq \frac{1}{\pi^2}\frac{M^{*2}C_1^2}{\hbar^3\rho c_s}\int_0^{q_m}\frac{q}{q_c^2}\,dq$$

$$= \frac{1}{\pi^2}\frac{M^{*2}C_1^2}{\hbar^3\rho c_s}\cdot\frac{q_m^2}{2q_c^2} = \frac{1}{\pi^2}\frac{M^{*2}C_1^2}{\hbar^3\rho c_s}\cdot\frac{q_m^2}{2}\cdot\frac{\hbar^2}{4M^{*2}c_s^2}$$

$$= \frac{1}{8\pi^2}\frac{C_1^2 q_m^2}{\hbar\rho c_s^3}.$$

$C_1 \sim 5\times 10^{-11}$ erg, $\quad q_m \sim 10^8\ \text{cm}^{-1}$,

$\rho \sim 5$, $\quad c_s \sim 5\times 10^5$ cm/sec,

$\langle N\rangle \sim 500 \gg 1$.

2. $$\Delta\epsilon = \sum_{\mathbf{q}}\frac{|\langle \mathbf{k}-\mathbf{q}, n_{\mathbf{q}}+1|H'|\mathbf{k}, n_{\mathbf{q}}\rangle|^2}{\varepsilon_{\mathbf{k}} - \varepsilon_{\mathbf{k}-\mathbf{q}} - \hbar\omega_{\mathbf{q}}}$$

where $H' = iC_1\displaystyle\sum_{\mathbf{k}'\mathbf{q}'}\sqrt{\frac{\hbar}{2\rho\omega_{\mathbf{q}'}}}\,|\mathbf{q}'|(a_{\mathbf{q}'} - a^{\dagger}_{-\mathbf{q}'})c^{\dagger}_{\mathbf{k}'+\mathbf{q}'}c_{\mathbf{k}'}$.

$$\Delta\epsilon = \frac{\hbar C_1^2}{2\rho c_s}\sum_{\mathbf{q}}\frac{q}{\dfrac{\hbar^2}{2m^*}\left(2\mathbf{k}\cdot\mathbf{q} - q^2 - \dfrac{2m^*}{\hbar}c_s q\right)}, \qquad q_c = 2m^* c_s/\hbar.$$

$$= \frac{C_1^2 m^*}{\hbar\rho c_s}\frac{2\pi}{(2\pi)^3}\int_0^{q_m}\int_{-1}^{1}\frac{q^2\,dq\,d\mu}{(2kq\mu - q - q_c)}$$

$$= \frac{C_1^2 m^*}{\hbar\rho c_s}\frac{1}{(2\pi)^2}\int_0^{q_m}\frac{q^2}{2k}\ln\left|\frac{2k - q - q_c}{2k + q + q_c}\right|dq.$$

Because the factor q^2, the contribution of the integrand comes mainly from large q, i.e., near q_m. For large q, we have

$$ln\left|\frac{2k-q-q_c}{2k+q+q_c}\right| = ln\left|\frac{1-\dfrac{2k}{q+q_c}}{1+\dfrac{2k}{q+q_c}}\right| = ln\left|\frac{1-x}{1+x}\right|,$$

where $x = \dfrac{2k}{q+q_c}$.

$$= -2\left(x+\frac{x^3}{3}+\cdots\right) \simeq -2x = -\frac{4k}{q+q_c}.$$

$$\Delta\epsilon \simeq \frac{C_1^2 m^*}{\hbar\rho c_s}\frac{1}{(2\pi)^2}\left\{\int_{q_0}^{q_m}\frac{-2q^2\,dq}{q+q_c} + \int_0^{q_0}\frac{q^2\,dq}{2k}ln\left|\frac{2k-q-q_c}{2k+q+q_c}\right|\right\}$$

where q_0 is chosen such that $q_0 > q_c$. For $q_m \gg k \gg q_c$,

$$2k \gg q - q_c, \quad ln\left|\frac{2k-q-q_c}{2k+q+q_c}\right| = ln\left|\frac{2k}{2k}\right| = 0.$$

The first integrand can be approximated as $-2q$. For $q_m \gg k > q_0$, then

$$\Delta\epsilon \simeq -\frac{C_1^2 m^*}{\hbar\rho c_s}\frac{q_m^2}{4\pi^2}.$$

With $C_1 \sim 5\times10^{-11}$ erg, $m^* \sim 0.9\times10^{-27}$ g, $q_m \sim 10^8$ cm^{-1}, $\rho \sim 5$, $c_s \sim 5\times10^5$ cm/sec,

$$\Delta\epsilon \sim \tfrac{1}{40}\times10^{-11}\,\text{erg} = 0.25\times10^{-12}\,\text{erg} \simeq 0.1\ \text{eV}.$$

4. $$H = \hbar\omega_l\sum_{\mathbf{q}} b_{\mathbf{q}}^\dagger b_{\mathbf{q}} + e\phi(\mathbf{x}_1) + e\phi(\mathbf{x}_2)$$

$$= \hbar\omega_l\sum_{\mathbf{q}} b_{\mathbf{q}}^\dagger b_{\mathbf{q}} - i4\pi Fe\sum_{\mathbf{q}}\frac{1}{q}\Big(b_{\mathbf{q}}e^{i\mathbf{q}\cdot\mathbf{x}_1} - b_{\mathbf{q}}^\dagger e^{-i\mathbf{q}\cdot\mathbf{x}_1}$$
$$+ b_{\mathbf{q}}e^{i\mathbf{q}\cdot\mathbf{x}_2} - b_{\mathbf{q}}^\dagger e^{-i\mathbf{q}\cdot\mathbf{x}_2}\Big).$$

Let $a_{\mathbf{q}} = ub_{\mathbf{q}} + v$, $a_{\mathbf{q}}^{\dagger} = u^* b_{\mathbf{q}}^{\dagger} + v^*$, and with $[a_{\mathbf{q}}, a_{\mathbf{q}'}^{\dagger}] = \delta_{\mathbf{q},\mathbf{q}'}$, we get $u^* u = 1$. Furthermore, from $[a_{\mathbf{q}}, H] = \lambda a_{\mathbf{q}}$, $[a_{\mathbf{q}}^{\dagger}, H] = -\lambda a_{\mathbf{q}}^{\dagger}$, we want to find u, v, λ.

$$\left[a_{\mathbf{q}}, H\right] = \hbar u \omega_l b_{\mathbf{q}} + \frac{i4\pi Fe}{q}\left(e^{-i\mathbf{q}\cdot\mathbf{x}_1} + {}^{-i\mathbf{q}\cdot\mathbf{x}_2}\right) = \lambda\left(ub_{\mathbf{q}} + v\right).$$

So, $\lambda = \hbar\omega_l$, $\quad v = \dfrac{i4\pi Fe}{\hbar\omega_l q}\left(e^{-i\mathbf{q}\cdot\mathbf{x}_1} + e^{-i\mathbf{q}\cdot\mathbf{x}_2}\right)$, $\quad u = 1$.

Consider

$$\hbar\omega_l a_{\mathbf{q}}^{\dagger} a_{\mathbf{q}} = \hbar\omega_l\left[b_{\mathbf{q}}^{\dagger} - \frac{i4\pi Fe}{\hbar\omega_l \mathbf{q}}\left(e^{i\mathbf{q}\cdot\mathbf{x}_1} + e^{i\mathbf{q}\cdot\mathbf{x}_2}\right)\right]$$
$$\cdot\left[b_{\mathbf{q}} + \frac{i4\pi Fe}{\hbar\omega_l q}\left(e^{-i\mathbf{q}\cdot\mathbf{x}_1} + e^{-i\mathbf{q}\cdot\mathbf{x}_2}\right)\right]$$
$$= \hbar\omega_l b_{\mathbf{q}}^{\dagger} b_{\mathbf{q}} - \frac{i4\pi Fe}{q}\left(-b_{\mathbf{q}}^{\dagger} e^{-i\mathbf{q}\cdot\mathbf{x}_1} - b_{\mathbf{q}}^{\dagger} e^{-i\mathbf{q}\cdot\mathbf{x}_2}\right.$$
$$\left. + b_{\mathbf{q}} e^{i\mathbf{q}\cdot\mathbf{x}_1} + b_{\mathbf{q}} e^{i\mathbf{q}\cdot\mathbf{x}_2}\right)$$
$$+ \frac{(4\pi Fe)^2}{\hbar\omega_l q^2}\left(2 + e^{i\mathbf{q}\cdot(\mathbf{x}_1 - \mathbf{x}_2)} + e^{-i\mathbf{q}\cdot(\mathbf{x}_1 - \mathbf{x}_2)}\right)$$

$$H = \hbar\omega_l \sum_{\mathbf{q}} a_{\mathbf{q}}^{\dagger} a_{\mathbf{q}} - \sum_{\mathbf{q}} \frac{2(4\pi Fe)^2}{\hbar\omega_l q^2} - \sum_{\mathbf{q}} \frac{2(4\pi Fe)^2}{\hbar\omega_l q^2} e^{i\mathbf{q}\cdot(\mathbf{x}_1 - \mathbf{x}_2)}$$

The last term is obtained by $\mathbf{q} \to -\mathbf{q}$ in the sum.

6. a. $\tilde{H} = e^{-S} H e^{S} = \left(1 - S + \tfrac{1}{2}S^2 - \cdots\right) H \left(1 + S + \tfrac{1}{2}S^2 + \cdots\right)$
$= H - SH + HS + \left(\tfrac{1}{2}S^2 H - SHS + \tfrac{1}{2}HS^2\right) + \cdots$
$= H + [H, S] + \tfrac{1}{2}[[H, S], S] + \cdots$

b. $H = H_0 + \lambda H'$

$\tilde{H} = H_0 + \lambda H' + [H_0 + \lambda H', S]$
$+ \tfrac{1}{2}[[H_0 + \lambda H', S], S] + \cdots$
$= H_0 + \lambda H' + [H_0, S] + \lambda[H', S]$
$+ \tfrac{1}{2}[[H_0, S], S] + \tfrac{1}{2}[[\lambda H', S], S] + \cdots$

If S is chosen, such that

$$\lambda H' + [H_0, S] = 0, \text{ then } \lambda H' + H_0 S - S H_0 = 0$$

implies $|S| \sim \lambda$.

All the terms in $\tilde{H}$ are either of the order of λ^0 or λ^n, where $n > 1$. From $SH_0 - H_0 S = \lambda H'$, we have

$$\langle n|SH_0|m\rangle - \langle n|H_0 S|m\rangle = \langle n|\lambda H'|m\rangle \qquad H_0|m\rangle = E_m|m\rangle$$

$$\langle n|S|m\rangle = \frac{\lambda\langle n|H'|m\rangle}{E_m - E_n} \qquad E_m \neq E_n.$$

$$\begin{aligned}\tilde{H} &= H_0 + \tfrac{1}{2}[[H_0, S], S] + \lambda[H', S] + 0(\lambda^3) \\ &= H_0 - \tfrac{1}{2}[\lambda H', S] + \lambda[H', S] + 0(\lambda^3) \\ &= H_0 + \tfrac{1}{2}[\lambda H', S] + 0(\lambda^3).\end{aligned}$$

c. $H = \hbar\omega a^\dagger a + \lambda(a^\dagger + a) = H_0 + \lambda H', \qquad H_0 = \hbar\omega a^\dagger a,$

$H' = (a^\dagger + a)$.

$$\langle n|\tilde{H}|n\rangle = \langle n|H_0|n\rangle + \frac{\lambda}{2}\langle n|H'S - SH'|n\rangle$$

$$= n\hbar\omega + \frac{\lambda}{2}\sum_m \{\langle n|H'|m\rangle\langle m|S|n\rangle - \langle n|S|m\rangle\langle m|H'|n\rangle\}$$

$$= n\hbar\omega + \frac{\lambda^2}{2}\sum_m \left\{\frac{|\langle n|H'|m\rangle|^2}{E_n - E_m} - \frac{|\langle n|H'|m\rangle|^2}{E_m - E_n}\right\},$$

result in b is used.

$$\langle m|(a^\dagger + a)|n\rangle = \delta_{m,n+1}\sqrt{n+1} + \delta_{m,n-1}\sqrt{n}.$$

$$\frac{\langle n|a^\dagger + a|m\rangle}{E_n - E_m} = \frac{\delta_{n,m+1}\sqrt{m+1}}{\hbar\omega} - \frac{\delta_{n,m-1}\sqrt{m}}{\hbar\omega}.$$

$$\langle n|\tilde{H}|n\rangle = n\hbar\omega + \frac{\lambda^2}{2}\left\{-\frac{(n+1)}{\hbar\omega} + \frac{n}{\hbar\omega} - \frac{(n+1)}{\hbar\omega} + \frac{n}{\hbar\omega}\right\} = n\hbar\omega - \frac{\lambda^2}{\hbar\omega}.$$

d. $\tilde{H} = H_0 + \frac{1}{2}[\lambda H', S] + 0(\lambda^3), \qquad H_0|0\rangle = E_0|0\rangle.$

$$\tilde{\Phi}_0 = |0\rangle + \sum_n \frac{|n\rangle\langle n|\frac{1}{2}[\lambda H', S]|0\rangle}{E_0 - E_n} + \cdots$$

$$= |0\rangle + \frac{\lambda^2}{2}\sum_n |n\rangle$$

$$\cdot\left\{\sum_m \left[\frac{\langle n|a^\dagger + a|m\rangle\langle m|a^\dagger + a|0\rangle}{(E_0 - E_n)(E_0 - E_m)}\right.\right.$$

$$\left.\left. - \frac{\langle n|a^\dagger + a|m\rangle\langle m|a^\dagger + a|0\rangle}{(E_m - E_n)(E_0 - E_n)}\right]\right\} + \cdots$$

for $\lambda \neq 0$, $\tilde{\Phi}_0$ does not have boson in first order of λ. Consider,

$$\Phi_0 = e^S\tilde{\Phi}_0 = \left(1 + S + \frac{1}{2!}S^2 + \cdots\right)\tilde{\Phi}_0$$

$$= \left(1 + S + \frac{1}{2!}S^2 + \cdots\right)\left(|0\rangle + \sum_n |n\rangle 0(\lambda^2) + \cdots\right)$$

$$= |0\rangle + S|0\rangle + \cdots$$

$$= |0\rangle + \sum_n |n\rangle\langle n|S|0\rangle + \cdots$$

$$= |0\rangle - \frac{\lambda}{\hbar\omega}|1\rangle + \cdots$$

Φ_0 has boson to 1st order of λ.

7. a. The electric field produced by the polarization is given by $\nabla \cdot (\mathbf{E} + 4\pi\mathbf{P}) = 0$.

$$\mathbf{E} = -\nabla\phi(z) = -\hat{z}\frac{\partial\phi}{\partial z}.$$

$$\mathbf{P} = \hat{z}P_z = \hat{z}C_p e_{zz} = \hat{z}C_p\frac{\partial R_z}{\partial z}.$$

$$\nabla \cdot \mathbf{E} = -\frac{\partial^2\phi}{\partial z^2} = -4\pi\nabla \cdot \mathbf{P} = -4\pi C_p\frac{\partial^2 R_z}{\partial z^2}.$$

So, $\phi(z) = 4\pi C_p R_z(z)$,

$$R_z = \sum_q \sqrt{\frac{\hbar}{2\rho\omega_q}} \left(b_q e^{iqz} + b_q^\dagger e^{-iqz} \right),$$

ρ is the density of the medium.

The electron-phonon interaction is $-e\phi(z)$.

$$H'(z) = -4\pi C_p e \sum_{\mathbf{q}} \sqrt{\frac{\hbar}{2\rho\omega_{\mathbf{q}}}} \left(b_{\mathbf{q}} e^{iqz} + b_{\mathbf{q}}^\dagger e^{-iqz} \right)$$
$$= -e\phi(z), \quad \mathbf{q} = q\hat{z}.$$

In second quantized form, we have

$$H' = \int \psi^*(\mathbf{x})(-e\phi(z))\psi(\mathbf{x})\, d^3x.$$

$$\psi(\mathbf{x}) = \sum_{\mathbf{k}} a_{\mathbf{k}} e^{i\mathbf{k}\cdot\mathbf{x}} \frac{1}{\sqrt{V}},$$

where V is the volume of the sample.

$$H' = -4\pi e C_p \sum_{\mathbf{q}} \sum_{\mathbf{k}} \sqrt{\frac{\hbar}{2\rho\omega_{\mathbf{q}}}} \left(a_{\mathbf{k}}^\dagger a_{\mathbf{k}-\mathbf{q}} b_{\mathbf{q}} + a_{\mathbf{k}}^\dagger a_{\mathbf{k}+\mathbf{q}} b_{\mathbf{q}}^\dagger \right).$$

$$H' \sim \sqrt{\frac{1}{\omega_{\mathbf{q}}}} \sim q^{-1/2}, \qquad \omega_{\mathbf{q}} = c_s q.$$

b. The mobility $\mu_{\mathbf{k}} = \dfrac{e\tau_{\mathbf{k}}}{m}$. The scattering rate due to electron-phonon interaction is

$$\frac{1}{\tau_{\mathbf{k}}} = \frac{2\pi}{\hbar} \sum_{\mathbf{q}} \Big\{ \left| \langle \mathbf{k}-\mathbf{q}, n_{\mathbf{q}}+1 | H' | \mathbf{k}, n_{\mathbf{q}} \rangle \right|^2 \delta(\varepsilon_{\mathbf{k}} - \hbar\omega_{\mathbf{q}} - \varepsilon_{\mathbf{k}-\mathbf{q}})$$
$$+ \left| \langle \mathbf{k}+\mathbf{q}, n_{\mathbf{q}}-1 | H' | \mathbf{k}, n_{\mathbf{q}} \rangle \right|^2 \delta(\varepsilon_{\mathbf{k}} + \hbar\omega_{\mathbf{q}} - \varepsilon_{\mathbf{k}+\mathbf{q}}) \Big\}$$
$$= \frac{2\pi}{\hbar} (4\pi e C_p)^2 \sum_{\mathbf{q}} \left(\frac{\hbar}{2\rho\omega_{\mathbf{q}}} \right) \Big\{ (n_{\mathbf{q}}+1)\, \delta(\varepsilon_{\mathbf{k}} - \hbar\omega_{\mathbf{q}} - \varepsilon_{\mathbf{k}-\mathbf{q}})$$
$$+ n_{\mathbf{q}}\, \delta(\varepsilon_{\mathbf{k}} + \hbar\omega_{\mathbf{q}} - \varepsilon_{\mathbf{k}+\mathbf{q}}) \Big\}.$$

$$\mu_{\mathbf{k}}^{-1} = \left(\frac{m}{e}\right)\frac{2\pi}{\hbar}(4\pi eC_p)^2\sum_{\mathbf{q}}\left(\frac{\hbar}{2\rho\omega_{\mathbf{q}}}\right)$$

$$\times\{(n_{\mathbf{q}}+1)\,\delta(\varepsilon_{\mathbf{k}}-\hbar\omega_{\mathbf{q}}-\varepsilon_{\mathbf{k}-\mathbf{q}})$$

$$+n_{\mathbf{q}}\,\delta(\varepsilon_{\mathbf{k}}+\hbar\omega_{\mathbf{q}}-\varepsilon_{\mathbf{k}+\mathbf{q}})\}.$$

$$n_{\mathbf{q}} = \frac{1}{e^{\hbar\omega_{\mathbf{q}}\beta}-1}, \qquad \beta = \frac{1}{k_BT}.$$

$$k_BT > \hbar\omega_{\mathbf{q}}, \qquad \text{then} \qquad n_{\mathbf{q}}+1 \simeq n_{\mathbf{q}} \sim \frac{k_BT}{\hbar\omega_{\mathbf{q}}}.$$

Also, $\varepsilon_{\mathbf{k}} \gg \hbar\omega_{\mathbf{q}}$,

$$\mu_{\mathbf{k}}^{-1} = \left(\frac{m}{e}\right)\left(\frac{2\pi}{\hbar}\right)(4\pi eC_p)^2\frac{1}{(2\pi)^3}\int d^3q\left(\frac{\hbar}{2\rho c_s q}\right)\left(\frac{k_BT}{\hbar\omega_{\mathbf{q}}}\right)$$

$$\times\left\{\delta\left(\frac{\hbar^2}{2m}q(2k\cos\theta-q)\right)+\delta\left(\frac{\hbar^2}{2m}q(-2k\cos\theta-q)\right)\right\}$$

$$\sim \frac{T}{k}.$$

In semiconductor, $\left\langle\dfrac{\hbar^2k^2}{2m}\right\rangle_T \sim k_BT, \qquad k \sim T^{1/2}.$

$$\mu_{\mathbf{k}}^{-1} \sim T^{1/2}, \qquad \mu_{\mathbf{k}} \sim T^{-1/2}.$$

CHAPTER 8

4. $N_{\mathbf{k}} = c_{\mathbf{k}}^{\dagger}c_{\mathbf{k}}.$

$$i\hbar\frac{\partial N_{\mathbf{k}}}{\partial t} = i\hbar c_{\mathbf{k}}^{\dagger}\frac{\partial c_{\mathbf{k}}}{\partial t} + i\hbar\frac{\partial c_{\mathbf{k}}^{\dagger}}{\partial t}c_{\mathbf{k}}.$$

$$i\hbar c_{\mathbf{k}}^{\dagger}\frac{\partial c_{\mathbf{k}}}{\partial t} = \varepsilon_{\mathbf{k}}c_{\mathbf{k}}^{\dagger}c_{\mathbf{k}} - c_{\mathbf{k}}^{\dagger}c_{-\mathbf{k}}^{\dagger}V\sum_{\mathbf{k}'}{}' c_{-\mathbf{k}'}c_{\mathbf{k}'}.$$

$$i\hbar \frac{\partial c_{\mathbf{k}}^{\dagger}}{\partial t} c_{\mathbf{k}} = -\varepsilon_{\mathbf{k}} c_{\mathbf{k}}^{\dagger} c_{\mathbf{k}} - c_{-\mathbf{k}} V \sum_{\mathbf{k}'}{}' c_{-\mathbf{k}'}^{\dagger} c_{\mathbf{k}'}^{\dagger} c_{\mathbf{k}}.$$

$$\begin{aligned} i\hbar \frac{\partial}{\partial t} \sum_{\mathbf{k}} N_{\mathbf{k}} &= -\sum_{\mathbf{k}} c_{\mathbf{k}}^{\dagger} c_{-\mathbf{k}}^{\dagger} V \sum_{\mathbf{k}'}{}' c_{-\mathbf{k}'} c_{\mathbf{k}'} - \sum_{\mathbf{k}} c_{-\mathbf{k}} V \sum_{\mathbf{k}'}{}' c_{-\mathbf{k}'}^{\dagger} c_{\mathbf{k}'}^{\dagger} c_{\mathbf{k}} \\ &= -\sum_{\mathbf{k}} c_{\mathbf{k}}^{\dagger} c_{-\mathbf{k}}^{\dagger} V \sum_{\mathbf{K}'}{}' c_{-\mathbf{k}'} c_{\mathbf{k}'} - \sum_{\mathbf{k}'}{}' c_{-\mathbf{k}'}^{\dagger} c_{\mathbf{k}'}^{\dagger} V \sum_{\mathbf{k}} c_{-\mathbf{k}} c_{\mathbf{k}} \\ &= -\sum_{\mathbf{k}} c_{\mathbf{k}}^{\dagger} c_{-\mathbf{k}}^{\dagger} V \sum_{\mathbf{k}'}{}' c_{-\mathbf{k}'} c_{\mathbf{k}'} + \sum_{\mathbf{k}'}{}' c_{\mathbf{k}'}^{\dagger} c_{-\mathbf{k}'}^{\dagger} V \sum_{\mathbf{k}} c_{-\mathbf{k}} c_{\mathbf{k}} = 0. \end{aligned}$$

5. $b_{\mathbf{k}}^{\dagger} = c_{\mathbf{k}\uparrow}^{\dagger} c_{-\mathbf{k}\downarrow}^{\dagger}, \qquad b_{\mathbf{k}} = c_{-\mathbf{k}\downarrow} c_{\mathbf{k}\uparrow}.$

$$\begin{aligned} \left[b_{\mathbf{k}}, b_{\mathbf{k}'}^{\dagger} \right] &= c_{-\mathbf{k}\downarrow} c_{\mathbf{k}\uparrow} c_{\mathbf{k}'\uparrow}^{\dagger} c_{-\mathbf{k}'\downarrow}^{\dagger} - c_{\mathbf{k}'\uparrow}^{\dagger} c_{-\mathbf{k}'\downarrow}^{\dagger} c_{-\mathbf{k}\downarrow} c_{\mathbf{k}\uparrow} \\ &= c_{-\mathbf{k}\downarrow} \left(\delta_{\mathbf{k}\mathbf{k}'} - c_{\mathbf{k}'\uparrow}^{\dagger} c_{\mathbf{k}\uparrow} \right) c_{-\mathbf{k}'\downarrow}^{\dagger} - c_{\mathbf{k}'\uparrow}^{\dagger} c_{-\mathbf{k}'\downarrow}^{\dagger} c_{-\mathbf{k}\downarrow} c_{\mathbf{k}\uparrow} \\ &= \left(1 - c_{-\mathbf{k}'\downarrow}^{\dagger} c_{-\mathbf{k}\downarrow} \right) \delta_{\mathbf{k}\mathbf{k}'} - c_{\mathbf{k}'\uparrow}^{\dagger} c_{\mathbf{k}\uparrow} c_{-\mathbf{k}\downarrow} c_{-\mathbf{k}'\downarrow}^{\dagger} \\ &\quad - c_{\mathbf{k}'\uparrow}^{\dagger} c_{-\mathbf{k}'\downarrow}^{\dagger} c_{-\mathbf{k}\downarrow} c_{\mathbf{k}\uparrow} \\ &= \left(1 - n_{\mathbf{k}\downarrow} \right) \delta_{\mathbf{k}\mathbf{k}'} - n_{\mathbf{k}\uparrow} \delta_{\mathbf{k}\mathbf{k}'} + c_{\mathbf{k}'\uparrow}^{\dagger} c_{-\mathbf{k}'\downarrow}^{\dagger} c_{-\mathbf{k}\downarrow} c_{\mathbf{k}\uparrow} \\ &\quad - c_{\mathbf{k}'\uparrow}^{\dagger} c_{-\mathbf{k}'\downarrow}^{\dagger} c_{-\mathbf{k}\downarrow} c_{\mathbf{k}\uparrow} \\ &= \left(1 - n_{\mathbf{k}\downarrow} - n_{\mathbf{k}\uparrow} \right) \delta_{\mathbf{k}\mathbf{k}'}. \end{aligned}$$

$$\begin{aligned} \left[b_{\mathbf{k}}, b_{\mathbf{k}'} \right] &= \left[c_{\mathbf{k}\downarrow} c_{\mathbf{k}\uparrow}, c_{-\mathbf{k}'\downarrow} c_{\mathbf{k}'\uparrow} \right] \\ &= c_{-\mathbf{k}\downarrow} c_{\mathbf{k}\uparrow} c_{-\mathbf{k}'\downarrow} c_{\mathbf{k}'\uparrow} - c_{-\mathbf{k}'\downarrow} c_{\mathbf{k}'\uparrow} c_{-\mathbf{k}\downarrow} c_{\mathbf{k}\uparrow} \\ &= c_{-\mathbf{k}'\downarrow} c_{\mathbf{k}'\uparrow} c_{-\mathbf{k}\downarrow} c_{\mathbf{k}\uparrow} - c_{-\mathbf{k}'\downarrow} c_{\mathbf{k}'\uparrow} c_{-\mathbf{k}\downarrow} c_{\mathbf{k}\uparrow} = 0. \end{aligned}$$

$$\begin{aligned} \{ b_{\mathbf{k}}, b_{\mathbf{k}'} \} &= 2 c_{-\mathbf{k}'\downarrow} c_{\mathbf{k}'\uparrow} c_{-\mathbf{k}\downarrow} c_{\mathbf{k}\uparrow} = 2 c_{-\mathbf{k}\downarrow} c_{\mathbf{k}\uparrow} c_{-\mathbf{k}'\downarrow} c_{\mathbf{k}'\uparrow} \\ &= 2 b_{\mathbf{k}} b_{\mathbf{k}'} \end{aligned}$$

for $\mathbf{k} \neq \mathbf{k}'$.

$$= 0 \quad \text{for } \mathbf{k} = \mathbf{k}', \text{ because } c_{\mathbf{k}\uparrow} c_{\mathbf{k}\uparrow} |0\rangle = 0.$$

$$\{ b_{\mathbf{k}}, b_{\mathbf{k}'} \} = 2 b_{\mathbf{k}} b_{\mathbf{k}'} \{ 1 - \delta_{\mathbf{k}\mathbf{k}'} \}.$$

6. $H_{\text{red}} = \sum_{\mathbf{k}} \varepsilon_{\mathbf{k}} (c_{\mathbf{k}}^{\dagger} c_{\mathbf{k}} + c_{-\mathbf{k}}^{\dagger} c_{-\mathbf{k}}) - V \sum_{\mathbf{k},\mathbf{k}'}{}' c_{\mathbf{k}'}^{\dagger} c_{-\mathbf{k}'}^{\dagger} c_{-\mathbf{k}} c_{\mathbf{k}}$. The ground state wavefunction is $|\Phi_0\rangle$ and the excited state is $|\Phi_e\rangle = \alpha_{\mathbf{k}}^{\dagger} \alpha_{\mathbf{k}'}^{\dagger} |\Phi_0\rangle$, where

$$c_{\mathbf{k}} = u_{\mathbf{k}} \alpha_{\mathbf{k}} + v_{\mathbf{k}} \alpha_{-\mathbf{k}}^{\dagger}, \qquad c_{-\mathbf{k}} = u_{\mathbf{k}} \alpha_{-\mathbf{k}} - v_{\mathbf{k}} \alpha_{\mathbf{k}}^{\dagger}.$$

$$c_{\mathbf{k}}^{\dagger} = u_{\mathbf{k}} \alpha_{\mathbf{k}}^{\dagger} + v_{\mathbf{k}} \alpha_{-\mathbf{k}}, \qquad c_{-\mathbf{k}}^{\dagger} = u_{\mathbf{k}} \alpha_{-\mathbf{k}}^{\dagger} - v_{\mathbf{k}} \alpha_{\mathbf{k}}.$$

Consider (i)

$$\langle\Phi_e|c^\dagger_{\mathbf{k}}c_{\mathbf{k}}|\Phi_e\rangle = \langle\Phi_0|\alpha_2\alpha_1\big(u_{\mathbf{k}}\alpha^\dagger_{\mathbf{k}} + v_{\mathbf{k}}\alpha_{-\mathbf{k}}\big)\big(u_{\mathbf{k}}\alpha_{\mathbf{k}} + v_{\mathbf{k}}\alpha^\dagger_{-\mathbf{k}}\big)\alpha^\dagger_1\alpha^\dagger_2|\Phi_0\rangle$$

For $\mathbf{k} \neq 1, 2$, only $v^2_{\mathbf{k}}\alpha_{-\mathbf{k}}\alpha^\dagger_{-\mathbf{k}}$ contributes, from Eq. 95.
$\mathbf{k} = 1$ or 2, $u^2_{\mathbf{k}}\alpha^\dagger_{\mathbf{k}}\alpha_{\mathbf{k}}$ is nonvanishing.
Combining the term of $c^\dagger_{-\mathbf{k}}c_{-\mathbf{k}}$,

$$\begin{aligned}\langle\Phi_e|\sum_{\mathbf{k}}\varepsilon_{\mathbf{k}}\big(c^\dagger_{\mathbf{k}}c_{\mathbf{k}} + c^\dagger_{-\mathbf{k}}c_{-\mathbf{k}}\big)|\Phi_e\rangle &= \sum_{\mathbf{k}\neq 1,2} v^2_{\mathbf{k}}\varepsilon_{\mathbf{k}} + u^2_1\varepsilon_1 + u^2_2\varepsilon_2\\ &= \sum_{\mathbf{k}} v^2_{\mathbf{k}}\varepsilon_{\mathbf{k}} - v^2_1\varepsilon_1 - v^2_2\varepsilon_2 + u^2_1\varepsilon_1 + u^2_2\varepsilon_2.\end{aligned}$$

Consider (ii)

$$\begin{aligned}\langle\Phi_e|c^\dagger_{\mathbf{k}'}c^\dagger_{-\mathbf{k}'}c_{-\mathbf{k}}c_{\mathbf{k}}|\Phi_e\rangle = \langle\Phi_0|\alpha_2\alpha_1&\big(u_{\mathbf{k}'}\alpha^\dagger_{\mathbf{k}'} + v_{\mathbf{k}'}\alpha_{-\mathbf{k}'}\big)\big(u_{\mathbf{k}'}\alpha^\dagger_{-\mathbf{k}'} - v_{\mathbf{k}'}\alpha_{\mathbf{k}'}\big)\\ &\times\big(u_{\mathbf{k}}\alpha_{-\mathbf{k}} - v_{\mathbf{k}}\alpha^\dagger_{\mathbf{k}}\big)\big(u_{\mathbf{k}}\alpha_{\mathbf{k}} + v_{\mathbf{k}}\alpha^\dagger_{-\mathbf{k}}\big)\alpha^\dagger_1\alpha^\dagger_2|\Phi_0\rangle.\end{aligned}$$

$\mathbf{k}', \mathbf{k} \neq 1, 2$, $\quad u_{\mathbf{k}'}v_{\mathbf{k}'}u_{\mathbf{k}}v_{\mathbf{k}}(\alpha_{-\mathbf{k}'}\alpha^\dagger_{-\mathbf{k}'}\alpha_{-\mathbf{k}}\alpha^\dagger_{-\mathbf{k}})$ contributes.
$\mathbf{k}' = 1$, $\mathbf{k} = 2$ or vice versa, we have

$$\begin{aligned}\langle\Phi_0|\alpha_2\alpha_1\big[&(-u_1v_1u_2v_2)\alpha^\dagger_1\alpha_1\alpha_{-2}\alpha^\dagger_{-2} + (u_1v_1u_2v_2)\alpha_{-1}\alpha^\dagger_{-1}\alpha_{-2}\alpha^\dagger_{-2}\\ &+(-u_1v_1u_2v_2)\alpha^\dagger_1\alpha_1\alpha^\dagger_2\alpha_2\\ &+(u_1v_1u_2v_2)\alpha_{-1}\alpha^\dagger_{-1}\alpha^\dagger_2\alpha_2\big]\alpha^\dagger_1\alpha^\dagger_2|\Phi_0\rangle = 0.\end{aligned}$$

Similarly, there is no contribution for $\mathbf{k}' = 1$, $\mathbf{k}$ arbitrary or vice versa,

$$\begin{aligned}\langle\Phi_e| - V\sum_{\mathbf{k},\mathbf{k}'}{}' c^\dagger_{\mathbf{k}'}c^\dagger_{-\mathbf{k}'}c_{-\mathbf{k}}c_{\mathbf{k}}|\Phi_e\rangle &= -V\sum_{\mathbf{k},\mathbf{k}'\neq 1,2}{}' u_{\mathbf{k}'}v_{\mathbf{k}'}u_{\mathbf{k}}v_{\mathbf{k}}\\ &= -V\sum_{\mathbf{k},\mathbf{k}'}{}' u_{\mathbf{k}'}v_{\mathbf{k}'}u_{\mathbf{k}}v_{\mathbf{k}} + 2Vu_1v_1\sum_{\mathbf{k}}{}' u_{\mathbf{k}}v_{\mathbf{k}}\\ &\quad + 2Vu_2v_2\sum_{\mathbf{k}}{}' u_{\mathbf{k}}v_{\mathbf{k}} - 2Vu_1v_1u_2u_2.\end{aligned}$$

$$\begin{aligned}\langle\Phi_e|H_{\text{red}}|\Phi_e\rangle - \langle\Phi_0|H_{\text{red}}|\Phi_0\rangle &= \big(u^2_1 - v^2_1\big)\varepsilon_1 + \big(u^2_2 - v^2_2\big)\varepsilon_2\\ &\quad + 2V(u_1v_1 + u_2v_2)\sum_{\mathbf{k}}{}' u_{\mathbf{k}}v_{\mathbf{k}} - 2Vu_1v_1u_2v_2\\ &= E.\end{aligned}$$

Let

$$u_{\mathbf{k}} = \cos\frac{\theta_{\mathbf{k}}}{2}, \quad v_{\mathbf{k}} = \sin\frac{\theta_{\mathbf{k}}}{2}.$$

$$u_1^2 - v_1^2 = \cos^2\theta_1/2 - \sin^2\theta_1/2 = \cos\theta_1 = \frac{\varepsilon_1}{\sqrt{\Delta^2 + \varepsilon_1^2}}.$$

$$2u_1 v_1 = 2\sin\theta_1/2\cos\theta_1/2 = \sin\theta_1 = \frac{\Delta}{\sqrt{\Delta^2 + \varepsilon_1^2}}.$$

$$\sum_{\mathbf{k}}{}' u_{\mathbf{k}} v_{\mathbf{k}} = \frac{1}{2}\sum_{\mathbf{k}}{}' \sin\theta_{\mathbf{k}} = \frac{\varepsilon_{\mathbf{k}}}{V}\tan\theta_{\mathbf{k}} = \frac{\Delta}{V}.$$

$$E = \frac{\varepsilon_1^2}{\left(\Delta^2 + \varepsilon_1^2\right)^{1/2}} + \frac{\varepsilon_2^2}{\left(\Delta^2 + \varepsilon_2^2\right)^{1/2}} + \frac{\Delta^2}{\left(\Delta^2 + \varepsilon_1^2\right)^{1/2}} + \frac{\Delta^2}{\left(\Delta^2 + \varepsilon_2^2\right)^{1/2}}$$

$$- \frac{2V}{4} \cdot \frac{\Delta}{\left(\Delta^2 + \varepsilon_1^2\right)^{1/2}} \cdot \frac{\Delta}{\left(\Delta^2 + \varepsilon_2^2\right)^{1/2}}$$

$$= \left(\varepsilon_1^2 + \Delta^2\right)^{1/2} + \left(\varepsilon_2^2 + \Delta^2\right)^{1/2} - \frac{V}{2} \cdot \frac{\Delta^2}{\left(\Delta^2 + \varepsilon_1^2\right)^{1/2}\left(\Delta^2 + \varepsilon_2^2\right)^{1/2}}$$

$$= \lambda_1 + \lambda_2 - \frac{V\Delta^2}{2\lambda_1\lambda_2} \simeq \lambda_1 + \lambda_2, \qquad \text{where } \lambda_i = \left(\varepsilon_i^2 + \Delta\right)^{1/2}.$$

In general, $|\Phi_e\rangle = \alpha_{\mathbf{k}'}^{\dagger}\alpha_{\mathbf{k}''}^{\dagger}|\Phi_0\rangle$, $E = \lambda_{\mathbf{k}'} + \lambda_{\mathbf{k}''}$.
For $\mathbf{k}' = -\mathbf{k}''$, $E = 2\lambda_{\mathbf{k}'}$.

8. $E_{\mathbf{k}} = (\varepsilon_{\mathbf{k}}^2 + \Delta^2)^{1/2}$. Let the normal electron density of state be $D_N(\varepsilon)$ and the superconducting electron density of state be $D_s(E)$; there is one-one correspondence between them,

$$D_s(E_{\mathbf{k}})\,dE_{\mathbf{k}} = D_N(\varepsilon_{\mathbf{k}})\,d\varepsilon_{\mathbf{k}}.$$

$$D_s(E_{\mathbf{k}}) = D_N(\varepsilon_{\mathbf{k}})\frac{d\varepsilon_{\mathbf{k}}}{dE_{\mathbf{k}}} = D_N(\varepsilon_{\mathbf{k}}) \Big/ \left(\frac{dE_{\mathbf{k}}}{d\varepsilon_{\mathbf{k}}}\right)$$

$$= \frac{D_N(\varepsilon_{\mathbf{k}})\left(\Delta^2 + \varepsilon_{\mathbf{k}}^2\right)^{1/2}}{\varepsilon_{\mathbf{k}}} = D_N(\varepsilon_{\mathbf{k}}) \cdot \frac{E_{\mathbf{k}}}{\varepsilon_{\mathbf{k}}}$$

$$= D_N\left(\sqrt{E_k^2 - \Delta^2}\right)\frac{E_{\mathbf{k}}}{\sqrt{E_k^2 - \Delta^2}}.$$

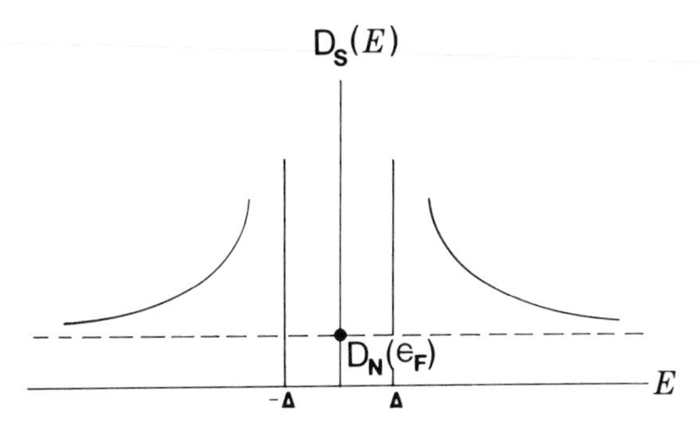

FIG. 8.1. $D_s(E)$ as a function of E.

For

$$E_{\mathbf{k}} < \Delta, \qquad D_s(E_{\mathbf{k}}) = 0,$$

$$E_{\mathbf{k}} > \Delta, \qquad D_s(E_{\mathbf{k}}) = \frac{D_N\left(\sqrt{E_{\mathbf{k}}^2 - \Delta^2}\right)E_{\mathbf{k}}}{\sqrt{E_{\mathbf{k}}^2 - \Delta^2}} \simeq \frac{D_N(\varepsilon_F)E_{\mathbf{k}}}{\sqrt{E_{\mathbf{k}}^2 - \Delta^2}}.$$

Since $\mathbf{k}$ is a region near the near k_F, we drop the subscript.

$$E < \Delta, \qquad D_s(E) = 0,$$

$$E > \Delta, \qquad D_s(E) = \frac{D_N(\varepsilon_F)E}{\sqrt{E^2 - \Delta^2}}.$$

CHAPTER 9

1. $\langle\phi|O_1|K\phi\rangle$. From Eq. (32), $\langle\phi|O_1|K\phi\rangle = (\langle\phi|K^{-1}O_1^\dagger K^{-1})|K\phi\rangle = \langle\phi|K^{-1}O_1^\dagger K^{-1}K|\phi\rangle = \langle\phi|K^{-1}O_1^\dagger|\phi\rangle = \langle\phi|O_1K^{-1}|\phi\rangle$, from $KO_1K^{-1} = 0_1^\dagger$. From Eq. (33), $K^{-1}|\phi\rangle = -K|\phi\rangle$, so, $\langle\phi|O_1|K\phi\rangle = -\langle\phi|O_1K|\phi\rangle = 0$.

2. $\langle K\phi|O_1|K\phi\rangle = (\langle\phi|K^{-1}O_1^\dagger K^{-1})K^2|\phi\rangle = -\langle\phi|K^{-1}O_1^\dagger K^{-1}|\phi\rangle = -\langle\phi|O_1K^{-1}K^{-1}|\phi\rangle = \langle\phi|O_1|\phi\rangle$, from prob. 1.

7. Eq. (49): $\langle\mathbf{k}|\mathbf{v}|\mathbf{k}\rangle = \langle\mathbf{k}|\mathbf{p}/m|\mathbf{k}\rangle = \dfrac{\hbar}{m}\sum_{\mathbf{G}}(\mathbf{k} + \mathbf{G})|\,f_{\mathbf{G}}(\mathbf{k})\,|^2$
$= \dfrac{1}{\hbar}\nabla_{\mathbf{k}}\varepsilon(\mathbf{k})$. $\mathbf{v} = \mathbf{p}/m, \qquad \langle\mathbf{k}|\mathbf{v}|\mathbf{k}\rangle = \langle\mathbf{k}|\mathbf{p}/m|\mathbf{k}\rangle$.
$|\mathbf{k}\rangle = e^{i\mathbf{k}\cdot\mathbf{x}}\sum_{\mathbf{G}} f_{\mathbf{G}}(\mathbf{k})\,e^{i\mathbf{G}\cdot\mathbf{x}}\sqrt{\Omega}$, where Ω is the volume of the sample.
We let it be 1.

$$\mathbf{p}|\mathbf{k}\rangle = \frac{\hbar}{i}\nabla\left(\sum_{\mathbf{G}} f_{\mathbf{G}}(\mathbf{k})\, e^{i(\mathbf{k}+\mathbf{G})\cdot\mathbf{r}}\right) = \hbar\sum_{\mathbf{G}}(\mathbf{k}+\mathbf{G}) f_{\mathbf{G}}(\mathbf{k})\, e^{i(\mathbf{k}+\mathbf{G})\cdot\mathbf{r}}.$$

$$\begin{aligned}\langle\mathbf{k}|\mathbf{p}/m|\mathbf{k}\rangle &= \frac{\hbar}{m}\sum_{\mathbf{G},\mathbf{G}'}(\mathbf{k}+\mathbf{G}) f^{*}_{\mathbf{G}'}(k) f_{\mathbf{G}}(\mathbf{k})\int e^{-i(\mathbf{k}+\mathbf{G}')\cdot\mathbf{r}} e^{i(\mathbf{k}+\mathbf{G})\cdot\mathbf{r}}\, d^3r \\ &= \frac{\hbar}{m}\sum_{\mathbf{G},\mathbf{G}'}(\mathbf{k}+\mathbf{G}) f^{*}_{\mathbf{G}'}(\mathbf{k}) f_{\mathbf{G}}(\mathbf{k})\,\delta(\mathbf{G}'-\mathbf{G}) \\ &= \frac{\hbar}{m}\sum_{\mathbf{G}}(\mathbf{k}+\mathbf{G})|\, f_{\mathbf{G}}(\mathbf{k})\,|^2.\end{aligned}$$

From Eq. (47),

$$\begin{aligned}&\nabla_{\mathbf{k}}\left\{\sum_{\mathbf{G}}\frac{\hbar^2}{2m}(\mathbf{k}+\mathbf{G})^2|\, f_{\mathbf{G}}(\mathbf{k})\,|^2 + \sum_{\mathbf{G}} f^{*}_{\mathbf{G}}(\mathbf{k})\sum_{\mathbf{g}} V(\mathbf{G}-\mathbf{g}) f_{\mathbf{g}}(\mathbf{k})\right\} \\ &\quad = \nabla_{\mathbf{k}}\left\{\varepsilon(\mathbf{k})\sum_{\mathbf{G}}|\, f_{\mathbf{G}}(\mathbf{k})\,|^2\right\}.\end{aligned}$$

Carrying out the gradient operation, we get

$$\begin{aligned}&\frac{\hbar^2}{m}\sum_{\mathbf{G}}(\mathbf{k}+\mathbf{G})|\, f_{\mathbf{G}}(\mathbf{k})\,|^2 + \sum_{\mathbf{G}}\frac{\hbar^2}{2m}(\mathbf{k}+\mathbf{G})^2\nabla_{\mathbf{k}}|\, f_{\mathbf{G}}(\mathbf{k})\,|^2 \\ &\quad + \sum_{\mathbf{G}}\nabla_{\mathbf{k}} f^{*}_{\mathbf{G}}(\mathbf{k})\sum_{\mathbf{g}} V(\mathbf{G}-\mathbf{g}) f_{\mathbf{g}}(\mathbf{k}) \\ &\quad + \sum_{\mathbf{G}} f^{*}_{\mathbf{G}}(\mathbf{k})\sum_{\mathbf{g}} V(\mathbf{G}-\mathbf{g})\nabla_{\mathbf{k}} f_{\mathbf{g}}(\mathbf{k}) \\ &\quad = (\nabla_{\mathbf{k}}\varepsilon(\mathbf{k}))\sum_{\mathbf{G}}|\, f_{\mathbf{G}}(\mathbf{k})\,|^2 + \varepsilon(\mathbf{k})\sum_{\mathbf{G}}\nabla_{\mathbf{k}}|\, f_{\mathbf{G}}(\mathbf{k})\,|^2.\end{aligned}$$

Because

$$\begin{aligned}\langle\varphi_{\mathbf{k}}|\varphi_{\mathbf{k}}\rangle = 1 &= \sum_{\mathbf{G},\mathbf{G}'} f^{*}_{\mathbf{G}}(\mathbf{k}) f_{\mathbf{G}'}(\mathbf{k})\int e^{-i(\mathbf{k}+\mathbf{G})\cdot r}\, e^{i(\mathbf{k}+\mathbf{G}')\cdot\mathbf{r}}\, d^3r \\ &= \sum_{\mathbf{G}}|\, f_{\mathbf{G}}(\mathbf{k})\,|^2,\end{aligned}$$

so,

$$\nabla_{\mathbf{k}} \sum_{\mathbf{G}} | f_{\mathbf{G}}(\mathbf{k}) |^2 = 0.$$

$$\sum_{\mathbf{G}} \frac{\hbar^2}{2m} (\mathbf{k} + \mathbf{G})^2 \nabla_{\mathbf{k}} | f_{\mathbf{G}}(\mathbf{k}) |^2$$

$$= \sum_{\mathbf{G}} \frac{\hbar^2}{2m} (\mathbf{k} + \mathbf{G})^2 \{ (\nabla_{\mathbf{k}} f_{\mathbf{G}}^*(\mathbf{k})) f_{\mathbf{G}}(\mathbf{k}) + f_{\mathbf{G}}^*(\mathbf{k}) \nabla_{\mathbf{k}} f_{\mathbf{G}}(\mathbf{k}) \}.$$

This result combines with

$$\sum_{\mathbf{G}} \nabla_{\mathbf{k}} f_{\mathbf{G}}^*(\mathbf{k}) \sum_{\mathbf{g}} V(\mathbf{G} - \mathbf{g}) f_{\mathbf{g}}(\mathbf{k}) + \sum_{\mathbf{G}} f_{\mathbf{G}}^*(\mathbf{k}) \sum_{\mathbf{g}} V(\mathbf{G} - \mathbf{g}) \nabla_{\mathbf{k}} f_{\mathbf{g}}(\mathbf{k})$$

to give

$$\sum_{\mathbf{G}} \varepsilon(\mathbf{k}) \{ \nabla_{\mathbf{k}} f_{\mathbf{G}}^*(\mathbf{k})) f_{\mathbf{G}}(\mathbf{k}) + f_{\mathbf{G}}^*(\mathbf{k}) \nabla_{\mathbf{k}} f_{\mathbf{G}}(\mathbf{k}) \}$$

$$= \sum_{\mathbf{G}} \varepsilon(\mathbf{k}) \nabla_{\mathbf{k}} | f_{\mathbf{G}}(\mathbf{k}) |^2,$$

so,

$$\frac{\hbar}{m} \sum_{\mathbf{G}} (\mathbf{k} + \mathbf{G}) | f_{\mathbf{G}}(\mathbf{k}) |^2 = \frac{1}{\hbar} \nabla_{\mathbf{k}} \varepsilon(\mathbf{k}).$$

8. From Eq. (49),

$$\frac{1}{\hbar} \nabla_{\mathbf{k}} \varepsilon(\mathbf{k}) = \frac{\hbar}{m} \sum_{\mathbf{G}} (\mathbf{k} + \mathbf{G}) | f_{\mathbf{G}}(\mathbf{k}) |^2.$$

$$\frac{1}{\hbar} \frac{\partial}{\partial k_\mu} \varepsilon(\mathbf{k}) = \frac{\hbar}{m} \sum_{\mathbf{G}} (k_\mu + G_\mu) | f_{\mathbf{G}}(\mathbf{k}) |^2.$$

$$\frac{1}{\hbar} \frac{\partial^2}{\partial k_\nu \partial k_\mu} \varepsilon(\mathbf{k}) = \frac{\hbar}{m} \sum_{\mathbf{G}} \left\{ \frac{\partial}{\partial k_\nu} (k_\mu + G_\mu) | f_{\mathbf{G}}(\mathbf{k}) |^2 + (k_\mu + G_\mu) \frac{\partial}{\partial k_\nu} | f_{\mathbf{G}}(\mathbf{k}) |^2 \right\}$$

$$= \frac{\hbar}{m} \left\{ \sum_{\mathbf{G}} \delta_{\mu\nu} | f_{\mathbf{G}}(\mathbf{k}) |^2 + \sum_{\mathbf{G}} (k_\mu + G_\mu) \frac{\partial}{\partial k_\nu} | f_{\mathbf{G}}(\mathbf{k}) |^2 \right\}.$$

From Eq. (46),

$$\langle \varphi_{\mathbf{k}}(\mathbf{x}) | \varphi_{\mathbf{k}}(\mathbf{x}) \rangle = 1 = \sum_{\mathbf{G},\mathbf{G}'} f^*_{\mathbf{G}'}(\mathbf{k}) f_{\mathbf{G}}(\mathbf{k}) \int e^{-i(\mathbf{k}+\mathbf{G}')\cdot\mathbf{r}} e^{i(\mathbf{k}+\mathbf{G})\cdot\mathbf{r}} \, d^3r$$

$$= \sum_{\mathbf{G}} | f_{\mathbf{G}}(\mathbf{k}) |^2 .$$

So,

$$\frac{\partial}{\partial k_\nu} \sum_{\mathbf{G}} | f_{\mathbf{G}}(\mathbf{k}) |^2 = 0.$$

$$\frac{1}{\hbar} \frac{\partial^2}{\partial k_\mu \, \partial k_\nu} \varepsilon(\mathbf{k}) = \frac{\hbar}{m} \delta_{\mu\nu} + \frac{\hbar}{m} \sum_{\mathbf{G}} G_\mu \frac{\partial}{\partial k_\nu} | f_{\mathbf{G}}(\mathbf{k}) |^2 .$$

From Eq. (50),

$$\left(\frac{1}{m^*} \right)_{\mu\nu} = \frac{1}{\hbar^2} \frac{\partial^2}{\partial k_\mu \, \partial k_\nu} \varepsilon(\mathbf{k}),$$

$$\left(\frac{1}{m^*} \right)_{\mu\nu} = \frac{1}{m} \left\{ \delta_{\mu\nu} + \sum_{\mathbf{G}} G_\mu \frac{\partial}{\partial k_\nu} | f_{\mathbf{G}}(\mathbf{k}) |^2 \right\} .$$

9. Eq. (56): $\left(\frac{m}{m^*} \right)_{\mu\nu} = \delta_{\mu\nu} + \frac{2\hbar^2}{m} \sum_{\delta}{}' \frac{\langle \gamma 0 | k_\mu | 0 \delta \rangle \langle 0 \delta | k_\nu | 0 \gamma \rangle}{\varepsilon_{\gamma 0} - \varepsilon_{\delta 0}} .$

$$\phi_{0\gamma} = \frac{1}{\sqrt{N}} \sum_j v_\gamma(\mathbf{r} - \mathbf{r}_j),$$

$$p_\nu \phi_{0\gamma} = \frac{1}{\sqrt{N}} \sum_j p_\nu v_\gamma(\mathbf{r} - \mathbf{r}_j).$$

$$\langle 0\delta | p_\gamma | 0\gamma \rangle = \frac{1}{N} \sum_{j,j'} \int v_\delta(\mathbf{r} - \mathbf{r}_{j'}) p_\nu v_\gamma(\mathbf{r} - \mathbf{r}_j) \, d^3r.$$

In the atomic limit, $\int v_\delta(\mathbf{r} - \mathbf{r}_{j'}) p_\nu v_\gamma(\mathbf{r} - \mathbf{r}_j) \, d^3r = \delta_{jj'} \langle v_\delta | p_\nu | v_\gamma \rangle$, where

$$\langle v_\delta | p_\nu | v_\gamma \rangle = \int v_\delta(\mathbf{r} - \mathbf{r}_j) p_\nu v_\gamma(\mathbf{r} - \mathbf{r}_j) \, d^3r = \int v_\delta(\mathbf{r}) p_\nu v_\gamma(\mathbf{r}) \, d^3r.$$

$$\langle 0\delta | p_\nu | 0\gamma \rangle = \langle v_\delta | p_\nu | v_\gamma \rangle = \langle \delta | p_\nu | \gamma \rangle .$$

$$\dot{r}_\nu = \frac{p_\nu}{m} = \frac{-i}{\hbar} [r_\nu, H].$$

$$\left\langle \delta \left| \frac{p_\nu}{m} \right| \gamma \right\rangle = -\frac{i}{\hbar}\langle \delta | [r_\nu, H] | \gamma \rangle = -\frac{i}{\hbar}(\varepsilon_\gamma - \varepsilon_\delta)\langle \delta | r_\nu | \gamma \rangle.$$

$$\left(\frac{m}{m^*}\right)_{\mu\nu} = \delta_{\mu\nu} + \frac{2}{m}\frac{1}{\hbar^2}\sum_\delta{}' \frac{m^2}{\varepsilon_\gamma - \varepsilon_\delta}(\varepsilon_\gamma - \varepsilon_\delta)^2 \langle \gamma | r_\mu | \delta \rangle \langle \delta | r_\nu | \gamma \rangle$$

$$= \delta_{\mu\nu} - \frac{2m}{\hbar^2}\sum_\delta{}' (\varepsilon_\delta - \varepsilon_\gamma) \left| \langle \gamma | r_\mu | \delta \rangle \right|^2.$$

$$\left(\frac{m}{m^*}\right)_{xx} = 1 - \frac{2m}{\hbar^2}\sum_\delta{}' (\varepsilon_\delta - \varepsilon_\gamma) \left| \langle \gamma | x | \delta \rangle \right|^2 = 0, \text{ from p. 301.}$$

$$T\varphi_{\mathbf{k}\gamma}(\mathbf{x}) = \varphi_{\mathbf{k}\gamma}(\mathbf{x} + \mathbf{t}_n) = \frac{1}{\sqrt{N}}\sum_j e^{i\mathbf{k}\cdot\mathbf{x}_j} v_\gamma(\mathbf{x} + \mathbf{t}_n - \mathbf{x}_j),$$

let $-\mathbf{x}_{j'} = \mathbf{t}_n - \mathbf{x}_j$,

$$= \frac{1}{\sqrt{N}}\sum_{j'} e^{i\mathbf{k}\cdot(\mathbf{t}_n + \mathbf{x}_{j'})} v_\gamma(\mathbf{x} - \mathbf{x}_{j'})$$

$$= e^{i\mathbf{k}\cdot\mathbf{t}_n}\frac{1}{\sqrt{N}}\sum_j e^{i\mathbf{k}\cdot\mathbf{x}_j} v_\gamma(\mathbf{x} - \mathbf{x}_j).$$

11. $$\varphi_{\mathbf{k}\gamma}(\mathbf{x}) = \varphi_{0\gamma}(\mathbf{x}) + \sum_\delta{}' \frac{\varphi_{0\delta}(\mathbf{x})}{\varepsilon_{\gamma 0} - \varepsilon_{\delta 0}}\langle 0\delta | \hbar\mathbf{k}\cdot\mathbf{p} | 0\gamma \rangle \frac{1}{m}.$$

$$\langle \varphi_{\mathbf{k}\gamma} | p_\mu | \varphi_{\mathbf{k}\gamma} \rangle = \langle \varphi_{0\gamma} | p_\mu | \varphi_{0\gamma} \rangle + \frac{1}{m}\sum_\delta{}' \frac{\langle 0\gamma | p_\mu | 0\delta \rangle \langle 0\delta | \hbar\mathbf{k}\cdot\mathbf{p} | 0\gamma \rangle}{\varepsilon_{\gamma 0} - \varepsilon_{\delta 0}}$$

$$+ \frac{1}{m}\sum_\delta{}' \frac{\langle 0\delta | \hbar\mathbf{k}\cdot\mathbf{p} | 0\gamma \rangle}{\varepsilon_{\gamma 0} - \varepsilon_{\delta 0}}\langle 0\gamma | p_\mu | 0\delta \rangle.$$

If the crystal has inversion symmetry, $\langle \varphi_{0\gamma} | p_\mu | \varphi_{0\gamma} \rangle = 0$.

$$\langle \varphi_{\mathbf{k}\gamma} | p_\mu | \varphi_{\mathbf{k}\gamma} \rangle = \sum_{\nu=1}^{3} \frac{\hbar k_\nu}{m} \cdot 2 \cdot \sum_\delta{}' \frac{\langle 0\gamma | p_\mu | 0\delta \rangle \langle 0\delta | p_\nu | 0\gamma \rangle}{\varepsilon_{\gamma 0} - \varepsilon_{\delta 0}}$$

from Eq. (56), $$= \sum_{\nu=1}^{3} \frac{\hbar k_\nu}{m} \cdot 2 \cdot \left\{ \left(\frac{m}{m^*}\right)_{\nu\mu} \cdot \frac{m}{2} - \frac{m}{2}\delta_{\mu\nu} \right\}.$$

$\mu \neq \nu,$ $$\langle \varphi_{\mathbf{k}\gamma} | p_\mu | \varphi_{\mathbf{k}\gamma} \rangle = \hbar k_\nu \left(\frac{m}{m^*}\right)_{\nu\mu}.$$

CHAPTER 10

1. Character table for Γ:

Table (1)

	e	C_{2z}	$2C_{4z}$	$m_{x,y}$	$m_{d,d'}$
Γ_1	1	1	1	1	1
Γ_2	1	1	1	-1	-1
Γ_3	1	1	-1	1	-1
Γ_4	1	1	-1	-1	1
Γ_5	2	-2	0	0	0

Character table for Δ

Table (2)

	e	m_y
Δ_1	1	1
Δ_2	1	-1

From Table (1):

$\Gamma_5|\ 2 \quad 0 \qquad \Gamma_5 = \Delta_1 + \Delta_2$

Similarly, $\quad \Gamma_5 = \Sigma_1 + \Sigma_2, \qquad M_5 = Z_1 + Z_2 \qquad M_5 = \Sigma_1 + \Sigma_2$

(M_5 is equivalent to Γ_5)

2. $D \rightarrow \mathbf{G}/2\pi = (0 \quad 1 \quad 0)$

$E \rightarrow \mathbf{G}/2\pi = (0 \quad \bar{1} \quad 0)$

$F \rightarrow \mathbf{G}/2\pi = (0 \quad 0 \quad 1)$

$G \rightarrow \mathbf{G}/2\pi = (0 \quad 0 \quad \bar{1})$

Construct the character table with D, E, F, G as basic functions.

Δ	e	$2C_{4x}$	C_{4x}^2	JC_{2y}, JC_{2z}	$2JC_2$
D, E, F, G	4	0	0	2	0

The character table for Δ

Δ	e	$2C_{4x}$	C_{4x}^2	JC_{2y}, JC_{2z}	$2JC_2$
Δ_1	1	1	1	1	1
Δ_2	1	-1	1	1	-1
Δ_3	1	-1	1	-1	1
Δ_4	1	1	1	-1	-1
Δ_5	2	0	-2	0	0

Δ	e	$2C_{4x}$	C_{4x}^2	JC_{2y}, JC_{2z}	$2JC_2$
$\Delta_1 + \Delta_2 + \Delta_5$	4	0	0	2	0

$$D + E + F + G = \Delta_1 + \Delta_2 + \Delta_5$$

CHAPTER 11

1. $$H = \frac{1}{2m}\left(\mathbf{p} - \frac{e}{c}\mathbf{A}\right)^2, \qquad \mathbf{A} = \left(-\frac{Hy}{2}, \frac{Hx}{2}, 0\right).$$

$$H = \frac{1}{2m}\left\{p_z^2 + \left(p_x + \frac{eHy}{2c}\right)^2 + \left(p_y - \frac{eHx}{2c}\right)^2\right\}.$$

$$i\hbar\dot{x} = [x, H] = \frac{i\hbar}{2m}\left\{2p_x + \frac{eHy}{c}\right\},$$

$$\dot{x} = \frac{1}{2m}\left(2p_x + \frac{eHy}{c}\right).$$

Similarly,

$$\dot{y} = \frac{1}{2m}\left(2p_y - \frac{eHx}{c}\right),$$

$$\dot{z} = \frac{p_z}{m}.$$

$$i\hbar\dot{p}_x = [p_x, H] = \frac{i\hbar}{2m}\left\{-\frac{e^2H^2}{2c^2}x + p_y\frac{eH}{c}\right\},$$

$$\dot{p}_x = \frac{1}{2m}\left\{p_y\frac{eH}{c} - \frac{e^2H^2}{2c^2}x\right\} = \frac{eH}{2c}\dot{y},$$

$$\dot{p}_y = \frac{1}{2m}\left\{-p_y\frac{eH}{c} - \frac{e^2H^2}{2c^2}y\right\} = -\frac{eH}{2c}\dot{x},$$

$\dot{p}_z = 0.$ $\qquad p_z$ is a constant of motion.

$$z = z(0) + \frac{p_z(0)}{m}t.$$

Let $\rho = x + iy$, then $\dot{\rho} = \dot{x} + i\dot{y} = \frac{1}{m}(p_x + ip_y) + \frac{\omega_c}{2}(y - ix)$, where $\omega_c = \frac{eH}{mc}$.

$$\ddot{\rho} = \frac{1}{m}(\dot{p}_x + i\dot{p}_y) + \frac{\omega_c}{2}(\dot{y} - i\dot{x}) = -i\omega_c\dot{\rho}.$$

This implies $\rho \sim e^{-i\omega_c t}$, $\ddot{\rho} = -\omega_c^2 \rho$. The motion in the $x-y$ plane is oscillatory.

$$\varphi \sim e^{ik_z z} \cdot \text{(harmonic oscillator function in } x-y \text{ plane)}.$$

$$\varepsilon_n = \frac{\hbar^2 k_z^2}{2m} + \left(n + \frac{1}{2}\right)\hbar\omega_c.$$

3. $\dot{\mathbf{p}} = -\dfrac{e}{c}\mathbf{v} \times \mathbf{H}$,

$\mathbf{H} = (0, 0, H_0)$.

Electron moves in an orbit $\perp$ to $\mathbf{H}$ in $\mathbf{k}$-space on a constant energy surface.

$$\dot{p}_x = -\frac{e}{c} v_y H_0, \qquad \dot{p}_y = \frac{e}{c} v_x H_0.$$

Let l be the distance between A and B (Fig. 11.1),

$$\frac{dl}{dt} = \sqrt{\dot{p}_x^2 + \dot{p}_y^2} = \frac{eH_0}{c}\sqrt{v_y^2 + v_x^2} = \frac{eH_0}{c} v_\perp .$$

$$dt = \frac{c}{eH_0}\frac{dl}{v_\perp}.$$

$$\int_0^T dt = T = \oint \frac{c}{eH_0}\frac{dl}{v_\perp}.$$

$$dl = \hbar\, dk_\parallel, \qquad v_\perp = \frac{1}{\hbar}\left(\frac{\partial \varepsilon(\mathbf{k})}{\partial \mathbf{k}}\right)_\perp, \text{ then}$$

$$T = \frac{c\hbar^2}{eH_0}\oint\left(\frac{\partial k}{\partial \varepsilon(\mathbf{k})}\right)_\perp dk = \frac{c\hbar^2}{eH_0}\frac{\partial}{\partial \varepsilon}\oint dS = \frac{c\hbar^2}{eH_0}\frac{\partial S}{\partial \varepsilon}.$$

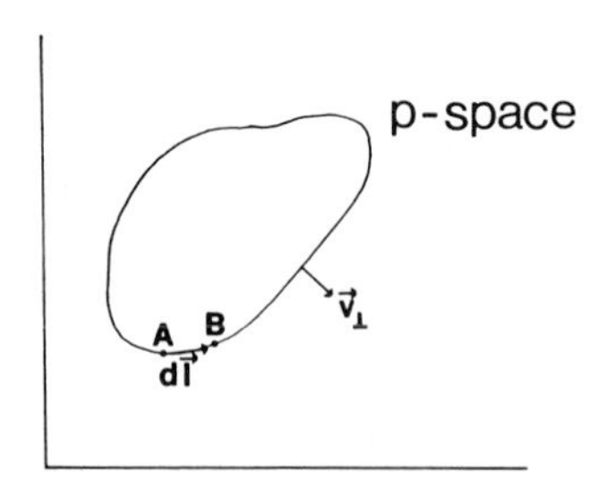

FIG. 11.1. The solid curve is a constant energy surface. $d\mathbf{l}$ is along the path between two points A, B. $\mathbf{v}_\perp$ is the velocity $\perp$ to the constant energy surface.

5. Use component form with the Landau gauge, $\mathbf{A} = (0, x, 0)H$.

$$\hbar k_x = p_x, \qquad \hbar k_y = p_y - \frac{e}{c}Hx, \qquad \hbar k_z = p_z.$$

$$[k_z, k_x] = 0, \qquad [k_y, k_z] = 0.$$

$$\hbar^2[k_x, k_y] = \left[p_x, p_y - \frac{e}{c}Hx\right] = \left[p_x, -\frac{e}{c}Hx\right]$$

$$= \frac{e}{c}H[x, p_x] = \frac{i\hbar e}{c}H.$$

$$[k_x, k_y] = k_x k_y - k_y k_x,$$

$$[k_z, k_x] = k_z k_x - k_x k_z, [k_y, k_z] = k_y k_z - k_z k_y.$$

These can be written as $\hbar^2 \mathbf{k} \times \mathbf{k} = \dfrac{i\hbar e}{c} H\hat{z}$.

$$\mathbf{k} \times \mathbf{k} = \frac{ie}{\hbar c}\mathbf{H}.$$

CHAPTER 12

1. Let the principal axes be 1, 2, and 3; the corresponding effective masses are $m_1 = m_2$, and m_3.

Let $\mathbf{H}$ be applied in the direction 3; the equations of motion in the transverse directions are

$$m_1 \frac{dv_1}{dt} + m_1 \frac{v_1}{\tau} = eE_1 + \frac{e}{c}v_2 H,$$

$$m_2 \frac{dv_2}{dt} + m_2 \frac{v_2}{\tau} = eE_2 - \frac{e}{c}v_1 H.$$

For steady state, $\dfrac{d\mathbf{v}}{dt} = 0$.

$$\begin{pmatrix} v_1 \\ v_2 \end{pmatrix} = \frac{1}{1 + \xi_1 \xi_2} \begin{pmatrix} \dfrac{e\tau}{m_1} & \xi_2 \dfrac{e\tau}{m_1} \\ -\xi_1 \dfrac{e\tau}{m_2} & \dfrac{e\tau}{m_2} \end{pmatrix} \begin{pmatrix} E_1 \\ E_2 \end{pmatrix},$$

where $\xi_1 = \dfrac{e\tau H}{m_1 c}$, $\quad \xi_2 = \dfrac{e\tau H}{m_2 c}$.

$$\begin{pmatrix} j_1 \\ j_2 \end{pmatrix} = -\frac{ne}{1+\xi_1\xi_2}\begin{pmatrix} \dfrac{e\tau}{m_1} & \xi_2\dfrac{e\tau}{m_1} \\ -\xi_1\dfrac{e\tau}{m_2} & \dfrac{e\tau}{m_2} \end{pmatrix}\begin{pmatrix} E_1 \\ E_2 \end{pmatrix}.$$

For $j_3 = j_2 = 0$, $\qquad E_2 = \xi_1 E_1$.

$$\begin{aligned} j_1 &= -\frac{ne}{1+\xi_1\xi_2}\left(\frac{e\tau}{m_1}E_1 + \xi_2\frac{e\tau}{m_1}E_2\right) \\ &= -\frac{ne}{1+\xi_1\xi_2}\left(\frac{e\tau}{m_1}\right)(1+\xi_1\xi_2)E_1 \\ &= -\frac{ne^2\tau}{m_1}E_1. \end{aligned}$$

The effective conductivity in direction 1 is independent of magnetic field, so the transverse magnetic resistance is zero.

In general, we should apply transformation to have the direction of **H** not along axis 3, and **E** not along axes 1 and 2.

2. For infinite circular cylinder, $m_1 = m_2 = m$, $m_3 \to \infty$. Let **H** be along axis 3, then the equations of motion are

$$m\frac{v_1}{\tau} = eE_1 + \frac{e}{c}v_2 H,$$

$$m\frac{v_2}{\tau} = eE_2 - \frac{e}{c}v_1 H,$$

$$m_3\frac{v_3}{\tau} = eE_3. \qquad v_3 = 0.$$

$$\begin{pmatrix} v_1 \\ v_2 \end{pmatrix} = \frac{e\tau/m}{1+\xi^2}\begin{pmatrix} 1 & \xi \\ -\xi & 1 \end{pmatrix}\begin{pmatrix} E_1 \\ E_2 \end{pmatrix}, \quad \xi = \frac{e\tau H}{mc} = \tau\omega_1, \omega_1 = \frac{eH}{mc}.$$

$$\sigma_{11} = \sigma_{22} = \frac{\sigma_0}{1+\xi^2}, \qquad \sigma_0 = ne^2\tau/m.$$

$$\sigma_{12} = -\sigma_{21} = \sigma_0\tau\omega_1/(1+\xi^2).$$

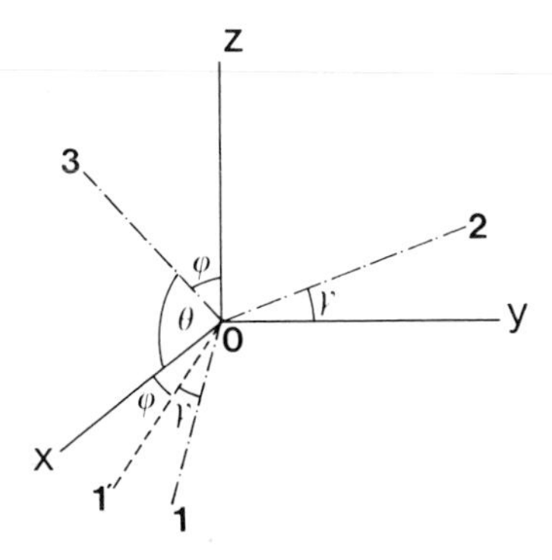

FIG. 12.1. The relative orientation of (1,2,3) with respect to (x, y, z).

The conductor does not carry any current along the axis of the cylinder. If **H** is perpendicular to axis 3, then the conduction property is independent of the field.

In general, $\hat{x}$, $\hat{y}$, and $\hat{z}$ do not coincide with 1, 2 and 3.

$$\begin{pmatrix} 1 \\ 2 \\ 3 \end{pmatrix} = \begin{pmatrix} \cos\gamma\sin\theta & \sin\gamma & -\cos\theta\cos\gamma \\ -\sin\gamma\sin\theta & \cos\gamma & \sin\gamma\cos\theta \\ \cos\theta & 0 & \sin\theta \end{pmatrix} \begin{pmatrix} x \\ y \\ z \end{pmatrix}$$

$H_3 = H\sin\theta$, $H_2 = H\cos\theta\sin\gamma$, $H_1 = -H\cos\theta\cos\gamma$. $E_3 = E_x\cos\theta + E_z\sin\theta$, $E_2 = -E_x\sin\gamma\sin\theta + E_y\cos\gamma + E_z\cos\theta\sin\gamma$, $E_1 = E_x\sin\theta\cos\gamma + E_y\sin\gamma - E_z\cos\theta\cos\gamma$.

$$m\frac{v_1}{\tau} = eE_x\sin\theta\cos\gamma + eE_y\sin\gamma - eE_z\cos\theta\cos\gamma + \frac{e}{c}v_2H\sin\theta - \frac{e}{c}v_3H\cos\theta\sin\gamma,$$

$$m\frac{v_2}{\tau} = -eE_x\sin\theta\sin\gamma + eE_y\cos\gamma + eE_z\cos\theta\sin\gamma - \frac{e}{c}v_3H\cos\theta\cos\gamma - \frac{e}{c}v_1H\sin\theta,$$

$$m_3\frac{v_3}{\tau} = eE_x\cos\theta + \frac{e}{c}v_1H\cos\theta\sin\gamma + \frac{e}{c}v_2H\cos\theta\cos\gamma. \qquad v_3 = 0.$$

$$v_1 = \frac{e\tau}{m} E_x \sin\theta\cos\gamma + \frac{e\tau}{m} E_y \sin\gamma + \frac{e\tau}{mc} H \sin\theta v_2 - \frac{e\tau}{m} E_z \cos\theta\cos\gamma,$$

$$v_2 = -\frac{e\tau}{m} E_x \sin\theta\cos\gamma + \frac{e\tau}{m} E_y \cos\gamma - \frac{e\tau}{mc} H \sin\theta v_1 + \frac{e\tau}{m} E_z \cos\theta\sin\gamma.$$

Let $\omega_1 = \dfrac{eH}{mc}$, $\quad \xi = \tau\omega_1 \sin\theta,$

$$\begin{pmatrix} v_1 \\ v_2 \end{pmatrix} = \frac{e\tau/m}{1+\xi^2} \begin{pmatrix} \sin\theta(\cos\gamma - \xi\sin\gamma) & \sin\gamma + \xi\cos\gamma & -\cos\theta(\cos\gamma - \xi\sin\gamma) \\ -\sin\theta(\xi\cos\gamma + \sin\gamma) & \cos\gamma - \xi\sin\gamma & \cos\theta(\xi\cos\gamma + \sin\gamma) \end{pmatrix} \begin{pmatrix} E_x \\ E_y \\ E_z \end{pmatrix}.$$

$$\begin{pmatrix} v_x \\ v_y \\ v_z \end{pmatrix} = \begin{pmatrix} \sin\theta\cos\gamma & -\sin\gamma\sin\theta & \cos\theta \\ \sin\gamma & \cos\gamma & 0 \\ -\cos\theta\cos\gamma & \sin\gamma\cos\theta & \sin\theta \end{pmatrix} \begin{pmatrix} v_1 \\ v_2 \\ v_3 \end{pmatrix}.$$

$$v_x = \frac{e\tau/m}{1+\xi^2}\left(\sin^2\theta E_x + \xi\sin\theta E_y - \sin\theta\cos\theta E_z\right),$$

$$v_y = \frac{e\tau/m}{1+\xi^2}\left(-\xi\sin\theta E_x + E_y + \xi\cos\theta E_z\right),$$

$$v_z = \frac{e\tau/m}{1+\xi^2}\left(-\cos\theta\sin\theta E_x - \xi\cos\theta E_y + \cos^2\theta E_z\right).$$

$$\sigma = \frac{\sigma_0}{1+\xi^2} \begin{pmatrix} \sin^2\theta & \xi\sin\theta & -\sin\theta\cos\theta \\ -\xi\sin\theta & 1 & \xi\cos\theta \\ -\sin\theta\cos\theta & -\xi\cos\theta & \cos^2\theta \end{pmatrix}.$$

$\theta \ll 1$, $\hat{z} \perp$ axis 3, $\sigma_{yy} \simeq \sigma_{zz}$, $\sigma_{yz} = -\sigma_{zx} \sim \omega_1\tau$. As H increases, $\theta \ll 1$, $\omega_1\tau\sin\theta \simeq 1$, $\sigma_{xy} \simeq -\sigma_{yx} < \omega_1\tau$, $\sigma_{yz} \sim 1$.

CHAPTER 13

2. $\psi_{\mathbf{k}} = \frac{1}{\sqrt{\Omega}} e^{i\mathbf{k}\cdot\mathbf{r}} - \sum_c \varphi_c \langle c|\mathbf{k}\rangle.$

$$\sum_c |\langle c|\mathbf{k}\rangle|^2 = \sum_c \langle \mathbf{k}|c\rangle\langle c|\mathbf{k}\rangle = \frac{1}{\Omega}\sum_c \int e^{-i\mathbf{k}\cdot\mathbf{x}}\varphi_c(\mathbf{x})\, d^3x$$

$$\cdot \int \varphi_c^*(\mathbf{x}')e^{i\mathbf{k}\cdot\mathbf{x}'}\, d^3x'$$

$$\leq \frac{1}{\Omega}\int_\Delta e^{-i\mathbf{k}\cdot\mathbf{x}}e^{i\mathbf{k}\cdot\mathbf{x}}\, d^3x = \frac{\Delta}{\Omega}.$$

3. $\varphi_{\mathbf{k}}(\mathbf{x}) = u_0(\mathbf{x})\Psi_{\mathbf{k}}(\mathbf{x}).$

a. $$\left\{-\frac{\hbar^2}{2m}\nabla^2 + V\right\}\varphi_{\mathbf{k}}(\mathbf{x}) = \varepsilon_{\mathbf{k}}\varphi_{\mathbf{k}}(\mathbf{x}),$$

$$\left\{-\frac{\hbar^2}{2m}\nabla^2 + V\right\}u_0(\mathbf{x})\Psi_{\mathbf{k}}(\mathbf{x}) = \varepsilon_{\mathbf{k}}u_0(\mathbf{x})\Psi_{\mathbf{k}}(\mathbf{x}).$$

Consider

$$p^2[u_0(\mathbf{x})\Psi_{\mathbf{k}}(\mathbf{x})] = \mathbf{p}\cdot\{[\mathbf{p}u_0(\mathbf{x})]\Psi_{\mathbf{k}}(\mathbf{x}) + u_0(\mathbf{x})\mathbf{p}\Psi_{\mathbf{k}}(\mathbf{x})\}$$

$$= [p^2u_0(\mathbf{x})]\Psi_{\mathbf{k}}(\mathbf{x}) + 2\mathbf{p}u_0(\mathbf{x})$$

$$\cdot\mathbf{p}\Psi_{\mathbf{k}}(\mathbf{x}) + u_0(\mathbf{x})p^2\Psi_{\mathbf{k}}(\mathbf{x}).$$

$$\left[-\frac{\hbar^2}{2m}\nabla^2 u_0(\mathbf{x})\right]\Psi_{\mathbf{k}}(\mathbf{x}) + Vu_0(\mathbf{x})\Psi_{\mathbf{k}}(\mathbf{x}) + 2\mathbf{p}u_0(x)\cdot\mathbf{p}\Psi_{\mathbf{k}}(\mathbf{x})\frac{1}{2m}$$

$$+ u_0(\mathbf{x})\frac{p^2}{2m}\Psi_{\mathbf{k}}(\mathbf{x}) = \varepsilon_{\mathbf{k}}u_0(\mathbf{x})\Psi_{\mathbf{k}}(\mathbf{x}).$$

But,

$$\left(-\frac{\hbar^2}{2m}\nabla^2 + V\right)u_0(\mathbf{x}) = \varepsilon_0 u_0(\mathbf{x}).$$

So,

$$u_0(\mathbf{x})\frac{p^2}{2m}\Psi_{\mathbf{k}}(\mathbf{x}) + (\varepsilon_0 - \varepsilon_{\mathbf{k}})u_0(\mathbf{x})\Psi_{\mathbf{k}}(\mathbf{x}) = -\frac{1}{m}\mathbf{p}u_0(\mathbf{x})\cdot\mathbf{p}\Psi_{\mathbf{k}}(\mathbf{x}).$$

b. $$u_0(\mathbf{x})\left[\frac{p^2}{2m} + (\varepsilon_0 - \varepsilon_{\mathbf{k}})\right]\Psi_N(\mathbf{x}) = -\frac{1}{m}\mathbf{p}u_0(\mathbf{x})\cdot\mathbf{p}\Psi_N(\mathbf{x}).$$

$u_0(\mathbf{x})$ is the same in each unit cell.
For s-state, $u_0(\mathbf{x})$ is large at the center of the cell. The continuity of the wavefunction at the cell boundary required $\mathbf{p}u_0(\mathbf{x}) = 0$ (Fig. 13.1).

c. In the interior of the cell, $\mathbf{x} \simeq 0, \Psi_N(\mathbf{x})$ is at an extremum, $\mathbf{p}\Psi_N(\mathbf{x}) \simeq 0$ (Fig. 13.2)

$$\left[\frac{1}{2m}p^2 + (\varepsilon_0 - \varepsilon_N)\right]\Psi_N(\mathbf{x}) = 0.$$

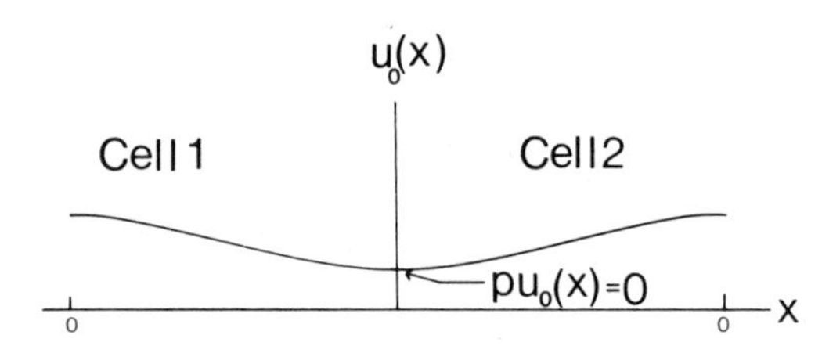

FIG. 13.1. Schematic illustration showing $u_0(x)$ as a function of x, and $pu_0(x) = 0$ at the cell boundary.

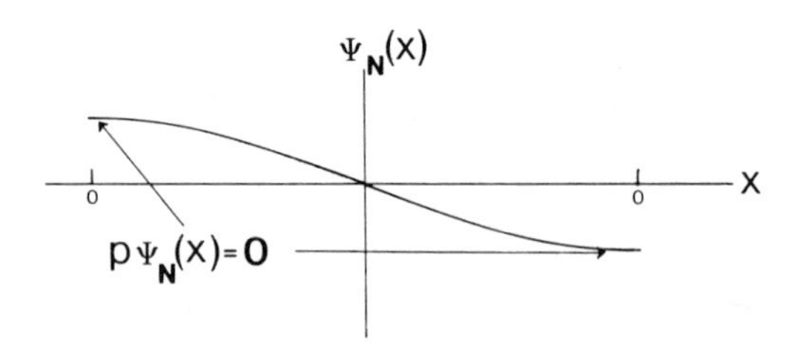

FIG. 13.2. Schematic illustration showing $p\Psi_N(x) = 0$ near the center of the cell.

$\Psi_N(\mathbf{x})$ can be described as free particle wavefunction,

$\Psi_N(\mathbf{x}) = e^{i\mathbf{k}_N \cdot \mathbf{r}}/\sqrt{\Omega}$, where Ω is the volume of the sample.

$$\varepsilon_N = \varepsilon_0 + \frac{\hbar^2 k_N^2}{2m}.$$

CHAPTER 14

3. a. $$D_{xy} = \frac{1}{m^2} \sum_\delta{}' \frac{\langle\gamma|p_x|\delta\rangle\langle\delta|p_y|\gamma\rangle}{\varepsilon_\gamma - \varepsilon_\delta}.$$

$$D_{yx} = \frac{1}{m^2} \sum_\delta{}' \frac{\langle\gamma|p_y|\delta\rangle\langle\delta|p_x|\gamma\rangle}{\varepsilon_\gamma - \varepsilon_\delta}.$$

Let R be the rotation operator which rotates by $\frac{\pi}{2}$ about $\hat{z}$-axis, then

$$R^{-1}D_{xy} = D_{xy}(R\mathbf{r}).$$

$$R\begin{pmatrix} x \\ y \end{pmatrix} = \begin{pmatrix} y \\ -x \end{pmatrix}.$$

$$R^{-1}D_{xy} = \frac{1}{m^2} \sum_\delta{}' \frac{\langle\gamma|p_y|\delta\rangle\langle\delta|p_{-x}|\gamma\rangle}{\varepsilon_\gamma - \varepsilon_\delta} = -D_{yx}.$$

From $\varepsilon(\mathbf{k}) = \sum_{\alpha\beta} D_{\alpha\beta} k_\alpha k_\beta$ and the energy is invariant under the symmetry operation,

$$D_{xy} = R^{-1}D_{xy},$$

so

$$D_{xy} = -D_{yx}.$$

$$D_{xy}^S = \tfrac{1}{2}\left(D_{xy} + D_{yx}\right) = 0.$$

$$D_{xy}^A = \tfrac{1}{2}\left(D_{xy} - D_{yx}\right) = D_{xy} \neq 0.$$

b. Without *s-o* interaction and from Eq. (32),

$$iD_{xy}^A = -\frac{1}{2m}\langle\gamma|L_z|\gamma\rangle.$$

$|\gamma\rangle$ can be one of $xf(r)$, $yf(r)$, and $zf(r)$ for the band edge at $\mathbf{k}=0$.

$$L_z = \frac{\hbar}{i}\left(x\frac{\partial}{\partial y} - y\frac{\partial}{\partial x}\right)$$

$\dfrac{\partial}{\partial y}f(r)$ gives an odd function of y.

$\dfrac{\partial}{\partial x}f(r)$ gives an odd function of x.

$\langle\gamma|L_z|\gamma\rangle = 0$ because we integrate the product of odd functions in space.

$D^A_{xy} = 0.$

4. $\varepsilon_{\mathbf{k}} = [A(k_x^2+k_y^2)+Bk_z^2]\begin{pmatrix}1 & 0\\ 0 & 1\end{pmatrix} + C(k_x\sigma_y - k_y\sigma_x)$ with $\hat{z}$-axis as symmetry axis. For a state with spin parallel to the $\hat{y}$-axis, we can write the wavefunction as $\begin{pmatrix}1\\0\end{pmatrix}$.

However, the symmetry axis is in the $\hat{z}$-direction, we transform $\begin{pmatrix}1\\0\end{pmatrix}$ by the rotation matrix

$$D^{1/2}\left(\frac{\pi}{2},\frac{\pi}{2},-\frac{\pi}{2}\right).$$

The wavefunction with the $\hat{z}$-axis as the symmetry axis is

$$\psi = \begin{pmatrix}1/\sqrt{2}\\ i/\sqrt{2}\end{pmatrix},$$

$$\sigma_y\begin{pmatrix}1/\sqrt{2}\\ i/\sqrt{2}\end{pmatrix} = \begin{pmatrix}1/\sqrt{2}\\ i/\sqrt{2}\end{pmatrix}.$$

So, for $k_y = 0$, the energy of the state with spin parallel to the $\hat{y}$-axis is

$$\varepsilon_{\mathbf{k}} = Ak_x^2 + Bk_z^2 + Ck_x = A\left(k_x + \frac{C}{2A}\right)^2 + Bk_z^2 - \frac{C^2}{4A}.$$

$$\varepsilon_{\mathbf{k}} + \frac{C^2}{4A} = Ak_x'^2 + Bk_z^2, \qquad k_x' = k_x + \frac{C}{2A}.$$

$$\mathbf{J} = ne\langle \mathbf{v} \rangle = \frac{ne}{\hbar}\sum_{\mathbf{k}} \nabla_{\mathbf{k}}\varepsilon_{\mathbf{k}}$$

$$v_x = \frac{1}{\hbar}\frac{\partial \varepsilon_{\mathbf{k}}}{\partial k_x} = \frac{1}{\hbar}(2Ak_x + C) = \frac{2A}{\hbar}\left(k_x + \frac{C}{2A}\right)$$

$$v_z = \frac{1}{\hbar}\frac{\partial \varepsilon_{\mathbf{k}}}{\partial k_z} = \frac{1}{\hbar}2Bk_z$$

$$\varepsilon_{\mathbf{k}} = \frac{\hbar^2}{4A}v_x^2 + \frac{\hbar^2}{4B}v_z^2 - \frac{C^2}{4A}$$

On the energy surface, for example, $\varepsilon_{\mathbf{k}} = 1 - \dfrac{C^2}{4A}$,

$$\frac{\hbar^2}{4A}v_x^2 + \frac{\hbar^2}{4B}v_z^2 = 1,$$

then $v_x, -v_x; v_z, -v_z$ are symmetric and $\langle \mathbf{v} \rangle = 0$.

5. $H = -\dfrac{\hbar^2}{2m}\nabla^2 + \sum_n [V_{\text{III}}(\mathbf{x} - \mathbf{x}_n) + V_V(\mathbf{x} - \mathbf{x}_n - \boldsymbol{\tau})]$; let the valence III atom sites be at the origin of the unit cell, while the valence V atom sites are at $\boldsymbol{\tau}$. $\mathbf{x}_n$ is the lattice vector.

$$\begin{aligned} H = -\frac{\hbar^2}{2m}\nabla^2 + \sum_n \Big[& V_{\text{III}}(\mathbf{x} - \mathbf{x}_n) + \frac{1}{2}V_{\text{III}}(\mathbf{x} - \mathbf{x}_n - \boldsymbol{\tau}) \\ & - \frac{1}{2}V_{\text{III}}(\mathbf{x} - \mathbf{x}_n - \boldsymbol{\tau}) + V_V(\mathbf{x} - \mathbf{x}_n - \boldsymbol{\tau}) \\ & + \frac{1}{2}V_V(\mathbf{x} - \mathbf{x}_n) - \frac{1}{2}V_V(\mathbf{x} - \mathbf{x}_n)\Big] \\ = -\frac{\hbar^2}{2m}\nabla^2 + \sum_n \Big\{ & \frac{1}{2}[V_{\text{III}}(\mathbf{x} - \mathbf{x}_n) + V_V(\mathbf{x} - \mathbf{x}_n)] \\ & + \frac{1}{2}[V_{\text{III}}(\mathbf{x} - \mathbf{x}_n - \boldsymbol{\tau}) + V_V(\mathbf{x} - \mathbf{x}_n - \boldsymbol{\tau})] \\ & + \frac{1}{2}[V_{\text{III}}(\mathbf{x} - \mathbf{x}_n) - V_V(\mathbf{x} - \mathbf{x}_n)] \\ & - \frac{1}{2}[V_{\text{III}}(\mathbf{x} - \mathbf{x}_n - \boldsymbol{\tau}) - V_V(\mathbf{x} - \mathbf{x}_n - \boldsymbol{\tau})]\Big\}. \end{aligned}$$

Define

$$V_+ = \frac{1}{2}[V_{\text{III}}(\mathbf{x}-\mathbf{x}_n) + V_V(\mathbf{x}-\mathbf{x}_n)],$$

$$V_- = \frac{1}{2}[V_{\text{III}}(\mathbf{x}-\mathbf{x}_n) - V_V(\mathbf{x}-\mathbf{x}_n)],$$

$$H = -\frac{\hbar^2}{2m}\nabla^2 + \sum_n \{V_+(\mathbf{x}-\mathbf{x}_n) + V_+(\mathbf{x}-\mathbf{x}_n-\boldsymbol{\tau}) + V_-(\mathbf{x}-\mathbf{x}_n) - V_-(\mathbf{x}-\mathbf{x}_n-\boldsymbol{\tau})\}.$$

Define

$$V^S = \sum_n \{V_+(\mathbf{x}-\mathbf{x}_n) + V_+(\mathbf{x}-\mathbf{x}_n-\boldsymbol{\tau})\},$$

$$V^A = \sum_n \{V_-(\mathbf{x}-\mathbf{x}_n) - V_-(\mathbf{x}-\mathbf{x}_n-\boldsymbol{\tau})\},$$

$$H = -\frac{\hbar^2}{2m}\nabla^2 + V^S + V^A.$$

The representation Γ_{15} are the antibonding states. The corresponding wave functions are

$$\Psi_{px}^+ = \frac{1}{\sqrt{2N}}\sum_n \{\varphi_{px}(\mathbf{x}-\mathbf{x}_n) + \varphi_{px}(\mathbf{x}-\mathbf{x}_n-\boldsymbol{\tau})\},$$

$$\Psi_{py}^+ = \frac{1}{\sqrt{2N}}\sum_n \{\varphi_{py}(\mathbf{x}-\mathbf{x}_n) + \varphi_{py}(\mathbf{x}-\mathbf{x}_n-\boldsymbol{\tau})\},$$

$$\Psi_{pz}^+ = \frac{1}{\sqrt{2N}}\sum_n \{\varphi_{pz}(\mathbf{x}-\mathbf{x}_n) + \varphi_{pz}(\mathbf{x}-\mathbf{x}_n-\boldsymbol{\tau})\}.$$

$\Gamma_{25'}$ are the bonding states, and

$$\Psi_{p\alpha}^- = \frac{1}{\sqrt{2N}}\sum_n \{\varphi_{p\alpha}(\mathbf{x}-\mathbf{x}_n) - \varphi_{p\alpha}(\mathbf{x}-\mathbf{x}_n-\boldsymbol{\tau})\},$$

where $\alpha = x, y, z$.

$$\langle \Psi_{p\alpha}^{+}|V^{A}|\Psi_{p\alpha}^{+}\rangle \simeq 0 \simeq \langle \Psi_{p\alpha}^{-}|V^{A}|\Psi_{p\alpha}^{-}\rangle.$$

$$\begin{aligned}\langle \Psi_{px}^{-}|V^{A}|\Psi_{px}^{+}\rangle &\simeq \frac{N}{2N}\int\left[\varphi_{px}^{*}(\mathbf{x}-\mathbf{x}_n)-\varphi_{px}^{*}(\mathbf{x}-\mathbf{x}_n-\boldsymbol{\tau})\right]\\ &\quad\times\left[V_{-}(\mathbf{x}-\mathbf{x}_n)-V_{-}(\mathbf{x}-\mathbf{x}_n-\boldsymbol{\tau})\right]\\ &\quad\times\left[\varphi_{px}(\mathbf{x}-\mathbf{x}_n)+\varphi_{px}(\mathbf{x}-\mathbf{x}_n-\boldsymbol{\tau})\right]d^3x\\ &\simeq \frac{1}{2}\int\{\varphi_{px}^{*}(\mathbf{x}-\mathbf{x}_n)V_{-}(\mathbf{x}-\mathbf{x}_n)\varphi_{px}(\mathbf{x}-\mathbf{x}_n)\\ &\quad+\varphi_{px}^{*}(\mathbf{x}-\mathbf{x}_n-\boldsymbol{\tau})V_{-}(\mathbf{x}-\mathbf{x}_n-\boldsymbol{\tau})\\ &\quad\times\varphi_{px}(\mathbf{x}-\mathbf{x}_n-\boldsymbol{\tau})\}\,d^3x\\ &=\Delta, \qquad \Delta \text{ is proportional to the strength of } V_{-}.\end{aligned}$$

$$\langle \Psi_{p\alpha'}^{-}|V^{A}|\Psi_{p\alpha}^{+}\rangle = 0, \qquad \alpha \neq \alpha'.$$

The matrix of V^A is:

$$\begin{array}{cccccc} x & y & z & x & y & z \end{array}$$
$$\begin{pmatrix} 0 & 0 & 0 & \Delta & 0 & 0\\ 0 & 0 & 0 & 0 & \Delta & 0\\ 0 & 0 & 0 & 0 & 0 & \Delta\\ \Delta & 0 & 0 & 0 & 0 & 0\\ 0 & \Delta & 0 & 0 & 0 & 0\\ 0 & 0 & \Delta & 0 & 0 & 0 \end{pmatrix}$$

This can be reduced to three identical eigenvalue problems for x, y, z.

$$\begin{pmatrix} -\lambda & \Delta \\ \Delta & -\lambda \end{pmatrix} = 0, \qquad \lambda_{\pm} = \pm\Delta.$$

$\lambda_{+} - (\lambda_{-}) =$ splitting between $\Gamma_{25'}$ and $\Gamma_{15} = 2\Delta$.

CHAPTER 15

1. The transition rate is

$$W_{fi} = \frac{2\pi}{\hbar}\,\delta(\varepsilon_{c\mathbf{k}} - \varepsilon_{v\mathbf{k}} - \hbar\omega)|P_{cv}|^2$$

where the dipole matrix $|P_{cv}|^2$ is assumed to be constant.

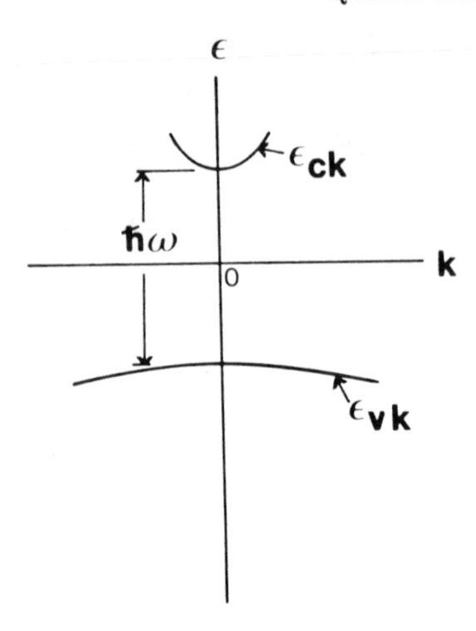

FIG. 15.1. The band structure near $\mathbf{k} = 0$.

The constant matrix element absorption coefficient is

$$\alpha \sim \sum_{\substack{\mathbf{k} \\ c,\, v}} \delta(\varepsilon_{c\mathbf{k}} - \varepsilon_{v\mathbf{k}} - \hbar\omega)$$

$$= \frac{\Omega}{(2\pi)^3} \int d^3k \sum_{c,\, v} \delta(\varepsilon_{c\mathbf{k}} - \varepsilon_{v\mathbf{k}} - \hbar\omega)$$

$$= \frac{\Omega}{(2\pi)^3} \int dS \int d\varepsilon_{cv} \frac{1}{|\nabla_{\mathbf{k}}\varepsilon_{cv}|} \delta(\varepsilon_{cv} - \hbar\omega),$$

where $\varepsilon_{cv} = \varepsilon_{c\mathbf{k}} - \varepsilon_{v\mathbf{k}}$; S is the constant energy surface of ε_{cv}. Near $\mathbf{k} \simeq 0$, ε_{cv} is a minimum. $\varepsilon_{cv} = \varepsilon_{cv}(0) + A_1 k_x^2 + A_2 k_y^2 + A_3 k_z^2$, A_1, A_2, and A_3 are positive and can be expressed in terms of the effective mass at $\mathbf{k} = 0$. Define the joint density of states

$$J_{cv}\, d\varepsilon_{cv} = \int dS \frac{1}{|\nabla_{\mathbf{k}}\varepsilon_{cv}|} d\varepsilon_{cv}.$$

Using the foregoing expression of ε_{cv}, for finite ε_{cv} the right-hand side gives the volume of an ellipsoid. For $d\varepsilon_{cv}$, we have

$$\mathcal{J}_{cv}\, d\varepsilon_{cv} = d\left\{\frac{4\pi}{3}\frac{1}{\sqrt{A_1A_2A_3}}\left(\varepsilon_{cv}-\varepsilon_{cv}(0)\right)^{3/2}\right\},$$

$$\mathcal{J}_{cv} = \frac{d}{d\varepsilon_{cv}}\left\{\frac{4\pi}{3}\frac{1}{\sqrt{A_1A_2A_3}}\left(\varepsilon_{cv}-\varepsilon_{cv}(0)\right)^{3/2}\right\}$$

$$= 2\pi\frac{1}{\sqrt{A_1A_2A_3}}\left(\varepsilon_{cv}-\varepsilon_{cv}(0)\right)^{1/2}, \qquad \text{for } \varepsilon_{cv} > \varepsilon_{cv}(0).$$

$$\alpha \sim \frac{\Omega}{(2\pi)^2}\frac{1}{\sqrt{A_1A_2A_3}}\left(\hbar\omega-\varepsilon_{cv}(0)\right)^{1/2}.$$

2. In uniaxial symmetry,

$$H = \frac{p_{e\|}^2}{2m_{e\|}} + \frac{p_{e\perp}^2}{2m_{e\perp}} + \frac{p_{h\|}^2}{2m_{h\|}} + \frac{p_{h\perp}^2}{2m_{h\perp}} - V(|\mathbf{r}_e - \mathbf{r}_h|).$$

$$\mathbf{R} = (X, Y, Z) = \left(\frac{m_{e\perp}x_e + m_{h\perp}x_h}{m_{e\perp}+m_{h\perp}}, \frac{m_{e\perp}y_e + m_{h\perp}y_h}{m_{e\perp}+m_{h\perp}}, \frac{m_{e\|}z_e + m_{h\|}z_h}{m_{e\|}+m_{h\|}}\right).$$

$$\mathbf{r} = \mathbf{x}_e - \mathbf{x}_h = (x_e - x_h,\, y_e - y_h,\, z_e - z_h).$$

$$H = \frac{p_{xe}^2 + p_{ye}^2}{2m_{e\perp}} + \frac{p_{ze}^2}{2m_{e\|}} + \frac{p_{xh}^2 + p_{yh}^2}{2m_{h\perp}} + \frac{p_{zh}^2}{2m_{h\|}} - V(r)$$

$$= -\frac{\hbar^2}{2M_\perp}\left(\frac{\partial^2}{\partial X^2} + \frac{\partial^2}{\partial Y^2}\right) - \frac{\hbar^2}{2M_\|}\frac{\partial^2}{\partial Z^2}$$

$$-\frac{\hbar^2}{2\mu_\perp}\left(\frac{\partial^2}{\partial x^2} + \frac{\partial^2}{\partial y^2}\right) - \frac{\hbar^2}{2\mu_\|}\frac{\partial^2}{\partial \mathfrak{z}^2} - V(r).$$

where

$M_\perp = m_{e\perp} + m_{h\perp}$, $m_\| = m_{e\|} + m_{h\|}$,

$$\frac{1}{\mu_\perp} = \frac{1}{m_{e\perp}} + \frac{1}{m_{h\perp}},$$

$$\frac{1}{\mu_\|} = \frac{1}{m_{e\|}} + \frac{1}{m_{h\|}}.$$

$$\psi = e^{i\mathbf{k}\cdot\mathbf{R}}u(x, y, \mathfrak{z}),$$

$$H\psi = \left\{ \frac{\hbar^2}{2M_\perp}\left(k_x^2 + k_y^2\right) + \frac{\hbar^2}{2M_\parallel}k_z^2 - \frac{\hbar^2}{2\mu_\perp}\left(\frac{\partial^2}{\partial x^2} + \frac{\partial^2}{\partial y^2}\right) - \frac{\hbar^2}{2\mu_\parallel}\frac{\partial^2}{\partial \mathfrak{z}^2} - V(r)\right\} e^{i\mathbf{k}\cdot\mathbf{R}}u(\mathbf{r})$$

$$= Ee^{i\mathbf{k}\cdot\mathbf{R}}u(\mathbf{r}).$$

$$\left\{ -\frac{\hbar^2}{2\mu_\perp}\left(\frac{\partial^2}{\partial x^2} + \frac{\partial^2}{\partial y^2}\right) - \frac{\hbar^2}{2\mu_\parallel}\frac{\partial^2}{\partial z^2} - V(r)\right\}\psi$$

$$= \left\{ E - \frac{\hbar^2}{2M_\perp}\left(k_x^2 + k_y\right) - \frac{\hbar^2}{2M_\parallel}k_z^2\right\}\psi.$$

Let $\mathfrak{z} = \sqrt{\frac{\varepsilon_\parallel}{\varepsilon_\perp}}\, z$, then consider

$$T = -\frac{\hbar^2}{2\mu_\perp}\left(\frac{\partial^2}{\partial x^2} + \frac{\partial^2}{\partial y^2}\right) - \frac{\hbar^2}{2\mu_\parallel}\frac{\partial^2}{\partial \mathfrak{z}^2}$$

$$= -\frac{\hbar^2}{2\mu_\perp}\left(\frac{\partial^2}{\partial x^2} + \frac{\partial^2}{\partial y^2}\right) - \frac{\hbar^2}{2\mu_\parallel}\left(\frac{\varepsilon_\perp}{\varepsilon_\parallel}\right)\frac{\partial^2}{\partial z^2}$$

Define

$$\nabla^2 = \frac{\partial^2}{\partial x^2} + \frac{\partial^2}{\partial y^2} + \frac{\partial^2}{\partial z^2}, \quad \text{and} \quad \frac{1}{\mu_0} = \frac{2}{3}\frac{1}{\mu_\perp} + \frac{1}{3}\frac{1}{\mu_\parallel}\cdot\frac{\varepsilon_\perp}{\varepsilon_\parallel},$$

$$\gamma = \frac{1}{\mu_\perp} - \frac{1}{\mu_\parallel}\cdot\frac{\varepsilon_\perp}{\varepsilon_\parallel},$$

then

$$T = -\frac{\hbar^2}{2\mu_0}\nabla^2 + \frac{\hbar^2}{2}\left(\frac{2}{3}\cdot\frac{1}{\mu_\perp} + \frac{1}{3}\frac{1}{\mu_\parallel}\frac{\varepsilon_\perp}{\varepsilon_\parallel}\right)\nabla^2$$

$$-\frac{\hbar^2}{2\mu_\perp}\left(\frac{\partial^2}{\partial x^2} + \frac{\partial^2}{\partial y^2}\right) - \frac{\hbar^2}{2\mu_\parallel}\left(\frac{\varepsilon_\perp}{\varepsilon_\parallel}\right)\frac{\partial^2}{\partial z^2}$$

$$= -\frac{\hbar^2}{2\mu_0}\nabla^2 - \frac{\hbar^2}{2}\frac{2\gamma}{3}\left(\frac{1}{2}\frac{\partial^2}{\partial x^2} + \frac{1}{2}\frac{\partial^2}{\partial y^2} - \frac{\partial^2}{\partial z^2}\right).$$

For the potential term,

$$\mathbf{D} = \varepsilon\mathbf{E} = \begin{pmatrix} \varepsilon_\perp & 0 & 0 \\ 0 & \varepsilon_\perp & 0 \\ 0 & 0 & \varepsilon_\parallel \end{pmatrix} \begin{pmatrix} E_x \\ E_y \\ E_z \end{pmatrix}, \qquad \mathbf{E} = -\nabla\phi,$$

then the potential energy $V(r) = -e\phi$. Assuming the hole is at the origin, then the potential at the electron satisfies

$$\nabla \cdot \mathbf{D} = \varepsilon_\perp \left(\frac{\partial^2\phi}{\partial x^2} + \frac{\partial^2\phi}{\partial y^2} \right) + \varepsilon_\parallel \frac{\partial^2\phi}{\partial \mathfrak{z}^2} = 4\pi e\, \delta(x)\, \delta(y)\, \delta(\mathfrak{z}).$$

Let $x' = x/\sqrt{\varepsilon_\perp}$, $\quad y' = y/\sqrt{\varepsilon_\perp}$, $\quad \mathfrak{z}' = \mathfrak{z}/\sqrt{\varepsilon_\parallel}$,

$$\nabla \cdot \mathbf{D} = \frac{\partial^2\phi}{\partial x'^2} + \frac{\partial^2\phi}{\partial y'^2} + \frac{\partial^2\phi}{\partial \mathfrak{z}'^2} = 4\pi \frac{e}{\varepsilon_\perp\sqrt{\varepsilon_\parallel}}\, \delta(x')\, \delta(y')\, \delta(\mathfrak{z}').$$

$$\phi = \frac{e}{\varepsilon_\perp\sqrt{\varepsilon_\parallel}\sqrt{x'^2 + y'^2 + \mathfrak{z}'^2}} = \frac{e}{\varepsilon_\perp\sqrt{\varepsilon_\parallel}\sqrt{\dfrac{x^2}{\varepsilon_\perp} + \dfrac{y^2}{\varepsilon_\perp} + \dfrac{\mathfrak{z}}{\varepsilon_\parallel}}}$$

$$= \frac{e}{\sqrt{\varepsilon_\perp \varepsilon_\parallel}\sqrt{x^2 + y^2 + \left(\dfrac{\varepsilon_\perp}{\varepsilon_\parallel}\right)\mathfrak{z}^2}} = \frac{e}{\sqrt{\varepsilon_\perp \varepsilon_\parallel}\sqrt{x^2 + y^2 + z^2}}$$

$$= \frac{e}{\varepsilon_0\sqrt{x^2 + y^2 + z^2}},$$

$$z = \sqrt{\frac{\varepsilon_\perp}{\varepsilon_\parallel}}\,\mathfrak{z}.$$

$$V(r) = -\frac{e^2}{\varepsilon_0\sqrt{x^2 + y^2 + z^2}}.$$

3. $$H_0 = -\frac{\hbar^2}{2\mu_0}\nabla^2 - \frac{e^2}{\varepsilon_0 r},$$

$$H' = -\frac{\hbar^2\gamma'}{3}\left(\frac{1}{2}\nabla^2 - \frac{3}{2}\frac{\partial^2}{\partial z^2} \right)$$

$$= -\frac{\gamma\hbar^2}{3\mu_0}\left(\frac{1}{2}\frac{\partial^2}{\partial x^2} + \frac{1}{2}\frac{\partial^2}{\partial y^2} - \frac{\partial^2}{\partial z^2} \right), \qquad \gamma = \gamma'\mu_0.$$

From $H_0\psi = E\psi$, (this is just the hydrogenic Schrödinger equation),

$$E_n = -E_1/n^2, \qquad E_1 = \mu_0 e^4/2\varepsilon_0^2\hbar^2.$$

The zero order energy for $n = 1$ and 2 states are

$$E_{1s} = E_g - E_1, \qquad \text{where } E_g \text{ is the gap energy.}$$
$$E_2 = E_g - E_1/4.$$

With H', we calculate the perturbed energy

$$E' = \langle 1s|H'|1s\rangle.$$

State $1s$ is spherically symmetric, so

$$\left\langle \frac{\partial^2}{\partial x^2} \right\rangle = \left\langle \frac{\partial^2}{\partial y^2} \right\rangle = \left\langle \frac{\partial^2}{\partial z^2} \right\rangle.$$
$$E' = 0.$$
$$E_{1s} = E_g - E_1.$$

For $n = 2$, we have fourfold degeneracies.

$$\psi_{200} = \frac{1}{4\sqrt{2\pi}} (a_0)^{-3/2} \left(2 - \frac{r}{a_0}\right) e^{-r/2a_0},$$

where a_0 is the effective Bohr radius,

$$a_0 = \frac{\hbar^2 \varepsilon_0}{\mu_0 e^2}.$$

$$\psi_{210} = \frac{1}{4\sqrt{2\pi}} (a_0)^{-3/2} e^{-r/2a_0} \left(\frac{r}{a_0}\right) \cos\theta,$$

$$\psi_{21\pm1} = \frac{1}{8\sqrt{\pi}} (a_0)^{-3/2} e^{-r/2a_0} \left(\frac{r}{a_0}\right) \sin\theta e^{\pm i\phi}.$$

The ϕ part of ψ is invariant under H', so there is no off-diagonal matrix element between ψ_{200}, ψ_{210}, to $\psi_{21\pm1}$ and between ψ_{211} and ψ_{21-1}. Furthermore, the spherical symmetry of ψ_{200} is invariant under H'; therefore ψ_{200} does not couple to ψ_{210} by H'. Similar to $|1s\rangle$, $\langle 2s|H'|2s\rangle = 0$.

$$E_{2s} = E_g - E_1/4.$$

The virial theorem gives $T = -V/2$, so

$$\langle 211| - \frac{\hbar^2}{2\mu_0}\nabla^2|211\rangle = -E_1/4.$$

Consider $\langle 211| \dfrac{\partial^2}{\partial z^2}|211\rangle$,

$$\frac{\partial}{\partial z}\left\{e^{-r/2a_0}\frac{r}{a_0}\sin\theta e^{i\phi}\right\} = -\frac{(x+iy)}{2a_0^2}e^{-r/2a_0}\cdot\frac{z}{r}.$$

$$\frac{\partial^2}{\partial z^2}\left\{e^{-r/2a_0}\frac{r}{a_0}\sin\theta e^{i\phi}\right\} = \frac{(x+iy)}{2a_0^2}e^{-r/2a_0}\left\{\frac{z^2}{2a_0r^2} - \frac{1}{r} + \frac{z^2}{r^3}\right\}$$

$$= \frac{e^{-r/2a_0}}{2a_0^2}\left\{\frac{r\cos^2\theta\sin\theta}{2a_0} - \sin^3\theta\right\}e^{i\phi}.$$

$$\langle 211|\frac{\partial^2}{\partial z^2}|211\rangle = -\frac{1}{20a_0^2}.$$

$$\frac{\hbar^2}{\mu_0 a_0^2} = 2E_1.$$

$$\langle 211|H'|211\rangle = -\frac{\gamma}{3}\langle 211|\frac{\hbar^2}{2\mu_0}\nabla^2 - \frac{3\hbar^2}{2\mu_0}\frac{\partial^2}{\partial z^2}|211\rangle$$

$$= -\frac{\gamma}{3}\left(-\frac{E_1}{4} + \frac{3}{20}E_1\right)$$

$$= \gamma\frac{E_1}{4}\left(\frac{2}{15}\right).$$

$$E_{211} = E_g - \frac{E_1}{4} + \frac{\gamma E_1}{4}\left(\frac{2}{15}\right)$$

$$= E_g - \frac{E_1}{4}\left(1 - \frac{2}{15}\gamma\right).$$

Similarly,

$$E_{21-1} = E_g - \frac{E_1}{4}\left(1 - \frac{2}{15}\gamma\right),$$

$$E_{210} = E_g - \frac{E_1}{4}\left(1 + \frac{4}{15}\gamma\right).$$

CHAPTER 16

1. The first step is to find ε_{xx}, ε_{yy}, ε_{xy}, and ε_{yx}; then use Eq. (79) to find ω. In a magnetic field, the equations of motion for the electron are:

$$\ddot{x} = -\frac{e}{m}E_x - \omega_c v_y, \qquad \omega_c = \frac{eH}{mc}.$$

$$\ddot{y} = -\frac{e}{m}E_y + \omega_c v_x.$$

$$v_x, v_y \sim e^{-i\omega t},$$

$$-\omega^2 x = -\frac{e}{m}E_x + i\omega\omega_c y,$$

$$-\omega^2 y = -\frac{e}{m}E_y - i\omega_c\omega x, \qquad \text{then } y = \frac{e}{m\omega^2}E_y + \frac{i\omega_c}{\omega}x.$$

$$-\omega^2 x = -\frac{e}{m}E_x + \frac{i\omega_c e}{m\omega}E_y - \omega_c^2 x.$$

$$\omega \ll \omega_c, \qquad x = -\frac{e}{m\omega_c^2}E_x + \frac{ie}{m\omega\omega_c}E_y, \qquad y = -\frac{ie}{m\omega_c\omega}E_x.$$

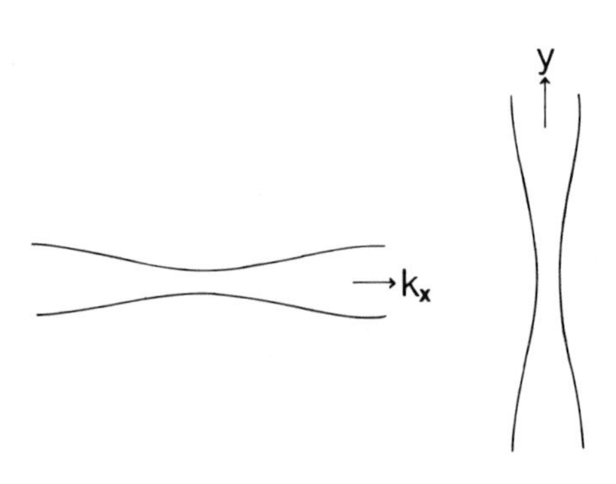

FIG. 16.1. The orbit in **k**-space and real space.

$$P_x = -nex = \frac{ne^2}{m\omega_c^2}E_x - \frac{ine^2}{m\omega\omega_c}E_y.$$

$$\varepsilon_{xx} = \frac{\omega_p^2}{\omega_c^2}, \qquad \varepsilon_{xy} = -\frac{i\omega_p^2}{\omega\omega_c}.$$

$$y = -\frac{ie}{m\omega\omega_c}E_x, \qquad P_y = \frac{ine^2}{m\omega\omega_c}E_x, \qquad \varepsilon_{yx} = \frac{i\omega_p^2}{\omega\omega_c}.$$

The current due to the electron motion along the open orbit is

$$j_{op} = -\frac{n_{op}e^2}{m}\tau E_y = -\sigma E_y.$$

$$\varepsilon_{yx} = i\frac{\omega_p^2}{\omega\omega_c}, \qquad \varepsilon_{yy} = \frac{i4\pi}{\omega}\sigma = -i\frac{\omega^{*2}\tau}{\omega}, \qquad \omega^{*2} = 4\pi n_{op}e^2/m.$$

$$\begin{vmatrix} \dfrac{c^2k^2}{\omega^2} - \dfrac{\omega_p^2}{\omega_c^2} & -i\dfrac{\omega_p^2}{\omega\omega_c} \\ i\dfrac{\omega_p^2}{\omega\omega_c} & \dfrac{c^2k^2}{\omega^2} - i\dfrac{\omega^{*2}\tau}{\omega} \end{vmatrix} = 0$$

$$\left(c^2k^2 - \frac{\omega_p^2\omega^2}{\omega_c^2}\right)\left(\frac{c^2k^2}{\omega^{*2}} - i\omega\tau\right) - \frac{\omega_p^4\omega^2}{\omega_c^2\omega^{*2}} = 0$$

$$\omega\tau \simeq 1, \qquad \omega_p^2\omega^2 \ll \omega_c^2, \qquad c^2k^2 \ll \omega^{*2}.$$

$$c^2k^2 \simeq i\frac{\omega_p^4\omega^2}{\omega_c^2\omega^{*2}}.$$

Consider the magnitude

$$\omega^2 \sim c^2k^2\omega_c^2\frac{\omega^{*2}}{\omega_p^4}.$$

$$\omega \sim ck\omega_c\omega^*/\omega_p^2.$$

3. $mv_x = p_x - \frac{e}{c}A_x, \qquad A_x = A_{\mathbf{q}} e^{iqz - i\omega t}.$

$$(j_{\mathbf{q}})_x e^{-i\omega t} = Tr\left(e^{-i\mathbf{q}\cdot\mathbf{r}} e\rho(t) v_x\right)$$

$$= Tr\left(e^{-i\mathbf{q}\cdot\mathbf{r}} e\left(\rho_0 + \delta\rho(t)\right)\left(p_x - \frac{e}{c}A_x\right)\right)\frac{1}{m}$$

$$= Tr\left(e^{-i\mathbf{q}\cdot\mathbf{r}}\left(-\frac{e^2}{c}\right)\rho_0 A_x\right)\frac{1}{m} + Tr\left(e^{-i\mathbf{q}\cdot\mathbf{r}}\,\delta\rho p_x\right)\frac{e}{m},$$

to first order of A_x.

$$= -\frac{e^2}{mc} Tr\left(\rho_0 A_{\mathbf{q}} e^{-i\omega t}\right) + Tr\left(e^{-i\mathbf{q}\cdot\mathbf{r}} e\delta\rho e^{-i\omega t} p_x\right)\frac{1}{m}.$$

$$i\hbar\dot{\rho} = [H, \rho] = i\hbar\delta\dot{\rho} = [H_0 + H', \rho_0 + \delta\rho]$$

$$= [H_0, \rho_0] + [H', \rho_0] + [H_0, \delta\rho].$$

$$i\hbar\frac{\partial}{\partial t}\langle\mathbf{k}+\mathbf{q}|\delta\rho(t)|\mathbf{k}\rangle = (\hbar\omega + i\eta)\langle\mathbf{k}+\mathbf{q}|\delta\rho e^{-i\omega t - \eta|t|/\hbar}|\mathbf{k}\rangle$$

$$= (f_{\mathbf{k}} - f_{\mathbf{k}+\mathbf{q}})\langle\mathbf{k}+\mathbf{q}| \times \left(-\frac{e}{mc}\right)\left(A_{\mathbf{q}} e^{iqz - i\omega t}\right)p_x|\mathbf{k}\rangle + (\varepsilon_{\mathbf{k}+\mathbf{q}} - \varepsilon_{\mathbf{k}})\langle\mathbf{k}+\mathbf{q}|\delta\rho|\mathbf{k}\rangle.$$

$$\langle\mathbf{k}+\mathbf{q}|\delta\rho|\mathbf{k}\rangle = \frac{(f_{\mathbf{k}} - f_{\mathbf{k}+\mathbf{q}})}{\varepsilon_{\mathbf{k}+\mathbf{q}} - \varepsilon_{\mathbf{k}} - \hbar\omega - i\eta}\langle\mathbf{k}+\mathbf{q}|\frac{e}{mc}A_{\mathbf{q}} e^{iqz}(\hbar k_x)|\mathbf{k}\rangle.$$

$$Tr\left(e^{-iqz} e\delta\rho p_x\right) = \sum_{\mathbf{k},\mathbf{k}'\mathbf{k}''}\langle\mathbf{k}|e^{-iqz}|\mathbf{k}'\rangle\langle\mathbf{k}'|e\delta\rho|\mathbf{k}''\rangle\langle\mathbf{k}''|p_x|\mathbf{k}\rangle$$

$$= \frac{e^2\hbar^2}{mc}A_{\mathbf{q}}\sum_{\mathbf{k}} k_x^2 \frac{f_{\mathbf{k}} - f_{\mathbf{k}+\mathbf{q}}}{\varepsilon_{\mathbf{k}+\mathbf{q}} - \varepsilon_{\mathbf{k}} - \hbar\omega - i\eta}$$

$$= \frac{e^2\hbar^2}{mc}\left(\frac{c}{i\omega}\right)E_{\mathbf{q}}\sum_{\mathbf{k}} k_x^2\left\{\frac{f_{\mathbf{k}}}{\varepsilon_{\mathbf{k}+\mathbf{q}} - \varepsilon_{\mathbf{k}} - \hbar\omega - i\eta} - \frac{f_{\mathbf{k}}}{\varepsilon_{\mathbf{k}} - \varepsilon_{\mathbf{k}-\mathbf{q}} - \hbar\omega - i\eta}\right\},$$

where $E_{\mathbf{q}}$ is electric field.

$$= \frac{e^2\hbar^2}{mc}\left(\frac{c}{i\omega}\right)E_{\mathbf{q}}\sum_{\mathbf{k}} k_x^2\left\{\mathscr{P}\frac{f_{\mathbf{k}}}{\varepsilon_{\mathbf{k}+\mathbf{q}} - \varepsilon_{\mathbf{k}} - \hbar\omega}\right.$$

$$+ i\pi\delta\left(\frac{\hbar^2}{m}kq\mu + \frac{\hbar^2}{2m}q^2 - \hbar\omega\right)$$

$$- \mathscr{P}\frac{f_{\mathbf{k}}}{\varepsilon_{\mathbf{k}} - \varepsilon_{\mathbf{k}-\mathbf{q}} - \hbar\omega}$$

$$- i\pi\delta\left(\frac{\hbar^2}{m}kq\mu + \frac{\hbar^2 q^2}{2m} - \hbar\omega\right)\Bigg\},$$

where $\mu = \cos\theta$, and $\mathscr{P}$ is the principal value.

$$Re\left\{Tr\left(e^{-iqz}e\delta\rho p_x\right)\right\}$$

$$= \frac{e^2\hbar^2}{mc}\left(\frac{c}{\omega}\right)E_{\mathbf{q}}\sum_{\mathbf{k}}k_x^2\left[\pi\delta\left(\frac{\hbar^2}{m}kq\mu + \frac{\hbar^2}{2m}q^2 - \hbar\omega\right) - \pi\delta\left(\frac{\hbar^2}{m}kq\mu - \frac{\hbar^2 q^2}{2m} - \hbar\omega\right)\right]$$

$$= \frac{e^2\hbar^2}{mc}\left(\frac{c}{\omega}\right)E_{\mathbf{q}}\frac{2\pi}{(2\pi)^3}\int k^2\, dk\, d\mu\, d\phi\, k^2(1-\mu^2)\cos^2\phi$$

$$\left\{\delta\left(\frac{\hbar^2}{m}kq\mu + \frac{\hbar^2}{2m}q^2 - \hbar\omega\right) - \delta\left(\frac{\hbar^2}{m}kq\mu - \frac{\hbar^2 q^2}{2m} - \hbar\omega\right)\right\}$$

$$= \frac{e^2\hbar^2}{mc}\left(\frac{c}{\omega}\right)E_{\mathbf{q}}\pi\frac{2\pi}{(2\pi)^3}\int k^4\, dk\,(1-\mu^2)\left[\delta\left(\frac{\hbar^2}{m}kqu + \frac{\hbar^2}{2m}q^2 - \hbar\omega\right) - \delta\left(\frac{\hbar^2}{m}kq\mu - \frac{\hbar^2 q^2}{2m} - \hbar\omega\right)\right]d\mu$$

$$= \frac{e^2\hbar^2}{mc}\left(\frac{c}{\omega}\right)E_{\mathbf{q}}\pi\frac{2\pi}{(2\pi)^3}\int k^4\, dk\,\frac{m}{\hbar^2}\frac{1}{kq}\left\{\left[1 - \left(-\frac{q}{2k} + \frac{m}{\hbar}\frac{\omega}{kq}\right)^2\right] - \left[1 - \left(\frac{q}{2k} + \frac{m}{\hbar}\frac{\omega}{kq}\right)^2\right]\right\}$$

$$= \frac{e^2\hbar^2}{mc}\left(\frac{c}{\omega}\right)E_{\mathbf{q}}\pi\frac{2\pi}{(2\pi)^3}\int k^4\, dk\,\frac{m}{\hbar^2}\frac{1}{kq}\left\{2\frac{m}{\hbar}\frac{\omega}{k^2}\right\}$$

$$= \frac{e^2\hbar^2}{mc}\left(\frac{c}{\omega}\right)E_{\mathbf{q}}\pi\frac{4\pi}{(2\pi)^3}\left(\frac{m^2}{\hbar^3}\right)\frac{\omega}{q}\int_0^{k_F} k\,dk$$

$$= \frac{e^2\hbar^2}{mc}\left(\frac{c}{\omega}\right)E_{\mathbf{q}}\pi\frac{4\pi}{(2\pi)^3}\left(\frac{m^2}{\hbar^3}\right)\frac{\omega}{q}\frac{k_F^2}{2}$$

$$= \frac{e^2 m\pi}{q\hbar}E_{\mathbf{q}}\frac{3}{4}\frac{n}{k_F} = \frac{3\pi n e^2 m}{4 m v_F q}E_{\mathbf{q}}, \qquad \hbar k_F = m v_F.$$

$$\mathscr{I}\left\{Tr\left(e^{-iqz}e\delta\rho p_x\right)\right\} = \frac{e^2\hbar^2}{mc}A_{\mathbf{q}}\sum_{\mathbf{k}} k_x^2\left\{\mathscr{P}\frac{f_{\mathbf{k}}}{\varepsilon_{\mathbf{k}+\mathbf{q}} - \varepsilon_{\mathbf{k}} - \hbar\omega} - \mathscr{P}\frac{f_{\mathbf{k}}}{\varepsilon_{\mathbf{k}} - \varepsilon_{\mathbf{k}-\mathbf{q}} - \hbar\omega}\right\}.$$

For $\frac{\hbar^2}{m}kq \gg \hbar\omega$,

$$\mathscr{I}\left\{Tr\left(e^{-iqz}e\delta\rho p_x\right)\right\} = \frac{e^2\hbar^2}{mc}A_{\mathbf{q}}\frac{2}{(2\pi)^3}\int k^4\,dk\,(1-\mu^2) \times d\mu\cos^2\phi\,d\phi\,\frac{-2q/2k}{\frac{\hbar^2}{m}kq\left[\mu^2 - \left(\frac{q}{2k}\right)^2\right]}$$

$$= \frac{e^2}{c}A_{\mathbf{q}}\frac{2\pi}{(2\pi)^3}\int k^2\,dk\,\frac{\mu^2 - 1}{\mu^2 - \left(\frac{q}{2k}\right)^2}\,d\mu$$

$$= \frac{e^2}{c}A_{\mathbf{q}}\frac{2\pi}{(2\pi)^3}\int k^2\,dk\int_{-1}^{1} d\mu\left[1 + \left(\frac{q}{4k} - \frac{k}{q}\right) \times\left(\frac{1}{\mu - q/2k} - \frac{1}{\mu + q/2k}\right)\right]$$

$$= \frac{e^2}{c}A_{\mathbf{q}}\frac{2\pi}{(2\pi)^3}\int k^2\,dk\left\{2 + \left(\frac{q}{4k} - \frac{k}{q}\right)2 \cdot ln\left|\frac{1 - q/2k}{1 + q/2k}\right|\right\} = I.$$

For $q/2k < 1$, $$I \simeq \frac{e^2}{c} A_{\mathbf{q}} \frac{2\pi}{(2\pi)^3} \int k^2\, dk \left\{ 2 + 2 \cdot \frac{k}{q} \cdot \frac{2q}{2k} \right\}.$$

For $q/2k > 1$, $$I \simeq \frac{e^2}{c} A_{\mathbf{q}} \frac{2\pi}{(2\pi)^3} \int k^2\, dk \left\{ 2 + 2 \cdot \frac{q}{4k} \cdot 2 \cdot \frac{2k}{q} \right\}.$$

So, for $q/2k \neq 1$,

$$I = \frac{e^2}{c} A_{\mathbf{q}} \frac{2\pi}{(2\pi)^3} \int dk\, k^2 \cdot 4 = \frac{e^2}{c} A_{\mathbf{q}} \frac{2}{(2\pi)^3} \cdot 4\pi \cdot \frac{k_F^3}{3} = \frac{e^2 n}{c} A_{\mathbf{q}}.$$

$$(j_{\mathbf{q}})_x = -\frac{ne^2}{mc} A_{\mathbf{q}} + \frac{ne^2}{mc} A_{\mathbf{q}} + \frac{3\pi ne^2}{4mv_F q} E_{\mathbf{q}} = \sigma_{\mathbf{q}} E_{\mathbf{q}}.$$

$$\sigma_{\mathbf{q}} = \frac{3\pi ne^2}{4v_F qm}, \qquad \varepsilon(\omega, \mathbf{q}) = \frac{4\pi i}{\omega} \sigma_{\mathbf{q}} = \frac{4\pi i}{\omega} \cdot \frac{3\pi ne^2}{4v_F qm}$$

CHAPTER 17

1. $$\lambda \sim \frac{cG_c}{eH}.$$

$G_c \sim 1 \times 10^{-19}$ g cm/sec for the dimension of the zone in the [0001] direction, and

$$f = \frac{v}{\lambda} \sim \frac{eHv}{cG_c} \sim \frac{5 \times 10^{-10} \times 10^3 \times 4 \times 10^5}{3 \times 10^{10} \times 10^{-19}} \sim 6 \times 10^7/\text{sec}.$$

CHAPTER 18

1. Eq. (30),

$$Z = \frac{2}{\pi} \sum_l (2l + 1)\eta_l(k_F).$$

Eq. (34),

$$\Delta\varepsilon_{\mathbf{k}} = \frac{h^2}{2m}(k'^2 - k^2) = \frac{\hbar^2}{2m}\left[\left(k - \frac{\eta_l}{R}\right)^2 - k^2\right] \simeq -\frac{\hbar^2 k \eta_l}{mR}.$$

DOS of l/unit wave number $= 2(2l+1)R/\pi$.

$$Z = \frac{2}{\pi}\sum_l (2l+1)\left(-\frac{mR}{\hbar^2 k}\right)\Delta\varepsilon_{\mathbf{k}}$$

$$= \frac{2}{\pi}\sum_l (2l+1)\left(-\frac{mR}{\hbar^2 k}\right)\langle \mathbf{k}|V_p|\mathbf{k}\rangle.$$

$$\rho_F = \frac{dn}{d\varepsilon} = \frac{dn}{dk}\frac{dk}{d\varepsilon} = \frac{dn}{dk}\bigg/\left(\frac{d\varepsilon}{dk}\right).$$

$$dn/dk = \sum_l 2(2l+1)R/\pi.$$

$$d\varepsilon/dk = \hbar^2 k/m.$$

$$\rho_F = \sum_l \frac{2}{\pi}(2l+1)\left(\frac{Rm}{\hbar^2 k}\right).$$

$$Z = -\rho_F\langle \mathbf{k}|V_p|\mathbf{k}\rangle.$$

2. $$F_\gamma(\varepsilon) = \int dE\,\frac{g_\gamma(E)}{\varepsilon - E}$$

$$= \int_0^{\varepsilon_1} dE\,\frac{g_0}{\varepsilon - E} = -g_0\,ln\left|\frac{\varepsilon_1 - \varepsilon}{\varepsilon}\right|.$$

$$V > 0, \qquad 1 + Vg_0 ln\left|\frac{\varepsilon_1 - \varepsilon_0}{\varepsilon_0}\right| = 0, \qquad ln\left|\frac{\varepsilon_1 - \varepsilon_0}{\varepsilon_0}\right| = -\frac{1}{Vg_0},$$

$$e^{-1/Vg_0} = \frac{\varepsilon_1 - \varepsilon_0}{\varepsilon_0},$$

$$\varepsilon_0 = \frac{\varepsilon_1}{e^{-1/Vg_0} + 1}.$$

$$V < 0, \qquad 1 - |V|g_0 ln\left|\frac{\varepsilon_1 - \varepsilon_0}{\varepsilon_0}\right| = 0.$$

$$\varepsilon_0 = \frac{\varepsilon_1}{e^{1/|V|g_0} + 1}.$$

3. a. From Eq. (88), $(\xi - \varepsilon_{\mathbf{k}})G^\sigma_{\mathbf{kk}} = 1 + V_{\mathbf{k}d}G^\sigma_{d\mathbf{k}}$. From Eq. (86),

$$G^\sigma_{\mathbf{k}d} = \frac{1}{\xi - \varepsilon_{\mathbf{k}}}V_{\mathbf{k}d}G^\sigma_{dd}.$$

$$G^\sigma_{d\mathbf{k}} = (G^\sigma_{\mathbf{k}d})^* = \frac{1}{\xi - \varepsilon_{\mathbf{k}}}V_{d\mathbf{k}}(G^\sigma_{dd})^*.$$

$$G^{\sigma}_{dd} = \frac{1}{\xi - E_\sigma - i\Delta}.$$

$$G^{\sigma}_{\mathbf{kk}} = \frac{1}{\xi - \varepsilon_{\mathbf{k}}} + \frac{1}{(\xi - \varepsilon_{\mathbf{k}})^2} \frac{|V_{d\mathbf{k}}|^2}{\xi - E_\sigma + i\Delta}.$$

b. $$\rho^{\sigma}_{\text{free}} = -\frac{1}{\pi}\left(\frac{1}{2\pi}\right)^3 \int d^3k\, \mathscr{I}(G^{\sigma}_{\mathbf{kk}}).$$

$$\frac{1}{\xi - \varepsilon_{\mathbf{k}}} = \frac{1}{\xi - \varepsilon_{\mathbf{k}} + i\delta} = \mathscr{P}\frac{1}{\xi - \varepsilon_{\mathbf{k}}} - i\pi\delta(\xi - \varepsilon_{\mathbf{k}}).$$

$$\frac{1}{(\xi - \varepsilon_{\mathbf{k}})^2} = -\frac{d}{\partial \xi}\left(\frac{1}{\xi - \varepsilon_{\mathbf{k}}}\right).$$

$$\rho^{\sigma}_{\text{free}} = -\frac{1}{\pi}\left(\frac{1}{2\pi}\right)^3 \int d^3k \left\{ -\pi\delta(\xi - \varepsilon_{\mathbf{k}}) + \pi\frac{\partial}{\partial \xi}[\delta(\xi - \varepsilon_{\mathbf{k}})] \times \frac{|V_{d\mathbf{k}}|^2(\xi - E_\sigma)}{(\xi - E_\sigma)^2 + \Delta^2} \right\}$$

$$= \rho^{\sigma}_0(\xi) + \frac{d\rho^{\sigma}_0}{\partial \varepsilon_{\mathbf{k}}} \frac{|V_{d\mathbf{k}}|^2(\xi - E_\sigma)}{(\xi - E_\sigma)^2 + \Delta^2}.$$

CHAPTER 19

1. $\varphi = e^{ikz} + \dfrac{\alpha}{r}e^{ikr}$, $e^{ikz} = e^{ikr\cos\theta}$,

$$\frac{1}{4\pi}\int d\Omega\, e^{ikr\cos\theta} = \frac{1}{4\pi}2\pi\int_{-1}^{1} d\mu\, e^{ikr\mu}$$

$$= \frac{1}{2}\frac{1}{ikr}\left(e^{ikr\mu}\right)\Big|_{-1}^{1} = \frac{\sin kr}{kr}.$$

$$\varphi_s = \frac{1}{kr}\sin kr + \frac{\alpha}{r}e^{ikr} = \frac{1}{2ikr}\left\{e^{ikr}(1 + 2ik\alpha) - e^{-ikr}\right\}$$

$$= e^{i\eta_0}\frac{\sin(kr + \eta_0)}{kr} = \frac{e^{i\eta_0}}{kr}\frac{1}{2i}\left(e^{ikr + i\eta_0} - e^{-ikr - i\eta_0}\right).$$

Identify the coefficient of e^{ikr}, $e^{i2\eta_0} = 1 + 2ik\alpha$,

$$\alpha = \frac{1}{2ik}(e^{i2\eta_0} - 1) = \frac{1}{k}e^{i\eta_0}(e^{i\eta_0} - e^{-i\eta_0})/2i$$
$$= \frac{1}{k}e^{i\eta_0}\sin\eta_0.$$

From Eqs. (20) and (21), $\alpha = -e^{i\eta_0}b$, $\eta_0 = -kb$, for s-wave $|kb| \ll 1$.

4. $\mathbf{P}_\perp = \hat{\mathbf{k}} \times [\mathbf{S}_l \times \hat{\mathbf{k}}] = \mathbf{S}_l - (\hat{\mathbf{k}} \cdot \mathbf{S}_l)\hat{\mathbf{k}}$.

$$\sum_\alpha P_\perp^\alpha P_\perp^\alpha = \sum_\alpha \{[S_l^\alpha - (\hat{\mathbf{k}} \cdot \mathbf{S}_l)\hat{\mathbf{k}}^\alpha][S_l^\alpha - (\hat{\mathbf{k}} \cdot \mathbf{S}_l)\hat{\mathbf{k}}^\alpha]\}$$
$$= \sum_\alpha \left\{S_l^\alpha S_l^\alpha - S_l^\alpha(\hat{\mathbf{k}} \cdot \mathbf{S}_l)\hat{\mathbf{k}}^\alpha - (\hat{\mathbf{k}} \cdot \mathbf{S}_l)\hat{\mathbf{k}}^\alpha S_l^\alpha + (\hat{\mathbf{k}} \cdot \mathbf{S}_l)^2\hat{\mathbf{k}}^\alpha\hat{\mathbf{k}}^\alpha\right\}$$
$$= \sum_{\alpha,\beta}\left(S_l^\alpha \delta_{\alpha\beta} S_l^\beta\right) - \sum_{\alpha,\beta}\left(\hat{\mathbf{k}}^\alpha S_l^\alpha\right)\left(\hat{\mathbf{k}}^\beta S_l^\beta\right)$$
$$= \sum_{\alpha,\beta}\left(\delta_{\alpha\beta} - \hat{\mathbf{k}}^\alpha\hat{\mathbf{k}}^\beta\right)S_l^\alpha S_l^\beta.$$

5. From Eq. (1),

$$\frac{d^2\sigma}{d\varepsilon\, d\Omega} = \frac{k'}{k}\left(\frac{M}{2\pi}\right)^2 |\langle \mathbf{k}'f|H'|i\mathbf{k}\rangle|^2\, \delta(\hbar\omega - \varepsilon_i + \varepsilon_f).$$
$$H' = n(\mathbf{r})b\frac{2\pi}{M}.$$

For elastic scattering $k' = k$,

$$\langle \mathbf{k}'f|H'|i\mathbf{k}\rangle = \int e^{-i\mathbf{k}'\cdot\mathbf{r}}\frac{2\pi}{M}bn(\mathbf{r})e^{i\mathbf{k}\cdot\mathbf{r}}\, d^3r$$

$$\frac{d^2\sigma}{d\varepsilon\, d\Omega} = b^2\,\delta(\hbar\omega - \varepsilon_i + \varepsilon_f)\left|\int e^{-i(\mathbf{k}'-\mathbf{k})\cdot\mathbf{r}}n(\mathbf{r})\, d^3r\right|^2$$
$$= b^2\,\delta(\hbar\omega - \varepsilon_i + \varepsilon_f)\left|\int e^{i\mathbf{K}\cdot\mathbf{r}}n(\mathbf{r})\, d^3r\right|^2, \qquad \mathbf{k} - \mathbf{k}' = \mathbf{K}.$$

$n(\mathbf{r}) = \text{const.}$, $\int e^{i\mathbf{K}\cdot\mathbf{r}}\, d^3r = \delta(\mathbf{K})$. This means that particle with incident momentum **k** is scattered only to momentum **k**. No scattering.

CHAPTER 20

1. $\frac{1}{3}K^2\langle u^2\rangle = \frac{K^2}{3}\frac{\hbar}{MN}\sum_{\mathbf{q}}\frac{(1/2+n_s)}{c_s q}$. The temperature dependent contribution comes from n_s,

$$\frac{1}{c_s}\sum_{\mathbf{q}}\frac{n_s}{q} = \frac{1}{c_s}\frac{V}{(2\pi)^3}4\pi\int_0^{q_m} q_2\,dq\frac{1}{e^{\beta\hbar\omega_q}-1}\frac{1}{q}$$

$$= \frac{1}{c_s}\frac{V}{2\pi^2}\frac{k_B^2T^2}{\hbar^2c_s^2}\int_0^{\Theta/T}\frac{x\,dx}{e^x-1}$$

$$= \frac{V}{2\pi^2}\frac{k_B^2T^2}{\hbar^2c_s^3}\int_0^{\Theta/T}\frac{x\,dx}{1+x-1}$$

$$= \frac{V}{(2\pi)^2}\frac{k_B^2T^2}{\hbar c_s^3}\left(\frac{\Theta}{T}\right), \qquad \text{for } T\gg\Theta.$$

$\frac{1}{3}K^2\langle u^2\rangle \sim T$, the proportionality constant is

$$\frac{K^2}{3}\frac{\hbar}{M}\Omega_{\text{cell}}\frac{1}{2\pi^2}\left(\frac{k_B^2\Theta}{\hbar^2c_s^3}\right), \qquad \text{where } \Omega_{\text{cell}} = V/N.$$

2. $I(E) = \frac{\Gamma}{4}\int_{-\infty}^{\infty} dt\, e^{-i(E-E_0)\frac{t}{\hbar}-\frac{\Gamma}{2}|t|-Q(t)}$.

$$Q(t) = \frac{1}{2}K^2\left\{\left\langle (u(t)-u(0))^2\right\rangle_T + [u(0),u(t)]\right\}.$$

For free atom, $u(t) = u(0)+vt$, $\Gamma = 0$.

$$\left\langle (u(t)-u(0))^2\right\rangle_T = \langle v^2\rangle_T t^2.$$

$$[u(0),u(t)] = [u(0),u(0)+vt] = [u(0),vt] = \left[u(0),t\frac{P}{M}\right]$$

$$= \frac{i\hbar t}{M}.$$

$$Q(t)=\frac{1}{2}K^2\left\{\langle v^2\rangle_T t^2+\frac{i\hbar t}{M}\right\}.$$

$$I(E)=\frac{\Gamma}{4}\int_{-\infty}^{\infty}dt\,e^{-i(E-E_0)\frac{t}{\hbar}-\frac{i\hbar K^2}{2M}t-\frac{K^2}{2}\langle v^2\rangle_T t2}.$$

Energy shift $=\dfrac{\hbar K^2}{2M}$. The line width is in Gaussian form $\sim\sqrt{K^2\langle v^2\rangle_T}$. Let energy of the γ-ray be 10^{-7} erg, then $\sqrt{K^2\langle v^2\rangle_T}\sim 10^{-15}\,\text{sec}^{-1}$.

$$\text{Width}=\frac{\hbar}{\tau}\sim 10^{-27}\times 10^{15}\sim 10^{-12}\,\text{erg}\sim 1\,\text{eV}.$$

CHAPTER 21

1. $$G(x,t)=-\frac{2i}{L}\int_0^{k_F}e^{ikx}e^{-i\left(\frac{\hbar^2}{2m}k^2\right)t}\,dk,\qquad t>0.$$

$$=-\frac{2i}{L}e^{imx^2/2\hbar^2t}\int_0^{k_F}e^{-i\frac{\hbar^2t}{2m}\left(k-\frac{mx}{\hbar^2t}\right)^2}dk$$

$$=-\frac{2i}{L}e^{imx^2/2\hbar^2t}\int_{-\frac{mx}{\hbar^2t}}^{k_F-\frac{mx}{\hbar^2t}}e^{-i\frac{\hbar^2t}{2m}k^2}dk$$

$$=-\frac{2i}{L}e^{imx^2/2\hbar^2t}\sqrt{\frac{2m}{\hbar^2t}}$$

$$\times\left\{\int_{-\sqrt{\frac{\hbar^2t}{2m}}\frac{mx}{\hbar^2t}}^{0}e^{-ik^2}dk+\int_0^{\sqrt{\frac{\hbar^2t}{2m}}\left(k_F-\frac{mx}{\hbar^2t}\right)}e^{-ik^2}dk\right\}$$

$$=-\frac{2i}{L}e^{imx^2/2\hbar^2t}\sqrt{\frac{2m}{\hbar^2t}}$$

$$\times\left\{\int_0^{\sqrt{\frac{\hbar^2t}{2m}}\left(k_F-\frac{mx}{\hbar^2t}\right)}e^{-ik^2}dk+\int_0^{\sqrt{\frac{\hbar^2t}{2m}}\frac{mx}{\hbar^2t}}e^{-ik^2}dk\right\}$$

$$=-\frac{2i}{L}e^{imx^2/2\hbar^2t}\sqrt{\frac{2m}{\hbar^2t}}$$

$$\times\left\{\int_0^{\sqrt{\frac{\hbar^2t}{2m}}\left(k_F-\frac{mx}{\hbar^2t}\right)}(\cos k^2-i\sin k^2)\,dk\right.$$

$$+\int_0^{\sqrt{\frac{\hbar^2 t}{2m}}\frac{mx}{\hbar^2 t}}(\cos k^2 - i\sin k^2)\,dk\Bigg\}$$

$$= -\frac{2i}{L}e^{imx^2/2\hbar^2 t}\sqrt{\frac{2m}{\hbar^2 t}}$$

$$\times\left\{\sqrt{\frac{\pi}{2}}\,C_1\left(\sqrt{\frac{\hbar^2 t}{2m}}\left(k_F - \frac{mx}{\hbar^2 t}\right)\right)\right.$$

$$-i\sqrt{\frac{\pi}{2}}\,S_1\left(\sqrt{\frac{\hbar^2 t}{2m}}\left(k_F - \frac{mx}{\hbar^2 t}\right)\right)$$

$$\left.+\sqrt{\frac{\pi}{2}}\,C_1\left(\sqrt{\frac{\hbar^2 t}{2m}}\,\frac{mx}{\hbar^2 t}\right) - i\sqrt{\frac{\pi}{2}}\,S_1\left(\sqrt{\frac{\hbar^2 t}{2m}}\,\frac{mx}{\hbar^2 t}\right)\right\}.$$

C_1 and S_1 are Fresnel integrals.

$$G(X, -|t|) = -\frac{2i}{L}\int_0^{k_F} e^{ikx}e^{i\left(\frac{\hbar 2}{2m}k^2\right)|t|}\,dk$$

$$= -\frac{2i}{L}e^{-imx^2/2\hbar^2|t|}\int_0^{k_F} e^{i\frac{\hbar 2|t|}{2m}\left(k+\frac{mx}{\hbar^2|t|}\right)^2}\,dk$$

$$= -\frac{2i}{L}e^{-imx^2/2\hbar^2|t|}\sqrt{\frac{2m}{\hbar^2|t|}}\left\{\int_0^{\sqrt{\frac{\hbar^2|t|}{2m}}\left(k_F+\frac{mx}{\hbar^2|t|}\right)} e^{ik^2}\,dk\right.$$

$$\left.-\int_0^{\sqrt{\frac{\hbar^2|t|}{2m}}\frac{mx}{\hbar^2|t|}} e^{ik^2}\,dk\right\}$$

$$= -\frac{2i}{L}e^{-imx^2/2\hbar^2|t|}\sqrt{\frac{2m}{\hbar^2|t|}}$$

$$\times\left\{\sqrt{\frac{\pi}{2}}\,C_1\left(\sqrt{\frac{\hbar^2|t|}{2m}}\left(k_F + \frac{mx}{\hbar^2|t|}\right)\right)\right.$$

$$+i\sqrt{\frac{\pi}{2}}\,S_1\left(\sqrt{\frac{\hbar^2|t|}{2m}}\left(k_F + \frac{mx}{\hbar^2|t|}\right)\right)$$

$$\left.-\sqrt{\frac{\pi}{2}}\,C_1\left(\sqrt{\frac{\hbar^2|t|}{2m}}\,\frac{mx}{\hbar^2|t|}\right) - i\sqrt{\frac{\pi}{2}}\,S_1\left(\sqrt{\frac{\hbar^2|t|}{2m}}\,\frac{mx}{\hbar^2|t|}\right)\right\}$$

2. $$\mathbf{j}_{\alpha\beta} = \frac{-e\hbar}{2mi}\left[\Psi_\alpha^\dagger \nabla \Psi_\beta - \left(\nabla \Psi_\alpha^\dagger\right)\Psi_\beta\right].$$

$$\mathbf{j}_{\alpha\beta}(\mathbf{x}'t', \mathbf{x}t) = \frac{-e\hbar}{2mi}\left[\Psi_\alpha^\dagger(\mathbf{x}'t')\nabla_{\mathbf{x}}\Psi_\beta(\mathbf{x}t) - \left(\nabla_{\mathbf{x}'}\Psi_\alpha^\dagger(\mathbf{x}'\mathbf{t}')\right)\Psi_\beta(\mathbf{x}t)\right]$$

$$\begin{aligned}\left\langle \mathbf{j}_{\alpha\beta}(\mathbf{x}t)\right\rangle &= \lim_{\substack{\mathbf{x}'\to\mathbf{x}\\ t'\to t}} \frac{-e\hbar}{2mi} \\ &\quad \times \langle \Psi_0 | \Psi_\alpha^\dagger(\mathbf{x}'t')\nabla_{\mathbf{x}}\Psi_\beta(\mathbf{x}t) - \left(\nabla_{\mathbf{x}'}\Psi_\alpha^\dagger(\mathbf{x}'\mathbf{t}')\right)\Psi_\beta(\mathbf{x}t) | \Psi_0 \rangle \\ &= -\frac{e\hbar}{2mi}(+i)\lim_{\substack{\mathbf{x}'\to\mathbf{x}\\ t'\to t}}(\nabla_{\mathbf{x}'} - \nabla_{\mathbf{x}})G_{\alpha\beta}(\mathbf{x}'t', \mathbf{x}t).\end{aligned}$$

$$\begin{aligned}\left\langle \mathbf{j}(\mathbf{x}, t)\right\rangle &= Tr\left\langle \mathbf{j}_{\alpha\beta}(\mathbf{x}t)\right\rangle \\ &= -\frac{e\hbar}{2m}\lim_{\substack{\mathbf{x}'\to\mathbf{x}\\ t'\to t}} Tr\left\{(\nabla_{\mathbf{x}'} - \nabla_{\mathbf{x}})G_{\alpha\beta}(\mathbf{x}'t', \mathbf{x}t)\right\} \\ &= -\frac{e\hbar}{m}\lim_{\substack{\mathbf{x}'\to\mathbf{x}\\ t'\to t}}(\nabla_{\mathbf{x}'} - \nabla_{\mathbf{x}})G(\mathbf{x}'t', \mathbf{x}t).\end{aligned}$$

$$G(\mathbf{x}'t', \mathbf{x}t) = G_{\alpha\alpha}(\mathbf{x}'t', \mathbf{x}t) = G_{\beta\beta}(\mathbf{x}'t', \mathbf{x}t).$$

Index

Abelian group, 179
Abragam, A., 388
Abrahams, E., 116
Abrikosov, A. A., 173, 397
Absorption, optical, 294
Acceleration theorems, 190
Acceptor states, 286
Acoustic attenuation, 326
Acoustic magnon, 53
Acoustic phonons, 12, 21
Adams, E. N., 193
Aigrain, P., 321
Akhiezer, A. I., 57
Allowed transitions, 294
Alloys, 338
Almost-free electron approximation, 254
Aluminum, 262
  fermi surface, 260
Analysis, graphical, 116
Anderson, P. W., 58, 62, 157, 163, 183, 351, 353
Annihilation, electron, 117
  hole, 117
Anomalous magnetic moment, 282
Anomalous skin effect, 308, 313
Antiferromagnet, heat capacity, 62
  temperature dependence of sublattice magnetization, 61
Antiferromagnetic magnons, 58
Aperiodic open orbits, 233, 234
Argyres, P. N., 193, 293
Atomic polyhedra, 251
Attenuation, magnetic field effects, 333
Austin, B. J., 262
Azbel, M. I., 315
Band edge, 268
Band structure, linear lattice, 256
Bands, degenerate, 189
  energy, 249
Bardeen, J., 130, 148, 150, 151, 172, 296, 315
Bar'yakhtar, V. G., 57
Basis vectors, 1
Baym, G., 397
BCS equation, 160, 414
  equation-of-the-motion method, 164
  spin-analog method, 157
BCS operators, 178
Beliaev, 397
Bethe, H. A., 207, 252
Blandin, A., 339, 360, 365
Blatt, F., 346
Blatt, J., 296
Bloch, C., 126
Bloch, F., 49
Bloch, M., 57
Bloch equation, 324
  functions, 179, 200
  theorem, 179
Blount, E. J., 194
Body-centered cubic lattice, Brillouin zone, 211
Bogoliubov, N., 23, 26, 59, 151
Bohm, H. V., 326
Boltzmann equation, 29, 313, 328
Born approximation, 344, 368
Born, M., 37
Boson gas, 23
Bouckaert, L., 206
Bound electron pairs, fermi gas, 153

Boundary condition, Wigner-Seitz, 253
Bowers, R., 320, 371
Bragg condition, 255, 374
Breakthrough, magnetic, 229
Brillouin zone, 199
 body-centered cubic lattice, 211
 face-centered cubic lattice, 213
 hexagonal close-packed structure, 214
 linear lattice, 202
 simple cubic lattice, 205
 simple square lattice, 203, 257
Brockhouse, B. N., 53, 371
Brooks, H., 249
Brown, E., 262
Brown, F. C., 141, 142
Brownian motion, 396
Brueckner, K. A., 24, 100, 418, 423
Brueckner method, 100
 technique, 417
Budd, H., 246
Burnham, D. C., 141
Button, K. J., 305

Cadmium mercury telluride, 284
Cadmium sulfide, 293, 305
Callaway, J., 249, 276
Canonical transformation, 148, 151
Casella, R. C., 305
Center-of-mass coordinates, 299
Cerenkov threshold, 136
Chambers, R. G., 232, 246
Classical skin depth, 308
Classification of plane wave states, 208
Clogston, A. M., 351, 359
Closed fermi surfaces, 238
Closed orbits, 229
Cohen, M., 26
Cohen, M. H., 215, 262, 267, 333
Coherence effects, 174
 length, 173
Coherent scattering, 373
Cohesive energy, 94, 115
Collision drag, 331
Compatibility relations, 204
Completeness relation, 272
Conduction electrons, *g* values, 284
Conductivity, effective, 312
 sum rule, 128
 tensor, 240
Conjugation, 184

Connected graph, 126
Constant, dielectric, 124, 126
Construction, fermi surfaces, 257
 Harrison, 259
Cooper, L. N., 150
Cooper pairs, 153, 177
Coordinates, center-of-mass, 299
Copper, 238
Correlation energy, 92, 99
 function, 368
  density, 370
  pair, 129
  two-electron, 94
Coulomb energy, 100
 ground-state, 93
Coulomb interaction, 117
 screened, 115
Creation, electron, 117
 hole, 117
Critical magnetic field, 162
Cross sections, scattering, 115, 380
Crystal, electron, 99
 momentum, 15
Cuprous oxide, 300, 302, 306
Curie point, 54, 366
Current density operator, 170
Cyclotron frequency, 218, 235, 236
 resonance, 217, 226, 234
  effective mass, 227
  metals, 315
  semiconductors, 279

Damon, R., 73
Damping, Landau, 104
Dangerous diagrams, 25
Deaton, B. C., 335
Debye temperature, 32
Debye-Waller factor, 379, 388
de Faget, 346
Deformation potential, 130, 134, 337
Degenerate bands, 189
Degenerate $\mathbf{k} \cdot \mathbf{p}$ perturbation theory, 188
de Gennes, P., 366
de Haas-van Alphen effect, 217, 220, 223, 229, 254, 261
de Launay, J., 30
Delta function representation, 4
Density correlation function, 370
 fluctuations, 111

Density correlation function, hamiltonian, 19
matrix, one-particle, 101
matrix operator, 79
momentum, 19
operator, current, 170
Dexter, D. L., 306
Diagrams, dangerous, 25
Goldstone, 117
Diamagnetic susceptibility, 225
Diamagnetism, Landau, 225
Diamond, 268, 269, 273
structure, 212
Dielectric anomaly, 319
constant, 103, 107, 124, 126, 319
Thomas-Fermi, 105
transverse, 324
response, 116
response analysis, 107
screening, 111
Diffraction, neutron, 24
Diffusion vector, 329
Dingle, R. B., 30
Direct optical transitions, 294
Dirty superconductors, 183
Dispersion relations, 405
Distribution function, pair, 110
Divalent metals, 242
Donath, W. E., 365
Donor states, 286
Doppler shift, 319
Double group, 216
Dresselhaus, G., 186, 215, 273, 277, 278, 301, 318
Drift velocity, 238, 239
Dynamic structure factor, 33, 110, 129
Dyson, F. J., 54, 324, 339
Dyson chronological operator, 8, 399
Dysonian line shape, 323
Dzyaloshinsky, E., 397

Easterling, V. J., 326
Eddy current equation, 308
Effective carrier charge, 336
Effective conductivity, 312
Effective mass, 292
mass, cyclotron resonance, 227
mass tensor, 185, 283
Elastic line, 13, 18
Electric dipole transitions, 302
Electrical resistivity, 345
Electric field, 190
hall, 241
Electrodynamics, metals, 308
superconductors, 169
Electron annihilation, 117
creation, 117
crystal, 99
-electron interaction, virtual phonons, 149, 151
-electron lifetime, 115
gas, 86, 118, 126
exchange integral, 91
perturbation theory, 417
self-energy, 93
-ion hamiltonian, metals, 144
operators, 84
orbits, 230, 232, 258
-phonon interaction, 130
metals, 142
Elliott, R. J., 215, 216, 274, 303
Empty lattice, 208
Energy-band data, semiconductor crystals, 271
Energy bands, 249, 268
Energy, cohesive, 94, 115
correlation, 92, 99
coulomb, 100
exchange, 67
-gap parameter Δ, 161
shift, 126
surfaces, spheroidal, 234
zero-point, 61
Equation-of-the-motion method—BCS equation, 164
Equations of motion, lagrangian, 18
Eshbach, J., 73
Exchange energy, 67
Hartree-Fock, 89
Exchange integral, electron gas, 91
Exchange interaction, indirect, 360, 364
Excited states, superconductors, 167
Excitons, 294
Frenkel, 298
ionic crystals, 306
longitudinal, 303
Mott, 298
transverse, 303
Extended orbits, 234
External lines, 124

Face-centered cubic lattice, Brillouin
zone, 213
Factor group splitting, 303
Falicov, L. M., 215
Fano, U., 44
Fermi, E., 112
Fermi gas, bound electron pairs, 153
surface, 222, 249
aluminum, 260
closed, 238
construction, 257
open, 232
stationary sections, 224
Fermion fields, 75
quasiparticles, 84
Ferrell, R. A., 105, 116, 127, 128
Ferromagnetic resonance, 323
scattering, 383
Feynman, R. P., 26, 141, 142
Feynman relation, 33
Feynman's theorem, 109
Field, fermion, 75
quantization, 19
variables, Holstein-Primakoff, 64
magnon, 64
Fletcher, P., 73
Fluctuations, density, 111
Form factor, magnetic, 381
Forward scattering, 421
Fourier lattice series, 3
series, 2
transforms, 401
Frauenfelder, H., 388
Frederikse, H. P. R., 220
Free electron, magnetic field, 217
Frenkel exciton, 298
Frequency, cyclotron, 218, 235, 236
distribution, phonons, 30
plasma, 36, 105
Friedel, J., 339, 346, 360, 365
Friedel, sum rule, 341, 343
f-sum rule, 128, 186, 198, 301

Galitskii, 397
Gas, boson, 23
Gauge invariance, 172
transformation, 172
Gavenda, J. V., 335
Gell-Mann, M., 423
Generalized OPW method, 254
Geometrical resonances, 334
Germanium, 234, 276, 305
Ginzburg, V. L., 173
Glauber, R. J., 392
Glide plane, 213
Glover, R. E., III, 128
Gold, A. V., 254
Goldstone diagrams, 117
Gorkov, L. P., 164, 397, 411
Graph, connected, 126
linked, 124, 125
ring, 129, 421, 424
Graphical analysis, 116
Green's functions, 397
perturbation expansion, 415
Grey tin, 277
Gross, E. F., 306
Ground state, superconducting, 155
Group, double, 216
theory, 199
wavevector, 201
Gurevich, V. L., 333
g values, conduction electrons, 284

Haering, R. R., 319
Hall, 296
Hall electric field, 241
Ham, F. S., 187, 249
Hamiltonian, magnetic field, 290
magnon-phonon, 74
Harman, T. C., 284
Harmonic oscillator, 218, 253, 293
Harrison, M. J., 333
Harrison, W. A., 149, 250, 255, 258, 260,
262, 265, 333, 335
Harrison construction, 259
Hartree approximation, 86, 113, 126,
410
Hartree-Fock approximation, 75, 80,
82, 86, 88, 410
Hartree-Fock exchange energy, 89
Heat capacity, antiferromagnet, 62
magnon, 55
Heine, V., 200, 262, 264, 267
Helicon modes, 321
Hellwarth, R. W., 142
Henshaw, D. G., 27, 33
Herman, F., 273, 274
Herring, C., 57, 215, 262
Herzfeld, K. F., 38

Hexagonal close-packed structure,
Brillouin zone, 214
High field limit, magnetoresistance, 241
Hole annihilation, 117
creation, 117
operators, 84
orbits, 230, 232, 258
Holstein-Primakoff transformation, 49, 58, 64
Holstein, T., 320, 329
Hopfield, J. J., 44, 216, 284, 304, 305, 306, 336
Huang, K., 37, 40
Hutson, A. R., 336

Iddings, C. K., 142
Impedance, surface, 309, 315
Impurity states, 286
Incoherent neutron scattering, 373
Indirect exchange interaction, 360, 364
Indirect optical transitions, 295
Indium antimonide, 277, 278, 283
Ineffectiveness concept, 311
Inelastic neutron scattering, 27, 53, 375
Interaction, coulomb, 117
deformation potential, 130, 134
electron-phonon, 130
magnon, 54
Ionic crystals, excitons, 306
Irreversible processes, thermodynamics, 240
Isotope effect, 150

James, R. W., 379
Jellium, 142
Jones, H., 200

Kadanoff, L. P., 397
Kaganov, M. I., 57, 68
Kahn, A. H., 220
Kane, E. O., 273, 277
Kaner, E. A., 315, 335
Keffer, F., 54
Kip, A. F., 273, 318
Kleinman, L., 255, 273, 276
Knox, R. S., 141
Kohn, W., 97, 113, 192, 237, 289, 293, 336, 348
Kohn effect, 113
Koopman's theorem, 83
Koster, G. F., 200, 351
Kothari, L. S., 371
k · **p** perturbation theory, 186, 278
Kramers-Kronig relation, 406
Kramers operator, 282
Kramers theorem, 183, 215
Kubo, R., 58
Kuhn, T., 250

Lagrangian equations of motion, 18
Landau, L. D., 141, 173
Landau damping, 104
diamagnetism, 225
gauge, 217
levels, 221, 286, 290
transitions, 296, 297
Langenberg, D., 318
Langer, J. S., 114
Lattice, empty, 208
reciprocal, 1
Lattice displacement operator, 23
Lattice series, fourier, 3
Laue theorem, 339
Lax, B., 298, 305
Lax, M., 216
Lead, 255
Lee, T. D., 141
Legendy, C., 320
Lehmann representation, 404
Leibfried, G., 16
Length, scattering, 371
screening, 112
Lifetime, electron-electron, 115
Lifshitz, I. M., 228, 237
Line shape, dysonian, 323
Linear lattice, band structure, 256
Brillouin zone, 202
Linked cluster theorem, 124
Linked graphs, 124, 125
Lipkin, H., 389
Liquid helium, second sound, 30
Liquid structure factor, 33, 129
Localized magnetic states, 353
London equation, 169, 172
Longitudinal attenuation phonon, 326
Longitudinal excitons, 303
Longitudinal optical phonons, 137
Loudon, R., 54, 216
Low, F. E., 141

Luttinger, J. M., 197, 237, 284, 293, 336
Lyddane, R. H., 38, 41

Macroscopic magnon theory, 63
Magnetic breakthrough, 229
Magnetic field, critical, 162
  electron dynamics, 217
  free electron, 217
  hamiltonian, 290
  orbits, 229
Magnetic field effects, attenuation, 333
Magnetic form factor, 381
Magnetic moment, anomalous, 282
Magnetic permeability, 323
Magnetic scattering of neutrons, 379
Magnetization, 223
  electromagnetic field, 46
  reversal, 56
  zero-point sublattice, 61
Magnetoabsorption, oscillatory, 296
Magnetoplasma propagation, 320
Magnetoresistance, 237
  high field limit, 241
Magnetostark effect, 307
Magnetostatic modes, 48, 72
Magnons, 47, 49
  acoustical, 53
  antiferromagnetic, 58
  field variables, 64
  heat capacity, 55
  interactions, 54
  optical, 53
  variables, 50, 59
Magnon-magnon scattering, 71
Magnon-phonon hamiltonian, 74
Magnon theory, macroscopic, 63
March, R. H., 371
Marshall, W., 371
Mass, polaron effective, 140
Matrix element, 174
Matthias, B. T., 351, 366
Mattis, D. C., 315, 318, 365
Mattuck, R. D., 339
Mean free path, 116
Meissner effect, 169
Metallic sodium, 115
Metals, electrodynamics, 308
  electron-ion hamiltonian, 144
Metals, electron-phonon interaction, 142
  velocity of sound, 143
Migdal, 397
Miller, A., 24
Miller, P. B., 319
Mirror planes, 213
Mitchell, A., 289
Modes, magnetostatic, 48, 72
Momentum, crystal, 15
  density, 19
  representation, 184
Monster, 261
Montroll, E., 30
Moore, 318
Morel, P., 163
Morse, M., 31
Mössbauer effect, 386
Mott exciton, 298

Nagamiya, T., 58
Nakamura, T., 58
Nearly free electrons, 255, 256
  model, 262
Néel temperature, 62
Nettel, S. J., 97
Neutron diffraction, 24, 368
  scattering, inelastic, 53
Normal process, 132
Normal product form, 78
Nozières, P., 24, 107, 114, 128, 397

One-particle density matrix, 101
Onsager, L., 225, 228
Open fermi surface, 232
Open-orbit resonance, 335
Operators, density matrix, 79
  dyson chronological, 8
  electron, 84
  hole, 84
  lattice displacement, 23
  particle density fluctuation, 95
  spin, 51
  statistical, 101
  time-ordering, 8
Optical absorption, 294
  magnons, 53
  phonons, 34, 37
  theorem, 347
Optical transitions, direct, 294
  indirect, 295

OPW method, 263
generalized, 254
Orbach, R., 164
Orbits, aperiodic open, 233, 234
closed, 229
electron, 230, 232, 258
extended, 234
hole, 230, 232, 258
magnetic field, 229
open, 230, 238, 244, 245, 247
periodic open, 232, 233, 234
quantization, 228
Orthogonalized plane wave, 263
Oscillations, free, 37
Oscillator, harmonic, 293
Oscillatory magnetoabsorption, 296
Overhauser, A. W., 97

Pair correlation function, 129
distribution function, 110
Pairs, Cooper, 153
Parallel pumping, 69
Paramagnetic scattering, 382
Parameter Δ, energy-gap, 161
Particle density fluctuation operator, 95
Pekar, S., 141
Periodic open orbits, 232, 233, 234
Perturbation expansion, Green's functions, 415
slowly varying, 196
theory, degenerate, 187
k · **p**, 188
electron gas, 417
k · **p**, 186, 278
time-dependent, 6
time-independent, 9
Peschanskii, V. G., 335
Phillips, J. C., 255, 273, 276
Phonons, 216
acoustic, 12, 21
amplification, 336
boson gas, 23
cloud, 134
electron-electron interaction, 151
frequency distribution, 30
longitudinal optical, 137
optical, 34, 37
Photon absorption, 294
Piezoelectric semiconductor, 149
Pincherle, L., 250
Pincus, P., 47
Pines, D., 24, 102, 107, 114, 128, 141, 397
Pippard, A. B., 169, 173, 220, 225, 229, 244, 311, 312, 322, 333
Plane wave states, classification of, 208
Plasma frequency, 36, 105
Plasmons, 34, 104, 105
dispersion relation, 36
Platzman, P. M., 142
Polarization, phonon, 16
waves, 34
Polaron, 130, 133
cloud, 139
effective mass, 140
Polyhedra, atomic, 251
Wigner-Seitz, 93
Potential, screened, 112, 116
Poynting vector, 310
Primitive basic vectors, 1
Privorotskii, I. A., 335
Process, normal, 132
umklapp, 133
Proton, 147
Pseudopotential, 361, 372
Pumping, parallel, 69

Quantization, field, 19
orbits, 228
Quantized flux, 175
Quantum equations, summary, 5
Quasimomentum, 15
Quasiparticles, fermion, 84
Quinn, J. J., 116

Raman processes, 28
Random phase approximation, 102, 113
Rare earth metals, 365
Reciprocal lattice, 1
Recoilless emission, 386
Reduced zone scheme, 209
Reflection, specular, 313
Reitz, J. R., 249
Relaxation time, 135
Representation, small, 201
Residual resistivity, 346
Resistivity, electrical, 345
residual, 346
surface, 311
tensor, 241

Resonance, cyclotron, 217, 226, 234
geometrical, 334
open-orbit, 335
Response, dielectric, 116
Reuter, G. E. H., 314
Reversal, magnetization, 56
Rigid band theorem, 344
Ring graphs, 129, 421, 424
Rocher, Y. A., 366
Rodriguez, S., 317
Rose, F., 320
Rosenstock, H. P., 31
Rosettes, 258
Roth, L., 298, 305
Roth relation, 283
Ruderman, M. A., 360

s spheres, 251
Sachs, R., 41
Sanders, T. M., 336
Sawada, K., 104
Scattering, coherent, 373
cross section, 115, 380
ferromagnetic, 383
forward, 421
incoherent, 373
inelastic lattice, 375
inelastic neutron, 27
length, 371
magnon-magnon, 71
of neutrons, magnetic, 379
paramagnetic, 382
Schrieffer, J. R., 150, 151, 172
Schultz, T. D., 141
Screened coulomb potential, 115
Screened potential, 112, 116
Screening, 118
charge density, 114
dielectric, 111
length, 112
Second sound, crystals, 28
liquid helium, 30
Seitz, F., 250, 266
Self-consistent field, 113
approximation, 103
method, 101
Semiconductors, 268
crystals, energy-band data, 271
cyclotron resonance, 279
piezoelectric, 149
Semiconductors, spin resonance in, 279
Sham, L. J., 262
Shinozaki, S., 56
Shirkov, D. V., 151
Shockley, W., 130, 148, 192
Shoenberg, D., 220
Shull, C. G., 371
Silicon, 234, 275
Simple cubic lattice, Brillouin zone, 205
Simple square lattice, Brillouin zones, 257
Singwi, K. S., 371
Skin depth, classical, 308
Skin effect, anomalous, 308, 313
Slater, J. C., 196, 250, 351
Small representation, 201
Smoluchowski, R., 206
Sodium, metallic, 115
Sondheimer, E. H., 314
Sound, velocity of, metals, 143
Space inversion, 183
Specular reflection, 313
Spheroidal energy surfaces, 234
Spin-analog method—BCS equation, 157
Spin operator, 51
Spin-orbit coupling, 181, 215, 279, 284
splitting, 271
Spin resonance, 323
in semiconductors, 279
Square lattice, Brillouin zone, 203
Stationary sections, fermi surface, 224
Statistical operator, 101
Stokes theorem, 228
Strong coupling limit, 177
Structure factor, dynamic, 33, 110, 129
liquid, 33, 129
Suhl, H., 351, 366
Sum rule conductivity, 128
f-, 186, 198, 301
Friedel, 341, 343
Superconducting, ground state, 155
Superconductivity, 411
Superconductors, electrodynamics of, 169
excited states, 167
Superfluidity, 26

Surface impedance, 309, 315
  resistivity, 311
Susceptibility, diamagnetic, 225

Teller, E., 41
Temperature, Debye, 32
  Néel, 62
  transition, 162, 163
Temperature dependence of sublattice magnetization, antiferromagnet, 62
Tetrahedral bonding orbitals, 268
Thermal averages, 407
Thermodynamics, irreversible processes, 240
Thomas, D. G., 112, 304, 305, 306
Thomas-Fermi, dielectric constant, 105
Threshold, Cerenkov, 136
Tight binding approximation, 197
  functions, 270
Time, relaxation, 135
Time-ordering operators, 8
Time reversal invariance, 214
  operator, 282
  symmetry, 182
Tinkham, M., 128, 151
Tolmachev, 151
Transformation, canonical, 148, 151
  gauge, 172
Transition element impurities, 356
Transitions, allowed, 294
  direct, 294
  electric dipole, 302
  indirect, 296
  Landau, 296, 297
  temperature, 162, 163
  vertical, 294
Transport equation, 246
Transverse dielectric constant, 324
  excitons, 303
  wave attenuation, 332
Tsukernik, V. M., 68
Tunneling, 178

Umklapp process, 133
Unlinked part, 124
U operator, 8, 119

Valence-band edge, 271, 284
Van Hove, L., 31, 110, 369, 371
Van Kranendonk, J., 57
Van Vleck, J. H., 57, 250, 360, 362
Vector representation, 186
Velocity of sound, metals, 143
Vertical transition, 294
Virtual phonons, electron-electron interaction, 149
Von der Lage, F. C., 207, 252
Vosko, S. H., 114, 348

Walker, L. R., 48, 73
Wannier, G. H., 194
Wannier effective wave equation, 279
  equation, 289, 292
  functions, 195, 270, 351
  theorem, 286
Ward, J. C., 30
Watanabe, 53
Wavevector, group of, 201
Weinreich, G., 336
Weinstock, R., 379
White, H. G., 336
Wick chronological operator, 399
Wigner, E., 206, 250, 266
Wigner limit, 99
Wigner-Seitz boundary condition, 253, 267
  method, 250
  polyhedra, 93
Wilks, J., 30
Wilson, A. H., 237
Wolff, P. A., 351
Woll, E. J., Jr., 113
Wollan, E. O., 371
Woodruff, T. O., 249
Woods, A. D. B., 371
Wurtzite, 305

Yafet, Y., 284
Yosida, K., 58, 360, 362
Yttrium iron garnet, 56

Zero-point energy, 61
  sublattice magnetization, 61
Ziman, J., 58, 237
Zinc blende, 271
  oxide, 304
Zwerdling, S., 298, 305

NO LONGER
PROPERTY OF
OLIN LIBRARY
WASHINGTON UNIVERSITY

NO LONGER
PROPERTY OF
OLIN LIBRARY
WASHINGTON UNIVERSITY